Mankiew
Mendelson

July 1986

THE ORIGIN OF MAJOR
INVERTEBRATE GROUPS

Dedicated to the memory of
Professor P. C. Sylvester-Bradley

Proceedings of a Symposium held at the University of Hull

THE SYSTEMATICS ASSOCIATION
SPECIAL VOLUME NO. 12

THE ORIGIN OF MAJOR INVERTEBRATE GROUPS

Edited by

M. R. HOUSE

Department of Geology, University of Hull, England

1979

Published for the
SYSTEMATICS ASSOCIATION
by
ACADEMIC PRESS · LONDON · NEW YORK · SAN FRANCISCO

ACADEMIC PRESS INC. (LONDON) LTD.
24–28 Oval Road
London NW1 7DX

U.S. Edition published by
ACADEMIC PRESS INC.
111 Fifth Avenue
New York, New York 10003

British Library Cataloguing in Publication Data
The origin of major invertebrate groups. –
(Systematics Association. Special volumes; no. 12).
1. Invertebrates – Evolution – Congresses
2. Phylogeny – Congresses
I. House, Michael Robert II. Series
592′.03′8   QL362   78-73884
ISBN 0-12-357450-1

Printed in Great Britain at the
University Press, Cambridge

List of Contributors

D. T. ANDERSON, F.R.S., *School of Biological Sciences, University of Sydney, Sydney, N.S.W. 2006, Australia*

M. D. BRASIER, *Department of Geology, University of Hull, Kingston upon Hull, HU6 7RX*

R. B. CLARK, *Department of Zoology, The University, Newcastle upon Tyne, NE1 7RU*

T. D. FORD, *Department of Geology, University of Leicester, Leicester, LE1 7RH*

A. GRAHAM, *Department of Zoology, The University, Whiteknights, Reading, RG6 2AJ*

C. H. HOLLAND, *Department of Geology, Trinity College, Dublin 2, Ireland*

M. R. HOUSE, *Department of Geology, University of Hull, Kingston upon Hull, HU6 7RX*

R. P. S. JEFFERIES, *Department of Palaeontology, British Museum (Natural History), Cromwell Road, London, SW7 5BD*

G. P. LARWOOD, *Department of Geology, Science Laboratories, South Road, Durham, DH1 3LE*

S. M. MANTON, F.R.S. *(Deceased)*

N. J. MORRIS, *Department of Palaeontology, British Museum (Natural History), Cromwell Road, London, SW7 5BD*

C. R. C. PAUL, *Department of Geology, University of Liverpool, Brownlow Street, Liverpool, L69 3BX*

R. B. RICKARDS, *Department of Geology, Sedgwick Museum, Downing Street, Cambridge, CB2 3EQ*

C. T. SCRUTTON, *Department of Geology, The University, Newcastle upon Tyne, NE1 7RU*

M. A. SLEIGH, *Department of Biology, Medical and Biological Sciences Building, University of Southampton, Southampton, Hampshire, SO9 3TU*

P. C. SYLVESTER-BRADLEY *(Deceased)*

P. D. TAYLOR, *Department of Geology, University College of Swansea, Singleton Park, Swansea, Wales, SA2 8PP*

(vi)

H. B. WHITTINGTON, F.R.S., *Department of Geology, Sedgwick Museum Downing Street, Cambridge, CB2 3EQ*

A. D. WRIGHT, *Department of Geology, The Queen's University, Belfast, BT7 1NN, N. Ireland*

E. L. YOCHELSON, *U.S. Geological Survey, Room 3-501, Museum of Natural History, Washington D.C., 20560, U.S.A.*

Preface

Since it was formed in 1937 the Systematics Association has been a forum for discussion of general theoretical and practical problems of taxonomy and systematics. Advantage has always been taken of the fact that venues in the British Isles normally allow a wide range of British specialists to attend symposia and to contribute to the discussions. Occasionally, finance allows other overseas speakers to be invited to attend and, in addition, overseas participants have always played a significant role in the Association functions.

The theme set for the meeting held from 19–21 April 1978 at the University of Hull, which is recorded in this publication, was a broad one, of interest to most zoologists and geologists. It was deliberately organized to bring neontologists and palaeontologists together, and although the latter outnumbered the former among the formal contributors, in discussion the balance was almost equal. Some selected points arising from the discussions are included in a final section based on recordings of the discussions and written contributions, but this cannot be more than a weak reflection of the informed, lively and witty exchanges which took place, nor does it enlarge on the numerous occasions when it had to be acknowledged that we were in ignorance of so much that was relevant.

How far, it may be asked, is the pursuit of certainty in matters relating to the origin of major invertebrate groups merely the pursuit of a will-o'-th-wisp? The chase, after all, has been going on for long enough, and this is but one of many contributions to it. I think the participants regarded the symposium more as an opportunity to discuss a review of current ideas on origins, and attenders were particularly indebted to the many speakers who were prepared to attempt a synthesis for their approval or criticism. In particular I am indebted to them for accepting my invitation to address the symposium and contribute to this volume.

The symposium was overshadowed by the death of Peter Sylvester-Bradley, who died the day before he was scheduled to give the opening address. This volume is dedicated to him. His loss meant not only that the discussion of Precambrian events lacked the main person who could have contributed so much to the subject, but the whole symposium felt the absence of one whose stimulating ideas and comments have done so much to enliven symposia of this sort in previous years. Dr T. D. Ford, at the shortest notice, gave an address of which he gives a brief report here, and the symposium was especially indebted to him.

The symposium was attended by about 125 participants from some 13 countries. Much of its success is due to the organizational work of Dr and Mrs M. D. Brasier and other staff of the Department of Geology and to the help of a range of others at the University of Hull. To these, and to the contributors and participants, are due the sincere thanks of the Association and not least to the members of Academic Press who have, once more, given great aid in the production of a symposium volume.

March 1979

M. R. HOUSE
President

Contents

1 | Precambrian Prelude

★P. C. SYLVESTER-BRADLEY

Abstract: Although the taxonomic division of all life into two kingdoms, animals and plants, has for a long time been abandoned, a physiological division into *heterotrophs* and *autotrophs* has taken its place. Recent research has shown that even this is an unjustifiable simplification. Since its origin and inception, life has developed more than two ways of capturing its energy.

In this paper, rival theories relating to the ancestry of animals are compared, and tested against available evidence. A model claiming methanogenic bacteria as the most primitive of all organisms is examined in the light of the geological record. Special attention is paid to the origin of sex and its importance with reference to models of diversification. Preference is given to what is termed the "dumb-bell" model (Fig. 1), in which the origin of life brings to an end a period of 1000 m.y. (million years) in which there was a multiplicity of protolife; it is followed by 3000 m.y. characterized by stasigenesis and relative uniformity. Then comes the third period, beginning in upper Proterozoic time and lasting another 1000 m.y., in which the origin of the nucleus is quickly succeeded by the origin of sex, of plants and of the Metazoa. The first example of adaptive radiation takes place, and this is followed by the Phanerozoic explosion of diversity.

THE MOST PRIMITIVE ORGANISMS: BIOLOGICAL EVIDENCE

The division of all life into two kingdoms, Animalia and Plantae, was always based more on their physiological distinction than on their phylogenetic record. For logically it seemed clear that if animals and plants were both derived from a common ancestor, then three kingdoms were necessary, not two, and a glance at any text-book of fifty years ago will show that both kingdoms were regarded as derived from the Protista, which could not therefore be included in either kingdom. "There are some

---

★ Professor Bradley died before the symposium began. This contribution represents the only part of his paper assembled for publication. To this has been appended a list of Professor Bradley's papers in this field. This has been assembled by Dr T. D. Ford.

Systematics Association Special Volume No. 12. "The Origin of Major Invertebrate Groups", edited by M. R. House, 1979, pp. 1–5, Academic Press, London and New York.

types among the simplest and smallest creatures which share animal and plant characteristics, being able both to take in solid food like a typical animal, and to build up food from simple inorganic substances like a green plant. Such examples only show the impossibility of drawing hard and

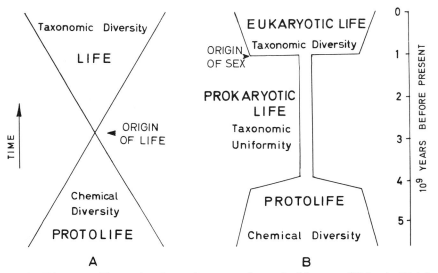

FIG. 1. Diagrams illustrating hypotheses on the early history of life. A, Pirie's hour-glass diagram. B, The dumb-bell model.

fast lines in Nature" (Haldane and Huxley, 1927, p. 5). The idea that the Protista should be regarded as a third kingdom had in fact already been proposed by Haeckel (1866, 1878) when he first introduced the term, and other naturalists of the same period suggested other names for the ancestral kingdom (e.g., "Protozoa", "Protoctista", "Acrita", "Primalia" etc.; see Whittaker, 1959, 1969).

Then, thirty years ago, it became apparent that a much more fundamental distinction could be drawn than that which exists between animals and plants, a distinction which applied not only to the physiology of the organisms concerned, but to the morphology of their very cell. These two divisions consist first of the eukaryotes, which include all animals, plants and protists, composed of individuals whose cells have a nucleus. Secondly, we have the prokaryotes, whose cells have no nucleus; these organisms are represented only by the bacteria and the blue-green algae. Moreover, the prokaryotes are considered not only to be more simply constructed than the eukaryotes, but to be more primitive and, in fact, ancestral to them. The taxonomic significance of this distinction was first recognized by Kluyver and van Niel (1936), and a kingdom for their reception, the "Monera", was adopted in the four-kingdom system of Copeland (1938), with the plants (or Metaphyta), the animals (or Metazoa) and the protoctists as the other three. Further elaborations have been proposed, but the five-kingdom system of Whittaker (1969) and Margulis (1970), in which the eukaryotes are divided into four kingdoms (plants, animals, fungi and protists) and the prokaryotes form just one (the Monera) has, until very recently, become generally accepted.

But now there is some unease. First, it seems that the distinction between the prokaryotes and the eukaryotes is greater than that which exists between any of the four eukaryotic kingdoms. The Monera are cut off from the other four by a gap greater than that which divides the plants from the animals, the fungi, or the protists. There is considerable debate about how this gap was bridged in evolutionary history. Margulis has put forward a convincing theory which postulates that all four eukaryotic kingdoms have been derived from a series of symbiotic associations of various prokaryotic ancestors (Margulis, 1970). Counter theories postulate that the ancestral eukaryote evolved from a single prokaryotic ancestor that was larger in size than normal (Raff and Mahler, 1972). The theories need testing against each other. It is crucial to understand the environment in which the transition from prokaryote to eukaryote took place. Fossil evidence comes down very strongly in favour of a long period of Precambrian time in which only prokaryotes exist. But the evidence is less strong when we come to the recognition of the first eukaryote.

Before we come to this evidence, it is necessary to understand the second cause for the unease in accepting the current division of life into five kingdoms. Should the prokaryotes properly be grouped together in just one kingdom? It is now becoming clear that within the prokaryotes there are several distinctly different groups of organism that differ both in metabolic function and morphological detail (in so far as morphology comprehends the molecular sequences of an organism's genome). Woese and Fox (1977) would separate all life into at least three "superkingdoms" (or "urkingdoms" as they prefer to call them), in which all eukaryotes are grouped together as the "Urkaryotes", and in which the prokaryotes are divided into two groups, the Archaebacteria (composed of the methanogens) and the Eubacteria (the rest of the prokaryotes). Woese and Fox go further. First, they suggest that the Eubacteria and Archaebacteria are equally old, "in that branchings between the two urkingdoms are comparably deep" (see also Fox *et al.*, 1977); secondly, they contend that they are so different from each other that, if they arose from a common ancestor, that ancestor was more primitive than any known prokaryote, and that the prokaryotic level of organization was independently achieved in each branch. Although there is as yet no fossil evidence for the existence of methanogenic bacteria in the Precambrian, the ecological significance of such a possibility is considerable. They are unique in that they are restricted to anaerobic habitats in which they obtain their energy from the oxidation of hydrogen and the simultaneous reduction of carbon dioxide to methane (Bryant, 1974). Comment on Woese and Fox's ideas suggests that the methanogens may have played an important part in the evolution of Precambrian ecosystems (Maugh, 1977; Wilkinson, 1978).

Maybe there are other divisions within the prokaryotes, as important as those which divide the eukaryotes into kingdoms. For example, Lewin (1976) has demonstrated that a newly discovered kind of alga, symbiotically associated with ascidians, although certainly prokaryotic, cannot be regarded as one of the cyanophytes (blue-green algae), as their chlorophyll constituents have more in common with the eukaryotic algae. He has proposed they should be regarded as a new class, the Prochlorophyta. But a new class of what phylum, what kingdom? Do they form a link between the prokaryotes and the eukaryotes? It is easier to pose the questions than to supply the answers. As Stanier and Cohen-Bazire (1977) have shown, we have something more exciting to debate than any taxonomic conclusion. We have evidence in the Prochlorophyta of the polyphyletic origin of chloroplasts. Perhaps the Chlorophyta and Rhodophyta represent two separate phyla of Protists that have been independently derived, through endosymbiosis, from two separate phyla of Eubacteria, one a prochlorophyte, the other, one of the Cyanobacteria.

These new discoveries have not only complicated our simple classical picture of life divided into just two categories, animals and plants. They have also demonstrated that we can no longer rely even on the simplicity of the physiological distinction between autotrophs and heterotrophs, for life has developed more than two ways of capturing its energy. There is some suggestion that even from the earliest days there was heterotrophy of at least two kinds, represented by the Archaebacteria and the Eubacteria. Likewise there seem to be at least two kinds of autotrophy, each using a different kind of chlorophyll. But these complications have been introduced by the study of living representatives of primitive organisms. It is time to turn to the geological evidence.

REFERENCES

BRYANT, M. P. (1974). Methane-producing bacteria. In "Bergey's Manual of Determinative Bacteriology" (R. E. Buchanan and N. E. Gibbons, eds) 8th edn, pp. 472–477. Williams and Wilkins, Baltimore.

COPELAND, N. F. (1938). The kingdoms of organisms. *Q. Rev. Biol.* **13**, 383–420.

FOX, G. E., MAGRUM, L. J., BALCH, W. E., WOLFE, R. S. and WOESE, C. R. (1977). Classification of methanogenic bacteria by 16 S ribosomal RNA characterization. *Proc. natn. Acad. Sci. USA.* **74**, 4537–4541.

HAECKEL, E. (1866). "Generalle Morphologie der Organismen." G. Reimer, Berlin.

HAECKEL, E. (1878). "*Das Protistenreich.*" E. Günther, Leipzig.

HALDANE, J. B. S. and HUXLEY, J. (1927). "Animal Biology." Oxford.

KLUYVER, A. J. and VAN NIEL, C. B. (1936). Prospects for a natural system of classification of bacteria. *Zentbl. Bakt., Abt. II,* **94**, 369–403.

LEWIN, R. A. (1976). Prochlorophyta as a proposed new division of algae. *Nature, Lond.* **261**, 697–698.

MARGULIS, L. (1970). "Origin of Eukaryotic cells." Yale Univ. Press. New Haven.

MAUGH, T. H., II (1977). Phylogeny: are Metazoans a third class of life? *Science, N.Y.* **198**, 812.

PIRIE, N. W. (1959). Chemical diversity and the origin of life. *In* "Origin of Life on Earth" (A. I. Oparin, ed.), pp. 76–83. Pergamon, London.

RAFF, R. A. and MAHLER, H. R. (1972). The non-symbiotic origin of mitochondria. *Science, N.Y.* **177**, 575–582.

STANIER, R. Y. and COHEN-BAZIRE, G. (1977). Phototropic prokaryotes: the Cyanobacteria. *A. Rev. Microbiol.* **31**, 225–274.

WHITTAKER, R. H. (1959). On the broad classification of organisms. *Q. Rev. Biol.* **34**, 210–226.

WHITTAKER, R. H. (1969). New concepts of kingdoms of organisms. *Science, N.Y.* **163**, 150–159.

WILKINSON, J. F. (1978). Merhanogenic bacteria: a new primary kingdom. *Nature, Lond.* **271**, 707.

WOESE, C. R. and FOX, G. E. (1977). Phylogenetic structure of the prokaryotic domain: the primary kingdoms. *Proc. natn. Acad. Sci. USA.* **74**, 5088–5090.

PUBLICATIONS BY P. C. SYLVESTER–BRADLEY
CONCERNING PRECAMBRIAN LIFE

1960. The nature and discovery of "missing links", *21st Int. geol. Cong.* Part 22, 111–112.

1963. Evidence for Abiogenic Hydrocarbons. *Nature, Lond.* **198**, 728–731 (with R. J. King).

1964. The Origin of Oil and Life. *Discovery* **25**, 37–42.

1967. Evolution versus Entropy. *Proc. Geol. Ass.* **78**, 137–148.

1968. The Science of Diversity. *Syst. Zoo.* **17**, 176–181.

1971. Environmental Parameters for the Origin of Life, *Proc. Geol. Ass.* **82**, 87–136.

1971. An evolutionary model for the Origin of Life. *In* "Understanding the Earth" (I. G. Gass, P. J. Smith and R. G. S. Wilson, eds), pp. 123–141. Artemis Press, London.

1971. Processes of geological evolution. *Jl geol. Soc.* **127**, 477–481.

1971. Carbonaceous chondrites and the prebiological origin of food. *In* "Molecular Evolution, 1: Chemical Evolution and the Origin of Life" (R. Buvet and C. Ponnamperuma, eds), pp. 62–94. N. Holland Pub. Co.

1972. The geology of juvenile carbon. In "Exobiology" (C. Ponnamperuma, ed.), pp. 62–94. N. Holland Pub. Co.

1975. The search for Protolife. *Proc. R. Soc., B.* **189**, 213–233.

1976. Evolutionary oscillation in prebiology: igneous activity and the origins of life. *Origins of Life*, **7**, 9–18.

1977. Evolution and growth. *Nature, Lond.* **265**, 492.

1977. Biostratigraphical tests of evolutionary theory. *In* "Concepts and Methods of Biostratigraphy" (E. G. Kauffman and E. Hazel, eds), pp. 41–63. Dowden, Hutchinson and Ross, Stroudsburg.

2 | Precambrian Fossils and the Origin of the Phanerozoic Phyla

TREVOR D. FORD

Geology Department, University of Leicester, England

Abstract: Spheroidal and filamentous algae, some found in stromatolites, demonstrate the presence of Protistan life forms throughout much of Precambrian time. None can be shown to be ancestral to the late Precambrian Ediacaran fauna. Now known from many trace-fossils this appears to include primitive coelenterates, annelids, arthropods and other organisms of uncertain affinity, though proof of the assignations to these phyla is lacking and links to their Cambrian descendants are minimal.

INTRODUCTION

The Phanerozoic Phyla must have had ancestry in the Precambrian, and the purpose of this paper is partly to update those of Glaessner (1962, 1966), Cloud (1968, 1976), Hofmann (1971), Grabert (1973) and Schopf (1975, 1977, 1978) and partly to provide a basis for the discussions which follow. This paper has been hurriedly prepared and I apologize for its shortcomings. Professor Peter Sylvester-Bradley, whose death robbed us of having his thoughts on the origin of the Phyla, had prepared but a few pages of his address, and these are produced as Chapter 1: they deal largely with the origin of life and particularly with the environments and significance of such prokaryotes as methanogenic bacteria and their successors, the early eukaryotes. The evidence is mostly biochemical or by analogy with modern bacteria and algae. My task is to look at the fossil evidence itself. Professor Sylvester-Bradley has been a stimulus for many years, and it is

Systematics Association Special Volume No. 12, "The Origin of Major Invertebrate Groups", edited by M. R. House, 1979, pp. 7–21, Academic Press, London and New York.

a matter of regret that pressure of work in other fields prevented me from approaching this review earlier and more thoroughly.

Some hundreds of Precambrian fossils have now been named and described in the literature, more are as yet nameless. They fall into the following categories:

1. Organic microfossils
2. Stromatolites
3. Trace fossils
4. Dubio-fossils
5. Pseudo-fossils

The last two categories can be disposed of rapidly; dubio-fossils are those features of rocks or rock surfaces where there is doubt as to whether the feature is of organic origin or not. If this can ever be provided then the fossil will be transferred to another category. Pseudo-fossils are those which are now regarded as of inorganic origin though they were once described, and sometimes named, as true fossils. The names may be in the literature but they no longer have significance.

Body fossils such as shells, carapaces and coral structures are so far absent from Precambrian records; we are thus largely restricted to a survey of organic microfossils, stromatolites and a miscellany of impressions, the traces of the former presence of soft-bodied life forms.

ORGANIC MICROFOSSILS

Spheroidal and filamentous nannofossils of organic composition are now known from all the continents and from almost the whole range of Precambrian time. Reviews have been provided by Downie (1967); J. M. Schopf (1969); J. W. Schopf (1975 and in Ponnamperuma, 1977) J. W. Schopf et al. (1971, 1973); Schopf and Oehler (1976). The earliest claimed so far are from cherts in the Onverwacht Formation of the Barberton Mountains in South Africa (Brooks and Shaw, 1971). Dated at some 3300 m.y. (million years before the present), they are very small spheroids only a few microns in diameter. Cloud (1976) is less certain of the organic origin (but see Muir et al. in Ponnamperuma, 1977). Slightly later rocks, The Fig Tree Series, in the same area, have yielded rod-shaped and filamentous bodies, now reasonably well accepted as organisms (Barghoorn and Schopf, 1966; Schopf and Baarghoorn, 1967). Dated at some 3100 m.y. they have been compared with algae and bacteria.

A much more diverse flora occurs in the Gunflint Formation of the Lake

Superior Region, dated at 2000 m.y. (Tyler and Barghoorn, 1954). Others occur in the Belcher Islands of Hudson Bay, dated at 1760 m.y. (Hofmann, 1976) and in the Beck Spring Formation (1300 m.y.) of California, the Nonesuch Shale of Michigan (1075 m.y.) (Moore *et al.*, 1969) the Bitter Springs cherts (900 m.y.) in Central Australia, the Brioverian of Northern France (700 m.y.) and the Kwagunt Formation of the Grand Canyon (800 m.y.). Schopf (1977) has shown that there appears to be a progressive increase in size and diversity, though delimiting taxa on the basis of size in featureless spheres and filaments is at best an arbitrary means of classification. Whilst most workers are agreed that the earliest forms are prokaryotes and that the latest include eukaryotes, there is considerable debate as to when eukaryotes first appeared, and as to how one may safely recognize them. Amongst the spheroids some appear to have a dark spot inside, which has been interpreted as a nucleus, though others (Awramik *et al.*, 1972), Hofmann (1976), Golubic and Hofmann (1976) have shown that it could be the inner lining of the cell wall which has shrunk into the middle. Similarly with the filamentous fossils, some have septa which appear to suggest cell division, typical of eukaryotes, though in some cases the apparent cell divisions have been shown to be due to the growth of diagenetic pyrite crystals. Taking a consensus view it seems likely that eukaryotes had evolved by about 1400 m.y. (Schopf and Oehler, 1976 and Schopf in Ponnamperuma, 1977; Peat *et al.*, 1978). Once this had occurred there was the possibility of much more rapid evolution with the onset of sex and its linked characteristics. The Bitter Springs and young microbiotas include clusters of cells, some apparently forming hollow spheres, while others form chains of cells. These provide the limited evidence of a progression from microfossils to Metazoa.

A variety of microfossils of acritarch type have been described from the Vendian by Timofeev (1969, 1973a, b). They appear to be forerunners of Palaeozoic forms (Downie, 1967).

Microbiologically degraded acritarchs, together with the attacking bacteria and fungi, have been identified in a shallow-water mud environment in the Nonesuch Shale of the Lake Superior region (Moore *et al.*, 1969).

STROMATOLITES

Stromatolites are layered, domed, columnar, digitate and cabbage-like structures in rocks, mostly carbonates, comparable with those built up by cyanophytes in shallow warm waters today. Though they are widespread in the Precambrian, only a few have yielded organic microfossils comparable with those building such structures today: the rest are devoid of microfossils though this is almost certainly due to obliteration by diagenetic changes. Thus, while it cannot be proved in many cases, most investigators accept that stromatolite structures in the Precambrian are of cyanophyte origin (Walter, 1972, 1976; Awramik in Ponnamperuma 1977). The oldest known stromatolites are about 3000 m.y. old, in the Archaean at Bulawayo (Schopf *et al.*, 1971); they are but 300 m.y. younger than the oldest nannofossils of the Onverwacht. Uncommon in the Archaean, stromatolites become abundant in the Proterozoic, and they have been used as zonal index fossils in Russia and Australia, though the validity of such claims has been challenged (see Walter, 1977). Those stromatolites which have yielded fossils have contained assemblages of both spheroidal and filamentous cyanophytes so that part of the problem is whether the individual stromatolite form-taxa are specific to particular assemblages or not. Stromatolites are less common in the Phanerozoic and Garrett (1970) has attributed this to the effect of grazers and burrowers from Cambrian times onwards. Modern stromatolites are largely, but not exclusively, confined to hypersaline waters where benthonic faunas are sparse. In the Precambrian, stromatolites seem to be generally indicative of shallow shelf seas, sometimes forming reefs, and there is no obvious change in form which could be associated with the change from anoxygenic to oxygenic atmospheres. There is, however, the occurrence of stromatolites in ferruginous and siliceous rocks of the Banded Iron Formations (about 2000 m.y.) with associated microbiotas in the Gunflint which may have formed an oxygen sink about this period of Earth history.

MEGASPHAEROMORPHS

It seems highly unlikely that evidence of the origin of the Phanerozoic phyla, of which most fossils are animal, will be found amongst the plants of the above categories, so one must turn to the record of the larger late Precambrian fossils. One group is that of the Megasphaeromorphs: these include spheroidal organic bodies, some of which were once thought to be primitive brachiopods (Walcott, 1899) and named *Chuaria circularis*. Re-assigned to the Megasphaeromorph acritarchs by Ford and Breed (1973) and regarded as forerunners to Palaeozoic Leiospheres by Jux (1977) these are thin-walled organic spheres comparable with microplankton but ranging up to 5 mm in diameter though occasionally much larger. Now known from the Grand Canyon, Utah, Canada, Sweden, France, Russia, Iran, India and Australia they are all in rocks difficult to date by isotopic techniques but in the range of 600–800 m.y. The Canadian specimens are immediately below the Cambrian (Gussow, 1973) but separated by an unconformity of unknown duration. Vidal (1976) has found megasphaeromorphs in association with a large assemblage of small acritarchs, with no distinct break in size range up to the 3 mm diameter. The small "medusoid" *Beltanelloides sorichevae* described from the Vendian of Russia by Sokolov (1973) has been re-interpreted as a large Megasphaeromorph allied to *Chuaria* (Sokolov, *in litt.*).

Whether the Megasphaeromorphs were prokaryote or eukaryote is not known, but their appearance seems to be more or less synchronous with that of metazoan fossils, and one cannot help speculating whether with giantism amongst algae was a stimulus by providing food for the appearance of the Metazoa or whether it was defensive reaction to the first animals with mouths.

THE EDIACARAN METAZOANS

Representatives of what is generally called the Ediacaran fauna of around 600–800 m.y. are now known from Charnwood Forest in Leicestershire (Ford, 1958, 1968; Boynton, 1978), from South Wales (Cope, 1977), Australia, Southwest Africa (= Namibia) and Russia, while dubio-fossil representatives have been reported from California and Alaska (Allison, 1975). Placed in the Phyla Coelenterata, Annelida and Uncertain by Glaessner and Wade (1966), most of the assignations can be questioned but they are the best available to date.

By far the most common of the Ediacara fauna are the medusoids: these are circular to ovate disc-like impressions with a variety of concentric, radial or lobate markings. Wade (1968, 1972a) has demonstrated how some of these could have arisen through sand invading the hollow gut of jellyfish, but with others there is no diagnostic evidence of medusoid affinities, and all one can say is that they are impressions of organisms with a moderately firm to leathery texture. Comparable impressions are known from various parts of the Phanerozoic (Häntzschel, 1962) but the same arguments as to whether or not they represent jellyfish can be applied. Even so, it is the most likely explanation of such impressions and supports the general evolutionary concept that the first animals were jellyfish.

The second most important group of the Ediacara fauna is that referred to as the pennatulid coelenterates by Glaessner and Wade (1966). These are frond-like impressions up to some 50 cm long. They have been assigned to the genera *Arborea, Charnia, Charniodiscus, Rangea* and *Pteridinium*. The definitions and relationships of these have recently been discussed by Jenkins and Gehling (1978). Broadly the impressions are of firm, perhaps leathery, leaf-shaped organisms with a central stem or rachis supporting a series of regular lobes on opposite sides. The lobes may or may not have constrictions suggesting some form of segmentation; the lobes also reduce in size towards the tip. The fossils also vary in the way the organisms were impressed, obverse (with segments) or reverse (without), with or without the stem in the plane of impression, with varying degrees of sharp demarcation of lobes from rachis. A few specimens have been found with fronds attached to the centres of disc impressions which on their own would have been called medusoids, suggesting that in some cases the discs were some sort of holdfast or anchorage for fronds. Glaessner and Wade (1966) have assigned the fronds to the pennatulids by comparison with modern forms, in which there are distinct differences: the lobes are separated not attached to each other, the segments are occupied by polyps, and there are no circular holdfast discs. Also pennatulids have a sparse geological history as proven fossils, only as far back as the Eocene. The Ediacaran fronds could well have had apertures for polyps but none have been preserved. So once again we are faced with a situation where the identification of these Precambrian fossils as pennatulids is the best we can do on present evidence. If Glaessner and Wade are correct, as seems possible, then it may be that these frondose organisms are colonial forerunners of corals and other Cnidaria. If they are not cnidarians, then what are they: a completely extinct group short-lived in Precambrian times only? This is possible but it would be just bad luck. Or, could they be plants, simply large algal fronds? Again, it would be logical to have large plant life before large animals appeared on the geological scene. The evidence is simply not fully convincing in any one direction.

The third group of Ediacaran fossils is that classified by Glaessner and Wade as Annelida. These include the elongate impression of a segmented animal with a crescentric cephalon, narrow abdominal axis and curved pleura, named *Spriggina floundersi*. There are several annelid worms living today which could have made comparable impressions, but none with such a large or distinctive head. Alternatively it could have been a much-elongated trilobite, or possibly a common ancestor to the annelids and trilobites. The annelid analogy seems best but the other possibilities must be borne in mind. Also in the Ediacaran annelid group are the several forms of *Dickinsonia*. Once regarded as a sole representative of an extinct group of coelenterates, the Dipleurozoa, these ovate biradiate impressions have been compared with the modern polychaete worm *Spinther* by Glaessner and Wade (1966) and Wade (1972b). The same problem as in other groups arises here—there is no fossil polychaete comparable with *Spinther* known in the Phanerozoic: *Spinther* could make impressions such as *Dickinsonia* but there is no certain evidence of affinity. The verdict is again—"not proven".

The Ediacara fauna includes three fossils left unassigned by Glaessner and Wade: *Parvancorina* has a semi-circular shaped cephalon (?) with an axis or tail and the sectors between containing the faintest suggestion of segmentation; the impression gained is of an overgrown trilobite larva. *Tribrachidium* is a disc-shaped object with a strong margin of radiate ridges surrounding a central area with three curved rays: it could be compared with a primitive coral or archeocyathid. Finally at Ediacara there is *Praecambridium*, an ovate impression of a small (5 mm long) membranous organism with markings suggesting muscles or, in the opinion of Glaessner and Wade (1971), segmentation of a primitive member of a Phylum Articulata, which could have been ancestral to both annelids and arthropods. Comparable with *Praecambridium* are the small impressions found in the Vendian of Russia. Rarely more than 4 mm in length these are ovate, apparently with a semi-circular head followed by an indistinct axis with five or six pairs of segmental lobes. Named *Vendia sokolovae*, *Onega stepanovi* and *Vendomia menneri* these are regarded as primitive arthropods (Sokolov, 1973; Keller and Fedonkin, 1976). A much larger form, 3 cm long, *Pseudovendia charnwoodensis*, has recently been found in Charnwood Forest (Boynton and Ford, in press).

A generalized deduction from these fossils *Praecambridium*, *Onega*, *Vendia* and *Vendomia* suggests the possible presence in late Precambrian times of a group of organisms with a single ovate carapace or shell covering a body indistinctly divided into perhaps five or six segments behind a semi-circular head. Allowing plenty of imagination one could speculate on this as a common ancestor to monoplacophoran molluscs, annelids and arthropods, but there is no proof. The contemporary *Spriggina* could then be a more advanced stage of somitic development.

An as-yet undescribed impression in the Charnian consists of an indistinct boss some 30 cm in diameter surrounded by a fringe of many small radiating fronds. Jenkins (*in litt.*) has said that these look like small *Rangea*-type fossils.

A wide variety of fossils have been described from strata contemporary with the Ediacara fauna in many parts of the world. Space does not allow a full discussion of these, and it should be noted that almost all the papers on the subject include speculative discussions on the interrelationships of these fossils. Perhaps the most intriguing is the large cup-shaped apparently colonial organism *Arumberia banksi* (Glaessner and Walter, 1975) from the Northern Territory of Australia. Apparently composed of a semi-rigid organic material these cup-shaped objects were sessile, and have been tentatively referred to as "probably of coelenterate grade" by Glaessner and Wade. These authors also drew comparisons with the bag-shaped *Baikalina sessilis* Sokolov 1972, the leaf-like *Nasepia altae* Germs 1972 and groups of tubes *Namalia villiersiensis* Germs 1972 all from the Kuibis Formation of the Nama group in Namibia.

The most problematic problematicum of all is *Xenusion auerswaldae* Pompeckj. Found in erratic quartzite boulders in the glacial deposits of north Germany, this was first referred to the Arthropoda, but more recently Halstead Tarlo (1967) has compared it with the pennatulids of the Ediacara fauna. In contrast the detailed studies of Jaeger and Martinsson (1967) suggest that it was a metameric organism of unknown affinities though with onychophoran appendages. They have identified the parent rock as lowermost Cambrian of Southeast Sweden.

Mention of the Nama Group draws one to the variety of fossils therein first described by Gürich (1930), with additions by Richter (1955), Glaessner (1963) and Germs (1963). These have been the subject of a series of highly speculative papers by Pflug (1970–73) who has assigned them to the group Petalonamae, emphasizing their basic foliate habit. Pflug has redefined Gürich's genus *Pteridinium* and has added several new genera including *Ernietta*, *Erniograndis* and *Petalostroma*. Formal objective descriptions are still lacking for some of the forms but Pflug has used a vivid imagination in his evaluation of the significance of the fossils. Beautifully drawn reconstructions have been used by Pflug to argue a case for these organisms being representatives of common ancestral forms for the Plant and Animal Kingdoms and ancestral to most of the Phanerozoic Phyla. For example a five-rayed foliate reconstruction was directly compared with a blastoid echinoderm. In my view this is imagination! Another bilateral foliate reconstruction was inverted and with the addition of eyes and mouth became an ancestral trilobite. I support Glaessner and Walter's comment (1975) that they "cannot accept some of his basic assumptions."

Taking a much more objective view of the Nama fauna, it seems likely that there are a variety of frondose impressions, in varying states of preservation, broadly comparable with those of the Ediacara fauna. Until much more objective study of these specimens is complete further comment is useless.

THE NEWFOUNDLAND FOSSILS

A further Precambrian fossil fauna awaiting detailed study is that noted in the Conception Group of Newfoundland by Anderson and Misra (1968), Misra (1969) and King *et al.* (1974). The rocks are broadly comparable in age and lithology with those of Charnwood Forest in England and both *Charnia* and *Charniodiscus* have been found. Among the many other specimens spread thickly on certain bedding planes are four groups as yet unnamed: type 1 of Misra (1969) is a spindle-shaped impression generally 15–25 cm long, pointed at both ends, with a thin median axis and tufted lobes along each side. There is no obvious mouth or anus and they are probably colonial coelenterates. Anderson has suggested (personal communication) that they might represent colonial hydrozoans with a float; compound forms are also present.

If punctured by a volcanic ash fall (thin ash layers occur in the sequence) such organisms could sink to the bottom in large numbers and be preserved by ashy sediment almost immediately. The second group (Misra's type 2) are leaf-like and directly comparable to *Charnia* and other similar fronds in the Ediacara fauna. Type 3 are large lobate impressions which could represent jellyfish, while type 4 are dendritic markings which could be either variants of the spindle-shaped organisms or damaged and distorted members of that group, though they may be distinct colonial organisms poorly preserved.

We await a full description of these intriguing fossils, but they emphasize the possible presence of both planktonic and benthonic organisms. A few trails of crawling organisms have also been found, but the showers of hot ash doubtless discouraged such "worms".

WORM TRAILS AND BURROWS

With the exception of the possible ancestral annelids or arthropods all the late Precambrian Metazoan fossils noted so far appear to have been filter-feeders on microplankton. One category of organisms is poorly represented but has given rise to controversy—detritus feeders. Considered under the broad heading of "worms", these are generally represented in the fossil record only by trails and burrows. These are sparsely present in Precambrian rocks but become abundant and with much greater diversity in the Cambrian, Seilacher (1956), Cowie (1967), Goldring (1967), Glaessner (1969) and Webby (1970) have reviewed the evidence. Among the problems is the identification of the organisms which made the trails and burrows. They have been variously referred to primitive molluscs, annelids, arthropods and "other organisms", but the evidence is far from clear. A further problem is whether the trails and burrows represent dwelling places for suspension feeders or whether they represent an infauna feeding on organic detritus, a probably ecological advance on plankton eaters. Most Precambrian trails and burrows are in very late Precambrian strata (e.g. Webby (1970); Squire (1973); Cloud *et al.* (1976) and Fedonkin (1977)) though Glaessner (1969) has suggested that there are some trails as old as 1000 m.y. Still older ones, noted in the Huronian of Canada by Hofmann (1967) were later re-interpreted as inorganic (Cloud, 1968; Hofmann, 1971), but those described recently by Clemmey (1976) from the Copperbelt dated at about 1400 m.y. are difficult to interpret as anything else but trails. If accepted they suggest that there may have been an infauna of Metazoans some 600 m.y. earlier than the Ediacara fauna, and long before other "worm" trails. However, the date of 1400 m.y. is the same as Schopf and Oehler's suggestion (1976) for the origin of the eukaryotes, and this alone must throw considerable doubt on the age of Clemmey's trails.

OTHER POSSIBLE PRECAMBRIAN FOSSILS

Among other Precambrian fossils there have been numerous claims of the discovery of sponge spicules (e.g. Dunn, 1964) but none confirmed so far

(Cloud, 1968). The various claims of Precambrian brachiopods have all been dismissed by Rowell (1971). Similarly claims of the discovery of primitive echinoderms, archeocyathids, foraminiferids and radiolarians are unsubstantiated. However, chitinozoan tubes, whatever the animal inhabiting them may have been, have recently been found in the uppermost Precambrian of the Grand Canyon (Bloeser *et al.*, 1977). They are clearly forerunners of forms otherwise known only from Upper Cambrian times onwards.

A possible primitive mollusc, *Wyattia reedensis*, of tube-like form in very late Precambrian limestones of California, is the only reasonably well substantiated mollusc in the Precambrian (Taylor, 1966).

A nodule with more or less regular arrangement of diagenetic pyrite crystals has been described as the oldest known organism from South African Archaean rocks (Pflug, 1976). In my opinion it can only be relegated to the category of pseudofossils.

CONCLUSION

With the majority of Precambrian fossils being of soft-bodied organisms difficult to classify in the recognized categories, they provide little firm evidence of the forerunners or common ancestors of Phanerozoic Phyla, indeed little even of the divergence of Plants and Animals. The fossil record is, however, very incomplete and our record of Precambrian fossils is restricted to a very limited group of sedimentary environments where soft-bodied organisms could be preserved. Who knows what other fossils will be discovered? As exploration of the Precambrian stratigraphic record continues to intensify, there is little doubt that more fossils will be found, but they will probably raise more problems than they solve.

REFERENCES

ALLISON, C. (1975). Primitive fossil flatworm from Alaska: new evidence bearing on the ancestry of the Metazoa. *Geology*, **3**, 649–652.

ANDERSON, M. M. and MISRA, S. B. (1968). Fossils found in the Precambrian Conception Group of Southeastern Newfoundland. *Nature, Lond.* **220**, 680–681.

AWRAMIK, S. M., GOLUBIC, S. and BARGHOORN, E. S. (1972). Blue-green algal cell degradation and its implication for the fossil record. *Geol. Soc. Am., Absts. with Progs.* **4** (7), 438.

BARGHOORN, E. S. and SCHOPF, J. W. (1966). Microorganisms Three Billion years old from the Precambrian of South Africa. *Science, N.Y.* **152**, 758–763.

BLOESER, B., SCHOPF, J. W., HORODYSKI, R. J. and BREED, W. J. (1977). Chitinozoans from the Late Precambrian Chuar Group of the Grand Canyon, Arizona. *Science, N.Y.* **195**, 676–679.

BOYNTON, H. E. (1978). Fossils from the Precambrian of Charnwood Forest, Leicestershire. *Mercian geol.* **6**, 291–296.

BOYNTON, H. E. and FORD, T. D. (1979). *Pseudovendia charnwoodensis*, a new Precambrian Fossil Arthropod from Charnwood Forest, Leicestershire. *Mercian geol.* **7** (In press.)

BROOKS, J. and SHAW, G. (1971). Evidence for life in oldest known sedimentary rocks in Onverwacht series chert, Swaziland System of South Africa. *Grana*, **11**, 1–8.

CLEMMEY, H. (1976). World's oldest animal traces. *Nature, Lond.* **261**, 576–578.

CLOUD, P. E. (1968). Pre-Metazoan evolution and the origins of the Metazoa. *In* "Evolution and Environment" (E. T. Drake, ed.), pp. 1–72. Yale Univ. Press.

CLOUD, P. (1976). Beginnings of biospheric evolution and their biogeochemical consequences. *Paleobiology*, **2**, 351–387.

CLOUD, P., WRIGHT, J. and GLOVER, L. (1976). Traces of animal life from 620 million year old rocks in North Carolina. *Am. Scient.* **64**, 396–406.

COPE, J. C. W. (1977). An Ediacara-type fauna from South Wales. *Nature, Lond.* **268**, 624.

COWIE, J. W. (1967). Life in Precambrian and early Cambrian times. *In* "The Fossil Record" (W. B. Harland, C. H. Holland, M. R. House, N. F. Hughes, A. B. Reynolds, M. J. S. Rudwick, G. E. Satterthwaite, L. B. H. Tarlo and E. C. Willey, eds), pp. 17–35. *Geol. Soc.* London.

DOWNIE, C. (1967). The geological history of the microplankton. *Rev. Palaeobot. Palynol.* **1**, 269–281.

DUNN, P. R. (1964). Triact spicules in Proterozoic rocks on the Northern Territory of Australia. *J. geol. Soc. Austr.* **11**, 195–7.

FEDONKIN, M. A. (1977). Precambrian–Cambrian ichnocoenoses of the east European platform. *In* "Trace Fossils 2" (T. P. Crimes and J. C. Harper, eds), pp. 183–194. Seel House Press, Liverpool.

FORD, T. D. (1958). Pre-Cambrian fossils from Charnwood Forest. *Proc. Yorks. Geol. Soc.* **31**, 211–217.

FORD, T. D. (1968). The Precambrian palaeontology of Charnwood Forest. *In* "Geology of the East Midlands" (P. C. Sylvester-Bradley and T. D. Ford, eds), pp. 12–14. Univ. Leicester Press.

FORD, T. D. and BREED, W. J. (1973). The problematical fossil *Chuaria*. *Palaeontology*, **16**, 535–550.

GARRETT, P. (1970). Phanerozoic stromatolites: noncompetitive ecologic restriction by grazing and burrowing animals. *Science, N.Y.* **169**, 171–173.

GERMS, G. J. B. (1972). The stratigraphy and palaeontology of the Lower Nama Group, South West Africa. *Bull. Chamber Mines Precambrian Res. Unit, Geol. Dept., Univ. Cape Town.* No. 12, 1–250.

GERMS, G. J. B. (1973). A re-interpretation of *Rangea schneiderhöhni* and the discovery of a related new fossil from the Nama Group, South West Africa. *Lethaia*, **6**, 1–10.

GLAESSNER, M. F. (1962). Precambrian fossils. *Biol. Rev.* **37**, 467–494.

GLAESSNER, M. F. (1963). Zur Kenntnis der Nama-fossilien Südwest-Afrikas. *Ann. Naturhist. Mus. Wien.* **66**, 113–120.

GLAESSNER, M. F. (1966). Precambrian palaeontology. *Earth Sci. Rev.* **1**, 29–50.

GLAESSNER, M. F. (1969). Trace fossils from the Precambrian and basal Cambrian. *Lethaia*, **2**, 369–393.

GLAESSNER, M. F. and WADE, M. (1966). The Late Precambrian fossils from Ediacara, South Australia. *Palaeontology*, **9**, 599–628.

GLAESSNER, M. F. and WADE, M. (1971). *Praecambridium*—a primitive arthropod. *Lethaia*, **4**, 71–77.

GLAESSNER, M. F. and WALTER, M. R. (1975). New Precambrian fossils from the Arumbera Sandstone, Northern Territory, Australia. *Alcheringa*, **1**, 11–28.

GOLDRING, R. (1967). Trace-fossils. *In* "The Fossil Record" (W. B. Harland *et al.* eds), pp. 622–623. Geol. Soc. London.

GOLUBIC, S. and HOFMANN, H. J. (1976). Comparison of holocene and mid-Precambrian Entophysalidaceae (Cyanophyta) in stromatolitic algal mats: cell division and degradation. *J. Paleont.* **50**, 1074–1082.

GRABERT, H. (1973). Die Biologie des Präkambrium. *Z. Geol. Pal.* **1**, 316–346.

GÜRICH, G. (1930). Der bislang ältesten Spuren von Organismen in Südafrika. *C.r. Int. Geol. Cong. XV (S. Africa 1929)*, **2**, 670–680.

GUSSOW, W. C. (1973). *Chuaria* sp. cf. *C. circularis* Walcott from the Precambrian Hector Formation, Banff National Park, Canada. *J. Paleont.* **47**, 1108–1112.

HALSTEAD TARLO, L. B. (1967). *Xenusion*—Onychopohoran or coelenterate? *Mercian geol.* **2**, 97–99.

HÄNTZSCHEL, W. (1962). Trace fossils and problematica. *In* "Treatise on Invertebrate Paleontology, Part W" (R. C. Moore, ed.), pp. 177–245. Geol Soc. Am. and Univ. Kansas Press.

HOFMANN, H. J. (1967). Precambrian fossils (?) near Elliot Lake, Ontario. *Science, N.Y.* **156**, 500–504.

HOFMANN, H. F. (1971). Precambrian fossils, pseudofossils and problematica in Canada. *Bull. geol. Surv. Can.* **189**, 1–146.

HOFMANN, H. J. (1976). Precambrian microflora, Belcher Islands, Canada: Significance and Systematics. *J. Paleont.* **50**, 1040–1073.

JAEGER, H. and MARTINSSON, A. (1967). Remarks on the problematic fossil *Xenusion auerswaldae*. *Geol. För. Stockh. Förh.* **88**, 435–452.

JENKINS, R. J. F. and GEHLING, J. G. (1978). A review of the frond-like fossils of the Ediacara assemblage. *Rec. S. Aust. Mus.* **17** (23), 347–359.

JUX, U. (1977). Über die wandstrukturen Sphaeromorpher acritarchen: *Tasmanites* Newton, *Tapajonites* Sommer and Van Boekel, *Chuaria* Walcott. *Palaeontographica*, A, **160**, 1–6.

KELLER, B. M. and FEDONKIN, M. A. (1976). New fossil finds in the Valdai Series (Precambrian). *Izv. Akad. Nauk*, Ser. Geol., No. 3, 38–44. (In Russian.)

KING, A. F., BRUECKNER, W. D., ANDERSON, M and FLETCHER, T. (1974). "Late Precambrian and Cambrian sedimentary sequences of Eastern Newfoundland. Ann. Mtg. Field Guide B-6." *Geol. & Miner. Ass. Can. St Johns*, 1–59.

MISRA, S. B. (1969). Late Precambrian (?) fossils from Southeastern Newfoundland. *Bull. geol. Soc. Am.* **80**, 2133–2140.

MOORE, L. R., MOORE, J. R. M. and SPINNER, E. (1969). A geomicrobiological study of the Precambrian Nonesuch shale. *Proc. Yorks. Geol. Soc.* **37**, 351–394.

PEAT, C. T., MUIR, M. D., PLUMB, K. A., McKIRDY, D. M. and NORWICK, M. S. (1978). Proterozoic microfossils from the Roper Group, Northern Territory, Australia. *BMR J. Austral. Geol. Geophys.* **3**, 1–17.

PFLUG, H. D. (1970a). Zur Fauna der Nama-Schichten in Südwest-Afrika. I. Pteridinia, Bau und systematische Zugehörigkeit. *Palaeontographia*, A, **134**, 226–262.

PFLUG, H. D. (1970b). Zur Fauna der Nama-Schichten in Südwest-Afrika. II. Rangeidae, Bau und systemetische Zugehörigkeit. *Palaeontographica*, A, **135**, 198–231.

PFLUG, H. D. (1971). Neue Fossilfunde im Jung-Präkambrium und ihre Aussagen zur Entstehung der höheren Tiere. *Geol. Tdsch.* **60**, 1340–1350.

PFLUG, H. D. (1972a). Systematik der jungpräkambrischen Petaloname Pflüg 1970. *Paläont. Z.*, **46**, 56–67.

PFLUG, H. D. (1972b). Zur Fauna der Nama-Schichten in Südwest-Afrika. III. Erniettomorpha, Bau und Systematik. *Palaeontographica*, A, **139**, 134–170.

PFLUG, H. D. (1973). Zur Fauna de Nama-Schichten in Südwest-Afrika. IV. Mikroskopische Anatomie der Petalo-Organismen. *Palaeontographica*, A, **144**, 166–202.

PFLUG, H. D. (1976). Strukturiet erhaltene Fossilien aus dem Archaikum von PONNAMPERUMA, C. (1977). "Chemical evolution of the early Precambrian". Academic Press, New York and London.

RICHTER, R. (1955). Die ältesten Fossilien Sud-Afrikas. *Senckenberg. leth.* **36**, 243–289.

ROWELL, A. J. (1971). Supposed Pre-Cambrian brachiopods. *Smithson. Cont. Palaeobiol.* **3**, 71–79.

SCHOPF, James M. (1969). Precambrian microfossils. In "Aspects of Palynology" (R. H. Tschudy and R. A. Scott, eds). Wiley, New York.

SCHOPF, J. W. (1975). Precambrian palaeobiology: problems jand perspectives. In: "Annual Review of Earth and Planetary Sciences."

SCHOPF, J. W. (1977). Biostratigraphic usefulness of stromatolitic Precambrian microbiotas: a preliminary analysis. *Precambrian Research*, **5**, 143–173.

SCHOPF, J. W. (1978). The evolution of the Earliest Cells. *Scientific American* **239**, 84–102.

SCHOPF, J. W. and BARGHOORN, E. S. (1967). Alga-like fossils from the early Precambrian of South Africa. *Science, N.Y.* **156**, 508–511.

SCHOPF, J. W. and OEHLER, D. Z. (1976). How old are the Eukaryotes? *Science, N.Y.* **193**, 47–49.

SCHOPF, J. W., OEHLER, D. Z., HORODYSKI, R. J. and KVENVOLDEN, K. A. (1971). Biogenicity and significance of the oldest known stromatolites. *J. Paleont.* **45**, 477–485.

SCHOPF, J. W., HAUGH, B. N., MOLNAR, R. E. and SATTERTHWAIT, D. F. (1973). On the development of Metaphytes and Metazoans. *J. Paleont.* **47**, 1–9.

SOKOLOV, B. S. (1973). Vendian of Northern Eurasia. *Mem. Am. Ass. Petrol. geol.* **19**, 204–218.

SEILACHER, A. (1956). Der Beginn des Kambriums als biologische Wende. *Neues Jb. Geol. Paläont.* **103**, 155–180.

SQUIRE, A. D. (1973). Discovery of Late Precambrian trace fossils in Jersey, Channel Isles. *Geol. Mag.* **110**, 223–226.

TAYLOR, M. E. (1966). Precambrian mollusc-like fossils from Inyo County, California. *Science, N.Y.* **153**, 198–201.

TIMOFEEV, B. V. (1969). "Proterozoic Sphaeromorphitidae". Akad. Nauk. USSR, Inst. Geol. Geophys. Precambrian. (In Russian.)

TIMOFEEV, B. V. (1973a). "Microscopic plant fossils of the Ukrainian Precambrian". Akad. Nauk. USSR. Inst. Geol. Geophys. Precambrian. (In Russian.)

TIMOFEEV, B. V. (1973b). Proterozoic and early Paleozoic microfossils. (In Russian, English Summary.) 7–12, Akad. Nauk. USSR, Siberian Branch, Inst. Geol. & Geophys., Proc. 3rd Int. Palynol. Congr. Microfossils of the Oldest Deposits.

TYLER, S. A. and BARGHOORN, E. S. (1954). Occurrence of structurally preserved plants in Pre-Cambrian Rocks of the Canadian Shield. *Science, N.Y.* **119**, 606–608.

VIDAL, G. (1976). Late Precambrian microfossils from the Visingsö Beds in southern Sweden. *Fossils and Strata*, No. 9, 1–57.

WADE, M. (1968). Preservation of soft-bodied animals in Precambrian sandstones at Ediacara, South Australia. *Lethaia*, **1**, 238–267.

WADE, M. (1972a). Hydrozoa and Scyphozoa and other medusoids from the Precambrian Ediacara fauna, South Australia. *Palaeontology*, **15**, 197–225.

WADE, M. (1972b). *Dickinsonia*: polychaete worms from the late Precambrian Ediacara fauna. *Mem. Qd. Mus.* **16**, 171–190.

WALCOTT, C. D. (1899). Precambrian fossiliferous formations. *Bull. geol. Soc. Am.* **19**, 199–244.

WALTER, M. R. (1972). Stromatolites and biostratigraphy of the Australian Precambrian and Cambrian. *Spec. Pap. Palaeontology*, **11**, 1–256.

WALTER, M. R. (1976). "Stromatolites." Elsevier.

WALTER, M. R. (1977). Interpreting stromatolites. *Am. Scient.* **65**, 563–571.

WEBBY, B. D. (1970). Late Precambrian Trace Fossils from New South Wales. *Lethaia*, **3**, 79–109.

Note added in proof:

While this chapter was in press a discovery of an even earlier assemblage of microfossils from the Archaean of Western Australia has been published by Dunlop *et al.* (1978).

DUNLOP, J. S. R., MUIR, M. D., MILNE, V. A. and GROVES, D. I. (1978). A new microfossil assemblage from the Archaean of Western Australia. *Nature* **274**, 676–678.

3 | Radiation of the Eukaryote Protista

M. A. SLEIGH

Department of Biology, University of Southampton, England

Abstract: The Protista evolved from prokaryote organisms and gave rise to higher animals, higher plants and higher fungi. The serial symbiosis theory of the origin of eukaryotes, which is supported in this article, suggests that the main radiation of eukaryotes took place among heterotrophic forms. Some of these heterotrophs fed on prokaryotic algae of various types, a few of which became endosymbionts and allowed the organisms to become autotrophic and to evolve into the various groups of algae and higher plants; the heterotrophic habit is retained in many of the protistan groups involved in this development. Other forms remained entirely heterotrophic and have evolved into such protistan groups as the various amoebae, flagellates (including lower fungi), ciliates, sporozoans and cnidosporans. From amongst the groups involved in this protistan radiation have evolved the sponges, fungi, cnidarians and the acoel ancestor of other metazoa, probably by four separate routes.

Among the Protista are many variants of nuclear structure, nuclear division, meiosis and life cycles, which have been taken as primary characters in suggesting a protistan phylogeny. Supporting evidence is derived particularly from flagellar characters, the nature of any pellicle or cell wall, mitochondrial structure and plastid structure and pigmentation.

INTRODUCTION TO THE PROTISTA

Living organisms are divided into prokaryotes and eukaryotes on the basis of the possession by the latter of a nuclear envelope which encloses the nuclear material, and membranous organelles such as mitochondria and plastids in the cytoplasm. Bacteria and blue-green algae are prokaryotes. The Protista are the simplest of the eukaryotes, being more complex than

Systematics Association Special Volume No. 12, "The Origin of Major Invertebrate Groups", edited by M. R. House, 1979, pp. 23–53. Academic Press, London and New York.

the prokaryotes but less complex than the land plants and multicellular animals. Protista evolved from prokaryotes and from the Protista these "higher" organisms evolved. Although most authors now reserve the term Protista for the simple eukaryotes, Haeckel (1866), had originally proposed the term for all lower organisms without tissues, and some authors still include prokaryotes within the Protista. In this account the term Protista will be used in its more restricted sense, to include only eukaryotes.

Amongst the more complex organisms it is an easy matter to distinguish animals from plants on the basis of nutritional characters. Although among the autotrophic higher plants there are a few heterotrophic forms that have abandoned photosynthesis for a saprophytic or parasitic mode of life, it is not difficult to allocate these forms appropriately to the Plant Kingdom and distinguish them from members of the Animal Kingdom, on the basis of a wealth of morphological characters. However, the limits of these two kingdoms among the simpler organisms has provoked considerable discussion in recent years, notably by Copeland (1956), Whittaker (1969) and Leedale (1974). (Whittaker's paper contains a valuable account of earlier theories as well as his own interpretation.)

The separation of prokaryotes from eukaryotes is now recognized as being more fundamental even than the division of higher animals from higher plants, and provides a separate kingdom of the prokaryotes, the Kingdom Monera. Copeland proposed a subdivision of living organisms into four kingdoms, Monera, Metaphyta, Metazoa and Protoctista (= Protista), and Whittaker has taken a further step by suggesting that the Fungi, which are plant-like in the possession of cell walls but animal-like in heterotrophic nutrition and lack of plastids, should qualify for recognition as a fifth kingdom of organisms. The Fungi, higher plants and higher animals all have their origins in the Protista, but are only three of many diverse lines of evolution of Protista after these had evolved from prokaryotes. In Whittaker's view the three kingdoms derived from the Protista could be separated on nutritional characters as photosynthetic (Kingdom Plantae), absorptive (Kingdom Fungi) and ingestive (Kingdom Animalia) (Fig. 1). Leedale points out that in Whittaker's scheme the Protista contains phyla more closely related to phyla in the three higher kingdoms than to other phyla in the Protista. It is also true that different groups within the Protista differ more profoundly from one another than any two metazoan animals, and in the view of some people more profoundly than higher animals differ from higher plants or Fungi. Leedale goes on to suggest that it would perhaps be more realistic to consider each major pattern of life as a kingdom, having its origins in the prokaryotes and developing through a protistan phase. This gives some 18 eukaryote kingdoms and is attractive because it separates such different groups as sponges from metazoan animals, red algae from brown algae and from higher green plants and also unicellular groups such as ciliates from

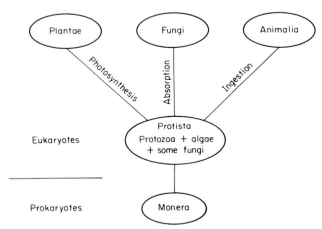

FIG. 1. The relationships of the five kingdoms of organisms discussed in the text.

sporozoans and from euglenids. However, the origins of eukaryotes from prokaryotes may not have been as polyphyletic as this scheme would suggest; in my opinion the concept of the Protista as an aggregation of lower eukaryotes remains valuable in discussion of the evolution of life, and in this account will be regarded as one of the five kingdoms, broadly in the sense used by Whittaker.

The Protista contains a diversity of radiating lines of evolution. Some lines are purely heterotrophic and recognizably animal (various Protozoa, in the strict sense). Others are purely autotrophic and are clearly lines of plant evolution which lead through various algal groups to multicellular photosynthetic plants, some of them phylogenetically distant from the line that led to the green land plants. Several lines contain both heterotrophic and autotrophic members and are now believed to be so scattered among the protistan radiation that it is not justifiable to attempt an extension of the separation of animal and plant kingdoms into the Protista. The Kingdom Protista includes a diversity of organisms showing a given level of organization rather than a common form of nutrition (as is also the case with the Kingdom Monera). The Protista is an important kingdom because from some branches of its radiation have emerged the three other recognized kingdoms of eukaryotes. It is also important because of the diversity of patterns of organization and functioning which show alternative life-forms of great variety, and because its members contribute in many significant ways to life on earth.

The organisms that comprise the Protista, in the sense used here, are classified by botanists and zoologists among the Algae, Protozoa and lower Fungi. It includes groups that are almost entirely unicellular, for example the euglenoid|flagellates, ciliate protozoa or chytrid fungi, as well as groups that contain both unicellular and multicellular forms. Because there is no recognizable boundary line, the red algae and brown algae, each of which show evolution from simple forms to macrophytic aquatic plants of considerable complexity, are all considered here as members of the Protista. These large algae are conventionally regarded as members of the plant kingdom, but the origins of the red and brown algae are from different lines of the protistan evolutionary radiation, and the differences of organization and reproduction of these plants, by comparison with each other and with land plants, suggest that they are better regarded as developments within the Kingdom Protista and not within the Kingdom Plantae. (They are of course prime candidates for consideration as separate kingdoms in the sense used by Leedale.) A similar justification could be applied to include the sponges as Protista rather than as members of the Kingdom Animalia. Various other groups amongst the Protista include colonial or multicellular organisms, so that, although the original Protista were unicellular forms, almost all patterns of organization that evolved have found benefits in aggregation of cellular units—colonial aggregates of prokaryotes also appear, of course, amongst both the bacteria and the blue-green algae. This account is concerned with those eukaryotes that are recognizably not metazoan animals, higher fungi or land plants, though the origins of these groups will also be considered; a list of the groups that may be considered to comprise the Protista is given in Table I.

It has been remarked that Charles Darwin wrote "The Origin of Species" without any mention of Protozoa (Sandon, quoted in Corliss 1974). However, with the acceptance of the concept of evolution and a better knowledge of single-celled animals it became natural to discuss the origin of Protozoa. The protozoologists of the late nineteenth and early twentieth centuries were divided as to whether amoeboid or flagellate forms of Protozoa were the more primitive; for example, Butschli (1882) favoured flagellates and Minchin (1912) favoured amoebae as ancestral

TABLE I. Groups that comprise the Eukaryote Protista

Informal name	Botanical name	Zoological name	Comment
*red algae	Rhodophyta	—	—
†*cryptophytes	Cryptophyta	Cryptomonadida	—
†*dinoflagellates	Dinophyta	Dinoflagellida	—
†*euglenids	Euglenophyta	Euglenida	—
†*chrysophytes	Chrysophyta	Chrysomonadida	—
*diatoms	Bacillariophyta	—	—
*brown algae	Phaeophyta	—	—
†*haptophytes	Haptophyta	—	coccolithophorids
†*green algae	Chlorophyta	Volvocida etc.	several groups?
† water moulds	Oomycetes	—	—
† chytrids	Chytridiomycetes	—	—
† collar flagellates	—	Choanoflagellida	—
† trypanosomes etc.	—	Kinetoplastida	—
† trichonymphids	—	Hypermastigida	—
† amoebae	—	Rhizopoda	polyphyletic?
† slime moulds	Myxomycetes	Mycetozoa	polyphyletic?
† foraminiferans	—	Foraminiferida	—
† radiolarians	—	Radiolaria	several groups?
† heliozoans	—	Heliozoia	several groups?
† sporozoans	—	Sporozoa	= Apicomplexa
† microsporans	—	Microspora ⎱	formerly
† myxosporans	—	Myxospora ⎰	Cnidospora
† ciliates	—	Ciliophora	—

The following additional flagellate groups are recognized: Eustigmatophyta*, Prasinophyta*, Xanthophyta*, Chloromonadophyceae*, Bicoecida†, Retortamonadida†, Oxymonadida†, Diplomonadida†, Trichomonadida†, Opalinata†. The amoeboid groups (amoebae to heliozoans, above) are often subdivided, the Acantharia† are usually separated from the radiolarians and the Labyrinthulida† are an enigmatic amoeboid group. Haplospora† may be separate from other cnidosporans.

* indicates autotrophic members.   † indicates heterotrophic members.

unicellular forms and hence as ancestors of all eukaryotes. This controversy has not been satisfactorily resolved, but it seems likely, as indicated below, that a prokaryote with phagocytic ability (and therefore an amoeboid form) may have developed into the first eukaryote before the development of eukaryote flagella, and that all, or practically all, of the eukaryotes subsequently evolved from this flagellate stock (see for example, Margulis 1970).

It is also interesting to speculate on whether the ancestry of all eukaryote organisms led through an autotrophic phase, in which photosynthetic ancestral forms underwent an adaptive radiation and gave rise not only to the diversity of autotrophic forms, but also to all heterotrophic eukaryotes, or whether there was a radiation of heterotrophic forms, a number of which became autotrophs. The author has previously written (Sleigh 1973; see also Cavalier-Smith 1975) in support of the first alternative because the autotrophic eukaryotes with the simplest plastids are the red algae, which also lack eukaryote $(9+2)$ flagella; these flagella are present in all the other major groups of eukaryotes (assuming a conventional classification of Fungi to include those lower forms with flagellate stages). It was therefore suggested that the red algae diverged from the eukaryote stock before the evolution of the $9+2$ flagellum, and that evolution among the autotrophic forms led first to the chromophyte algae (brown, yellow, etc.) which radiated to give a diversity of descendants including all protozoan and other animal groups, and also gave rise to the chlorophyte algae (green) from which the land plants evolved (Fig. 2). However, it is also possible to assume that the ancestors of the red algae possessed flagella, but lost them, and that the ability to perform photosynthesis was acquired secondarily and independently by a diversity of heterotrophic eukaryotes; on present evidence this appears the more satisfactory hypothesis and will be considered in more detail below.

In the context of this Symposium on the origins of invertebrate groups, particular attention should be paid to the animal representatives among the Protista and to the origins of multicellular forms from protistan ancestors. In this article, the main pattern of radiation of Protista will be considered and evidence for the relationships will be discussed; however, the dispersed occurrence of autotrophy and the generally more complete information available about members of groups containing autotrophs make it essential to include them in the discussion. Indeed, I am convinced that this is a genuinely biological (as opposed to botanical or zoological) area of interest, and cannot be approached from one side alone. In his recent monumental survey of interrelationships of lower animals, Hanson (1977) only includes those groups that are entirely animal-like; he neglects thereby a large part of the evidence that can be brought to bear on the phylogeny of Protista, and even of its animal groups.

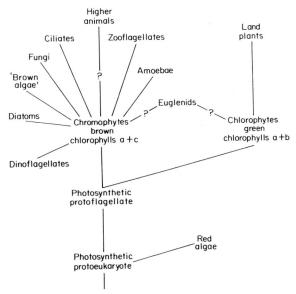

FIG. 2. A phylogenetic scheme of Protista which assumes that radiation among autotrophic forms led to the origins of present-day heterotrophic groups.

THE NATURE OF THE EVIDENCE FOR RELATIONSHIPS

The most direct evidence for the history of a group of organisms can usually be obtained from the fossil record. The majority of Protista lack hard parts that produce clear fossils, and there is at present very little evidence to suggest which members of the group may have evolved first. Microfossils that resemble filaments of blue-green algae are known from middle and late Precambrian rocks (Barghoorn and Schopf, 1965; Nagy, 1974; Schopf and Barghoorn, 1967), but Precambrian fossils that are supposedly eukaryotic are not easily assigned to any present-day group. Characteristic microfossils called acritarchs appeared abundantly during the Precambrian and persisted through several later periods, but although they *may* be dinoflagellates or related forms, the earliest fully accepted

TABLE II. Some characteristics of protistan and related groups (from references in text)

Group of organisms	Photosynthetic pigmentation, colour green (G), brown (B), red (R), type of chlorophylls (a, b, c_1, c_2) and presence of phycobilins (p)	Approximate proportion of autotrophic members	Cell surface naked (N), pellicle (P), wall of cellulose (Wc) of silica (Ws), scales (Sc), shell (Sh)	Flagella, number, mastigonemes (M), intraflagellar rod (R), hairs (H), naked (N)	Meiosis zygotic (Z), gametic (G), intermediary (I), believed absent (X), uncertain position (?)	Kinetochores present (✓) or absent (X)	Centrioles (C), nuclear plaques (Pn), cytoplasmic non-centriole plaques (Pc), no centriolar structures (X)	Nuclear spindle external (E), internal (I) or both	Nuclear envelope in mitosis open (O), closed (C) or fenestrated (F)	Chromosomes att. to nuc. memb. in mitosis	Chromosomes condensed in interphase
Dinophyta	B,a c_2	$\frac{1}{2}$	P	2,RH	N	X	X	E	C	✓	✓
Syndiniophyceae	—	0	N	2,RH?	?	✓	C	E	C	✓	✓
Hypermastigid Zooflag.	—	0	N	many,N	Z,G	✓	'C'	E	C	✓	✓
Kinetoplastida	—	0	P	2,R(H)	X	?	X	I	C	✓	?
Euglenophyta	G,ab	1/3	P	2,RH	X	✓	'C'	I	C	X	✓
Sporozoa	—	0	P	(2,N)	N	✓	Pn(C)	I(both)	C(F)	X	X
Ciliophora	—	0	P	many,N	G	✓	Pn	I	C	X	X
Myxospora	—	0	N	X	G	✓	Pn	I	C	X	X

Microspora	X	X	C	I	Pn	√	X	X	N	0	—
Fungi (higher groups)	X	X	C(F)	I	(Pn)	(√)	Z,I,G	X	Wc	0	—
Chytrids	X	X	F	both	C	√	Z,I	1	Wc?	0	—
Oomycetes	X	X	F(C)	both	Pn(C)	√	G?	2,M	Wc	0	—
Zooflagellates (other)	X	X	C,O	I,both	?	?	Z,G	1–many,N	N	0	—
Radiolaria	X	X	C	I	Pn(C)	√	?	(√)	N	0	—
Foraminiferida	X	X	C,O	I,both	Pn	?	I	(2)(3)	Sh	0	—
Amoebae, Schizopyrenida	X	?	C	I	X	X	X	(√)	N	0	—
Amoebae (some others)	X	X	(O)	I,both	?	X?	X	X	(N,Sh,Sc)	0	—
Heliozoa	X	X	O	both	Pc	?	G	(√)	N(Sc)	0	—
Chrysophyta	X	X	F,O	both	'C'	X	X	2,M	N(Sc)	$\frac{3}{4}$	B,a$c_1$$c_2$
Bacillariophyceae	X	X	O	both	Pc	?	G	(√)	Ws	All	B,a$c_1$$c_2$
Phaeophyta	X	X	F	both	C	?	I,G	2,M	Wc	All	B,a$c_1$$c_2$
Cryptophyta	X	X	O	both	X	X	X	2,H	(P)	$\frac{1}{2}$	Varies, ac_2p
Chlorophyta	X	X	F(O)	both	C(Pc)	(√)	Z,I	2–many	Wc	most	G,ab
Rhodophyta	X	X	F	both	Pc	√	Z,I	X	Wc	All	R,ap
Metaphyta	X	X	O	both	Pc	√	I	(many)	Wc	most	G,ab
Metazoa	X	X	O	both	C	√	G	1–many	N	0	—

When a symbol is enclosed by brackets, only some members of the group are included. 'C' indicates that some structure associated with the centriole is involved.

Note. Many records in this table are based on single observations and need confirmation.

dinoflagellate fossils appear in Silurian deposits (Sarjeant, 1974). Among other Protista that might be expected to leave good fossils, the foraminiferans first appeared in Cambrian and Ordovician rocks, radiolarians in the Ordovician or possibly the Cambrian, tintinnid ciliates in the Ordovician, coccolithophores in the upper Cambrian, diatoms in the Jurassic and silicoflagellates in the Cretaceous (Moore, 1964). Members of the majority of protistan groups might not be expected to leave such well preserved fossils as these, and even those that are well preserved provide few clues about relationships. However, the study of many types of microfossils is at an early stage, and there is yet hope for valuable evidence from fossil material.

Corliss (1974) has pointed out that it is not only the relative lack of fossil evidence that makes it difficult to give confident support to phylogenetic schemes involving protistan organisms—he was referring to Protozoa. In addition, the characteristic nature of Protozoa, including their microscopic size and subcellular differentiation, implying the need to study features at the ultrastructural or even molecular level, their common lack of sexuality, and their apparent cosmopolitan distribution, preventing any zoogeographical considerations, all make phylogenetic comparisons more difficult. Further, our knowledge of the organisms is still both elementary and very uneven—there is a high proportion of undescribed species, and in few groups have more than one or two species been adequately described even ultrastructurally, let alone in terms of physiology, morphogenesis and ecology—so that the data is too scanty for a really adequate comparison between groups.

In spite of these justified comments about our thin and uneven knowledge of the protistan groups, a number of authors have considered such information as is available, and it is possible to make preliminary comparisons of a number of structural, biochemical and physiological features. Since almost all of the groups involved have unicellular members, and all members of some groups are unicellular, the principal comparative characteristics to be studied are at the cellular and subcellular level. It is believed that a variety of features of chromosomes, nuclei and nuclear division appeared during the early evolution of the eukaryotes, and a number of combinations of these features have persisted in different groups. In addition, the nature of the body surface and the form of the flagella are often distinctive. Such features provide the data for Table II; this information will be used extensively and further explained in subsequent sections of this article, and is combined with other information to support a phylogenetic scheme for the Protista.

THE SYMBIOTIC THEORY OF THE ORIGIN OF EUKARYOTES

The most satisfying theory of the origin of eukaryotes suggests that two
or more prokaryotic organisms came together in a symbiotic relationship
to form an ancestral eukaryote. The arguments for this have been

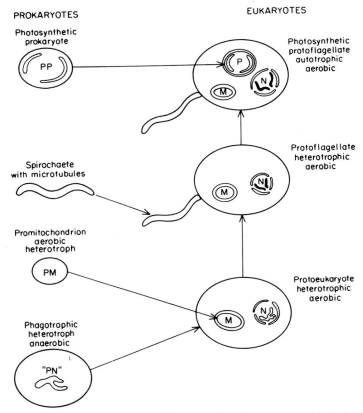

FIG. 3. The serial symbiosis theory of the origin of eukaryotes (modified from
Margulis, 1970). A phagotrophic form with "pronucleus" (PN) engulfed
a "promitochondrion" (PM) and became associated with a spirochaete to
form a protoflagellate with mitochondrion (M) and nucleus (N).
Autotrophic eukaryotes arose by symbiosis with a "proplastid" (PP)
which became integrated in the cell as a plastid (P).

explained in detail by Margulis (1970, 1975), and subsequent evidence,
referred to below, has continued to strengthen the support for the theory.
However, the theory has not been without critics, for example Raff and
Mahler (1975) and Mahler and Raff (1975) gave biochemical evidence for

the view that the mitochondrion is not a symbiont but a "complex supra-molecular structure," and Cavalier-Smith (1975) has suggested that the membranous organelles of the cell could have been formed from the photosynthetic membrane systems of ancestral blue-green algae; more substantial and wide ranging evidence is required if these proposals are to find wide support among workers on algae and protozoa. Symbiotic associations involving relationships at the cellular level between members of a great diversity of protistan and prokaryotic groups abound; many appear to be stable symbioses and whilst some are supposedly ancient others appear to be recent. If such symbioses are so frequent and successful today, there can be little reason to doubt that symbioses occurred in the early history of life on earth and the likelihood of these having persisted in the form of the eukaryotic cell appears very strong.

As presented here, the theory (Fig. 3) suggests that a Precambrian phagotrophic prokaryote, that was heterotrophic but unable to use oxygen for energy releasing reactions, engulfed an aerobic heterotrophic pro-karyote like the *Paracoccus* of today, which can perform ATP synthesis using oxygen in the same way as a mitochondrion (John and Whatley, 1975; Whatley *et al.*, 1978). If the aerobic form was maintained as an intracellular symbiont within the phagotroph, the composite "organism" could survive as an aerobic heterotroph, at a level of organization that bears comparison with the present-day amoeba *Pelomyxa palustris*, which is believed to be a rather simple phagotroph with prokaryotic endosymbionts and without true mitochondria. In time the aerobic endosymbiont is assumed to have become fully integrated into the biochemical systems of the cell as a mitochondrion, improving its exchange of metabolites and nucleotides with the host cell and losing some of its genetic information to the host cell nucleus. (Margulis (1975) has pointed out that since phagocytosis is virtually unknown in present-day prokaryotes, the alter-native of active invasion of an anaerobic prokaryote by a penetrative *Bdellovibrio*-like "promitochondrial", prokaryote may provide a more acceptable explanation.)

The host cell nucleus is assumed to have become more complex in the early eukaryote by separation of the genetic material to form several chromosomes and by segregation of cellular membrane material to form a nuclear envelope. The "chromosome" of bacterial prokaryotes is often associated with the cell membrane, and in some of the Protista the chromosomes retain a persistent association with the inner membrane of the nuclear envelope which is most evident at nuclear division. The eukaryote chromosomes have also become more complex in several ways, of which the following two are the most obvious: (i) the synthesis of DNA in bacterial "chromosomes" is continuous, but in most eukaryotes it is confined to a specific period (the S period) of the cell cycle; (ii) the DNA of the bacterial chromosome is not associated with histone proteins whilst these proteins normally form part of the chromatin structure of eukaryote chromosomes.

In general, prokaryotes lack microtubules, but these are characteristic of eukaryote cells. It has recently been shown that the large pillotina-type spirochaetes (prokaryotes) contain genuine microtubules (Margulis, 1978a, b); spirochaetes can form symbiotic associations with protozoan cells and by their motility propel the protozoan (Cleveland and Grimstone, 1964). Margulis (1970) has suggested that the origin of flagellar and mitotic microtubule systems of eukaryotes may be from a symbiotic association of spirochaetes with early eukaryote cells that gave rise to the ancestral flagellated eukaryote (Fig. 3). (The "flagellum" of a bacterium is morphologically, biochemically and physiologically quite different from the eukaryote 9 + 2 flagellum, and could not have directly given rise to the eukaryote flagellum.) Microtubules seem generally to be formed in association with specific types of microtubule organizing centres (MTOC's), the most important of which is the flagellar basal body (functionally and morphologically equivalent to, and probably ancestral to, the centriole); such MTOC's may be supposed to have had their origin from the motile spirochaete symbiont, whose genetic capacity to produce tubulin, the microtubule protein, has been transferred to the nucleus of the host cell. The acquisition of microtubules allowed the development of a more precise, fibre-based, system for regulating chromosome separation at nuclear division (mitosis) to replace the membrane-associated system that had been inherited from prokaryotes. Such a flagelleted, microtubule-producing, eukaryote would appear to be ancestral to all eukaryote groups, except possibly some amoebae of the *Pelomyxa* type. It is assumed here that it was the source of the main radiation of eukaryote Protista.

A third step in the serial symbiosis is supposed to have involved the phagocytic uptake of a photosynthetic prokaryote by a flagellated eukaryote (Fig. 3). Thus Margulis (1970) suggested that a blue-green algal symbiont taken up by an amoebo-flagellate would have given rise to an early plant flagellate, with the alga developing into the plastid of the plant cell and giving up some of its genetic control to the nucleus of the host cell. From a blue-green prokaryote ancestor the logical development would be the production of a red alga for in both cases the only chlorophyll pigment present is chlorophyll a, and in both groups phycobilic proteins are present in phycobilisomes located around the membranous sacs in which the chlorophyll is bound. The red algae also contain the simplest type of plastids with separate thylakoids (the chlorophyll-containing lamellate sacs), that could well be derived from the single thylakoids of blue-green algae.

The recent discovery of *Prochloron*, a different type of photosynthetic prokaryote which has grouped thylakoids containing chlorophylls a and b (Lewin, 1976, Whatley, 1977), suggests that different types of prokaryotes containing true chlorophylls have evolved, not just the blue-green type. An alga of the *Prochloron* type could well have been the ancestor of the plastid symbiont of some green algal groups (Chlorophyta and Euglenophyta), in which the pigments include chlorophylls a and b. It seems reasonable to speculate that yet more chlorophyll-containing prokaryotes evolved, and that among them were forms with chlorophylls a and c, as well as others with phycobilic proteins in addition; these "prochromophyte" prokaryotes could have provided the symbionts ancestral to the plastids of the various groups of brown algae (Chromophyte phyla).

The evolution of different combinations of photosynthetic pigments in these prokaryotic ancestors provides a satisfying explanation of the appearance of a diversity of plastid types within the eukaryote radiation. According to this explanation, a Precambrian prokaryote, which had already developed chlorophyll a evolved along various lines in which different additional pigments (among them phycobilins and chlorophylls b and c) were developed. The resulting prokaryote forms provided a diversity of potential symbionts which could be taken up by heterotrophic forms from different parts of the radiation of Protista and give similar plastid symbionts in dissimilar host cells (e.g. Chlorophyta and Euglenophyta) (see Fig. 6, below). If this polyphyletic origin of the plastid symbionts in different groups of Protista is correct, then plastid characters and pigmentation should only be used with great caution when inferring relationships between various major photosynthetic groups.

The conclusion from this discussion is that an ancestral flagellated eukaryote gave rise to a variety of heterotrophic forms with different characteristics. The diet of many of these forms would have included photosynthetic prokaryotes, and in some of the flagellates prokaryotic organisms became established as plastid symbionts. Some of the resulting photosynthetic eukaryotic organisms have retained the ability to feed heterotrophically, but others have become pure autotrophs and have often developed cell walls that prevent further phagocytosis. Evolution of the heterotrophs, the heterotroph–autotrophs ("plant-animals") and the autotrophs has proceeded from these ancestral forms in many directions and provides the diversity of eukaryotic forms that we see today, as well as, presumably, a number of lines of evolution that have become extinct. It is important to recognize that unicellular heterotrophic flagellates occupy a central place in this scheme of protistan evolution. The evidence that supports this view of the radiation of Protista will be discussed in the next section.

SOME FEATURES OF PROTISTAN GROUPS AND SUGGESTIONS CONCERNING THEIR PHYLOGENY

The earliest protistan is assumed to have been a protoeukaryote, formed by a symbiotic association between an anaerobic, heterotrophic prokaryote and a "promitochondrial" prokaryote, which was beginning to develop eukaryote nuclear features. At this stage the chromosomes of the proto-eukaryote probably underwent continuous DNA synthesis throughout the cell cycle, lacked associated histone proteins and remained in the condensed configuration during the interphase stage of division. A nuclear envelope had been developed but during the nuclear division the nuclear envelope remained intact, extending and pinching into two as the divided chromosomes separated into two groups within it. The chromosomes probably remained associated with the nuclear envelope at all times, and separation of the daughter chromosomes involved membrane activity.

Throughout the transition to a protoflagellate, during which microtubules and the capacity to develop flagella were gained, possibly by symbiosis with a spirochaete, the body surface probably retained its generalized phagocytic function. Thus the body surface was formed by a naked membrane capable of forming phagocytic vacuoles at any or many parts of the surface.

With the acquisition of microtubules the form of mitosis could develop to something more conventional. In one pattern the microtubules become associated with the outside of the nuclear envelope in places where chromosomes are attached on the inside, but the nuclear envelope remains "closed", as in many dinoflagellates. In many other Protista the microtubules form a mitotic spindle within the closed nuclear envelope and regulate the movement of chromosomes which have become detached from the nuclear envelope. In another series of groups openings develop at the poles of the nuclear envelope, or it breaks down completely, to give an "open" mitosis in which only microtubules remain to control the positions and motion of the chromosomes. Centrioles may or may not be present. Meiosis, though relatively uniform in character, occurs erratically through the groups involved in these developments; it is not yet clear whether it evolved early and was lost in some groups, or whether a similar form of meiosis was developed independently by different groups; detailed work on the structural aspects of the process is even more difficult than with mitosis, and comparative information at this level, which is necessary for phylogenetic purposes, is not yet available. It appears that the primitive form of sexual life cycle was a haploid one with meiosis of the zygote following quickly after formation of the nuclear synkaryon at fertilization. In various groups the meiosis has been delayed, giving a prolonged diploid phase, first with intermediary meiosis separating two substantial phases, one diploid, one haploid, and eventually with a principal diploid phase being followed by meiosis in the formation of gametes, and the haploid phase being confined to the gametes.

The evolution of the protoflagellate seems to have led along two main routes, one in which the nuclear envelope remains closed at mitosis, at least until a much later stage of evolution, and the other in which the nuclear envelope becomes perforated at the poles or breaks down to a greater or lesser extent during an open mitosis; the distribution of these features in flagellates has been reviewed by Taylor (1976), and the evolution of the mitotic spindle has been reviewed by Kubai (1975). Open mitosis appears to be associated with a retention, for some time at least, of a naked cell surface capable of widely distributed phagocytosis, while in the forms with closed mitosis phagocytic activity was restricted to a single cytostome area and the body surface became modified to form a pellicle over the remaining area. Groups which develop cell walls and lose phagocytic ability are present in both sections. Much of the data that are used in the following paragraphs are summarized in Table II; where detailed references are not given, the relevant information is drawn from Dodge (1973), Grell (1973), Sleigh (1973) or Taylor (1976). Relationships inferred in these discussions are integrated in the schemes presented in Figs 4, 5, and 6.

The most primitive nuclear organization among eukaryotes appears in dinoflagellates (reviewed by Loeblich, 1976). Within this group all forms appear to have condensed chromosomes throughout interphase and histones are only present in small quantities, the relevant basic proteins apparently being qualitatively different from those in higher eukaryotes. There is some evidence that the chromosomes contain significant amounts of repeated sequences of DNA, which is a eukaryote feature. It is reported that DNA synthesis in the primitive dinoflagellate *Prorocentrum* is continous throughout the cell cycle, as in prokaryotes, although some restriction of the period of synthesis has been found in some other dinoflagellates. However, reports that microtubules are not associated with nuclear division of *Prorocentrum* have been disproved by more recent work. In all dinoflagellates the nuclear division is of a primitive, closed type with the chromosomes remaining (or becoming) attached to the persistent nuclear envelope during division. In the majority of dinoflagellates in which the dividing nucleus has been studied, the nucleus is penetrated by many membrane-lined tunnels containing extranuclear microtubules. In *Cryp-thecodinium* Kubai and Ris (1969) reported that these microtubules do not make contact with the membranes but pass straight through and remain of constant length so that they merely provide a framework through the nucleus, whose polarity can direct the membrane-mediated separation of the chromosomes. In other forms some of the microtubules terminate on the outer nuclear membrane at sites where chromosomes are attached internally (with or without kinetochores). Centrioles are not present in these dinoflagellates. In the parasitic *Syndinium* species (at present classed as dinoflagellates, but placed in a separate subsection by Loeblich, 1976) the microtubules terminate at centrioles in the cytoplasm and pass into a single membrane-lined tunnel through the nucleus to attach to the nuclear envelope at the sites where chromosomes are anchored by kinetochores— there are few chromosomes in this type, many in other species studied. Separation of daughter chromosomes is associated with elongation of pole-to-pole microtubules, but not of those attached to the kinetochores (Ris and Kubai, 1974). Microtubule elongation is usual in eukaryote mitoses. In all dinoflagellates for which evidence is available, the normal cells are haploid and the zygote undergoes a two-division meiosis. The nuclear characters of the dinoflagellates appear rather intermediate between prokaryotes and typical eukaryotes; the most acceptable explanation for this appears to be that a heterotrophic dinoflagellate ancestor diverged from the main line of eukaryote evolution after the appearance of meiosis, but before the full development of eukaryote nuclear characters.

Other groups showing primitive nuclear features are the euglenids, the kinetoplastids and the polymastigid flagellates. In the last, particularly in

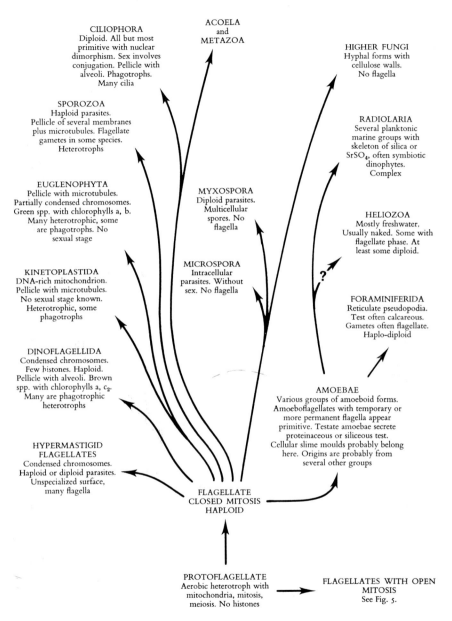

CILIOPHORA
Diploid. All but most
primitive with nuclear
dimorphism. Sex involves
conjugation. Pellicle with
alveoli. Phagotrophs.
Many cilia

ACOELA
and
METAZOA

HIGHER FUNGI
Hyphal forms with
cellulose walls.
No flagella

SPOROZOA
Haploid parasites.
Pellicle of several membranes
plus microtubules. Flagellate
gametes in some species.
Heterotrophs

RADIOLARIA
Several planktonic
marine groups with
skeleton of silica or
SrSO$_4$, often symbiotic
dinophytes.
Complex

EUGLENOPHYTA
Pellicle with microtubules.
Partially condensed chromosomes.
Green spp. with chlorophylls a, b.
Many heterotrophic, some
are phagotrophs. No
sexual stage

MYXOSPORA
Diploid parasites.
Multicellular
spores. No
flagella

HELIOZOA
Mostly freshwater.
Usually naked. Some with
flagellate phase. At
least some diploid.

KINETOPLASTIDA
DNA-rich mitochondrion.
Pellicle with microtubules.
No sexual stage known.
Heterotrophic, some
phagotrophs

MICROSPORA
Intracellular
parasites. Without
sex. No flagella

FORAMINIFERIDA
Reticulate pseudopodia.
Test often calcareous.
Gametes often flagellate.
Haplo-diploid

DINOFLAGELLIDA
Condensed chromosomes.
Few histones. Haploid.
Pellicle with alveoli. Brown
spp. with chlorophylls a, c$_2$.
Many are phagotrophic
heterotrophs

AMOEBAE
Various groups of amoeboid forms.
Amoeboflagellates with temporary or
more permanent flagella appear
primitive. Testate amoebae secrete
proteinaceous or siliceous test.
Cellular slime moulds probably belong
here. Origins are probably from
several other groups

HYPERMASTIGID
FLAGELLATES
Condensed chromosomes.
Haploid or diploid parasites.
Unspecialized surface,
many flagella

FLAGELLATE
CLOSED MITOSIS
HAPLOID

PROTOFLAGELLATE
Aerobic heterotroph with
mitochondria, mitosis,
meiosis. No histones

FLAGELLATES WITH OPEN
MITOSIS
See Fig. 5.

FIG. 4. Some features and interrelationships amongst those groups of protistans
with closed mitosis.

hypermastigids (symbionts in insects) the chromosomes tend to remain condensed in interphase, the spindle is extranuclear with centrioles and the chromosomes are attached to the nuclear envelope at the sites of spindle attachment. Kubai (1975) suggests that the microtubules only become attached to kinetochores in the membrane after the initial separation of daughter kinetochores by a membrane-mediated mechanism, and that only the later stages of formation of daughter nuclei involve activity of microtubules.

The euglenids also have partially condensed chromosomes and a closed mitosis, but in this case the mitotic spindle is intranuclear, the spindle fibres terminating at the internal nuclear membrane. Here the chromosomes do not appear to possess kinetochores and are not attached to either the microtubules or the membrane; the separation of daughter chromosomes during mitosis is not well synchronized, but the suspicion remains that the microtubules play some part in their movement (see note, p. 54).

In the kinetoplastid flagellates (*Trypanosoma, Bodo* and allied forms) the chromatin is not in the form of continuously condensed chromosomes, but remains associated with the nuclear envelope. At division of the closed nucleus an internal mitotic spindle is formed as in euglenids. The separation of daughter chromosomes appears to involve membrane-mediated activities, distant from the spindle, at least initially, though kinetochore-like structures appear on the spindle during anaphase and could imply attachment of chromosomes to the spindle in the later stages of mitosis (Vickerman, 1974). The kinetoplastids are unique in the possession of a large DNA body (the kinetoplast) within the single mitochondrion.

Meiosis does not appear to occur in either euglenids or kinetoplastids; in the hypermastigid *Trichonympha* the cells are haploid with zygotic meiosis, but some related genera are diploid with gametic meiosis, and among these *Urinympha* shows a curious one-division meiosis of a type found in a few other symbiotic flagellates (Oxymonadida). It is interesting that the euglenids, kinetoplastids and dinoflagellates share another unique feature in the striated intraflagellar rod structure lying alongside the axoneme (9 + 2 bundle of fibres). They also resemble one another in that the body surface is specialized to a greater or lesser extent to form a pellicle (within the surface membrane), a substantial component of which is provided by arrays of microtubules.

Although euglenids and dinoflagellates share a number of similar features of cellular organization, their plastids are very different. Of 38 genera of euglenids, 11 contain autotrophic members, the others being phagotrophic or "osmotrophic" (Leedale, 1967). The plastids of euglenids are green, containing chlorophylls a and b, suggesting that an ancestral euglenid may have taken in a *Prochloron*-like symbiont. Photosynthetic dinoflagellates are brown, with chlorophylls a and c, derived presumably from a symbiotic "prochromophyte". The normal pigment composition is chlorophylls a and c_2, but *Peridinium foliaceum* contains chlorophylls a, c_1 and c_2, a pattern that is characteristic of chrysophytes and diatoms, suggesting that this species of *Peridinium* has acquired its plastids independently. In fact, it is believed to have taken in a eukaryote symbiont from one of these other chromophyte groups because the thylakoid pattern of the plastids is unlike that of dinoflagellates but typical of chrysophytes and diatoms, and because the plastids are associated with a second nucleus of a more typical eukaryote type with chromosomes unlike those in the dinoflagellate nucleus of the main cell (Loeblich, 1976). Phagotrophy is very widespread in dinoflagellates, even among some species that are autotrophic; the relatively recent acquisition of such a symbiont should not be surprising. Both euglenids and dinoflagellates are here regarded as having evolved as heterotrophic groups, and the independent acquisition of plastids in these two otherwise fairly closely related groups seems the most acceptable explanation of the origins of their quite different photosynthetic features.

Two other groups which show a closed mitosis with intranuclear spindles, in these cases terminating at plaques on the nuclear envelope, are the ciliates (micronuclei) and the wholly parasitic sporozoans. A few sporozoa have fenestrate mitotic nuclei with spindles that extend through these polar openings from centrioles in the cytoplasm (see Vivier and Vickerman, 1974). Kinetochores have been reported in both groups. The Sporozoa are haploid with zygotic meiosis, but the ciliates are diploid (in their micronuclei which contain the genetic material concerned with sexual events) and undergo meiosis in the maturation of the gametic nuclei. Ciliates normally possess many $9+2$ organelles, but Sporozoa lack them except for a few with flagellate gametes. These two groups share with dinoflagellates a pellicular structure in which the surface membrane is characteristically underlain by a single layer of flattened alveoli (membranous sacs); in Sporozoa the three membranes (one surface and two alveolar) are normally pressed closely together, but in ciliates the alveoli often remain fluid-filled as a layer of "cushions" on the cell surface, and in dinoflagellates the cellulose pellicular stiffening plates of the theca are laid down within the alveoli (Dodge and Crawford, 1970; Sleigh, 1973).

Although the typical modern ciliate has a unique nuclear dimorphism, this is now regarded as having developed within the group (see e.g. Corliss, 1975). The most primitive of present-day ciliates is *Stephanopogon* (Primociliatida), whose nuclei are of only one diploid type, but which have a characteristic ciliate life-form in many other respects. An intermediate stage is represented by several families of lower ciliates (Karyorelictida, e.g. *Loxodes*) which possess diploid macronuclei that never divide, and synthesize RNA but not DNA, as well as more typical ciliate micronuclei which contain no RNA and synthesize only DNA—all new nuclei are therefore formed by division of micronuclei. (An almost identical nuclear dualism also occurs in foraminiferans.)

The Microspora and Myxospora (formerly the two subsections of the Cnidospora, but now regarded as independent) are specialized parasites with closed mitosis, intranuclear spindles that terminate at plaques on the nuclear envelope and with kinetochores (at least in microsporans, which are better known). The myxosporans appear to be diploid with a gametic meiosis, which apparently takes place in a single division (Schulman and Semenovitch, 1973). Flagella are not present in either group and the body surface is naked except in the spore stages. While these groups are probably close to the other groups with closed mitosis, more definite relationships cannot yet be established.

In the majority of zygomycete, ascomycete and basidiomycete fungi the mitosis is closed with intranuclear spindles. According to Kubai (1975) and Fuller (1976) mitotic patterns in these fungi are very variable. In some cases the nuclear division is comparable with that of euglenids in the lack of kinetochores and absence of contact between the chromosomes and microtubules. In other cases kinetochores are present and spindle attachments occur, as in Sporozoa and ciliates. The ends of the microtubules at the nuclear envelope may or may not be embedded in dense plaques. In some Basidiomycetes the nuclear envelope is fenestrate or may break down at metaphase. None of these "higher" fungi possess flagella, and should not be confused with such "lower fungi" as Oomycetes and chytrids, which probably arose at different parts of the evolutionary radiation of Protista. Oomycetes have two flagella of the typical heterokont type (see below), while chytrids have a single posterior flagellum; centrioles are normally present, as are kinetochores, but the nucleus sometimes appears to remain closed, even in the presence of centrioles in the cytoplasm outside e.g. *Catenaria* while in other cases the nuclear envelope is fenestrate during mitosis. Myxomycetes are probably more closely related to amoeboid or

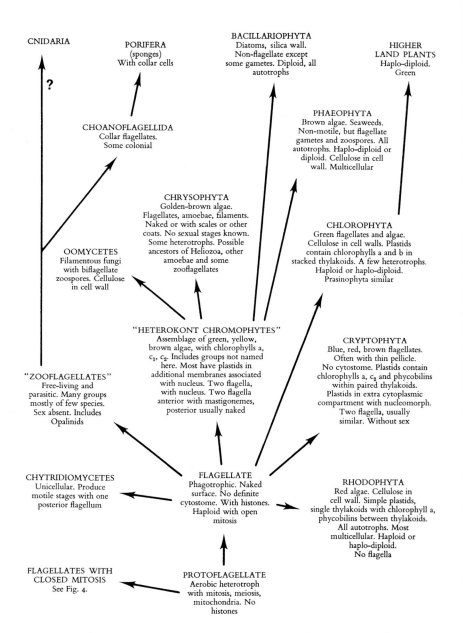

CNIDARIA

?

PORIFERA
(sponges)
With collar cells

BACILLARIOPHYTA
Diatoms, silica wall.
Non-flagellate except
some gametes. Diploid, all
autotrophs

HIGHER
LAND PLANTS
Haplo-diploid.
Green

CHOANOFLAGELLIDA
Collar flagellates.
Some colonial

PHAEOPHYTA
Brown algae. Seaweeds.
Non-motile, but flagellate
gametes and zoospores. All
autotrophs. Haplo-diploid or
diploid. Cellulose in cell
wall. Multicellular

CHRYSOPHYTA
Golden-brown algae.
Flagellates, amoebae, filaments.
Naked or with scales or other
coats. No sexual stages known.
Some heterotrophs. Possible
ancestors of Heliozoa, other
amoebae and some
zooflagellates

CHLOROPHYTA
Green flagellates and algae.
Cellulose in cell walls. Plastids
contain chlorophylls a and b in
stacked thylakoids. A few heterotrophs.
Haploid or haplo-diploid.
Prasinophyta similar

OOMYCETES
Filamentous fungi
with biflagellate
zoospores. Cellulose
in cell wall

"HETEROKONT CHROMOPHYTES"
Assemblage of green, yellow,
brown algae, with chlorophylls a,
c_1, c_2. Includes groups not named
here. Most have plastids in
additional membranes associated
with nucleus. Two flagella,
with nucleus. Two flagella
anterior with mastigonemes,
posterior usually naked

CRYPTOPHYTA
Blue, red, brown flagellates.
Often with thin pellicle.
No cytostome. Plastids contain
chlorophylls a, c_2 and phycobilins
within paired thylakoids.
Plastids in extra cytoplasmic
compartment with nucleomorph.
Two flagella, usually
similar. Without sex

"ZOOFLAGELLATES"
Free-living and
parasitic. Many groups
mostly of few species.
Sex absent. Includes
Opalinids

CHYTRIDIOMYCETES
Unicellular. Produce
motile stages with one
posterior flagellum

FLAGELLATE
Phagotrophic. Naked
surface. No definite
cytostome. With histones.
Haploid with open
mitosis

RHODOPHYTA
Red algae. Cellulose in
cell wall. Simple plastids,
single thylakoids with chlorophyll a,
phycobilins between thylakoids.
All autotrophs. Most
multicellular. Haploid or
haplo-diploid.
No flagella

FLAGELLATES WITH
CLOSED MITOSIS
See Fig. 4.

PROTOFLAGELLATE
Aerobic heterotroph
with mitosis, meiosis,
mitochondria. No
histones

FIG. 5. Some features and interrelationships amongst those groups of protistans
with open mitosis.

flagellate Protozoa than to the higher fungi and here again several groups with different origins may have been put together; open and closed mitoses may occur at different parts of the same life cycle in *Physarum*, according to Aldrich (1969), so it is clear that nuclear factors must be considered alongside other phylogenetic indicators.

The amoeboid Protozoa are probably polyphyletic, but closed mitoses are frequent among them, and at least some amoebae may be assumed to belong in this section of the radiation of Protista. A group of small "limax-amoebae" (the order Schizopyrenida, Page, 1976) includes the amoeboflagellates *Tetramitus* (which Margulis (1970) regarded as primitive on nuclear characters) and *Naegleria*. In *Naegleria* the mitotic nucleus remains closed, with an intranuclear spindle, no obvious kinetochores and no plaques at the membrane (see Vivier and Vickerman, 1974). In the foraminiferans and radiolarians Hollande (1972) has described a closed mitosis with kinetochores and prominent plaques at the ends of the intranuclear spindle—in the radiolarian *Collozoum* centrioles were present in the cytoplasm outside the nucleus and near the plaques, as in the oomycete *Catenaria*. Both foraminiferans and radiolarians have flagellate gametes. Some other amoeboid forms will be considered later.

Organisms on the other main evolutionary route typically show a more or less open mitosis, though often the nuclear membrane only becomes fenestrate at the poles allowing the spindle microtubules to extend into the nucleus from MTOC's in the cytoplasm. Closed mitoses have also been reported in some of these groups, but mostly in cases where the intranuclear microtubules appear continuous with extranuclear micro-tubules, often associated with a centriole.

It is assumed that primitively the flagellate ancestor of this branch of protistan evolution had a naked surface and was capable of phagocytosis over a considerable part of its surface. Several of the autotrophic groups in this section have developed cell walls after the acquisition of plastids and became incapable of phagocytic heterotrophy. The red algae appear to provide a good example of this development. It is assumed that a naked phagotrophic protoflagellate engulfed a blue-green alga, which became a symbiotic plastid, since the plastids of red algae contain phycobilic proteins and chlorophyll a in a distribution only found elsewhere in blue-green algae. Subsequent to the acquisition of the photosynthetic symbionts the red algae are assumed to have abandoned phagotrophy, developed cellulose cell walls (cell walls are formed outside the cell membrane) and lost their flagella, to give the characteristic modern rhodophyte pattern.

The two main lines of the group have developed multicellular forms with complex life cycles. The nuclear membrane is fenestrate at mitosis (*Membranoptera*) with a spindle extending between cytoplasmic plaques, and with kinetochores; some red-algae are haploid, while others have intermediary meiosis and alternating haploid and diploid phases. Not to be confused with red algae are various symbiotic associations involving blue-green algal "cyanelles", that are apparently of a more recent origin. The most interesting of these is probably *Cyanophora paradoxa* in which the host cell may be a cryptophyte or a dinoflagellate, and there are other examples (e.g. *Glaucocystis*, *Glaucosphaera*) in which the host may be of chlorophyte origin—neither cryptophytes nor chlorophytes appear today to be phagotrophic, which presumably makes them unlikely hosts. Some have regarded *Glaucocystis* as a representative of a flagellate group close to red algae.

Green algae (Chlorophyta) have followed a similar route, but in this case the autotrophic prokaryote was presumably a "prochlorophyte" (like *Prochloron*) containing chlorophylls a and b, rather than a blue-green alga. Chlorophytes have evolved in various directions and their origins may well have been polyphyletic, for there is considerable diversity in nuclear patterns, reproduction and life forms. Normally a cellulose cell wall is present, and the diversity of forms includes solitary and colonial flagellates (some of which have become secondarily heterotrophic), filamentous forms lacking any flagellate stage and with direct conjugation as a sexual process (these especially may have had an independent origin), filamentous and thalloid forms with both flagellate gametes and flagellate asexual zoospores, and the line that led, presumably from near charophytes, to the green land plants that have flagellate male gametes in most groups. In this group the nucleus at mitosis is fenestrate or open, with centrioles in some members and cytoplasmic polar plaques in others; in general, the lower forms are haploid and higher forms have intermediary meiosis.

The cryptophytes have a curious form of plastid organization (Coombs and Greenwood, 1976). In the plastid itself the thylakoid membranes carry chlorophylls a and c_2 (as in dinoflagellates), but the thylakoids are usually paired and contain phycobilisomes—not found within the thylakoids of any other group. The plastids are isolated from the cytoplasm and nucleus of the cryptophyte by a second membrane which also encloses a cytoplasmic compartment containing a nucleomorph (a reduced eukaryote nucleus) (cf. *Peridinium foliaceum*, above). This suggests that the ancestral cryptophyte established a symbiotic association with an eukaryote which itself contained a symbiotic plastid derived from a "prochromophyte" prokaryote with an unusual combination of photosynthetic pigments. (A further stage of symbiosis is seen in the "red-tide" ciliate *Mesodinium rubrum* which contains a symbiotic cryptophyte, so that *Mesodinium* contains macro- and micro-nuclei of the ciliate, a large cryptophyte nucleus, the symbiont nucleomorph and plastid DNA in four separate concentric compartments (Hibberd, 1977).) Cryptophytes have an unspecialized surface or a thin pellicle, and present-day heterotrophic members of the group, a few species of which are extremely common, are not phagotrophic. The flagella usually carry hairs, but of a unique pattern which does not suggest any particular relationship. The nuclear membrane breaks down in mitosis and the chromosomes become arranged at the metaphase plate of a spindle that lacks centrioles or obvious plaques but kinetochores appear to be absent; no sexual stages are known.

A different form of "prochromophyte" symbiont is supposed to have provided the plastids of the chromophyte groups of algae which contain chlorophylls a, c_1 and c_2 and usually vary in coloration from golden-brown to dark brown. The most complex of three important groups here are the multicellular Phaeophyta (brown seaweeds) which have developed cellulose in their cell walls but reproduce by asexual flagellate zoospores and by flagellate gametes. They have intermediary or gametic meiosis and have nuclear spindles with centrioles, the microtubules from which enter the nucleus through polar fenestrations. The Chrysophyta are an important group of mostly biflagellate organisms characterized, like the flagellate stages of phaeophytes, by the presence of one posterior naked flagellum and one anterior flagellum with stiff lateral hairs (mastigonemes), arising side-by-side from the cell surface; this is known as the heterokont pattern, and is also found in oomycete fungi and xanthophytes. Chrysophytes have an open or fenestrate mitotic nucleus, the spindle being associated with flagellar basal bodies or structures attached to them; no sexual activity has been found in the group. Some of the chrysophytes secrete silica extracellularly, notably in loricas or cyst walls, but the related diatoms (Bacillariophyta) secrete silica cell walls and are entirely autotrophic diploid forms with an open mitosis and sometimes flagellate male gametes. Chrysophytes frequently secrete proteinaceous scales which may cover the otherwise naked cell surface, and others secrete a cellulose lorica or cup in which the cell lies. The group contains flagellate and amoeboid phagotrophic forms, some of which are pure heterotrophs while others are plant-animals, practising both autotrophy and heterotrophy simultaneously, e.g. *Ochromonas*; other forms are flagellate, amoeboid, coccoid or filamentous autotrophs. Other small chromophyte groups with heterokont flagellation and chlorophylls, a, c_1 and c_2 are xanthophytes, chloromonads and eustigmatophytes, which differ in a number of internal characters but do not appear to lead to any other groups. The haptophytes are a group of small algae that have two similar naked flagella and usually a pseudoflagellar organelle called a haptonema that contains about seven microtubules, shows motility unlike that of a flagellum and appears to be adhesive; members of this group often have dimorphic life cycles and many secrete scales, some of which are calcified as coccoliths and are abundant as microfossils.

The fact that some chrysophytes have an amoeboid form might be taken as an indication that some amoebae may have evolved from this chromophyte group; the secretion of silica in some testate amoebae, heliozoa, radiolaria and a few foraminiferans might also suggest such a relationship. In general the amoeboid life-form is characterized by a paucity of structural features that could be valuable phylogenetic indicators, and it seems very probable that it is a life-form that could have arisen many times during the evolution of the Protista, making the amoebae highly polyphyletic. Nuclear characters would seem to support this. For example, the nucleus of *Pelomyxa palustris* is thought to be "primitive"—such a form, lacking mitochondria but with other prokaryote symbionts, could have been derived from an early protoeukaryote and persisted in anaerobic sediments ever since, or it could have been derived from a much later flagellate by loss of mitochondria (it is reputed to carry occasional flagella); when more is known of its nuclear characteristics and division it may be possible to narrow down its ancestry. The occurrence of closed mitosis and intranuclear spindles in some amoeboid forms has already been mentioned. In *Amoeba proteus* gaps appear in the nuclear envelope but the spindle still appears intranuclear; in *Chaos* the nuclear envelope breaks down further and in heliozoans there is an open mitosis with polar plaques. In some amoebo-flagellates, notably the helioflagellate *Dimorpha*, as in some chrysophytes, the basal bodies of the flagella, or structures attached to them, act as the poles of the nuclear spindle, suggesting again that some amoebae and probably most heliozoan forms, may be derived from chrysophytes. Some heliozoan forms carry flagella or have biflagellate stages (*Clathrulina*) and it is known that some of the larger Heliozoa are diploid. It may be concluded that at least some of the considerable diversity of amoeboid groups were derived from chrysophyte flagellates or their ancestors but that others may have had their origins at one or several much earlier stages in the lineage of eukaryotes. Our ignorance of the features of the majority of amoebae, particularly the naked forms, which are in general less well known than flagellates because they are difficult even to identify, does not at present allow any better interpretation.

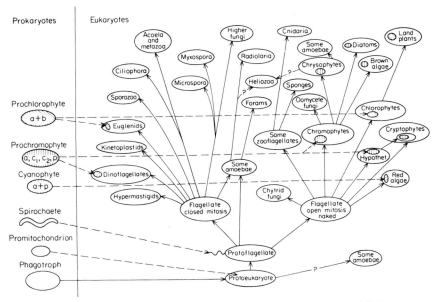

FIG. 6. A scheme of protistan radiation showing the distribution of different types of autotrophic symbionts.

There is also a heterogeneous collection of "zooflagellates" that have not yet been accounted for in this evolutionary scheme. The most interesting and characteristic group are the collar flagellates (choanoflagellates) which have often been supposed to have originated from chrysophytes, although Leadbeater and Manton (1974) suggest that the evidence of mitochondrial structure makes this very unlikely. This group is most notable because it is supposed to have led to the sponges, the only other group with a well developed collar of slender pseudopodia around the flagellar base. However our knowledge of the ultrastructural and nuclear features of both collar flagellates and sponges is inadequate for further conclusions about their relationships. It seems more certain that the heterotrophic bicoecid flagellates are derived from chrysophytes. The remaining parasitic multiflagellate forms, retortamonads, diplomonads and oxymonads, have a fenestrated nuclear envelope during mitosis, or a closed mitosis with intranuclear spindle (oxymonads) but many of these forms live in a highly peculiar and isolated environment. There exist other enigmatic flagellates whose classification is obscure, but have occasionally been studied, such as *Apusomonas* (Vickerman, 1974) whose nuclear division has similarities with the chrysomonad type.

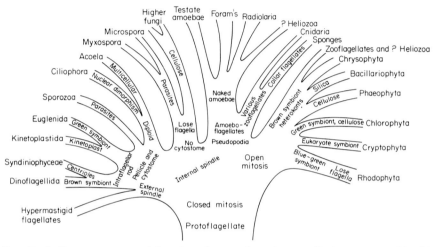

FIG. 7. A diagram summarizing the scheme of protistan radiation proposed in this article.

In this discussion considerable emphasis has been placed on nuclear characters, but other authors have drawn upon other data. Thus Taylor (1976) notes that the mitochondrial forms in euglenids and kinetoplastids are each unique, while cryptophytes and chlorophytes (and perhaps choanoflagellates (Leadbeater and Manton, 1974)) with flattened cristae can be distinguished from the tubular (or tubular-vesicular) cristae of chrysophytes, phaeophytes, diatoms and dinoflagellates (and which other authors have recorded in amoebae, ciliates, sporozoans and fungi); members of the multiflagellate parasitic groups lack mitochondria. The significance of mitochondrial characters is not clear since some features vary with the state of the cell. Flagellar characters have proved of phylogenetic interest in some cases, but not all protistans possess flagella. Characters of plastid structure, pigmentation and storage products, upon which much emphasis has been placed in the past, may be expected to depend more upon the photosynthetic symbiont than upon the host cell, so must be used cautiously when inferring phylogenetic relationships of the host cells. There are still numerous other features to be explored, for example the similarities of RNA sequences or cytochromes could be most informative, and the distribution of less common molecules like hydroxyproline or of particular biochemical pathways, should also provide some helpful data. The tentative conclusions drawn by the author on the basis of the information which has been discussed are shown as a dendrogram in Fig. 7. It is hoped that others will be stimulated to bring forward new data that will improve the accuracy of such syntheses, for many suggestions made here are based on only one or two reports, and some will certainly require revision when more species are studied.

THE ORIGINS OF MULTICELLULAR ANIMALS

If anything, discussions of this theme are fraught with even more difficulties than relationships within the Protista, for the sorts of feature which are now becoming better documented among the Protista are poorly known in the lower multicellular animals. Much is known about the structure and biochemistry of rat liver cells, but very little about the cells of simpler cnidarians and flatworms that may be closest to the protistan ancestry. A thorough survey of the previous literature on the subject is given by Hanson (1977) and will not be repeated here, where the concern is merely to suggest which groups of the Protista may have provided starting points for the evolution of multicellular groups.

It seems clear that the sponges represent an independent line of descent, and that this may be assumed to be an evolutionary dead end. The protistan choanoflagellate is assumed to have evolved a colonial habit like *Proterospongia* and further differentiated to form a sponge (see discussion by Hanson, 1977).

Hanson has also discussed the origin of the Cnidaria and remarks that their origin "seems at present to be totally obscure". Features which could indicate an ancestry among certain protistans are the presence of only one flagellum/cilium per cell, the absence of cell walls, the amoeboid phagocytosis and the nematocysts (similar structures to which occur in dinoflagellates and Myxospora). It appears that extrusive organelles may have been evolved several times in different groups, for somewhat simpler structures are widespread in Protista, so that nematocysts should be ignored in seeking an ancestor. The other three features can be found in colonies of colourless chrysomonads, *Oikomonas* for example, which could conceivably have developed into a planula stage and on to such simple polyps as *Otohydra*, which is ciliated all over and which Hanson regards as a possible primitive cnidarian.

Turbellarian flatworms have many cilia per cell and the smaller acoelans live much as ciliates do. The discovery that ciliates such as *Stephanopogon* have a single type of nucleus allows one to believe that such a ciliate, which is normally multinucleate and undergoes multiple fission, could have divided into many cells but not separated, so that it could have formed a small ciliated mass like a simple acoel. Although the cells of embryonic stages of acoels possess an open mitosis with centrioles, it is claimed from light microscopy that the mitoses of tissue cells in adult flatworms have a closed mitosis without centrioles—as in ciliates (see Hanson, 1977). It seems that the most plausible ancestry for the main line of metazoan evolution is from a stem form of the ciliate line which evolved into an acoel flatworm.

REFERENCES

ALDRICH, H. C. (1969). The ultrastructure of mitosis in myxamoebae and plasmodia of *Physarum flavicomum*. *Am. J. Bot.* **56**, 290–299.

BARGHOORN, E. S. and SCHOPF, J. W. (1965). Microorganisms from the late Precambrian of Central Australia. *Science, N.Y.* **150**, 337–339.

BUTSCHLI, O. (1882). Protozoa. *In* "Klassen und Ordnungen des Thier-Reichs" (H. G. Bronn, ed.), Vol. 1. C. F. Winter, Leipzig.

CAVALIER-SMITH, T. (1975). The origin of nuclei and of eukaryotic cells. *Nature, Lond.* **256**, 463–8.

CLEVELAND, L. R. and GRIMSTONE, A. V. (1964). The fine structure of the flagellate *Mixotricha paradoxa* and its associated microorganisms. *Proc. R. Soc. Lond. B*, **159**, 668–685.

COOMBS, J. and GREENWOOD, A. D. (1976). Compartmentation of the photosynthetic apparatus. *In* "The intact chloroplast" (J. Barber, ed.), Vol. 1, pp. 1–52. Elsevier/North Holland, Amsterdam.

COPELAND, H. F. (1956). "The Classification of Lower Organisms." Pacific Books, Palo Alto.

CORLISS, J. O. (1974). Time for evolutionary biologists to take more interest in protozoan phylogenetics? *Taxon*, **23**, 497–522.

CORLISS, J. O. (1975). Nuclear characteristics and phylogeny in the protistan phylum Ciliophora. *BioSystems*, **7**, 338–349.

DODGE, J. D. (1973). "The Fine Structure of Algal Cells." Academic Press, London and New York.

DODGE, J. D. and CRAWFORD, R. D. (1970). A survey of thecal fine structure in the Dinophyceae. *Bot. J. Linn. Soc.* **63**, 53–67.

FULLER, M. S. (1976). Mitosis in Fungi. *Int. Rev. Cytol.* **45**, 113–153.

GRELL, K. G. (1973). "Protozoology." Springer-Verlag, Berlin.

HAECKEL, E. (1866). "Generelle Morphologie der Organismen." Reimer, Berlin.

HANSON, E. D. (1977). "The origin and early evolution of animals." Pitman, London.

HIBBERD, D. J. (1977). Observations on the ultrastructure of the cryptomonad endosymbiote of the red-water ciliate *Mesodinium rubrum*. *J. mar. biol. Ass. U.K.* **57**, 45–61.

HOLLANDE, A. (1972). Le déroulement de la cryptomitose et les modalités de la ségrégation des chromatides dans quelques groupes de protozoaires. *Ann. Biol.* **11**, 427–466.

JOHN, P. and WHATLEY, F. R. (1975). *Paracoccus denitrificans*: A present-day bacterium resembling the hypothetical free-living ancestor of the mitochondrion. *Symp. Soc. Exp. Biol.* **29**, 39–40.

KUBAI, D. F. (1975). The evolution of the mitotic spindle. *Int. Rev. Cytol.* **43**, 167–227.

KUBAI, D. F. and RIS, H. (1969). Division in the dinoflagellate *Gyrodinium cohnii* (Schiller). A new type of nuclear reproduction. *J. Cell Biol.* **40**, 508–528.

LEADBEATER, B. S. C. and MANTON, I. (1974). Preliminary observations on the chemistry and biology of the lorica in a collared flagellate (*Stephanoeca diplocostata* Ellis). *J. mar. biol. Ass. U.K.* **54**, 269–276.

LEEDALE, G. F. (1967). "Euglenoid Flagellates". Prentice-Hall, Englewood Cliffs.

LEEDALE, G. F. (1974). How many are the kingdoms of organisms? *Taxon*, **23**, 261–270.

LEWIN, R. A. (1976). Prochlorophyta as a proposed new division of algae. *Nature, Lond.* **261**, 697–698.

LOEBLICH, A. R. III (1976). Dinoflagellate evolution: speculation and evidence. *J. Protozool.* **23**, 13–28.

MAHLER, H. A. and RAFF, R. A. (1975). The evolutionary origin of the mitochondrion: a non-symbiotic model. *Int. Rev. Cytol.* **43**, 1–124.

MARGULIS, L. (1970). "Origin of eukaryotic cells". Yale University Press, New Haven.

MARGULIS, L. (1975). Symbiotic theory of the origin of eukaryotic organelles: criteria for proof. *Symp. Soc. Exp. Biol.* **29**, 21–38.

MARGULIS, L. (1978a). Microtubules in prokaryotes. *Science, N.Y.* **200**, 1118–1124.

MARGULIS, L. (1978b). Biology of host-associated spirochaetes. *Proc. R. Soc. Lond. B.* (In Press.)

MINCHIN, E. A. (1912). "An introduction to the study of the Protozoa". Edward Arnold, London.

MOORE, R. C. (1964). Editor. "Treatise on invertebrate palaeontology", Parts C, D. Geol. Soc. Amer. Univ. Kansas Press.

NAGY, L. (1974). Transvaal stromatolite: first evidence for the diversification of cells about $2 \cdot 2 \times 10^9$ years ago. *Science, N.Y.* **183**, 514–516.

PAGE, F. C. (1976). A revised classification of the Gymnamoebia (Protozoa: Sarcodina). *Zool. J. Linn. Soc.* **58**, 61–77.

Raff, R. A. and Mahler, H. A. (1975). The symbiont that never was: an enquiry into the evolutionary origin of the mitochondrion. *Symp. Soc. Exp. Biol.* **29**, 41–92.

Ris, H. and Kubai, D. F. (1974). An unusual mitotic mechanism in the parasitic protozoan *Syndinium sp. J. Cell Biol.* **60**, 702–720.

Sarjeant, W. A. S. (1974). "Fossil and living dinoflagellates." Academic Press, London and New York.

Schopf, J. W. and Barghoorn, E. S. (1967). Algal–like fossils from the early Precambrian of South Africa. *Science, N.Y.* **156**, 508–512.

Schulman, S. S. and Semenovitch, V. N. (1973). The life cycle in Myxosporidia and the taxonomic position of the Cnidosporidia in the animal kingdom. *Proc. IV Int. Cong. Protozool., Clermont Ferrand*, p. 368.

Sleigh, M. A. (1973). "The Biology of Protozoa". Edward Arnold. London.

Taylor, F. J. R. (1976). Flagellate phylogeny: a study in conflicts. *J. Protozool.* **23**, 28–40.

Vickerman, K. (1974). Nuclear division in lower zooflagellates. *Actualités Protozoologiques*, **1**, 368. (Université de Clermont.)

Vivier, E. and Vickerman, K. (1974). Divisions nucleaires chez les protozoaires. *Actualités Protozoologiques*, **1**, 161–177. (Université de Clermont.)

Whatley, J. M. (1977). The fine structure of *Prochloron. New Phytol.* **79**, 309–313.

Whatley, J. M., John, P. and Whatley, F. R. (1978). From extracellular to intracellular: the establishment of mitochondria and chloroplasts. *Proc. R. Soc. Lond. B.* (In Press.)

Whittaker, R. H. (1969). New concepts of the kingdoms of organisms. *Science, N.Y.* **163**, 150–159.

Note added in proof:

Gillott and Triemer (1978) have found that during the mitosis of *Euglena* flagellar bases lie adjacent to the outside of the nucleus at both ends of the intranuclear spindle, and that the chromosomes are attached to spindle microtubules by definite kinetochores.

Gillott, M. A. and Triemer, R. E. (1978). The ultrastructure of cell division in *Euglena gracilis. J. Cell Sci.* **31**, 25–36.

4 | Radiation of the Metazoa

R. B. CLARK

Department of Zoology, University of Newcastle upon Tyne, England

Abstract: The major taxa are defined by anatomical features and this review of metazoan evolution is concerned with the emergence of difference styles of body architecture in animals. Despite the diversity of views about the origin of the Metazoa, only two alternative structures have been proposed for the stem group from which subsequent radiation took place: a solid bodied acoeloid/planuloid, or a trimerous coelomate. The latter hypothesis implies that the radiation of the "lower" Metazoa involved a steady loss of structural complexity (loss of trimery, obliteration of the coelom) and it is difficult to offer a functional explanation of the development of coelomic cavities so early in metazoan evolution. The former hypothesis entails an increase in structural complexity with increasing size and the colonization of new habitats, and avoids these difficulties. The radiation of acoelomate surface-creeping animals embraces the Platyhelminthes, Nemertea and Mollusca. Burrowing in the substratum by animals too large to exist as meiofauna was associated with the evolution of a hydrostatic skeleton in the form of a coelom. Three types of functional organization of early coelomates are visualized: unsegmented sedentary, metameric vagrant and oligomeric tubicolous. These three types of organization may have evolved independently from acoelomate predecessors. Pseudocoelomates in some respects parallel these developments, but this heterogeneous assemblage of animals may include secondarily simplified as well as primitive innovations in body architecture. The successful radiation of metameric metazoans followed the evolution of appendages, but the origin of the arthropodan phyla presents some unexplained features. The trimeric organization, represented in Pterobranchia and Phoronidea has produced two important radiations of the deuterostome and lophophorate phyla, respectively.

Systematics Association Special Volume No. 12, "The Origin of Major Invertebrate Groups", edited by M. R. House, 1979, pp. 55–101. Academic Press, London and New York.

INTRODUCTION

In considering metazoan radiation we are concerned with the evolution of the major styles of animal architecture. Animal classification has a purely anatomical basis and the major taxa are defined by their anatomical features. Biochemical, physiological, ecological, behavioural, or any other innovation is assumed to be reflected in anatomical change (if it is not, it is not usually recognized in the classificatory system) and so these are essentially subsidiary matters. They cannot be entirely neglected on that account. It is possible that non-morphological characters are useful in indicating relationships between different groups of animals (Løvtrup (1977) goes so far as to claim that *only* non-morphological characters can be used in considerations of phylogeny above the level of phyla) and any hypothesis that satisfies anatomical requirements must also be consistent with biochemical, physiological and other considerations as far as they are known. But, ultimately, it is the history of anatomical innovations that has to be reconstructed.

If there is rarely a complete concensus about phylogeny and radiation within individual phyla, areas of disagreement are usually well defined and the limited number of available options is reasonably clear. The situation is very different with the evolution of the major structural plans represented in the 30 or so separate phyla. At this level there are almost as many views as there have been writers, and interpretations range from a steady, progressive, almost linear evolution of the phyla (Hadzi, 1953) to the independent origin of almost every kind of metazoan from protistans (Nursall, 1962). This variety is a reflection of the equivocal nature of the evidence—largely that of comparative anatomy and comparative embryology—on which reconstructions of the major directions of animal evolution are based. Furthermore, the highest taxa have been consciously defined to include all animals between which there is any discernable relationship and, with few exceptions, to set a profound gulf between different phyla.

At various times more or less plausible arguments have been presented to demonstrate an evolutionary link between almost any phylum and almost any other. It has been postulated, for example, that vertebrates evolved from Nemertini (Willmer, 1970, 1974),* Annelida (Gutmann,

---

* In this paper, to keep the bibliography within bounds, I have generally cited only literature published within the last 25 years. This reflects recent or current views but does less than justice to the originators of them. I have also often cited comprehensive and later works of authors rather than papers in which a theory was first suggested.

1966, 1967), Mollusca (Sillman, 1960), Arachnida (Løvtrup, 1977), Echinodermata (Jeffries, 1968), Hemichordata (Bone, 1960) or Urochordata (Berrill, 1955). Most of these theories have a venerable history, but all have been canvassed, in one form or another, in recent years. Despite a revival of interest in evolutionary biology its latest manifestations have been more concerned with the mechanisms of evolution or with experiments in techniques to establish relationships between animals than with phylogeny, and have made relatively little contribution to solving these problems. It is therefore unrealistic to expect much immediate improvement in this chaotic situation.

Rather than catalogue the many phylogenies that have been proposed, still less to add to their number, I shall try to identify a number of key areas in this continuing debate and indicate what appear to be the alternative solutions that exist, and their implications. Later papers in this symposium will be concerned with detailed arguments about the origin, affinities and radiation of individual phyla, so in discussing the radiation of the Metazoa, I shall direct attention to broader considerations.

METHODOLOGY

In the absence, for the most part, of a useful fossil record, studies of metazoan phylogeny have historically been based on the classical evidence of comparative anatomy and comparative embryology. The use of anatomical characters of existing and, where available, fossil animals to indicate relatedness depends upon identifying homologous structures. Since these, by definition, are structures in existing animals which were transformed from the same structure in a common ancestral animal, there is inevitably a degree of circularity in this process. In practice, the identification of supposedly homologous structures depends on more that simple anatomical similarity, and Remane (1952) codified current practice in a number of criteria which, so far as they can be satisfied, increase the probability that two structures are, indeed, homologous. Despite the theoretical unsatisfactoriness of this approach there is surprising concensus about many of the homologies that have been proposed.

This has become less true of the findings of comparative embryology. Stemming from von Baer's conclusion that the early stages of ontogeny of related animals are more alike than later stages, considerable weight was commonly placed on resemblances between developmental and larval stages as evidence of phyletic relatedness between animals. This can be a two-edged weapon, however, and the same evidence may be interpreted as indicating recapitulation (following Haeckel) or paedogenesis (following Garstang). Although it is recognized that developmental processes may be subject to a variety of changes in the course of evolution (Jägersten, 1968), this approach tends to underestimate the importance of larval or other developmental adaptations which might result in convergent features in embryonic or larval stages of unrelated or only distantly related animals. Although there can be no doubt that phylogenetic relationships may be reflected in developmental similarities, understanding of them demands a more detailed and penetrating view of embryological or larval resemblances between different animals, as Jägersten (1968) and Anderson (1973) have shown, than the rather superficial comparisons that have often been made in the past.

The first objective in these enquiries, as well as of the more formalized approaches of numerical taxonomy (Sneath and Sokal, 1973) and cladism (Hennig, 1966), is to establish degrees of relatedness between groups of animals in the form of a phyletic tree or a cladistic dendrogram. But beyond this, if we are to reconstruct the evolutionary history of a group, it is necessary to invent hypothetical ancestral animals and offer some explanation of the sequence of events which gave rise to their modern descendants. Too often in the past this was an excuse for almost unbridled speculation. It frequently degenerated into a mere juggling with existing or idealized morphological types into a tidy series which was then assumed to represent the course of evolution of these animals.

In fact, there are a number of constraints on speculation (Clark, 1977).

1. The hypothetical animals must have been real animals living in a real world. In many respects that world was not different from our own, and physical and chemical constraints and ecological and physiological principles which have been found to apply in existing animals would have applied with equal force to their ancestors.
2. Animals function as an integrated whole. (There are some obvious qualifications to this, particularly among the lower Metazoa.) Evolutionary change in one organ system must therefore have been accompanied by greater or smaller associated changes in other organ systems, and throughout any transitional evolutionary changes (represented by hypothetical "intermediate" types) the organisms must have continued to be functional entities.
3. Jägersten (1968) has argued that the developmental stages, as well as adult structures, must be viewed in a similar light and the evolution of

the whole life cycle must show an intelligible transition from one form to another.

4. Each new evolutionary development must be presumed to have had conferred a selective advantage on its possessor over what existed before. This advantage is measured in improved or different functional capabilities in the animal.

It is now axiomatic that structure cannot be considered except in relation to its function and the achievements of functional morphology in the last fifty years have been such that it is often possible to stipulate with confidence the minimal conditions (or sometimes alternative sets of conditions) which must be satisfied for an animal to perform a particular task (Clark, 1964, 1978a). Although these conditions have been discovered by the study of living animals, they must have applied also in extinct forms. Comparable adaptations leading to close similarity in essential structure have undoubtedly appeared several times in quite unrelated animals and if this type of convergence is evident in existing animals there can be no reason to suppose it was not a feature of earlier stages of evolution. The strong correlation that can be demonstrated between particular types of structure and particular functions may suggest phyletic pathways, but to distinguish related features from those that are simply convergent it is necessary to turn to classical anatomical, embryological and, indeed, any other evidence of relationships between existing animals. A consideration of the functional attributes of particular structures does, however, offer some guide as to the selective advantage of postulated evolutionary changes and allows a more realistic view to be taken of the course of metazoan evolution.

METAZOAN STEM GROUPS

Essentially two types of structure have been postulated for the early metazoans, although there are several different views of how these structures evolved from protistans. These structures are a hollow gastraea or a solid acoeloid-planuloid. Either may be envisaged as being pelagic or benthic, or something approaching both types may be represented as pelagic and benthic stages, respectively, in the life cycle of early metazoans, but whichever view is taken there appears to have been little scope for morphological innovation and the evolution of radically different types of body architecture except for animals living on the substratum.

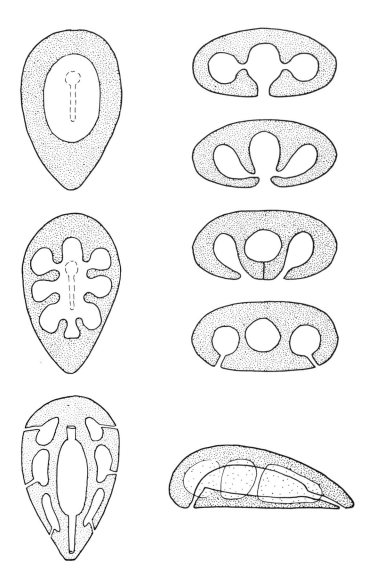

FIG. 1. Bilaterogastraea theory of Jägersten (1955). Diagrammatic plans and transverse sections of stages in the derivation of coelomic pouches from gastric pockets. Bottom right, lateral view of the hypothetical early oligomerous coelomate from which all Metazoa except coelenterates are derived.

1. *Gastraea-enterocoel ("Archicoelomate") theories*

Modern versions of the gastraea theory are linked with the enterocoel theory (Remane, 1950; Jägersten, 1955; Marcus, 1958; Siewing, 1969). It is envisaged that at an early stage in its history, the gastraea developed a small number of gastric pouches (4, 5, 6 or 8 in various statements of this theory) which then separated from the gut to form coelomic pouches (Figs 1, 2). It follows that the Cnidaria, Ctenophora and, in some versions of the theory, the Porifera also, are derived from the gastraea, but that all other Metazoa are coelomates or evolved from coelomates. It is therefore necessary to postulate that many groups suffered regressive evolution and loss of structural complexity before the start of their subsequent radiation. Whatever may be thought of these phylogenetic implications of so early an appearance of the coelom, there are two fundamental objections to the gastraea-enterocoel theory.

The first relates to the size of the animals in which these events are supposed to have taken place. Small animals can swim by cilia, creep on the substratum, or live as meiofauna, but once their size exceeds a very few millimetres, both swimming (except with the aid of special developments, as in Ctenophora) and burrowing require more powerful forces than can be generated by cilia, and other locomotory specializations become necessary. Metazoan cells do not vary greatly in size and small animals (whether primitively or secondarily small) are composed of fewer cells than their larger relatives and have a correspondingly simpler structure. Although none of the expositions of the gastraea theory indicate the size of this organism, it is clear that it is envisaged as being small: it either swims by cilia or is a ciliated benthic animal, and the simplicity of its postulated structure is consistent with small size and precludes large size. An examination of the functions of coelomic structures in existing animals suggests that their prime role and the functional advantage they conferred on early coelomates was as a hydrostatic skeleton (Clark, 1978a). But animals the size of the gastraea gain no benefit from the existence of a separate liquid-filled compartment because the mechanical properties of a hydrostatic skeleton can be as well satisfied by deformation of the cellular contents of the body.

The variation in the postulated number of gastric/coelomic pouches in the gastraea relates entirely to their supposed subsequent fate. Eight pouches relate to the existence of eight mesenteries in the Octocorallia,

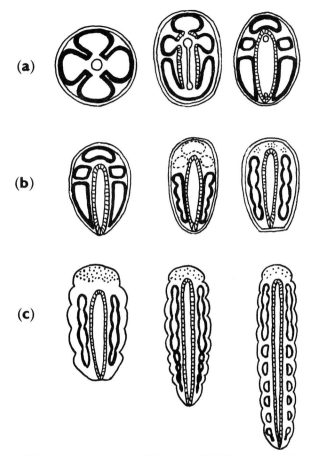

FIG. 2. Archicoelomate theory of Remane (1950). (a) Transformation of gastric pockets of a medusoid into the coelomic pouches of an oligomerous coelomate, (b) reduction of proto- and mesocoels and primary segmentation of the metacoel, (c) formation of secondary segments by proliferation.

which then become a stem group for the Cnidaria; at least two are suppressed in all coelomates (Fig. 2b, c), and all are suppressed in non-coelomate triploblastic metazoans. Five or six gastric/coelomic pouches relate to the embryonic or adult organization of the coelom in echinoderms and oligomerous animals (prejudicially known as the "Archicoelomata"). Four pouches refer to the tetraradiate symmetry of many larvae, the posterior coelomic pouch later dividing, also resulting in a typical oligomerous organization of the coelom and body (Fig. 2a). In no case is it suggested that the development of a particular number of coelomic pouches had any functional significance in the organisms in which this event took place. The fact that an innovation subsequently proved to have great potential for development and exploitation is not an explanation of its initial appearance. It is necessary to suggest the selective advantage such a development afforded the animals in which it occurred. In fact, there is no such selective advantage.

If the gastraea theory is divorced from the enterocoel theory, we are left with Haeckel's recapitulatory interpretation which now has few, if any, advocates.

2. *Acoeloid-Planuloid Theories*

The acoeloid-planuloid group of theories of the origin of the Metazoa has implications only for the phylogenetic position of the Cnidaria, Cteno- phora and, possibly, the Porifera, although the last are usually regarded as having an independent origin from the Protista. In several schemes, the earliest metazoans had a structure comparable to that of the Acoela and gave rise directly to the Platyhelminthes. They evolved from either a plasmodial (Hadzi, 1953, 1963; Hanson, 1958, 1977; Steinböck, 1958) or amoeboid (Reutterer, 1969; Boaden, 1975) protistan stock. Coelenterates are then regarded as being derived from Platyhelminthes (Hadzi, Reutterer) or independently of the acoeloid from protistan stocks (Boaden, Hanson, Steinböck). In the planuloid versions of these theories, early metazoans are envisaged as being comparable to the cnidarian planula larva, from which both the coelenterates and Turbellaria diverged (Hyman, 1959; Hand, 1959).

Thus, at present there is a variety of possibilities for the early history of the Metazoa. Multicellular animals may be monophyletic, diphyletic or polyphyletic, the protistan stock or stocks from which they emerged may have been amoeloid, ciliate or flagellate. The structure of the earliest metazoans may have been a gastraea, plakula, acoeloid or planuloid. The coelenterate phyla and Porifera may have independent origins from the Platyhelminthes, may have given rise to them, or may be derived from them. A common feature of all these views of early metazoan phylogeny, with the exception of the gastraea-enterocoel theory, is that the stem group for subsequent metazoan evolution was an organism of comparable structure to simple Turbellaria.

While there is a general concensus about the radiation of the Turbellaria and Nemertini as surface-creeping, carnivorous forms and the evolution of the endoparasitic habit in the Platyhelminthes with profound consequences for their morphology, physiology and reproductive biology, there are two areas of dispute about the early radiation of triploblastic Metazoa. The first concerns their habitat and the second, the phylogenetic position of the Mollusca.

1. Habitat of the Early Metazoa

It is generally assumed that the earliest Turbellaria lived on the substratum or were of such dimensions that they could swim with the aid of cilia. Whether or not they were initially pelagic, it is evident from physical considerations that larger members of this group would have been obliged to adopt a creeping habit on the substratum. An alternative view stems from the discovery of a rich variety of organisms, with representatives of 12–18 phyla, living in the anoxic, sulphide biome below the surface of marine sediments (Fenchel and Riedl, 1970). While many organisms have clearly become secondarily adapted as meiofauna and a few have penetrated the sulphide biome, Fenchel and Riedl thought it possible that some of the primitive taxa were relics of "the oldest biosystem on earth which preceded the aerobic biosphere". Among the Turbellaria, the two most primitive orders (according to Ax, 1963), the Acoela and Catenulida, and the Phylum Gnathostomulida are represented in the "thiobenthos" and may be such survivors. Boaden (1975) supports this view and suggests that much of the early radiation of the Metazoa occurred in the meiobenthos, initially in the sulphide biome and later in the oxidized layers above it. This would have the effect of backdating much of the early evolution of the Metazoa to at least the mid-Precambrian and before oxidizing conditions were fully established.

As Boaden points out, there is at present no conclusive evidence for or against such an extensive Precambrian metazoan evolution and a search for fossil remains of, for example, the genital armament of turbellarians and nematodes, or the jaws of gnathostomulids in Precambrian anoxic sediments would be valuable. The merits of this hypothesis therefore depend at present on the balance of indirect evidence for it or for alternative hypotheses. The maximum depth to which meiobenthic taxa

occur in marine sediments, in relation to deoxygenation, can be related
to their supposed phylogenetic position (Fig. 3), sometimes with the most
primitive surviving groups in a phylum living at the greatest depth
(Boaden, 1975). The Cnidaria, arthropods and the deuterostomes do not

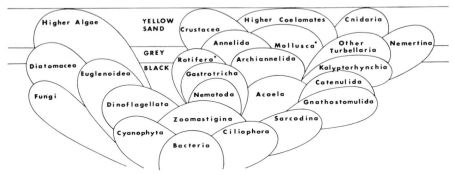

FIG. 3. Distribution of meiofaunal groups in relation to oxygenation of the
substratum. The names are at the level of the maximum depth at which
each group occurs and may reflect a relationship between depth and
evolutionary origin. (From Boaden, 1975.)

occur in anoxic sediments and are, by this hypothesis, of later origin than
the early metazoan radiation in the thiobenthos. The very wide
geographical distribution of many meiofaunal taxa is also regarded as
evidence of their great age (Sterrer, 1973). The inability of some aerobic
Turbellaria, as well as parasitic Platyhelminthes, to synthesize sterols and
polyunsaturated fatty acids was interpreted by Meyer and Meyer (1972)
as a secondary loss, but could equally well be interpreted as a reflection
of an origin of this phylum in anaerobic conditions (Boaden, 1975). The
existence of epidermal microvilli in gnathostomulids and retronectid
Catenulida may also reflect an ability of the early Metazoa to absorb
through ectodermal surfaces dissolved and particulate organic matter
which would have been readily available in the reducing conditions of their
thiobenthic habitat (Boaden, 1975).

Whatever views are held about the habitat of the early Metazoa, it is
beyond dispute that the meiobenthos has been secondarily colonized many
times by members of almost all phyla. Adaptations to this environment
are well known (Ax, 1963; Swedmark, 1964) and lead to a remarkable
degree of convergence among quite unrelated meiobenthic animals.
Among these adaptations is the evolution of small size, or more particularly

small diameter, with a consequent simplification of structures compared with corresponding features in larger relatives of the meiobenthic taxa. This adds to the uncertainty of deciding which are primitive members of a group because simplicity, whether primary or secondary, often appears primitive. An additional difficulty of the thiobenthos hypothesis is that nearly all groups of animals represented in the meiobenthos have larger relatives which live on or in the surface of the substratum. In the terms of this hypothesis, their evolution can only have taken place after oxidizing conditions were firmly established at the surface, and Boaden (1975) envisages that most of the spiralian metazoan phyla had already appeared before then. While it is possible to offer an explanation of the selective factors influencing metazoan radiation in large animals, factors determining a metazoan radiation in the meiobenthos have not been suggested and are not at all evident. It is possible that some such radiation did take place, in this environment, whether or not life originated in the thiobenthos, but there are unsolved difficulties in supposing that it was as extensive as Boaden suggests.

2. Origin of Mollusca

Confusion about the origin of the Mollusca has been created by the supposition that they are closely related to the Annelida (because of the developmental similarities between molluscs and polychaetes) coupled with the view that molluscs are coelomate and even segmented. The discovery of *Neopilina* allowed a brief revival by Lemche (1959) of the early view that molluscs are segmented animals and, because of the small size of the coelomic compartments, have a close affinity with arthropods. However, on examination, it is difficult to sustain the view that molluscs were derived from or gave rise to metameric coelomates and it has generally been concluded that such repetition of structures as occurs in more primitive molluscs is pseudometamerism comparable to that seen in some turbellarians and nemerteans (Salvini-Plawen, 1969).

The view that molluscs are coelomate depends partly on the fact that the pericardial cavity conforms with a minimum definition of a coelom (a cavity within the mesoderm bounded by an epithelium), but more because the gonads are primitively associated with the rostral part of the pericardium. The pericardium has been regarded as homologous with the annelid coelom even by those who accept that the cavity was never spaceous, because of their adherence to the gonocoel theory of the origin of the coelom (e.g. Fretter and Graham, 1962). The pericardium of molluscs is an expansion chamber which permits pulsation of the compact heart; it may have originated as a dilation of the genital ducts, but there is no reason to relate it to the annelid coelom. Indeed, there are formidable difficulties to supposing that the mollusc pericardium is either a reduced coelom or a cavity which later gave rise to the spaceous coelom. The view that it is of quite independent origin and unrelated to the coelom of annelids or any other coelomates poses few difficulties and now has wide acceptance (Clark, 1964; Vagvolgyi, 1967; Salvini-Plawen, 1968, 1969).

Disposal of the notion that the morphology of molluscs is related to that of segmented coelomates removes any obstacles to regarding them as part of a radiation of early acoelomate animals. The development of a locomotory ventral foot with associated mucus glands in the Caudofoveata and their equivalent in Solenogastres (Salvini-Plawen, 1969) as well as in other molluscs which may be regarded as primitive (Placophora, Monoplacophora), suggests that the emergence of molluscs if not the whole of the turbellarian–nemertean–molluscan radiation, took place among animals creeping on the surface of the substratum.

PSEUDOCOELOMATES

Little sense can be made of the "pseudocoelomate" phyla at present. The term refers to a grade of construction, not a phyletic entity, although there is generally a strong presumption that these animals have, if not a common origin, at least an ancestry among a group of related animals. In fact, although described as a pseudocoelom, the nature of the body cavity is unclear and may not be the same in all these animals. Their relationship with one another as well as to other animals is uncertain and it is not even settled which groups should be included among the pseudocoelomates. The Rotifera, Gastrotricha, Kinorhyncha, Nematoda and Nematomorpha may be regarded as separate pseudocoelomate phyla or somewhat loosely associated as classes of a single phylum Aschelminthes or Nemathelminthes. The Acanthocephala, Entoprocta (Kamptozoa) and Priapulida may be included in the Aschelminthes or, more commonly, be regarded as separate phyla. On the other hand, the Priapulida, Entoprocta and Gastrotricha are sometimes not considered to be pseudocoelomate at all. The lack of consensus is so great that practically no two writers on the subject in recent years has treated the affinities of the "pseudocoelomate" groups in precisely the same way.

Pseudocoelomate animals are usually regarded as being of a simpler and inferior grade of construction to coelomates and are sharply distinguished from them on the grounds that the body cavity is not a "true" coelom but a persistent blastocoel (Hyman, 1959). In theory, this is a precise definition which categorizes an otherwise somewhat heterogeneous assemblage of animals. But, in fact, most pseudocoelomates are small and have compact developmental stages in which the blastocoel becomes severely reduced in more advanced stages or fails to appear at all. There is no continuity between adult body cavities and an embryonic blastocoel. This definition has therefore an uncertain justification and little practical value. To decide whether a cavity is coelomic or pseudocoelomic in a particular instance, it is commonly necessary to rely on criteria derived from features of adult structures.

Remane (1963), when considering this problem, suggested that a coelom is typified as a body cavity lined with an epithelium which contributes a musculature, the cavity receives the gametes, contains nephridial funnels and is partitioned by mesenteries. However, as Remane indicated, undoubted coelomates do not necessarily satisfy all these criteria while pseudocoelomates may satisfy some of them. There is therefore no sharp anatomical distinction between the two types of body cavity and this, coupled with the confusing relationships between groups conventionally regarded as pseudocoelomate, questions the validity of the pseudocoelomate condition as a grade of construction with any phylogenetic significance.

The difficulties created by this unsatisfactory situation are illustrated by the present status of groups widely regarded as pseudocoelomate.

Macrodasyid Gastrotricha contain three parallel body compartments although these are not liquid-filled cavities but are filled with other tissues. The central compartment is occupied by the intestine and mature gametes, the two lateral compartments by the gonads and immature gametocytes and the longitudinal muscles. Remane (1963) believed that the lateral compartments might be coelomic, though reduced in size and extent. Subsequent ultrastructural investigations have not clarified this question. Rieger et al. (1974) have been unable to confirm the existence of an extracellular membrane or endothelial lining of the lateral compartments which are confined to the posterior part of the body. These authors conclude that these compartments are not coelomic but are merely areas delimited by the configuration of the circular muscles. Against this, Teuchert (1977) concludes that since the lateral compartments contain gonads which are associated with mesothelial cells, they are coelomic. No such compartments are visible in the chaetonotid gastrotrichs.

Nematomorpha have three body compartments comparable to those of gastrotrichs and presumably any conclusion drawn from the structure of the latter will apply also to nematomorphs.

The existence of a membrane and cellular epithelium lining a body cavity, which are evidently lacking or very incomplete in gastrotrichs, has several times been regarded as of critical importance in distinguishing a coelom from a pseudocoel. There is such a membrane in some Rotifera, though it lacks nuclei and the body cavity is therefore regarded as pseudocoelomic (Remane, 1963). A similar situation was thought to exist in the Priapulida (Lang, 1953), but with the discovery of nuclei associated with the membrane lining the body cavity (Shapeero, 1961), this group was accepted as coelomate. This repeats the experience with the dinophilid archiannelids in which the body cavity was thought to be bounded by a non-cellular membrane and therefore a haemocoel (= blastocoel, pseudo-coel) until Jägersten (1944) showed that the membrane contained nuclei.

Vaguely coelomic features can also be discerned in the body cavities of nematodes and Acanthocephala. Remane (1963) drew attention to the structure of the body wall of nematodes in which the muscle cells, particularly the platymyarian type in free-living forms, are arranged like an epithelium and he concluded that this is more like a typical outer lining of a coelom than is found in many coelomates. Furthermore, the muscle cells are derived from two entomesodermal bands comparable to the mesodermal bands of annelids and molluscs. The body cavity of Acanthocephala is not a blastocoel, but appears as a split in the inner cellular mass in the larva, and so resembles a schizocoel (Nicholas and Hynes, 1963).

It is evident that there is no satisfactory way of defining a pseudocoel in embryological or anatomical terms, such that it can be distinguished from a coelom.

Siewing (1976) grasps this nettle firmly by claiming that all Aschelminthes are coelomate. As a supporter of the "archicoelomate" theory, he regards all triploblastic metazoa as derived from an oligomerous coelomate stock and hence, that all these body cavities represent different stages of reduction of a coelom. This interpretation renders the concept of a coelom so vague that it embraces any body cavity other than the gut, and even no cavity at all if the space occupied by developing gametes in gastrotrichs is regarded as coelomic, and this undermines the strict homology of body cavities described as coelomic, on which much stress is laid by proponents of the "archicoelomate" theory.

ORIGIN AND RADIATION OF COELOMATES

1. Functions of the Coelom

In existing coelomates, a spaceous coelom has multiple functions: as a circulatory system for the transport of respiratory gases and metabolites, as a repository for accumulating gametes whether attached to the gonad or floating in the coelomic fluid, as a space in which mobile organs can move independently of the body wall, and as a hydrostatic skeleton for the body-wall musculature. In none of these respects is the coelom unique. Body cavities of a variety of morphological natures, the pseudocoel, the haemocoel or even the cnidarian coelenteron, may sometimes serve the same purposes (Clark, 1978a).

Because of its multiple functions there may well have been considerable selective advantage in the evolution of a body cavity distinct from the gut by the early metazoans, but it cannot be assumed that all these functions were equally important selective factors in the first appearance of a secondary body cavity. The only consistent correlation between function and the size and organization of the coelom in existing coelomates is in its role as a hydrostatic skeleton. With the exception of a few explicable anomalies (soft-bodied arthropods, echinoderms), when the behaviour of the animal is such as to require a mechanically efficient hydrostatic skeleton for the body-wall musculature, the coelom is spaceous, when there is not it is reduced or highly modified. If a large coelom does exist, it has in varying degrees other functions as well, but if it does not, the other functional requirements are satisfied in different ways.

Since the prime function of the coelom in existing animals is as a hydrostatic skeleton for the body-wall musculature, it is likely that this was its role in the earliest coelomates.

2. Mechanical Capabilities of Precoelomates and Coelomates

Some idea of the mechanical properties of the immediate forerunners of coelomates can be gained from a consideration of the behaviour of turbellarians and nemerteans (Clark, 1964). These worms have circular and longitudinal muscle fibres in the body wall and are able to produce reversible changes of shape and, if of appropriate dimensions, peristaltic movements in which the hydrostatic skeleton is provided by the parenchymatous tissue occupying the space between the gut and the body wall. Although a tissue of this nature is capable of transmitting fluid pressures, it damps them down, and changes of shape which involve parts of the body-wall musculature behaving antagonistically are slow and weak. Such animals, if they are small enough, can swim by cilia or live as meiofauna, or if larger can creep on the substratum by cilia or muscular waves (cf. Tricladida, Polycladida) but they cannot generate the more powerful forces needed for burrowing.

Some nemerteans, although theoretically of comparable mechanical structure to the Turbellaria, provide an illuminating exception. Proboscis eversion, which is rapid and involves a considerable transfer of volume, does involve the use of a liquid hydrostatic skeleton in the form of the rhynchocoel. Many nemerteans can also burrow into soft substrata with the aid of peristaltic locomotory movements. In them, the cellular component of the parenchymatous tissue is much reduced and the volume of intercellular fluid increased: this is an approach to a liquid hydrostatic skeleton for the body-wall musculature and its mechanical superiority over the turbellarian parenchyma in this respect is reflected in the greater locomotory capabilities of nemerteans.

The evolution of a coelom as a hydrostatic skeleton in early metazoans with a structure mechanically equivalent to that of Turbellaria would have permitted them to perform the same changes of body shape with the same musculature as before, but to perform them more powerfully. In particular, it would have allowed them to occupy a new ecological niche as burrowers in the substratum. Most theories of the origin of the coelom (gonocoel, nephrocoel, schizocoel, enterocoel, etc.) are concerned with the structures from which a coelom was derived or the manner in which it is formed in ontogeny. If, however, the coelom had primarily a mechanical, hydrostatic function, any spaceous, liquid-filled compartment, whatever its morphological origin, would have served this purpose. Comparative embryological evidence suggests a polyphyletic origin of the coelom and at least three basic types of functional organization can be detected among coelomates. While this does not rule out a monophyletic origin, in which case all coeloms are homologous, it is not inconsistent with a radiation among precoelomates in which coelomic hydrostatic skeletons were evolved independently several times as adaptations to different life styles (Clark, 1978a).

3. Unsegmented Coelomates

The most direct adaptation is that found in unsegmented coelomates such as Echiura. The large coelom is used as a hydrostatic skeleton in the performance of peristaltic burrowing and in perstaltic irrigation of the burrow. While the presence of a single, unobstructed coelom allows considerable pressures to be concentrated at a point by contraction of most of the body-wall muscles and this may be important in constructing and widening a burrow in compacted substrata, the worms do not appear to be capable of long-sustained burrowing. Almost all worms with an unobstructed hydrostatic skeleton, whether this condition is primary as in Echiura or secondary as in many polychaetes, are sedentary. The Priapulida are exceptional, but there is some evidence that the body-wall musculature responsible for the burrowing movements is physiologically specialized for rapid and sustained activity (Mattisson *et al.*, 1974).

4. Metameric Coelomates

The metamerism of the oligochaete coelom appears to be an adaptation to vagrant burrowing. The existence of intersegmental septa prevents the transmission of fluid pressures from one part of the body to another (Fig. 4), and the longitudinal and circular muscles of a segment behave approximately as antagonists only to each other. As a result, during peristaltic locomotion body-wall muscles are concerned solely with developing locomotory forces and are not required to prevent inappropriate deformation of the body caused by pressure changes elsewhere in it; this is a function of the septal musculature. This mechanical improvement over non-segmented coelomates permits long periods of peristaltic burrowing without fatigue. Intersegmental septa have other functions, particularly in bracing the body wall and preventing deformation of the gut (Mettam, 1969), but the fact that, although perforated, oligochaete septa prevent the transmission of fluid during active movements by closure of the septal foramina, indicates the importance of their role as hydrostatic barriers.

5. Oligomerous Coelomates

Oligomerous coelomates, such as Phoronidea, Pterobranchia and Pogonophora, have a small number of coelomic compartments and are adapted as tubicolous, tentaculate feeders. One coelomic compartment, the first in

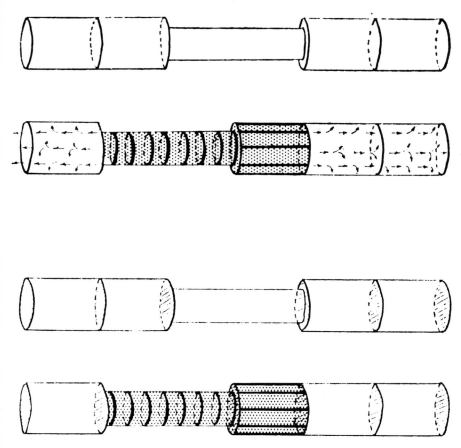

FIG. 4. Effect of intersegmental septa in limiting the transmission of pressure changes during the passage of a peristaltic wave. (a) As one region of the body is shortened and the next elongated by contraction of the body-wall muscles, the increase in fluid pressure in these regions (stippled) is dissipated and transmitted to all other parts of the body wall. (b) In a septate worm, pressure changes are confined to segments undergoing change of shape. (From Clark, 1963.)

Pogonophora, the second in Pterobranchia and probably the Phoronidea, is a hydrostatic organ for the tentacular feeding structure, another coelomic compartment provides a general perivisceral cavity and functions much as the undivided coelom of unsegmented worms. There may be an additional coelomic compartment serving as a hydraulic organ for a prehensile (the pre-oral lobe of pterobranchs) or other structure (the frenulum of pogonophores), and in the Pogonophora there are a few coelomic compartments of unknown function in the opisthosoma. Related animals to these oligomerous phyla which have no such tubicolous habit show a radical modification of coelomic organization.

6. Monophyletic or Polyphyletic Origin of the Coelom?

The question whether or not these three separate adaptations with strikingly different organization arose independently from acoelomate worms has not been resolved. The general conviction that all coeloms are homologous has resulted in numerous attempts to derive one type of organization from another, but with little success. There are difficulties in deriving metameric from unsegmented coelomates or vice versa. Despite claims to the contrary, there is no satisfactory evidence that the unsegmented Echiura evolved from metameric worms (Clark, 1969, 1978b) and, since septa cannot perform their hydrostatic isolating role unless complete, it is difficult to explain how a metameric coelom could have evolved in gradual steps in an unsegmented coelomate (Mettam, 1969). Thus, these two adaptations to burrowing, one sedentary, the other vagrant, may be separate developments among acoelomate, surface-creeping worms. This view has no implications about the antecedent structures from which the undivided or metameric coelom arose.

The gastraea (or bilaterogastraea)–enterocoel theories of Remane (1950), Jägersten (1955) and Marcus (1958) suppose the coelom and its first appearance to be related to the trimeric structure of oligomerous animals, though not specifically as an adaptation to a tubicolous existence. If anything, it was apparently envisaged that the evolution of the coelomic compartments was not accompanied by any change of habit or habitat. There is no functional justification for the appearance of coelomic compartments at the early stage in metazoan radiation that they propose. It is difficult to envisage a transition from an acoelomate creeping organism directly to an oligomerous tubicolous animal and the tubicolous adaptation is perhaps more likely to have taken place in animals that were already coelomate and living in the substratum. At present, it is difficult to resolve the question whether this represents a third independent development from acoelomate forerunners or was a result of radiation of unsegmented or, less likely, metameric coelomates represented today by the Echiura and Annelida.

Proponents of the gastraea–enterocoel theory suppose that existing unsegmented and metameric spiralian coelomates were derived from "archicoelomates" by suppression of the pro- and mesocoels and a secondary development of metameric segments in or behind the metacoel. This view does not pose problems on functional grounds. The absence of a tentacular structure and virtual obliteration of the mesocoel of enteropneusts (cf. Pterobranchia) shows that these structures are not immutable. Such an interpretation, however, supposes a more complicated phylogeny than the independent origin of a secondary body cavity in several groups of early coelomates, there is no evidence to support this complication and the interpretation is subject to the same ontological uncertainties as any other view which links spiralian coelomates and deuterostomes.

RADIATION OF METAMERIC COELOMATES

Metameric spiralian coelomates include the Annelida and arthropod phyla. Problems connected with the evolution of the Uniramia and other arthropods are touched on only briefly here since they are discussed in detail elsewhere in this symposium (Manton and Anderson, Chapter 10).

1. Origin of Annelida

Views of the origin of polychaetes, with their segmental parapodia, are confused. Polychaetes are often supposed to be derived from archiannelids which are still commonly regarded as the most primitive annelids. Archiannelids lack parapodia and live as meiofauna, yet metamerism is regarded as primarily an adaptation to sinusoidal swimming. Parapodia, themselves, are usually regarded as primarily swimming appendages (though clearly in many polychaetes they have become severely modified to serve other functions), but occasionally as walking appendages. It is impossible to reconcile all these views and, in fact, none of them appears to be well founded.

At various times, resemblances have been described between archiannelids and various polychaete families: Spionidae (Orrhage, 1964, 1974), Opheliidae, Oweniidae (Bubko and Minichev, 1972; Bubko, 1973), Histriobdellidae (Sveshnikov, 1958; Gelder and Jennings, 1975) and a number of families which possess a ventral proboscis or buccal organ (Dales, 1962). These anatomical similarities may well indicate a relationship between some of these families and archiannelids and it has often been concluded that archiannelids with their simple structure are a primitive annelid stock from which polychaetes evolved, a view supported in recent years by Sveshnikov (1958), Dales (1962, 1977), Bubko (1973), Mileikovsky (1977) and others.

The difficulty with this interpretation is that it entails both the coelom and metamerism making a gradual appearance among such small animals as archiannelids living as a meiofauna. (That this probably also implies that the annelid coelom is not homologous with that of most other coelomates appears not to have been remarked upon.) If the primary function of a metamerized coelom is as a hydrostatic skeleton used in peristaltic burrowing (Clark, 1964, 1978a), it is not likely that this structure evolved in small animals for which such an adaptation is unnecessary. Although Gray (1969) described leech-like muscular creeping in the Protodrilidae —which he suggested might illustrate how interstitial animals may have developed muscular locomotion with a concomitant development of musculature and coelom while still small in size—most anatomical features of archiannelids are more intelligibly viewed as the result of a simplification of structure as a result of their becoming small. The more specialized features of archiannelids are typical adaptations to life in the mesopsammon and these, as well as the small size and simplified structure, are exactly comparable to adaptations by members of several other phyla which have colonized this environment (Swedmark, 1964; Laubier, 1967; Fauchald, 1974). The view that archiannelids are interstitial polychaetes (Hermans, 1969; Clark, 1969; Fauchald, 1974; Orrhage, 1974; Westheide, 1977) is consistent with what is known of them and avoids the difficulties associated with the view that they are a stem group for the Annelida. The relationships that have been suggested between archiannelids and certain polychaete families may be well founded but the direction of evolution that has been proposed for them is not.

If the earliest annelids evolved as burrowing worms it is most likely that they initially retained complete circular and longitudinal muscle coats which are present in almost all worm-like animals. They would therefore have had a musculo-hydrostatic system comparable to that of oligochaetes, though no doubt without the specializations of reproductive and excretory systems of modern oligochaetes, which are associated with their colonization of freshwater and terrestrial habitats. The polychaete development of lateral appendages, entailing a wholesale reorganization of the structure and functions of the body-wall musculature and, generally, a loss of a metamerized coelom, is a later event in the history of the Annelida (Clark, 1969).

The anatomy of polychaete segments suggests that parapodia evolved as lateral appendages. The attempt by Storch (1968) to derive the segmental musculature of all polychaetes from that in the Aphroditidae in which, unusually among polychaetes, the parapodia are directed ventro-laterally and are used for walking on a substratum, presents serious difficulties. Storch took no account of the parapodial musculature which has a profound effect on the anatomy of the body-wall muscles (Mettam, 1971), and, as Fauchald (1974) pointed out, Storch's use of Remane's homology criteria coupled with comparisons of similarity characteristics by the methodology of numerical taxonomy, almost inevitably results in the most complicated organization being selected as the primitive stem from which all others are derived. Far from aphroditids indicating the structure and habit of the earliest animals with segmental appendages, everything about them points to their being very advanced errant polychaetes.

Sharov (1966) also considered the earliest annelids to be surface-creeping polychaetes, though in his scheme, parapodia were supposed to be the modified oral tentacles of a ctenophore ancestor. The Spintheridae represent the most primitive annelids. This proposal is based on the most superficial considerations of anatomy and has been effectively demolished by Manton (1967) and others.

In other views of polychaete origins, parapodia are regarded as primarily lateral structures, generally as swimming appendages. The idea that the metamerism of protostome worms originated as an adaptation to undulatory swimming (Hyman, 1951; Bonik *et al.*, 1976) and that parapodia evolved as a further adaptation in this direction is not supported by what is known of the kinetics of undulatory swimming. All long, narrow animals swim in essentially the same way by propagating a lateral or dorso-ventral sinusoidal wave along the body (Taylor, 1952) and there is no evidence to suggest that swimming ability or efficiency is related to segmentation of the body (Clark, 1976). The locomotory waves travel from head to tail in forward swimming by smooth-bodied animals (the majority) or in the reverse direction if the body is sufficiently rough (as in most swimming polychaetes). The opheliids *Armandia* and *Ammotrypane* with diminutive parapodia and the archiannelid *Polygordius* with none are smooth enough to swim by the former method (Clark and Hermans, 1976). Rough-bodied swimming, with reversed locomotory waves, is not efficient because of the large drag component created by the body waves and, although theoretically possible, it is doubtful if any polychaetes swim by this method (Clark and Tritton, 1970) (Fig. 5). In most cases that have

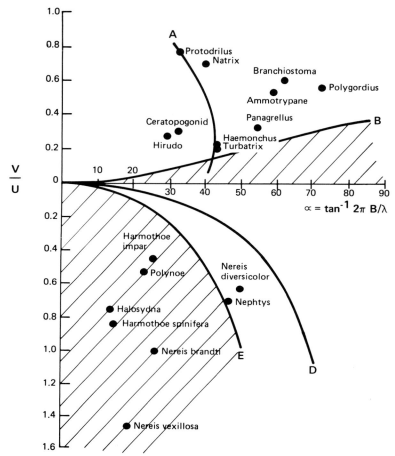

Fig. 5   Relationship between ratio of velocity of swimming (V) to that of body
waves (U) and pitch of the undulations (α) in sinusoidal swimming by
long thin animals. For rough-bodied animals the direction of propagation
of the undulations is reversed (U negative). Line A, condition for
maximum speed for a given energy output. In the shaded areas the body
waves produce no locomotory thrust (only drag). Between lines D and
E the undulations produce increasing drag and decreasing thrust and
undulatory swimming is inefficient. (After Clark, 1976.)

been investigated, the body undulations produce no locomotory thrust,
only drag which has to be overcome by the parapodial beat, and it is
noteworthy that in fast-swimming heteronereids and some polynoids, the
body undulations are reduced to a small amplitude and are almost

eliminated (Clark and Tritton, 1970; Clark, 1976). The development of lateral appendages in a formerly smooth-bodied animal, far from enhancing swimming ability would have impaired it, and there are no grounds for relating the evolution of annelid metamerism or of segmental parapodia to undulatory swimming.

The development of parapodia, and the disruption of circular and longitudinal muscle layers of the body wall that this entails, is incompatible with peristaltic burrowing and is unrelated to swimming. It is most likely that the evolution of lateral appendages in annelids was related to locomotion in a surface detrital ooze too soft to permit this burrowing and not fluid enough for swimming. This is a similar habit and habitat to that postulated by Whittington (1975) for the mid-Cambrian *Opabinia regalis* which in some respects had a structure comparable to that envisaged in the early annelids.

2. Uniramia

There is convincing morphological (Manton, 1977) and embryological (Anderson, 1973) evidence that onychophorans, myriapods and hexapods are related and Manton includes them in a single arthropodan phylum, the Uniramia. Anderson has shown that the pattern of embryological development in the Onychophora is the basic theme for all the Uniramia and the mode of formation and subsequent development of presumptive areas of the blastoderm in onychophorans in a functional modification of that in oligochaetes. The oligochaete pattern of development, in turn, can be regarded as a modification of that retained by polychaetes.

Manton has shown that the musculature of the onychophoran limb is constructed and functions on a different principle from that of the polychaete parapodium. It is evident that segmental appendages and their musculature evolved independently and as adaptations to quite different methods of locomotion in the Onychophora and Polychaeta.

These morphological and embryological observations, together with conclusions about relationships within the Annelida, can best be reconciled if it is supposed that the earliest annelids had a developmental pattern more like that of polychaetes than oligochaetes, but an adult morphology like that of oligochaetes, without parapodia. The Polychaeta are a morphologically progressive group in that they evolved segmental parapodia, but are embryologically conservative. Oligochaetes have retained the basic anatomical features of early annelids but show developmental modifications that are reflected also in the Onychophora which, in turn, evolved walking legs.

One remaining puzzling feature of this evolutionary sequence is the fate of the coelom in onychophorans. Both they and polychaetes stemmed from burrowing worms with a metameric coelom. The functional advantage of a compartmented coelom disappeared when new methods of locomotion were adopted. In all annelids there appears to be an advantage in providing for communication between the coelomic compartments (even in oligochaetes the septa are perforated, though the foramina can be sealed by sphincter muscles) and in polychaetes the intersegmental septa are often severely reduced. Onychophora have the ability to perform considerable changes of shape in order to squeeze through crevices and the different gaits used in locomotion at different speeds involve a change in length of the animal. However, in onychophorans, the need for a longitudinal communication in the hydrostatic skeleton has not been met simply by reducing the septa, as in polychaetes, but by an expansion of the blood vascular system at the expense of the coelom. The haemocoel now provides the hydrostatic skeleton. For some reason, the simple expedient of reducing the septa to allow longitudinal transmission of pressure changes in the coelomic fluid was not available to, or not adopted by them. This remains an unexplained feature of the emergence of the Uniramia.

3. Crustacea

Manton (1977) has given convincing morphological evidence that the Crustacea are not closely related to the Uniramia and that they acquired an arthropod structure independently of them. Anderson (1973) has given powerful support to this view by demonstrating that the presumptive fate map of the crustacean blastula cannot be derived from those of the Annelida and Uniramia, although the basic pattern of development in Crustacea includes a modified spiral cleavage. He therefore concludes that this phylum diverged from an early spiralian group independently of the Annelida. Since the Spiralia include unsegmented and non-coelomate groups such as molluscs, nemerteans and polyclads, this implies that crustacean metamerism and possibly also the coelom evolved independently of those of annelids and, presumably, as quite different adaptations. The coelomic compartments of crustacean somites are never more than vestigial and the development of the gonads is not closely associated with them. There is little evidence that crustaceans evolved from animals with large coelomic compartments and the functional significance of the initial appearance of these cavities and indeed of crustacean metamerism remains unsolved.

OLIGOMEROUS COELOMATE STEM GROUPS

The Pogonophora, Phoronidea and Pterobranchia all correspond approximately in their gross structure to that envisaged in the early oligomerous coelomates. This is true whether oligomery is supposed to have originated in a small creeping benthic organism (following the gastraea-enterocoel theories) or in a coelomate or acoelomate worm that adopted a tubicolous habit. A close relationship of these phyla near the stem of the radiation of oligomerous animals is therefore commonly proposed. There are, however, difficulties in this view. The Pogonophora appear to be far removed from either of the other oligomerous tubicolous worms, the position of the Phoronidea is not beyond dispute, and the relationship between Phoronidea and Hemichordata is not clear.

1. *Pogonophora*

After an initial period when pogonophorans were considered on the most superficial evidence to be aberrant polychaetes, there was a general consensus that they are oligomerous deuterostomes (Southward, 1963, review). While this view is still maintained by some (e.g. Siewing, 1975a), it has been increasingly questioned in recent years. The metamerism of the pogonophoran opisthosoma now appears to agree in most essentials with that of annelids (Southward, 1975a). The lack of excretory organs and gonads in the opisthosomal segments is intelligible in view of their small size, if for no other reason, and is common in annelid terminal segments; the absence of a gut, as a characteristic of pogonophorans, is immaterial. The nervous system, previously regarded as dorsal, now seems more likely to be ventral (Nørrevang, 1970; Ivanov, 1975), and divergent views about the method of coelom formation make it uncertain if the pogonophore coelom is enterocoelous (Ivanov, 1975) or schizocoelous (Nørrevang, 1970).

These observations have allowed a revival of the view that pogonophorans and Vestimentifera (if these worms are to be separated from the Pogonophora) are aberrant annelids (Land and Nørrevang, 1975) or at least are allied to annelids (Southward, 1975a), and A. J. Southward (1975) suggests that they may have originated as neotenous lecithotrophic polychaete larvae.

Much depends on the interpretation placed on the metamerism of the opisthosoma. Siewing (1975a), impressed by the close structural similarities of the pro-, meso- and metasomes of pogonophorans and the three body regions of "archicoelomates", concludes that the opisthosoma originated as a metamerism of the terminal region of the body of an oligomerous animal. Southward (1971, 1975a, b), who believes the opisthosomal segments are homologous with annelid segments, argues that the opisthosoma is a burrowing organ. In view of its small size and small number of segments, it is not clear that the hydrostatic isolation of the coelomic compartments of the opisthosoma makes much contribution to mechanical efficiency (Clark, 1978b). If this proves to be so, the independent origin of a metamerized coelom (Siewing, 1975a) with this mechanical function seems less likely that the retention of an almost vestigial organ in a highly modified annelid (Land and Nørrevang, 1975). The latter interpretation implies that the Pogonophora show a remarkable convergence in the structure of the anterior and major part of the body to phoronids and pterobranchs. That this is convergence and not homology is supported by the different functions of the pro- and mesosomes in these three groups of oligomerous worms.

Ivanov's (1975) separation of the Pogonophora from all other coelomates on the grounds of the unique absence of a gut, and division of the Coelomata into Deuterostomia, Pogonophora, Chaetognatha and Trochozoa reflects a taxonomic rather than a phylogenetic view and the implications of it for the evolution of coelomates have not been considered.

2. Phoronidea

The matter of debate about the Phoronidea concerns the nature of the epistome. According to the "archicoelomate" theory, phoronids are composed of three coelomate body regions, the protosome being much reduced and represented by the epistome, the lophophore being borne on the mesosome. However, the functional coelomic organization of phoronids is commonly considered to consist of only two compartments: the anterior compartment providing a hydraulic organ for the lophophore and the posterior compartment forming the perivisceral coelom. Nichols

(1966) in fact suggested that the phoronid organization is derived from that of sipunculids in which a small coelomic compartment serves the circum-oral ring of tentacles and the second, and main, coelom occupies the remainder of the body. The gut of sipunculids, like that of phoronids, is reflexed with the anus in an anterior position.

The view that the epistome is the protosome and the cavity within it a protocoel has often been questioned (Hyman, 1959). The pre-oral region of the larva is shed at metamorphosis and the epistome appears only later as a small fold above the mouth. The cavity within it was long considered to be blastocoelic rather than coelomic, although the existence of cilia on the protonephridia in the preseptal space of the actinotroch larva suggests that in fact it is a coelom. Recent studies of the development and adult anatomy of several species of phoronid support the view that the epistomial cavity is coelomic. Zimmer (1973) has found that mesodermal cells budded from the archenteric wall line the blastocoelic cavity in the developing pre-oral hood and that in the six species of phoronid he studied, a typical pre-oral hood coelom is always formed, though in some is transitory. Investigations of the adult anatomy of several poronids by Emig and Siewing (1975) and Siewing (1975b) have shown that there is a separate epistomial cavity which sends branches to the lophophore and tentacles, accompanying branches of the mesocoel. These are interpreted as having a hydraulic function and the slits between blood vessels, placing the protocoel in communication with the mesocoel, are interpreted as valves allowing an exchange of fluid pressures.

Resolution of this matter reinforces the view that the Phoronidea have retained the body form and tubicolous habit of early oligomerous worms. There is a clear tendency among other lophophorates for a reduction and elimination of the epistome and its small size in phoronids suggests that even in them it is a reduced structure. If this is so and the protosome was formerly a larger structure, the fundamentally trimeric organization of the phoronid body becomes more evident.

3. Phoronid-pterobranch affinities

The existence of a large pre-oral lobe and trimerous organization of the body, coupled with their tubicolous habit, suggests that pterobranch hemichordates have suffered less modification than phoronids have and are therefore closer than them in their gross structure to the body architecture of the early oligomerous worms. Whether the similarity in anatomical organization between the Phoronidea and Pterobranchia implies related-ness or is the result of convergence (as is probably the case with the Pogonophora) requires examination.

Formerly, a close relationship between phoronids and hemichordates was discounted on the grounds that, while the latter were clearly deuterostome, the Phoronidea display some protostome features during their ontogeny, and the actinotroch larva is a modified trochophore (Hyman, 1959). Zimmer (1973) has complained at the prejudicial nature of a search for resemblances between actinotroch and trochophore larvae and he suggests that, instead, the question should be posed "do the larvae of lophophorates have more in common with protostomatous or deuterostomatous larvae?" He concludes that the only feature which any lophophorate irrefutably shares with protostomes is the derivation of the adult mouth from the blastopore. The presence of solenocytes in phoronid larvae has parallels with protostomes rather than deuterostomes, and the sub-epidermal position of the nervous system and prototroch-like corona of the larvae in Ectoprocta (if related to phoronids) are also protostomian. Against this, Zimmer claims no less than eighteen significant similarities between lophophorates and deuterostomes and a relationship between them seems confidently assured.

RADIATION OF THE LOPHOPHORATES

The Ectoprocta (Bryozoa) and Brachiopoda are usually regarded as closely related to the Phoronidea and the three phyla are accordingly linked in a single group, the lophophorates. All have the characteristic mesosomal lophophore around the mouth; it is horseshoe-shaped in phoronids, phylactolaematous ectoprocts and very small brachiopods and has this form as a developmental or growth stage in gymnolaematous ectoprocts and larger brachiopods. The protosomal epistome appears in phoronids and phylactolaemes but is not clearly differentiated in gymnolaemes or brachiopods. Both the Ectoprocta and Brachiopoda have undergone extensive radiation and have clearly evolved quite different structure from any conceivable common ancestral lophophorate. The Phylactolaemata, despite their highly specialized development associated with their restriction to freshwater habitats, have an adult structure which is generally regarded as the most primitive among living Ectoprocta. The zooid is cylindrical and has circular and longitudinal muscles in the body wall, the lophophore is horseshoe-shaped and, alone among ectoprocts, the phylactolaemes have an epistome and three clearly identified body regions. In all these respects they approach the gross structure of phoronids and are not far removed from the postulated structure of a basic oligomerous worm.

It is therefore possible to envisage the radiation of Phoronidea, Ectoprocta and Brachiopoda, with increasing modification of the basic body plan in Ectoprocta and Brachiopoda, from an early trimerous, tubicolous coelomate and this is the view that is widely adopted. But while the position of the Brachiopoda in this scheme is uncontroversial, the Ectoprocta present certain difficulties.

Supporters of a link between the Ectoprocta and Phoronidea are usually emphatic that the Entoprocta, being pseudocoelomate, are not related to the Ectoprocta and that the old phylum Bryozoa in which the two are united as constituent classes has no validity (e.g. Brien, 1960; Ryland, 1970). However, there are similarities between entoprocts and ectoprocts which cannot be ignored and Brien and Papyn (1954) and Hyman (1959), took the view that they were early (Entoprocta) and late (Ectoprocta) offshoots of the line leading to the Annelida. These views are difficult to reconcile with any tenable view of phoronid origins if, as these authors have maintained, the Ectoprocta are related to them. Marcus (1958), one of the few authorities on Bryozoa to include both Ectoprocta and Entoprocta in the phylum, considered the latter to be derived from attached larvae of ectoprocts.

Among more recent authors, Nielsen (1971) argues strongly in favour of a close relationship between the Ectoprocta and Entoprocta and against a link between either with the Phoronidea. The most critical reasons advanced for separating ectoprocts and entoprocts are that the body cavity is a coelom in the former but a pseudocoel in the latter, the anus is outside the lophophore in ectoprocts but within it in entoprocts, as their names suggest, and that the functioning of the lophophore is different in each.

Nielsen points out that the origin of the mesoderm is obscure and there is nothing in the larval development of either group to suggest a coelom. In the Entoprocta the space between the mesoderm cells is occupied by "haemolymph" which, in some, is circulated by a special system of star cells. In the Ectoprocta, the body cavity is not a typical coelom. At larval metamorphosis the enclosed regions of the corona and the entire cyphonautes alimentary canal degenerate leaving only one layer of cells, except in the area of the future polypide bud where it is lacking. This is not much different from the metamorphosis of some Entoprocta in which the prototroch becomes enclosed at metamorphosis and degenerates. Nielsen therefore concludes that although more extensive and more fully developed as a hydrostatic skeleton for eversion and retraction of the polypides in Ectoprocta, the body cavity of the two groups is comparable and probably not coelomic.

The arrangement of cilia on the tentacles is very similar in the two groups and the anus lies outside the ciliary girdle in Entoprocta, although within the area embraced by the lophophore, as it does in Ectoprocta, but the function of the ciliated tentacular apparatus appears to be quite different. In the Entoprocta the lateral cilia of the tentacles create a current of water towards the centre of the lophophore and out over the atrium; the same cilia trap particles and transfer them to the frontal cilia which carry them to the mouth. In the Ectoprocta, the lateral cilia produce a current towards the mouth, and then out between the tentacles; denser food particles are then deposited on the mouth or the ciliated groove between the tentacles.

Apart from this difference in food collecting methods, the differences between Ectoprocta and Entoprocta do not appear to Nielsen to reflect a great phyletic separation. In particular, the ectoproct "coelom" is interpreted as no more than an enlargement of the entoproct pseudocoel associated with the evolution of a retractable polypide.

The conventional view of the relationship of the Ectoprocta with the two other lophophorate phyla results to a considerable extent from a consideration of adult characters. Jägersten (1968) argued that this is an insufficient basis for a reconstruction of phylogeny and the evolution of the entire life cycle must be taken into account. Innovations at larval or other developmental stages must be accounted for and reconciled with both the life cycle and adult structure of existing animals in any phyletic scheme. Jägersten applied this thesis to a detailed consideration of the evolution of the lophophorates (Fig. 6) with a result which is in substantial agreement with the views of most anatomists. Nielsen (1971), however, using a similar approach to Jägersten found serious difficulties in deriving the life cycle of ectoprocts from that of phoronids. The principle objection to a close relationship between them arises from the timing, in both ontogeny and phylogeny, of the appearance of tentacles and buds. In the Phoronidea the tentacles arise from the metatroch of the larva or in close association with it, but in the Ectoprocta this region is completely enveloped by an overgrowth of the episphere which makes contact with the substratum all round the settled larva. The episphere gives rise to the

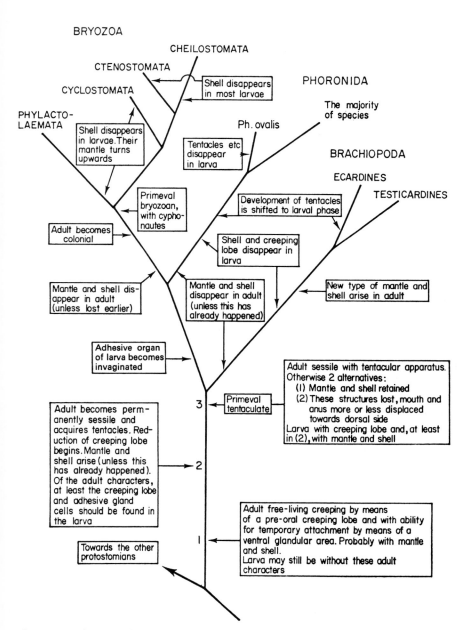

BRYOZOA

CHEILOSTOMATA

CTENOSTOMATA

CYCLOSTOMATA

PHORONIDA

PHYLACTO-
LAEMATA

Shell disappears
in most larvae

The majority
of species

Shell disappears
in larvae.Their
mantle turns
upwards

Ph. ovalis

Tentacles etc
disappear
in larva

BRACHIOPODA

ECARDINES

TESTICARDINES

Primeval
bryozoan,
with cypho-
nautes

Development of tentacles
is shifted to larval phase

Adult becomes
colonial

Shell and creeping
lobe disappear in
larva

Mantle and shell dis-
appear in adult
(unless lost earlier)

Mantle and shell
disappear in adult
(unless this has
already happened)

New type of mantle and
shell arise in adult

Adhesive organ
of larva becomes
invaginated

Adult sessile with tentacular apparatus.
Otherwise 2 alternatives:
  (1) Mantle and shell retained
  (2) These structures lost, mouth and
      anus more or less displaced
      towards dorsal side
Larva with creeping lobe and, at least
in (2), with mantle and shell

3

Primeval
tentaculate

Adult becomes perm-
anently sessile and
acquires tentacles. Red-
uction of creeping lobe
begins. Mantle and
shell arise (unless this
has already happened).
Of the adult characters,
at least the creeping lobe
and adhesive gland
cells should be found in
the larva

2

Adult free-living creeping by means
of a pre-oral creeping lobe and with ability
for temporary attachment by means of a
ventral glandular area. Probably with mantle
and shell.
Larva may still be without these adult
characters

1

Towards the other
protostomians

FIG. 6. Radiation of lophophorates indicated the postulated sequence of appearance of developmental and adult anatomical features. (From Jägersten,
1968.)

buds which then become tentaculate. Nielsen argues that if a phoronid-like "protobryozoon" had a metamorphosis like that of extant ectoprocts, its feeding structure would have been covered by the episphere. If, to escape this difficulty, it is proposed that budding had already been developed in this phoronid-like ancestor (a much earlier stage than that envisaged by Jägersten) then the buds must have arisen from the narrow space inside the ring of tentacles, corresponding to the episphere in phoronids. Budding from this area is incompatible with what is known of phoronid organization. In enumerating the similarities and differences between Ectoprocta and Phoronidea, Nielsen (1977) finds the feeding pattern as the only significant feature held in common between them. In most other respects, Ectoprocta resemble Entoprocta.

Nielsen's interpretation entails a reconsideration of bryozoan phylogeny. He suggests that the Ectoprocta are a specialized group in which the larval gut is not retained and all adults arise through budding from the metamorphosed larva. The Ectoprocta are, in principle, colonial animals and solitary forms are regarded as colonies in which the stolons break very easily. Thus, the Gymnolaemata are the more primitive; the Phylactolaemata and Cyclostomata are the more specialized forms. This view of relationships within the Ectoprocta has been strongly contested. Jebram (1973) points out that the existence of the epistome and protocoel, and the U-shaped lophophore in phylactolaemes but not gymnolaemes link the former with other lophophorate phyla and indicate a more intelligible evolutionary pathway. The orally directed growth and budding of phylactolaemes also appears more primitive than the anally directed budding of gymnolaemes, and there is greater difficulty in deriving the former from the latter than the reverse.

ORIGIN AND RADIATION OF DEUTEROSTOMES

The principle deuterostome phyla are clearly related to one another (only the affinities of the supposedly deuterostome Chaetognatha are obscure). The Echinodermata and Hemichordata have convincing embryological and larval similarities and share an obviously trimeric basic organization of the coelom, although this has become highly modified in echinoderms. The propensity of the central nervous system, when it sinks beneath the epidermis, to be folded inwards in these animals results in a neural tube in enteropneusts, urochordate larvae and the other chordates. The existence of characteristic pharyngeal gill clefts in enteropneusts, urochordates and the lower chordates also link these phyla. Any interpretation of deuterostome, or more particularly vertebrate, origins must therefore offer a plausible explanation of the evolution of both the invertebrate and vertebrate phyla in this group. The first task is to decide the structure of the earliest deuterostome and postulate its derivation from other metazoans. Vertebrates, echinoderms and hemichordates have all been cast in this role and it is necessary to consider the merits and implications of all three interpretations.

1. Vertebrata

One group of theories derives vertebrates from animals that were already metamerically segmented: molluscs (Lemche, 1959, who considered this phylum to be metamerically segmented), annelids (Gutmann, 1966, 1967) or arthropods (Løvtrup, 1977). Although only Gutmann among recent authors is explicit on the matter, it is implied in all these theories that non-segmented deuterostomes have secondarily lost a metameric organization and have evolved a simpler structure than their proto-vertebrate relations.

The first question to be resolved in assessing these theories is whether or not vertebrate metamerism is homologous with that of annelids. In existing annelids and vertebrates, segmentation is certainly morphologically different and serves different functions (Clark, 1964). In annelids it is primarily a metamerism of the coelomic compartments and is associated with peristaltic burrowing; the musculature is not segmented. In vertebrates it is the longitudinal musculature and not the coelom that is segmented and is an adaptation to producing lateral swimming flexures of the body in animals with an axial skeleton.

During the embryonic development of the mesodermal blocks in cephalochordates, segmented coelomic pouches appear although they soon lose their metameric organization, and in lower vertebrates there is a transitory segmentation of anterior parts of the coelom during ontogeny. This may be interpreted as recapitulatory: the last vestige of annelid coelomic metamerism in animals in which the nature and function of segmentation has changed. Equally, though, it is possible that this transitory segmentation of the coelom is a function of the time of appearance of a coelomic cavity in relation to the segmentation of the mesoderm and formation of myotomal rudiments. If the mesoderm cavitates to form a coelom at a time when the mesoderm is segmenting, a segmented coelom will inevitably result whether this feature has phylogenetic significance or not. The same argument applies to other coelomic and mesodermal components that are metameric in lower chordates.

Metamerism of the body-wall musculature is rare in annelids (Clark, 1964; Clark and Richardson, 1967). In arthropods it is related to the existence of an exoskeleton which must be articulated if the animals are to flex, and if articulated, the musculature must be related to the sclerites. In Chordates the segmented musculature is evidently related to the production of lateral swimming flexures, but all long, ribbon-shaped animals swim in the same manner whether they are segmented or not and the existence of a segmented musculature appears to make no difference to their swimming ability (Clark, 1976). The only mechanical difference between vertebrates and invertebrates which might account for the development of a segmented musculature in the former is the presence of an axial skeleton which evidently made its first appearance as a column of turgid, vacuolated cells enclosed in a connective tissue sheath.

Such a structure resists longitudinal compressive forces during swimming and prevents shortening of the body. This essential function is performed by the hydrostatic skeleton (coelom, pseudocoel, parenchymatous tissue etc.) in invertebrates and appears an insufficient explanation of the evolution of a notochord in animals that were already coelomate. An additional function of the notochord is that it provides stiffness, and its resistance to lateral deformation appears to be more significant than its resistance to longitudinal compressive forces. In *Branchiostoma*, the stiffness of the notochord is adjustable and this is considered to be an important adaptation to fast swimming and burrowing (Guthrie and Banks, 1970; Webb, 1973). The muscular architecture is clearly designed to apply torsional forces to the notochord (Nursall, 1956) and appears to explain the metamerism of vertebrate, but not invertebrate, musculature as an adaptation to swimming.

The only swimming chordate without a functionally segmented musculature is the ascidian larva, but a comparable situation exists in dogfish embryos when a "precocious band" of unsegmented muscle is formed on either side of the notochord and produces lateral undulations of the body analogous to swimming movements (Harris and Whiting, 1954). Whether the condition in ascidian larvae is primitive or secondary, it is evidently related to the small size of the organism and weakness of the forces required to flex the tail. The role of this notochord is as compression strut in an organism in which there are no alternative structures to resist compression, and as a supporting structure for a propulsive tail which would otherwise droop.

These considerations, while not precluding the evolution of vertebrates from metameric invertebrates, offer no support to such a view. The only evidence in favour of this hypothesis is the equivocal embryological evidence of transitory metamerism of the coelom in cephalochordates. In addition to the rather slight nature of the evidence supporting them, these theories necessarily suppose that non-vertebrate deuterostomes evolved from an early metameric chordate. The arguments for this, though tenable, are the reverse of those used to account for the progressive evolution from invertebrate deuterostomes and introduce a complexity which it is not necessary to postulate.

The Nemertini are the only unsegmented animals outside the Deuterostomia that have been considered as possible sources of the vertebrates. Willmer (1974) has reviewed the considerable number of nemertean features which might be compared with vertebrate structures, including the pituitary complex, cephalic sense organs, excretory and blood vascular systems and, most importantly, the notochord. This, it is claimed, is derived from the nemertean proboscis (or in earlier versions of the nemertean hypothesis, from the rhynchocoel sheath and contents). Most, if not all, the alleged resemblances between nemerteans and vertebrates appear to be superficial and of very doubtful validity and Willmer's functional interpretation of the proposed evolutionary changes is equally unconvincing. These developments are seen as an adaptation to swimming, despite the fact that many nemerteans swim well and it is not at all clear how modifications of the proboscis apparatus as a skeletal element would have assisted this process. In Willmer's version of the hypothesis, the paired lateral diverticula of the nemertean intestine become the myotomal musculature and cephalochordates, urochordates, petromyzonts, myxinoids and gnathostome fish are all conceived as having an independent origin from a heterogeneous nemertean stock, but these matters are not crucial to the hypothesis.

2. Echinodermata

If the vertebrates did not evolve directly from metameric invertebrates the search for the earliest deuterostomes turns to the Echinodermata and Hemichordata. (The chordate features of the urochordate larva are such that whether regarded as primitive or regressive, the Urochordata have never been considered to be close to the origin of the deuterostome radiation.) At one time the Hemichordata were considered only slightly further removed than urochordates from chordates because of the existence of a hollow, dorsal neural tube, pharyngeal gills and a supposed notochord (the stomochord) in enteropneusts. Attention then focused on the Echinodermata as the only indubitably non-chordate deuterostomes.

Although the Echinodermata have a number of features which may be regarded as primitive (the nervous system largely in the form of a subepidermal plexus, etc.), they have a basic asymmetry and a number of unique features, such as the water vascular system, which make it difficult to derive the echinoderm structure directly from that of adult protostomes. Instead, the dipleurula larva took its place at the stem of the deuterostome radiation.

The dipleurula theory, with its strong overtones of Haeckel's recapitulation theory, envisaged the ancestral deuterostome as little more than the lowest common denominator of echinoderm larvae. In the pentactula theory the appearance of a coelomic tentacular structure is envisaged at a somewhat earlier stage than in the dipleurula theory, but the two differ from one another less than appears at first sight, and, indeed they were reconciled in a common theory by MacBride (1896). In one form or another the dipleurula/pentactula theory enjoyed almost universal acceptance for many years and is still repeated in many textbooks, but, in fact, the structure of the supposed ancestral form and subsequent evolutionary developments in it involve the most arbitrary postulates and there is no functional justification for any aspect of the theory (Clark, 1964).

Nichols (1967) avoided the worst aspects of the dipleurula theory by proposing that echinoderms evolved from coelomates with a structure comparable to that of sipunculids. The circum-oral tentacular ring with its separate coelom and compensation sacs in sipunculids is equated with the arms, hydrocoel and polian vesicles in a hypothetical primitive echinoderm. The recurved gut of echinoderms, which is responsible for surprising contortions in the dipleurula/pentactula theory, is already present in the Sipuncula. The middle Cambrian carpoids, such as *Gyrocystus*, were interpreted by Nichols as being at a grade of construction little advanced beyond his supposed early echinoderm. Concentration on the origin of the water vascular system, as in Nichols' theory, leaves unanswered the questions of the origin of the clearly trimeric organization of the echinoderm coelom. The very great modification of the coelomic compartments during the ontogeny of existing echinoderms suggests that trimery did not originate in echinoderms (characterized in particular by the possession of the unique specialization of one coelomic compartment as the water vascular system), but that this type of coelomic organization was inherited from oligomeric ancestors.

3. Hemichordata

The only deuterostomes which have a functional trimerous organization are the Hemichordata, represented by the Pterobranchia and Enteropneusta. The former are tubicolous polyps and the latter burrowing worms, and the body organization serves somewhat different functions in each. The protocoel provides a hydrostatic skeleton for the mobile pre-oral lobe or proboscis which, in different ways, is used in locomotion in both pterobranchs and enteropneusts. The metacoel is the general perivisceral coelom for the greater part of the body, though small and largely occluded by muscle and connective tissue in enteropneusts where it has little hydrostatic function. The function of the mesocoel, however, differs in the two classes of hemichordates. In pterobranchs it is a hydrostatic organ for the lophophore; in enteropneusts it is the collar coelom, consisting of a number of ill defined spaces and, so far as one can tell, it is functionless. Although Gutmann (1969, 1970) interprets it as an incipient mesocoel and believes that the enteropneust organization preceded and gave rise to that of pterobranchs, a more likely explanation is that with the adoption of a burrowing existence, the lophophore became useless as a feeding and respiratory organ, hindered burrowing, was lost and its coelomic hydrostatic organ, the mesocoel, became occluded as a result and persists only as a vestige. To compensate for the loss of the chief respiratory organ of the body, pharygeal gill slits became multiplied and elaborated.

The fact that the body architecture of pterobranchs is comparable to one of the basic types of coelomate functional organization is consistent with the view that they are not far removed from a stem group of the deuterostomes.

4. Deuterostome Radiation

If, as it appears, deuterostomes evolved from oligomerous, tubicolous worms, the broad directions of the subsequent radiation are clear although there is room for a variety of alternative interpretations of the details of this process.

The Pterobranchia retain much of the fundamental structure of the

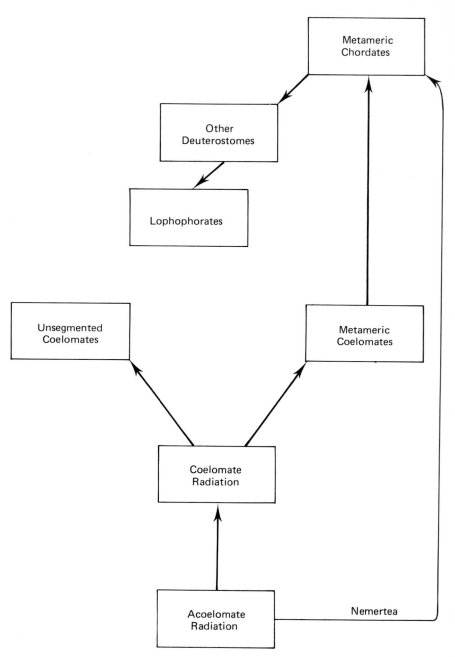

Fig. 7. Outline of metazoan phylogeny, assuming vertebrate metamerism is derived from that of spiralians or that vertebrates are derived from other unsegmented spiralians.

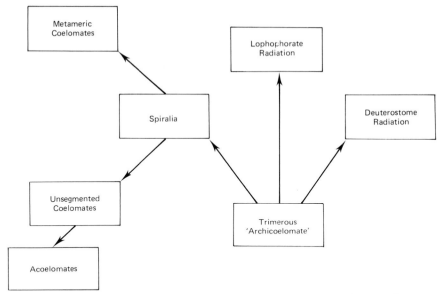

Fɪɢ. 8. Outline of metazoan phylogeny following the archicoelomate theory.

earliest deuterostomes although the sessile habit and colonial structure of most pterobranchs need not be regarded as features of the earliest hemichordates. The related Enteropneusta display the morphological changes associated with the adoption of a burrowing existence to which we have already referred.

The derivation of echinoderms from an organism comparable to a pterobranch is foreseen in Grobben's (1923) theory of the origin of echinoderms and this does not differ greatly from the pentactula theory, save that the organism is thought to be sedentary rather than ciliated and swimming.

The Chordata emerged from this deuterostome radiation by the evolution of a notochord and, possibly later, a metameric musculature. There is no shortage of views as to precisely how and in what organism these events took place and I shall not discuss the selection between them further here.

CONCLUSION

Although there are widely differing views about the phyletic position and relationships of individual phyla, there is less conflict than appears at first sight about the main features of metazoan radiation. The principal differences between the schemes shown in Figs 7–9 are as follows.

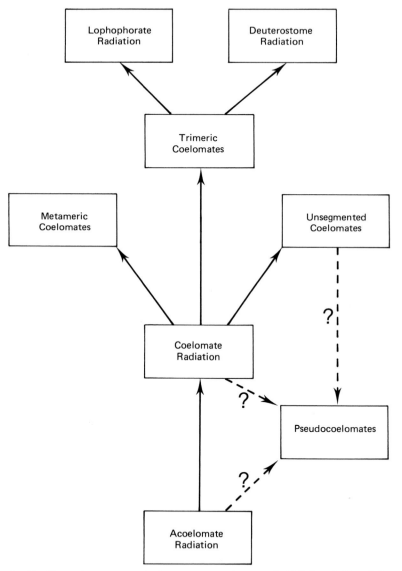

FIG. 9. Outline of metazoan phylogeny assuming acoeloid/planuloid origins and polyphyletic origin of the coelom.

1. The trimeric coelomate organization is the basis of the radiation of lophophorates and deuterostomes, except in theories which derive chordates directly from (generally metameric) protostomes (Fig. 7). In that case, both lophophorates and non-chordate deuterostomes are derived from early segmented chordates in a progressive reduction of their structure.

2. The first appearance of the trimeric coelomate organization may be early in the history of the Metazoa (Fig. 8). It follows from this that the Spiralia (protostomes) have suffered a loss of the proto- and mesocoels and acoelomates show a further progressive simplification of structure. Alternatively (Fig. 9) the trimeric coelom appears as part of the radiation of early coelomates in which the unsegmented, metameric and trimeric types of coelomate organization appeared as different adaptations to life in, rather than on the substratum.

The phyla described as pseudocoelomate (indicated only in Fig. 9) occupy an uncertain position and may be derived from acoelomates, coelomates or represent a separate radiation comparable to that of the early coelomates.

These general schemes are unaffected by changing views about the position of individual phyla (e.g. whether or not molluscs are regarded as coelomate or metamerically segmented).

REFERENCES

ANDERSON, D. T. (1973). "Embryology and Phylogeny in Annelids and Arthropods." Pergamon Press, Oxford.

AX, P. (1963). Die Ausbildung eines Schwanzfadens in der interstitiellen Sandfauna und die Verwandtbarkeit von Lebensform-charakteren für die Verwandtschaftsforschung. *Zool. Anz.* **171**, 51–76.

BERRILL, N. J. (1955). "The Origin of Vertebrates". Clarendon Press, Oxford.

BOADEN, P. J. S. (1975). Anaerobiosis, meiofauna and early metazoan evolution. *Zool. Scripta*, **4**, 21–24.

BONE, Q. (1960). The origin of chordates. *J. Linn. Soc. (Zool.)*, **44**, 252–269.

BONIK, K., GRASSHOFF, M. and GUTMANN, W. F. (1976). Die Evolution der Tierkonstruktion. III. Vom Gallertoid zur Coelomhydraulik. *Natur Mus.* **106**, 178–188.

BRIEN, P. (1960). Bryozoaires. *In* "Traité de Zoologie" (P. P. Grasse, ed.), Vol. 5 (2), pp. 1053–1335. Masson, Paris.

BRIEN, P. and PAPYN, L. (1954). Les Endoproctes et la classe des Bryozoaires. *Annls Soc. Zool. Belg.* **85**, 59–87.

BUBKO, O. V. (1973). On the systematic position of the Oweniidae and Archiannelida (Annelida). (In Russian, English summary.) *Zool. Zh.* **52**, 1286–1296.

BUBKO, O. V. and MINICHEV, Y. S. (1972). The nervous system of Oweniidae (Polychaeta). (In Russian, English summary.) *Zool. Zh.* **51**, 1288–1299.

CLARK, R. B. (1964). "Dynamics in Metazoan Evolution, the origin of the coelom and segments." Clarendon Press, Oxford.

CLARK, R. B. (1969). Systematics and phylogeny: Annelida, Echiura, Sipuncula. *In* "Chemical Zoology" (M. Florkin and B. T. Scheer, eds), Vol. 4, pp. 1–68. Academic Press, London and New York.

CLARK, R. B. (1976). Undulatory swimming in polychaetes. *In* "Perspectives in Experimental Zoology" (P. Spencer-Davis, ed.), Vol. 1, pp. 437–446. Pergamon, Oxford.

CLARK, R. B. (1977). Phylogenetic reconstruction. *Verh. Dtsch Zool. Ges.* 1977, 175–183.

CLARK, R. B. (1978a). Functional correlates of the coelom. *Zool. Jb.* (*Anat.*) (In Press.)

CLARK, R. B. (1978b). Composition and relationships. *In* "Physiology of the Annelida" (P. J. Mill, ed.). Academic Press, London and New York. (In Press.)

CLARK, R. B. and HERMANS, C. O. (1976). Kinetics of swimming in some smooth-bodied polychaetes. *J. Zool., Lond.* **178**, 147–160.

CLARK, R. B. and RICHARDSON, M. E. (1967). Apodemes and tonofibrillae in polychaetes. *Bull. Soc. zool. Fr.* **92**, 207–212.

CLARK, R. B. and TRITTON, D. J. (1970). Swimming mechanisms in some nereidiform polychaetes. *J. Zool., Lond.* **161**, 257–271.

DALES, R. P. (1962). The polychaete stomodeum and the inter-relationships of the families of Polychaeta. *Proc. zool. Soc. Lond.* **139**, 389–428.

DALES, R. P. (1977). The polychaete stomodeum and phylogeny. *In* "Essays on Polychaetous Annelids in Memory of Dr Olga Hartman" (D. J. Reish and K. Fauchald, eds), pp. 525–546. Allan Hancock Fdn., Los Angeles.

EMIG, C. C. and SIEWING, R. (1975). The epistome of *Phoronis psammophila* (Phoronida). *Zool. Anz.* **194**, 47–54.

FAUCHALD, K. (1974). Polychaete phylogeny: a problem in protostome evolution. *Syst. Zool.* **23**, 493–506.

FENCHEL, T. M. and RIEDL, R. J. (1970). The sulfide system: a new biotic community underneath the oxidized layer of marine sand bottoms. *Mar. Biol.* **7**, 225–268.

FRETTER, V. and GRAHAM, A. (1962). "British Prosobranch Molluscs." Ray Soc., London.

GELDER, S. R. and JENNINGS, J. B. (1975). The nervous system of the aberrant symbiotic polychaete *Histriobdella homari* and its implications for the taxonomic position of the Histriobdellidae. *Zool. Anz.* **194**, 293–304.

GRAY, J. S. (1969). A new species of *Saccocirrus* (Archiannelida) from the West coast of North America. *Pacif. Sci.* **23**, 238–251.

GROBBEN, K. (1923). Theoretische Erörterungen begriffend die phylogenetische

Ableitung der Echinodermen. *S. B: Akad. wiss. Wien, Math. Naturw. Kl.* **132**, 263–290.

GUTHRIE, D. M. and BANKS, J. R. (1970). Observations on the function and physiological properties of a fast paramyosin muscle—the notochord of amphioxus (*Branchiostoma lanceolatum*). *J. exp. Biol.* **52**, 125–138.

GUTMANN, W. F. (1966). Coelomgliederung, Myomerie und die Frage der Vertebraten-Antezedénten. *Z. zool. Syst. Evolutionsforsch.* **4**, 13–57.

GUTMANN, W. F. (1967). Nachtrag zur "Wurmtheorie" der Vertebraten-Evolution. *Z. zool. Syst. Evolutionsforsch.* **5**, 314–332.

GUTMANN, W. F. (1969). Acranier und Hemichordaten, ein Seitenast der Chordaten. *Zool. Anz.* **182**, 1–26.

GUTMANN, W. F. (1970). Die Entstehung des Muskelapparates der Hemichordaten. *Z. zool. Syst. Evolutionsforsch.* **8**, 139–154.

HADZI, J. (1953). An attempt to reconstruct the system of animal classification. *Syst. Zool.* **2**, 145–154.

HADZI, J. (1963). "The Evolution of the Metazoa." Pergamon, Oxford.

HAND, C. (1959). On the origin and phylogeny of the coelenterates. *Syst. Zool.* **8**, 191–202.

HANSON, E. D. (1958). On the origin of the Eumetazoa. *Syst. Zool.* **7**, 16–47.

HANSON, E. D. (1977). "Origin and Early Evolution of Animals." Wesleyan Univ. Press, Middletown, Conn.

HARRIS, J. E. and WHITING, H. P. (1954). Structure and function of the locomotory system of the dogfish embryo. The myogenic stage of movement. *J. exp. Biol.* **31**, 501–524.

HENNING, W. (1966). "Phylogenetic Systematics." Univ. Illinois Press, Urbana.

HERMANS, C. O. (1969). The systematic position of the Archiannelida. *Syst. Zool.* **18**, 85–102.

HYMAN, L. H. (1951). "The Invertebrates: Platyhelminthes and Rhyncocoela". McGraw-Hill, New York.

HYMAN, L. H. (1959). "The Invertebrates: Smaller Coelomate Groups." McGraw-Hill, New York.

IVANOV, A. V. (1975). Embryonalentwicklung der Pogonophora und ihre Systematische Stellung. *Z. zool. Syst. Evolutionsforsch.* (Special issue.) The phylogeny and systematic position of the Pogonophora, 10–44.

JÄGERSTEN, G. (1944). Zur Kenntnis der Morphologie, Enzystierung und Taxonomie von *Dinophilus*. *K. svenska Vetensk Akad. Handl.*, iii, **21** (2), 1–90.

JÄGERSTEN, C. (1955). On the early phylogeny of the Metazoa. The bilaterogastraea theory. *Zool. Bidr. Upps.* **30**, 321–354.

JÄGERSTEN, G. (1968). "Livscykelns Evolution hos Metazoa. En generell teori." Lärodmedels-förlagen, Stockholm. (In Swedish. English edition: "Evolution of the Metazoan Life Cycle." Academic Press, London and New York.

JEBRAM, D. (1973). The importance of different growth directions in Phylactolaemata and Gymnolaemata for reconstructing the phylogeny of the Bryozoa. *In* "Living and Fossil Bryozoa Recent Advances in Research" (G. P. Larwood, ed.), pp. 565–576. Academic Press, London and New York.

JEFFRIES, R. P. S. (1968). Some fossil chordates with echinoderm affinities. *Symp. zool. Soc. Lond.* **20**, 163–208.

LAND, A. VAN DER and NØRREVANG, A. (1975). The systematic position of *Lamellibranchia* (Annelida, Vestimentifera). *Z. zool. Syst. Evolutionsforsch.* (Special issue.) The phylogeny and systematic position of the Pogonophora, 86–101.

LANG, K. (1953). Die Entwicklung des Eies von *Priapulus caudatus* und die systematische Stellung der Priapuliden. *Ark. Zool.* ii, **5**, 321–348.

LAUBIER, L. (1967). Adaptations chez les Annélides Polychètes interstitielles. *Année biol.* **43**, 1–16.

LEMCHE, H. (1959). Protostomian relationships in the light of *Neopilina*. *Proc. XV int. Congr. Zool.* 381–389.

LØVTRUP, S. (1977). "The Phylogeny of Vertebrata." Wiley, London.

MACBRIDE, E. W. (1896). The development of *Asterina gibbosa*. *Quart. Jl microsc. Sci.* **38**, 339–411.

MANTON, S. M. (1967). The polychaete *Spinther* and the origin of the Arthropoda. *J. nat. Hist.* **1**, 1–22.

MANTON, S. M. (1977). "The Arthropoda, Habits, Functional Morphology and Evolution." Clarendon, Oxford.

MARCUS, E. (1958). On the evolution of the animal phyla. *Q. Rev. Biol.* **33**, 24–58.

MATTISSON, A., NILSSON, S. and FÄNGE, R. (1974). Light microscopical and ultrastructural organization of muscles of *Priapulus caudatus* (Priapulida) and their responses to drugs, with phylogenetic remarks. *Zool. Scripta*, **3**, 209–218.

METTAM, C. (1969). Peristaltic waves of tubicolous worms and the problem of irrigation in *Sabella pavonina*. *J. Zool., Lond.* **158**, 341–356.

METTAM, C. (1971). Functional design and evolution of the polychaete *Aphrodite aculeata*. *J. Zool., Lond.* **163**, 489–514.

MEYER, F. and MEYER, H. (1972). Loss of fatty acid biosynthesis in flatworms. *In* "Comparative Biochemistry of Parasites." (van der Bossche, ed.). Academic Press, London and New York.

MILEIKOVSKY, S. A. (1977). On the systematic interrelationships within the Polychaeta and Annelida—an attempt to create an integrated system based on their larval morphology. *In* "Essays on Polychaetous Annelids in Memory of Dr. Olga Hartman" (D. J. Reisch and K. Fauchald, eds), pp. 503–524. Allan Hancock Fdn., Los Angeles.

NICHOLAS, W. L. and HYNES, H. B. N. (1963). Embryology, post-embryonic development and phylogeny of the Acanthocephala. *In* "The Lower Metazoa" (E. C. Dougherty, Z. N. Brown, E. D. Hanson and W. D. Hartman, eds), pp. 385–402. Univ. Calif. Press, Berkeley, Los Angeles.

NICHOLS, D. (1966). "Echinoderms", 2nd edn. Hutchinson, London.

NICHOLS, D. (1967). The origin of echinoderms. *Symp. zool. Soc. Lond.* **20**, 209–229.

NIELSEN, C. (1971). Entoproct life-cycles and the entoproct/ectoproct relationship. *Ophelia*, **9**, 209–341.

NIELSEN, C. (1977). The relationships of Entoprocta, Ectoprocta and Phoronida. *Am. Zool.* **17**, 149–150.

NØRREVANG, A. (1970). On the embryology of *Siboglinum* and its implications for the systematic position of the Pogonophora. *Sarsia* **42**, 7–16.

Nursall, J. R. (1956). The lateral musculature and the swimming of fish. *Proc. zool. Soc. Lond.* **126**, 127–143.

Nursall, J. R. (1962). On the origins of the major groups of animals. *Evolution*, **16**, 118–123.

Orrhage, L. (1964). Anatomische und morphologische Studien über die Polychaetenfamilien Spionidae, Disomidae und Poecilochaetidae. *Zool. Bidr. Upps.* **36**, 335–405.

Orrhage, L. (1974). Über die Anatomie, Histologie und Verwandtschaft der Apistobranchidae (Polychaeta Sedentaria) nebst Bemerkungen über die systematische Stellung der Archianneliden. *Z. Morph. Tiere*, **79**, 1–45.

Remane, A. (1950). Die Entstehung der Metamerie der Wirbellosen. *Verh. Dtsch zool. Ges.* 1949, 16–23.

Remane, A. (1952). "Die Grundlagen der natürlichen Systems, der vergleichenden Anatomie und der Phylogenetik." Akad. Verlagsgesellschaft, Leipzig.

Remane, A. (1963). The systematic position and phylogeny of pseudocoelomates. *In* "The Lower Metazoa." (E. C. Dougherty, Z. N. Brown, E. D. Hanson and W. D. Hartman, eds), pp. 247–255. Univ. Calif. Press, Berkeley, Los Angeles.

Reutterer, A. (1969). Zum Problem der Metazoenabstammung. *Z. zool. Syst. Evolutionsforsch.* **14**, 198–226.

Rieger, R. M., Ruppert, E., Rieger, G. E. and Schoepfer-Sterrer, C. (1974). On the fine structure of gastrotrichs with description on *Chordodasys antennatus* n.sp. *Zool. Scripta*, **3**, 219–237.

Ryland, J. S. (1970). "Bryozoans." Hutchinson, London.

Salvini-Plawen, L. von (1968). Die "Funktions-Coelomtheorie" in der Evolution der Mollusken. *Syst. Zool.* **17**, 192–208.

Salvini-Plawen, L. von (1969). Solenogastres und Caudofoveata (Mollusca, Aculifera): Organisation und phylogenetische Bedeutung. *Malacologia*, **9**, 191–216.

Shapeero, W. L. (1961). Phylogeny of Priapulida. *Science, N.Y.* **133**, 879–880.

Sharov, A. G. (1966). "Basic Arthropodan Stock with Special Reference to Insects." Pergamon, Oxford.

Siewing, R. (1969). "Lehrbuch der vergleichenden Entwicklungsgeschichte der Tiere." Paray, Hamburg, Berlin.

Siewing, R. (1975a). Thoughts about the phylogenetic-systematic position of Pogonophora. *Z. zool. Syst. Evolutionsforsch.* (Special issue.) The Phylogeny and Systematic position of the Pogonophora, 127–138.

Siewing, R. (1975b). Gliederung des Phoronidenkörpers. *Verh. Dtsch zool. Ges.* 1974, 116–121.

Siewing, R. (1976). Probleme und neuere Erkenntnisse in der Grosssystematik der Wirbellosen. *Verh. Dtsch zool. Ges.* 1976, 59–83.

Sillman, L. R. (1960). The origin of vertebrates. *J. Paleont.* **34**, 540–544.

Sneath, P. H. A. and Sokal, R. R. (1973). "Numerical Taxonomy. The Principles and Practice of Numerical Classification." Freeman, San Francisco.

Southward, A. J. (1975). On the evolutionary significance of the mode of feeding of Pogonophora. *Z. zool. Syst. Evolutionsforsch.* (Special issue.) The Phylogeny and Systematic position of the Pogonophora, 77–85.

SOUTHWARD, E. C. (1963). Pogonophora. *Oceanogr. Mar. Biol. Ann. Rev.* 1, 405–428.

SOUTHWARD, E. C. (1971). Recent researches on the Pogonophora. *Oceanogr. Mar. Biol. Ann. Rev.* 9, 193–220.

SOUTHWARD, E. C. (1975a). Fine structure and phylogeny of the Pogonophora. *Symp. zool. Soc. Lond.* 36, 235–251.

SOUTHWARD, E. C. (1975b). A study of the structure of the opisthosoma of *Siboglinum fjordicum. Z. zool. Syst. Evolutionsforsch.* (Special issue.) The Phylogeny and Systematic position of the Pogonophora, 64–76.

STEINBÖCK, O. (1958). Zur Phylogenie der Gastrotrichen. *Zool. Anz., Suppl.* 21, 128–169.

STERRER, W. (1973). Plate tectonics as a mechanism for dispersal and speciation in interstitial sand fauna. *Netherl. J. Sea Res.* 7, 200–222.

STORCH, V. (1968). Zur vergleichenden Anatomie der segmentalen Muskelsysteme und zur Verwandtschaft der Polychaeten-Familien. *Z. Morph. Tiere,* 63, 251–342.

SVESHNIKOV, V. A. (1958). On the morphology of larvae of some Eunicemorpha (Polychaeta). (In Russian, English summary.) *Dokl. Akad. Nauk SSSR.* 121, 565–568.

SWEDMARK, B. (1964). The interstitial fauna of marine sand. *Biol. Rev.* 39, 1–42.

TAYLOR, G. (1952). Analysis of the swimming of long and narrow animals. *Proc. R. Soc., A,* 214, 158–183.

TEUCHERT, G. (1977). Liebeshöhlenverhältnisse von dem marinen Gastrotrich *Turbanella cornuta* Remane (Ordnung Macrodasyoidea) und eine phylogenetische Bewertung. *Zool. Jb. Anat.* 97, 586–596.

VAGVOLGYI, J. (1967). On the origin of molluscs, the coelom, and coelom segmentation. *Syst. Zool.* 16, 153–168.

WEBB, J. E. (1973). The role of the notochord in forward and reverse swimming and burrowing in the amphioxus (*Branchiostoma lanceolatum*). *J. Zool., Lond.* 170, 325–338.

WESTHEIDE, W. (1977). The geographical distribution of interstitial polychaetes. *Microfauna Meeresboden,* 61, 287–302.

WHITTINGTON, H. B. (1975). The enigmatic animal *Opabinia ragalis*, Middle Cambrian, Burgess Shale, British Columbia. *Phil. Trans. R. Soc. B,* 271, 1–43.

WILLMER, E. N. (1970). "Cytology and Evolution", 2nd edn. Academic Press, London and New York.

WILLMER, E. N. (1974). Nemertines as possible ancestors of the vertebrates. *Biol. Rev.* 49, 321–363.

ZIMMER, R. L. (1973). Morphological and developmental affinities of the lophophorates. *In :* "Living and Fossil Bryozoa, Recent Advances in Research" (G. P. Larwood, ed.), pp. 593–599. Academic Press, London and New York.

5 | The Cambrian radiation event

M. D. BRASIER

Department of Geology, University of Hull, England

Abstract: The fossil record indicates that the metazoan phyla originated over a period of from 200 to 500 million years in a series of radiations. Of these the Cambrian radiation event (taken to include latest Vendian and early Cambrian) was apparently the most dramatic. The broad stratigraphic framework of this radiation is outlined and the fossil record of participating groups reviewed. Studies of the phyletic, skeletal, trophic, size-related and environment-related changes over the critical time interval reveal the all-embracing nature of this radiation event. Although eutrophication of the ecosystem and increased cropping pressures are implicated they do not explain the timing of the radiation. The "cause" is most likely entwined in the evidence for planetary upheaval that brought about the associated transgression, evidence that has yet to be properly unearthed and evaluated.

INTRODUCTION

Although microscopic algal and bacterial remains are known from rocks over $3 \cdot 0 \times 10^9$ years old, the remains of animals are not well known before about 680 m.y. (million years before the present day), whilst plants and animals with preserved hard parts did not become established until about 570 m.y. The latter event, sometimes called the "Cambrian explosion" was a spectacular adaptive radiation heralding the end of the age of "first life" or Proterozoic aeon and the dawn of the age of "revealed life" or Phanerozoic aeon (*c.* 570 m.y.). The resultant contrast between Precambrian rocks almost barren of fossils and Cambrian rocks in which they may abound has prompted many ingenious explanations. The nature of this

Systematics Association Special Volume No. 12, "The Origin of Major Invertebrate Groups", edited by M. R. House, 1979, pp. 103–159. Academic Press, London and New York.

Cambrian radiation event is still insufficiently understood, however, to demand that we study its nature rather than its "cause".

THE STRATIGRAPHIC AND PALAEONTOLOGICAL FRAMEWORK

The biostratigraphic control of late Precambrian and early Cambrian rocks is notoriously shaky because the fossils are usually scarce, facies-controlled and provincial in their distribution. Most workers acknowledge the existence of two late Precambrian stratigraphic units, here treated as systems within the Proterozoic aeon. The Riphean system ranged from about 1600 to 680 m.y. and can be correlated by means of the varied columnar stromatolites (Raaben, 1969), microphytolites (Kolosov, 1975) and microfloras, which contain unicellular and multicellular eukaryotes, prokaryotes and the problematic acritarchs (Schopf, 1977). The Vendian system ranged in time from about 680 to 570 m.y. (Sokolov, 1972). At the base it contains deposits of the Varangian glaciation which may be used as marker horizons (Harland, 1974). Subdivisions of the Vendian on the Russian platform include, in ascending order, the Volhyn (or Mogilev-Podolian) series, the Redkino series and the Valdai series. In this paper the Volhyn, Mogilev and Redkino series are informally referred to as "lower Vendian" and the Valdai series as "upper Vendian". This system contains invertebrate trace fossils and soft body fossils of "Ediacara type" which may be used for correlation (Glaessner, 1971a) along with acritarchs and microphytolites (Semikhatov et al., 1970, Vidal, 1976a). If the fossiliferous Nama Group of South Africa is truly of Vendian age (as the latest

FIG. 1. A tentative correlation of some fossiliferous rocks of late Proterozoic and early Cambrian age. In each column is noted the approximate order of appearance of various fossil groups, including trace fossil genera. Data and earlier references can be found in the following sources: 1, Brasier et al. (1978); 2, Poulsen (1976), Brasier (unpublished); 3, Sokolov (1973, 1976), Fedonkin (1977); 4, Bergström (1970), Martinsson (1974), Cowie and Glaessner (1975), Brasier (unpublished); 5, Banks (1970); 6, Anderson and Misra (1968), North (1971), Dr T. Fletcher (personal communication, 1978); 7, Rozanov and Debrenne (1974), Brasier (1976b), Crimes et al. (1977); 8, Cowie and Spencer (1970); 9, 10, Young (1972); 11, Alpert (1975a, b, 1977), Wigett (1977); 12, Rozanov et al. (1969), Cowie and Rozanov (1974); 13, Cowie and Glaessner (1975), Matthews and Missarzhevsky (1975); 14, Glaessner (1969), Rozanov and Debrenne (1974); Cowie and Glaessner (1975); Brasier (1976a); 15, Cowie and Glaessner (1975); 16, Germs (1972a).

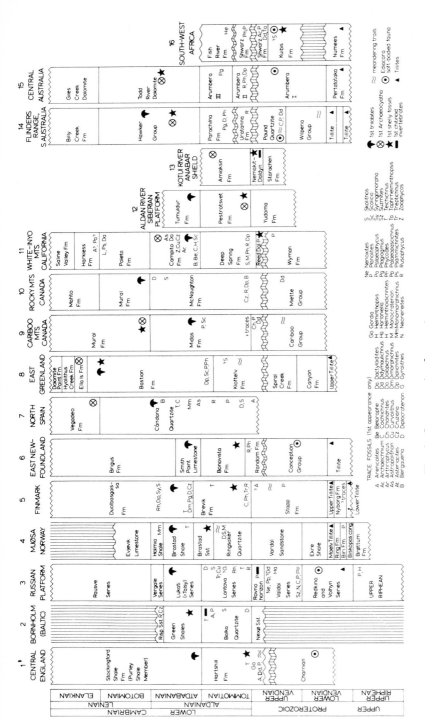

Fig. 1. See opposite for caption.

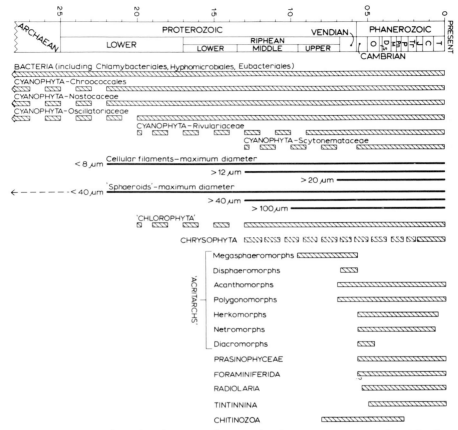

FIG. 2. Stratigraphic distribution of microfossil groups connected with the Cambrian radiation event. Data from sources in text.

palaeontological work seems to affirm) then invertebrate skeletal fossils also first appeared in Vendian times.

The succeeding Rovno horizon of the Russian platform and the Nemakit-Daldyn horizon of the Anabar shield, Siberia, are both fossiliferous transitional units between undoubted Vendian and Cambrian, but here will be considered to be "uppermost (or latest) Vendian" in age. Simple tubular skeletons occur in these sequences.

The Cambrian system which followed (age *c.* 570 to 500 m.y.), is currently subdivided into lower, middle and upper units but the lower Cambrian is the main interest of this paper. The lower Cambrian is further

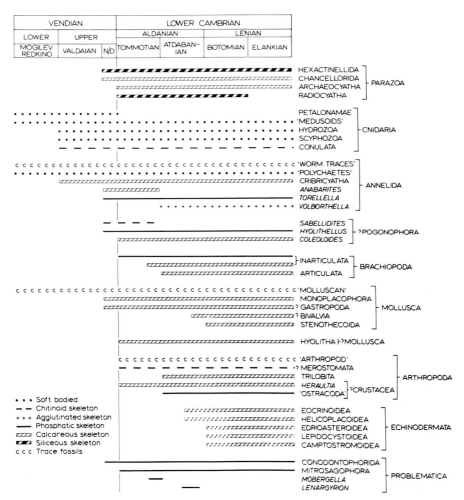

FIG 3. Stratigraphic distribution of invertebrate macrofossils in the Vendian and early Cambrian, with an indication of their skeletal composition. Data from sources in the text.

subdivided on the basis of trilobite biozones (Cowie *et al.*, 1972; Fritz, 1972) or on archaeocyathid biozones (Rozanov and Debrenne, 1974). The latter scheme recognizes four stages: the Tommotian, Atdabanian, Lenian and Elankian; because of recent confusion over the term "Lenian", however, the term "Botomian" will be retained instead.

In Fig. 1, the writer has tentatively correlated some major sequences

across the Precambrian–Cambrian boundary, using the available information on acritarchs, microproblematica, trace fossils, "Ediacara type"
fossils, trilobites, archaeocyathids and other data. From this the general
sequence of fossil appearances can be reconstructed (Figs 2, 3 and ensuing
discussion). This review of the fossil record is followed by a brief analysis
of the phyletic, skeletal, trophic, size-related and environmental changes
observed over the late Proterozoic to early Cambrian interval and
concludes with a list of attributes of the event.

THE FOSSIL RECORD

1. Prokaryotes and Stromatolites

A great deal has now been written about fossil prokaryotes and the
organo-sedimentary structures they accreted in Proterozoic times (see for
example the reviews by Schopf, 1971; Cloud, 1976). Most examples of
cellular preservation are known from silicified banded-iron formations,
silicified carbonates and silicified stromatolites, with relatively few from
silicified oncolithic-oolitic carbonates and phosphorite nodules. Preservation in other lithologies such as unsilicified carbonates and shales is poorer,
usually yielding kerogenous, graphitic or pyritic sheaths and degraded cell
walls and contents. For this reason they are less well known from such
facies and may receive convenient rag-bag names as "acritarchs" and
"microphytolites". Furthermore, the energy devoted to the study of
Proterozoic stromatolite microbiotas has barely extended to the early
Palaeozoic ones, hence the stromatolite-dominated view of Proterozoic
life, with its monotonous prokaryote microbiotas, cannot be said to give
the true picture of the Precambrian "prelude" until the degree of biotic
change across the Vendian and Cambrian is better known.

Late Riphean prokaryote microbiotas probably included Eubacteriales,
Oscillatoriaceae, Nostocaceae, Rivulariaceae and Chroococcaeae (Schopf,
1971) and we may infer that the sulphur, iron and manganese bacteria were
also present at this time (Cloud, 1976; Karkhanis, 1976). Branched and
tufted colonies of oncolith-building cyanophytes of early Cambrian age
(e.g. Brasier, 1977; Edhorn and Anderson, pp. 113–123 in Flügel, 1977)
may represent scytonemataceans and rivulariaceans. The *Anacystis*-like
chroococcalean *Renalcis* first appeared in upper Vendian strata but thrived
in early Cambrian times (Kolosov, 1975; Brasier, 1977).

A change in both the gross morphology and microfabric of stromatolites is discernible through the late Proterozoic so that strata can be correlated by this means (Walter, 1972). Some of these changes may relate to the increasing participation of red and green algae in stromatolite and oncolite formation during Riphean to early Palaeozoic times.

Thrombolites are organic calcareous mounds having a clotted or nebulous, rather than a laminar, microstructure (Aitken, 1967) and were primarily constructed by *Renalcis*, *Epiphyton* (? a branched red alga) and by *Girvanella* (? a tangle of cyanophyte filaments). They are first reported from possible Vendian strata in Morocco (Schmitt and Monninger, pp. 80–85 in Flügel, 1977) but became a very important component of subtidal platform deposits in early Cambrian times, providing a habitat for archaeocyathids, sponges and molluscs.

Endolithic cyanophytes, chlorophytes and rhodophytes are known as fossils from their minute borings into calcareous shells. Although they occur in the early Cambrian carbonates of Labrador (James and Kobluk, 1978) it is difficult to demonstrate their presence before this time owing to the lack of calcareous shells. One could view the problem differently, though, relating the lack of calcareous shells in Proterozoic strata to their destruction by the omnipresent cyanophytes. There certainly seems little reason to believe that the endolithic habit was not present in Proterozoic algae.

2. Eukaryote Protists, Acritarchs and Chitinozoans

Only those Protista with resistant organic, agglutinated, calcareous or siliceous skeletons leave a viable fossil record. Even when preserved, it can be difficult to distinguish simple protistan skeletons from other structures built by "higher" organisms and, of course, to decide to which group a protist fossil (if such it be) may belong. This vexing problem has been temporarily set aside by the use of the term "Acritarcha" for those microscopic organic-walled hollow vesicles of uncertain biological affinity (Evitt, 1963).

Acritarchs are generally released from their rock matrices by maceration and digestion in strong acids, a process which, unfortunately, may also break up plant tissues, filaments and colonies into small clusters or discrete vesicles and thereby cause confusion. Nonetheless, there is a degree of similarity between most Palaeozoic acritarchs in having single walls of a sporopollenin-like material, often provided with an irregular opening (Downie, 1973). Some acritarchs may prove to be peridinalean dinoflagellate cysts (e.g. Dale, 1976) whilst others could be the cysts of gymno-dinalean or dinophysialean dinoflagellates (Lister, 1970) or the cysts of unicellular prasinophyceans (Wall, 1962).

Solitary, single-walled and formerly spherical vesicles, called "sphaeromorphs", predominate in Proterozoic acritarch assemblages. There is a tendency for their maximum diameter to increase throughout this long period, reaching 40 μm by middle Riphean times and 100 μm or more by late Riphean times (Schopf, 1977). There may also be a trend towards increasing morphological diversity within this group during the late Proterozoic (Timofeev, 1959). Giant "megasphaeromorphs" some 500 to 5000 μm in diameter, such as *Chuaria*, are characteristic of late Riphean and Vendian rocks (Ford and Breed, 1973; Schopf, 1977) but their affinities are uncertain. Other smaller (i.e. < 100 μm) sphaeromorph acritarchs of late Proterozoic age probably include coccoid cyanophytes and cyanophytes spores (e.g. Binda, 1977; Vidal 1976a), decomposed filamentous cyanophytes (e.g. Hofmann, 1975), eukaryote spores and resting cysts of planktonic protists or larger marine algae (e.g. Cloud and Germs, 1971). Some late Proterozoic acritarchs that occur in loose, monospecific clusters may represent the remains of colonial protists, of spore clusters or the dissaggregated vegetative cells of larger algae.

On the Russian platform, sphaeromorph acritarchs predominate in the Valdai series and the Rovno series but are diminished in importance in the succeeding Lontowa series (correlated with the Tommotian stage of Siberia), although *Tasmanites*, a probable prasinophycean cyst, appears there (Volkova, 1971). Double walled vesicles ("disphaeromorphs"), sometimes provided with openings, were characteristic of late Proterozoic assemblages and may well have been the resting cysts of unicellular algae (Downie, 1967).

Acritarchs with spiny processes ("acanthomorphs") such as *Micrhystridium* and *Baltisphaeridium*, flourished at the expense of sphaeromorphs in early Cambrian times, including horizons now thought to be of Tommotian age (e.g. Timofeev, 1959; Volkova, 1971; Downie, 1967, 1974). Possible acanthomorphs such as *Eomicrhystridium* are nevertheless reported by Deflandre (1968) from the early Proterozoic Gunflint Formation and the Vendian rocks of Brittany. There are also spinose vesicles (e.g. *Pteinosphaeridium*) from possible Vendian rocks in Sweden (Vidal, 1976a) whilst Volkova (1971) mentions scarce *Micrhystridium* in the Valdai series of Russia and Poland.

Acritarchs with fewer polygonal arms ("polygonomorphs") are not reported from the Riphean but occur in the Vendian of Sweden (Vidal, 1976a) and in lower Cambrian strata (e.g. Volkova, 1971; Rudavskaja, 1973) becoming common from this time onwards in the early Palaeozoic. Novel acritarch forms that appear first in the early Cambrian include crested forms with perforate walls of the *Cymatiogalea* group ("herkomorphs"), fusiform vesicles of the *Leiofusa* group ("netromorphs") and elongate vesicles with polar sculpture of the *Acanthodiacrodium* group ("diacromorphs", see Downie, 1967, 1973).

Kazmierczak (1976) has likened the early Proterozoic microfossil *Eosphaera* to *Volvox*-like microfossils in Devonian rocks, although the former are smaller and the comparison has been dismissed by Cloud (1976) along with other contemporaneous candidates for eukaryote organization. A structure resembling a chrysophyte cyst has been noted in the middle Riphean Beck Spring microbiota (Cloud, 1976) whilst a calcareous seam within the late middle Riphean Nonesuch Shale yielded siliceous scales comparable with coccoliths (Jost, 1968) but more akin to the scales of Recent *Chrysochromulina*. Other than these reports, chrysophytes are unknown with certainty until coccolithophorids appeared in late Triassic times and diatoms and chrysomonads in the Cretaceous. Pyrrhophyta are not certainly known until Silurian times and remained scarce until the Jurassic, whilst ebridians did not appear until the Palaeocene (see Brasier, 1979).

Rhizopod protists are not known with certainty from the Proterozoic, records of this age having been dismissed by Deflandre (1950) mostly because they proved to be organic-walled acritarchs. Although Zhamoyda (1968) refers to nine families of Cambrian Radiolaria, Zhuravleva (1970a) dismisses some of the early Cambrian ones as vegetable or mineral in origin. Entactiniid Radiolaria (suborder Spumellaria) have since been reported from middle and upper Cambrian strata by Nazarov (1975) whilst Holdsworth (1977) considers that other Spumellaria may have been present at this time.

Foraminiferida are reported from Proterozoic strata by Pflug (1965) but the evidence is not convincing. Glaessner (1963a, 1975) and Føyn and Glaessner (1978) interpret the Tommotian and younger agglutinated tubes of *Platysolenites* and *Spirosolenites* as early benthonic foraminiferids resembling *Bathysiphon*, although Hamar (1967) formerly regarded them as of worm origin. Simple ammodiscacean benthonic foraminifera resembling *Psammosphaera* and *Hippocrepina* have been recorded from the lower Cambrian of Greenland and Labrador (Poulsen, 1932; Howell and

Dunn, 1942) but these reports seem doubtful. Recollecting has not confirmed them and at least some specimens appear to be inorganic ooliths (J. W. Cowie, *in litt.*, 1975). The suggestion that *Renalcis* had foraminiferid affinities (Riding and Brasier, 1975) has been dismissed for the early Cambrian records by Brasier (1977). Early Cambrian foraminifera described by Reitlinger (1948) and Vologdin (1958) have likewise not received acceptance (Loeblich and Tappan, 1964). Pflug (1965) has reported a biserial calcareous test called *Scaniella* from upper Cambrian rocks, regarded by Glaessner (1975) as a Cretaceous contaminant. To these records may be added finds of adherent calcareous organisms resembling *Tolypammina* in the early Cambrian archaeocyathid bioherms of Labrador (Dr D. Kobluk, *in litt.*, 1977). In a recent review, Conkin and Conkin (1977) make no mention of the above foraminiferid records, considering the group to have appeared in mid Ordovician times with simple ammodiscaceans such as *Bathysiphon* and *Hyperammina*.

Although Recent Ciliata are numerous and widely distributed, only the Tintinnina have a fossil record, supposedly ranging from Ordovician times, but they are unknown from many periods (Tappan and Loeblich, 1968). Flask or bottle-shaped hollow organic vesicles of chitin-like material ("Chitinozoa") may represent the tests of ciliate protists but could also have belonged to other protistan groups or even to Metazoa (Jansonius, 1970). Although formerly thought to range from early Ordovician (i.e. Tremadocian) to Devonian times, Bloeser *et al.* (1977) have recently reported chitinozoans from the late Riphean Chuar Group. The pseudochitin nature of the wall still needs to be demonstrated, however.

3. Larger Algae and Vascular Plants

The ecological relationship between larger metaphytes and metazoans and its evolutionary significance in Proterozoic to early Palaeozoic times has already been emphasized by Schopf *et al.* (1973) but the evidence is meagre. Calcified metaphytes are rare in Cambrian rocks and almost unknown in Proterozoic ones, whilst organic compressions referable to algae are extremely rare in both.

Perhaps the oldest metaphytes are the ribbon-like "*Helminthoidichnites*" and phylloid *Beltina* from the middle Riphean Greyson Shale of Montana (*c.* 1300 m.y., Walter *et al.*, 1976). Curved drag marks about 760 m.y. old that may have been made by algae in Central Australia are reported by

Milton (1966). The Vendian deposits of the Redkino and Valdai series also contain several examples, such as the ribbon-like ?brown algae *Vendotaenia* and *Tyrasotaenia* (formerly called *Laminarites*) found near Leningrad (Gnilovskaja, 1971), Siberia (Sokolov, 1973, 1976) and Sweden (Vidal, 1976b). *Papillomembrana*, a minute (< 0·5 mm diameter) organic structure from the Vendian of Norway has been likened to a dasycladalean alga (Spjeldnaes 1963) but this similarity is uncertain and the fossil may actually be of Riphean age (N. Spjeldnaes, personal communication, 1978). In lower Cambrian rocks, calcareous red algae may be represented by the fossils *Epiphyton* and *Bija*, dasycladalean green algae by *Cambroporella* and *Vologdinella* and codiacean green algae by *Proaulopora* (Johnson, 1966).

The middle Cambrian lycopod reported by Kryshtofovitch (1953) may be a case of mistaken identity (? dasycladalean). Some Proterozoic and early Cambrian acritarchs were reported as bearing trilete marks (e.g. Timofeev, 1959) but inferences regarding terrestrial vegetation cannot be drawn here, as J. M. Schopf (1969) has shown, and the existence of these trilete marks may be questioned (G. Vidal, personal communication, 1978).

4. Parazoa

(a) *Porifera*. The Proterozoic spicules of *Eospicula* referred to by de Laubenfels (1955) are now thought to be of inorganic origin (Schindewolf, 1956). Some early Cambrian sponges are reviewed by Sdzuy (1969) and Finks (1970). The oldest spicules occur in Nemakit-Daldyn and Tommotian rocks in Siberia (Zhuravleva, 1975) and Tommotian rocks in England (Brasier *et al.*, 1978) and include calcareous triactines, stauracts (cf. those of *Protospongia* and *Calcihexactina* of the class Hexactinellida) and phosphatized heptactines (?Hexactinellida). Calcareous monaxons, triacts, pentacts, hexacts and decacts occur in Atdabanian–Botomian archaeocyathid biostromes in Spain (A. Perejon, personal communication, 1978). Possible boring sponges associated with hexacts and megascleres occur in the Green Shales of Bornholm (Poulsen, 1967) of probable Atdabanian age. Although the first certain demosponges occur in the middle Cambrian Burgess Shale, monaxons found in lower Cambrian rocks could belong to this class (Finks, 1970).

(b) *Chancelloriida (or Heteractinellida)*. The curious sponge-like organism *Chancelloria* was first described from the middle Cambrian Burgess Shale by Walcott (1920) but its hollow calcareous, rosette-like spicules are known from the Nemakit-Daldyn horizon and earliest Tommotian rocks in Siberia (Cowie and Rozanov, 1974; Zhuravleva, 1975). Although Doré and Reid (1965) and Sdzuy (1969) have argued for the sponge affinities of the Chancelloriida, others are less convinced. Missarzhevsky and Rozanov (1965) considered some specimens to have had bryozoan affinites whilst Goriansky (1973) suggested echinoderm affinities.

(c) *Radiocyatha*. Subglobular and conical sponge-like fossils with ?siliceous skeletons composed of stellate rays have been called Radiocyatha by Debrenne *et al.* (1970, 1971) who regarded them as more primitive than true sponges. They occur in Tommotian to Botomian rocks and resemble, to some extent, the Chancelloriida and Archaeocyatha.

(d) *Archaeocyatha*. Archaeocyatha left behind them the most diverse and promising fossil record of any group in early Cambrian times and yet were largely restricted to that epoch. Their skeletons were calcareous, perforate and generally conical with two walls connected by septa, tabulae or both, but the morphology is highly variable (see Hill, 1972). Their lack of spicules contrasts with most sponge groups, hence they have usually assumed the status of an independent phylum (Hill, 1972) or taken their place in a subkingdom Archaeozoa (Zhuravleva, 1970b) or a kingdom Archaeata (Zhuravleva, 1974). Handfield and McKinney (1975) even raised the possibility of algal affinities. Ziegler and Rietschel (1970), however, contested the separation of archaeocyathids from calcisponges whilst Brasier (1976a) deduced that the former were invertebrates with a somatic integration at least of sponge grade, considering the perforate forms to have been sessile suspension feeders.

Reports of Proterozoic examples (Haughton, 1960) have since been reinterpreted as worm burrows (Glaessner, 1963b) whilst the similarity of the sac-like *Namalia* from South Africa (Germs, 1968) probably has no significance. The oldest archaeocyathids are therefore those used to define the base of the Tommotian stage in the Aldan River area of Siberia (Rozanov *et al.*, 1969) although they are not older than Atdabanian elsewhere in the world.

(e) *Stromatoporoida*. The calcareous skeletons of the stromatoporoids were probably built by sponges (Stearn, 1975). There is some similarity between irregular archaeocyathids and Cambrian stromatoporoids, which could be taken to indicate phyletic relationships (Nestor, 1966). Early Cambrian stromatoporoids have been reported by Yavorsky (1963) but the validity of these reports is uncertain according to Zhuravleva (1970a).

(*f*) *Anzalia sp.* Non-calcareous bioherms with a porous, vesicular structure built of fine detrital sediment, called *Anzalia*, occur in littoral facies in the lower Cambrian of Morocco. They have been compared with colonial sponges (Termier and Termier, 1964) but their affinities remain obscure.

(*g*) *Bag-shaped organisms.* Bag-shaped and conical impressions of soft-bodied organisms with flexible outer walls have been described from the upper Vendian of Central Australia (*Arumberia* of Glaessner and Walter, 1975), the upper Vendian of Siberia (*Baikalina* of Sokolov, 1972, 1973) and the Vendian Nama Group of South Africa (*Namalia* of Germs, 1968). *Arumberia*-like structures have been shown to the writer in the Long-myndian of England by Dr B. Bland. *Arumberia* was supposedly benthonic and colonial with radially grooved sides, the overturned bags readily becoming aligned by currents (Glaessner and Walter, 1975) but this "organism" is probably a pseudofossil caused by turbid water flow (c.f. Dzulyński and Walton, 1965, pp. 61–81).

5. *Radiata*

(*a*) *Petalonamae.* The Petalonamae of Pflug (1972a, 1972b) are in extinct group of feather-shaped soft bodied organisms preserved as impressions with a central stem that bears pinnule-like elements. Early Vendian examples include *Charnia* from England (Ford, 1958), *Charnia* and *Pteridinium* from the Mogilev-Redkino suite of the Russian platform and *Glaessnerina*, *Pteridinium* and *Nasepia* from the lower Yudomian of Siberia (Sokolov, 1972, 1973, 1976). Late Vendian examples include *Pteridinium*, *Glaessnerina* (= "*Rangea*") and *Arborea* from South Australia (Glaessner and Wade, 1966), "*Rangea*" from Central Australia (Cowie and Glaessner, 1975), "*Charnia*" from Newfoundland (see Anderson and Misra, 1968) and *Rangea*, *Ernietta*, *Ernionorma*, *Erniobaris* and probably *Nasepia* from the Nama Group of South Africa (Pflug, 1972b; Germs, 1973a). *Xenusion*, now thought to come from the *Diplocraterion*-bearing Kalmersund sandstone of Sweden (Jaeger and Martinsson, 1967) has also been compared with the petalonamaeans (Tarlo, 1967).

The age range of this petalonamaean fauna must be in the region of 680–570 m.y. (Glaessner, 1971a). Unfortunately, the biology of these important creatures remains obscure. Early finds were compared with algae (Ford, 1958) and later with pennatulacean octocorals (Glaessner and Wade, 1966). Pflug (1972a) regarded them as an ancestral metazoan stock that arose from the organization of protistan colonies. If Pflug's models are correct then the Petalonamae were colonies of suspension-feeding organisms with, in the more "advanced" forms, tissues arranged around a central cavity. Not yet satisfactorily explained, however, is the attachment of *Charnia* to the discoidal medusoid *Charniodiscus*, first reported in England by Ford (1958) and now also known from Newfoundland (T. P. Fletcher, personal communication, 1978). This association suggests that the Petalonamae could have been cnidarian polyps with a float structure for buoyancy in the water column.

(*b*) *Medusoids.* Discoidal and conical moulds and compressions generally thought to be of cnidarian medusae occur in many Vendian rocks (e.g. Glaessner, 1971a). Early Vendian examples include *Charniodiscus* from England (Ford, 1958), *Cyclomedusa* and *Kimberella* from Siberia and *Beltanella*, *Beltanelloides*, *Bronicella*, *Charniodiscus*, *Conomedusites*, *Cyclomedusa* and *Medusinites* from the Mogilev-Redkino suite of Russia (Sokolov, 1976). Younger Vendian medusoids include over a dozen genera at Ediacara (Glaessner and Wade, 1966; Wade 1972a) of which Wade thought that *Eoporpita* was hydrozoan, *Cyclomedusa* was possibly hydrozoan and *Brachina* and *Kimberella* may have been scyphozoan. The medusoids *Hallidaya* and *Skinnera* from similar horizons in Central Australia resemble Scyphozoa in complexity but are not tetrameral in symmetry (Wade, 1969). *Conomedusites* was later placed by Glaessner (1971b) in the Conulariida, an extinct order of benthonic polyps with organic or phosphatic exoskeletons and an organisation intermediate between primitive Hydrozoa and more complex Scyphozoa. *Beltanelloides* is a compressed, spherical fossil resembling a large *Chuaria* that occurs en masse on bedding planes in the Redkino series of Russia (e.g. Sokolov, 1973) and in Spain (Perejon and Brasier, unpublished). It could, perhaps, have been a medusoid structure.

Early Cambrian medusoids such as *Velumbrella* from Poland (Stasinska, 1966) are uncommon, possibly because their remains were consumed by scavengers such as trilobites.

(*c*) *Polypoids.* Although *Conomedusites* and the Petalonamae may have been Vendian polyps, the first abundant examples of soft-bodied benthonic polyps are the early Cambrian impressions and vertical burrows of *Bergaueria*, *Astropolithon* and *Dolopichnus*, which were probably made by actinians (Alpert, 1973; Alpert and Moore, 1975), and *Anthoichnites* (Melendez, 1966). Sokolov (1976) reports the presence of *Bergaueria* in the upper Vendian of Russia but this is not confirmed by Fedonkin (1977).

Cnidarians with mineralized hard parts are unknown in upper Protero-zoic rocks and rare or disputed in lower Cambrian ones. *Palaeoconularia*, for example, is a possible conulariid from the Botomian of the USSR (Tchudinova, 1959) and the calcareous tubular clusters of *Bija* have been interpreted as tabulate coral skeletons by Bondarenko (1966). Rugosa-like cups from the lower Cambrian rocks were even referred by Korde (1963) to a new class, the Hydroconozoa. Jell and Jell (1976) recently discovered fossils in the lower middle Cambrian of Australia which they interpret as tabulate coral skeletons. There is, however, no compelling evidence for the importance of polyps with mineralized skeletons until Ordovician times.

6. Bilateria

(*a*) *Acoelomates* and *Pseudocoelomates*. The preservation potential of these small creatures is low (Fig. 4) so no certain examples of Vendian or Cambrian age are known. The turbellarian-like fossil called *Brabbinthes* from ?Proterozoic strata in Alaska (Allison, 1975) now appears to be a section through a hexactinellid sponge spicule of Cambrian age (Cloud *et al.*, 1976). *Dinomischus*, a compression fossil from the middle Cambrian Burgess Shale, has been likened to the Entoprocta by Conway Morris (1977a).

(*b*) *Priapulida*. Fossil priapulid worms are an important constituent of the middle Cambrian Burgess Shale fauna (Conway Morris, 1977b) and, like recent species, they were soft-bodied, infaunal and predatory. Of these only *Selkirkia columbia* dwelt in a tube, a rigid annulated structure probably of organic composition. Although their activities must have left behind trace fossils these have not yet been recognized.

(*c*) *Sipunculida*. The middle Cambrian *Ottoia* and *Banffia* were compared by Walcott (1911) with the Sipunculida but *Ottoia* is now considered to have been a priapulid (Conway Morris, 1977b). Runnegar *et al.* (1975) likened the Hyolitha (lower Tommotian–Permian) to sipunculids but preferred a distinct phylum for them. Vendian and Cambrian examples of sipunculids are otherwise unknown.

(d) *Mollusca*. Although mollusc shells are unknown below Nemakit-Daldyn rocks, mollusc-like trials are widely reported in Vendian strata. *Bunyerichnus*, a locomotion trail that occurs only a short distance above the Upper Tillite in South Australia may have been made by a primitive mollusc (Glaessner, 1969). *Scolicia*-like trails (mostly formed at present by gastropods) and *Buchholzbrunnichnus* occur in the Vendian of South Africa (Germs, 1972b, 1973b) whilst a ?molluscan *Plagiogmus*-like trace is known in the late Vendian of Russia (Sokolov, 1976). Bilobed trails of *Didymaulichnus* that could also be molluscan are common in Vendian rocks (e.g. Young, 1972) as well as in early Cambrian ones. Despite all this evidence of "molluscan" activity, though, none of the soft-bodied Ediacara fauna are thought to have been molluscan.

The first molluscan shells occur in the Nemakit-Daldyn and Tommotian rocks of Siberia and by that time had already developed a wide range of forms. The cap-shaped and univalved Helcionellacea (e.g. *Bemella*) have variously been interpreted as gastropods (Rozanov *et al.*, 1969) and as cyrtonellid monoplacophorans (Runnegar and Pojeta, 1974; Runnegar and Jell, 1976). Tryblidiid monoplacophora with reduced metameric features are also known from lower Cambrian rocks (Missarzhevsky, 1976).

The laterally compressed *Anabarella* from the Siberian Nemakit-Daldyn and Tommotian rocks is interpreted by Runnegar and Pojeta (1974) as a monoplacophoran ancestral to rostroconch molluscs via the intermediate form "*Heraultia*." The latter genus, however, appears to have been a bivalved crustacean (Müller, 1975).

Fordilla from upper Atdabanian rocks may be amongst the oldest of fossil bivalves (Pojeta *et al.*, 1973) but its association with a small taxodont bivalve in the ?Atdabanian Green Shales of Bornholm (Poulsen, 1967) should also be noted.

Helically coiled shells such as *Aldanella* and *Pelagiella* occur in Tommotian rocks and are interpreted as the oldest gastropods by Runnegar and Pojeta (1974) and Matthews and Missarzhevsky (1975) and as worms or Mollusca *Incertae sedis* by Yochelson (1975). The molluscan radula reported from the lower Cambrian of California by Firby and Durham (1974) has since proved to be clusters of the "worm" tube *Volborthella* (Glaessner, 1976; Yochelson *et al.*, 1977).

The brachiopod-like bivalved molluscs of the class Stenothecoida Yochelson (Yochelson, 1969) are still known only from lower to middle Cambrian rocks (Runnegar and Pojeta, 1974). The more frequently encountered Palaeozoic group of conical tubes with opercula, referred to as Hyolitha, are regarded as a molluscan class by Marek and Yochelson (1976) but as a distinct phylum with both molluscan and sipunculid affinities by Runnegar *et al.* (1975). Hyoliths occur in the oldest Tommotian rocks of Siberia (Rozanov *et al.*, 1969) but the conical calcareous tubes of *Wyattia* from probable latest Vendian or early Tommotian in California (Taylor, 1966) could also have been an ancestral hyolith stock.

Cephalopoda, Scaphopoda, Aplacophora and Polyplacophora are not known in either lower or middle Cambrian rocks (see Runnegar and Pojeta, 1974).

(e) *Annelida.* Of the many late Proterozoic and Cambrian trace fossils now known it is difficult to say which are certainly of annelid origin. Burrows that were apparently made by coelomates moving and feeding in waves appear to be present in Riphean rocks (e.g. Clemmey, 1976) and in Vendian ones (e.g. *Cylindrichnus* of Glaessner, 1969 and Sokolov, 1976). Many other Vendian and Cambrian deposit and suspension feeding traces could also be ascribed, questionably, to the Annelida (e.g. *Helminthoidichnites, Planolites, Chondrites, Treptichnus, Teichichnus, Asterosoma, Asterichnus, Phycodes, Arenicolites, Skolithos, Monocraterion* and *Diplocraterion*).

Segmented annelid-like worms are known as soft-bodied impressions and compressions in both Vendian and Cambrian strata. The flat, *Spinther*-like *Dickinsonia* is reported from both the Mogilev-Redkino suite of Russia (Sokolov, 1976) and from the upper Vendian of South Australia (Glaessner and Wade, 1966; Wade, 1972b). *Spriggina* from the latter horizon (Glaessner and Wade, 1966) resembles more the Recent pelagic polychaete *Tomopteris* although the cephalization suggests it may have been close to the root stock of the trilobite arthropods. Cloud *et al.* (1976) also consider that *Vermiforma* from the late Vendian of Carolina is the impression of a soft-bodied tubular worm, possibly an annelid.

Fossil polychaetes and the annelid *Palaeoscolex* occur together in the high lower Cambrian Emu Bay Shales (Glaessner, 1976). The latter genus, which may be closer to the ancestral oligochaete group, also occurs in the high lower Cambrian to middle Cambrian Kinzer Shale (see Conway Morris, 1977b, p. 87; Resser and Howell, 1938) and in the middle Cambrian Spence Shale (Robison 1969). *Palaeoscolex* is unknown in the middle Cambrian Burgess Shale but there are remains of scaly and setiferous annelids of uncertain systematic position, such as *Wiwaxia*, *Pollingeria*, *Canadia* and *Worthenella* (e.g. Walcott, 1911). Errant polychaetes with chitinous jaws ("scolecodonts") are unknown until Ordovician times according to Howell (1962).

Tubular fossils from the Vendian and Cambrian are often referred to the Polychaeta. The oldest example may be *Cloudina*, a calcareous, chambered, tubular structure from the Nama Group of South Africa (Germs, 1972a, Glaessner, 1976). Similar fossils known as Cribricyathida are typical of the lower Cambrian in Siberia (Yankauskas, 1972). Somewhat similar are the agglutinated *Volborthella* and the calcareous or calcareous-agglutinated *Salterella* (Glaessner, 1976) known from rocks of mid to late early Cambrian age although Yochelson (1977) has argued for their placement in an extinct phylum, the Agmata. These two genera may represent the same organisms preserved in different lithologies (Yochelson *et al.*, 1977). Another agglutinated fossil is the tubular *Platysolenites*, known from the Tommotian of the Baltic region and northern Scandinavia. Hamar (1967) regarded this as a polychaete tube but Glaessner (1963a) considers it foraminiferid. Also questionably annelid is the phosphatic tube of *Torellella* and the calcareous tube of *Anabarites*, known from Nemakit-Daldyn and younger horizons (Razanov *et al.*, 1969).

(*f*) *Onycophora and Tardigrada*. The earliest onycophoran–like organism appears to be *Aysheaia*, a compression fossil from the middle Cambrian Burgess Shale, but it may also be compared with the Tardigrada (Whittington, Chapter 9, this volume). More doubtful is the soft–bodied impression of *Xenusion* from the Tommotian Kalmarsund sandstone (Jaeger and Martinsson, 1967).

(*g*) *Pogonophora*. Several long and narrow tubes from Vendian and Cambrian deposits have been assigned to the phylum Pogonophora: the annulated organic tubes of *Sabellidites*, *Saarina* and *Sokolovia* from the Rovno horizon (highest Vendian) on the Russian platform, for example, and *Calyptrina* and *Paleolina* from the Nemakit-Daldyn horizon (highest Vendian) of Siberia (Sokolov, 1965). *Sabellidites* also occurs in the Tommotian Lontowa series (Volkova, 1971) but the specimens reported from the upper Riphean of Turukhan Territory now appear to be of the alga *Vendotaenia* (Sokolov, 1976). The phosphatic tubes of *Hyolithellus*, which occur in Nemakit–Daldyn and early Cambrian strata (Matthews and Missarzhevsky, 1975) could also have been pogonophoran skeletons (see Poulsen, 1963) as could the calcareous tubes of *Coleoloides* found in Tommotian and younger rocks (Brasier and Hewitt, 1979).

(*h*) *Arthropoda*. The oldest reported "arthropods" are very small, well preserved black-brown specimens with a continuous carapace, found in boreholes in the Mogilev–Redkino suite of the Vendian (Sokolov, 1976) but the interpretation requires confirmation. Of similar age but better known are the trilobite-like impressions of *Vendia* (Sokolov, 1976) very similar in form to the late Vendian *Praecambridium* from South Australia (Glaessner and Wade, 1971). *Parvancorina* from the latter rocks might be compared with the protaspis stage of younger trilobite arthropods (see Glaessner and Wade, 1966) but no associated arthropod tracks are known. In fact, arthropod tracks are entirely lacking from undisputed Vendian strata. Although this seems to cast doubt on the arthropod nature of the above, it might also be argued that such protoarthropods were largely pelagic and left no tracks.

Small impressions resembling the arthropod resting trail *Isopodichnus* occur just below ?early Tommotian shells in England (Brasier *et al.*, 1978). Trilobite scratch marks, made by laterally grazing trilobites or merostomes (i.e. the trace fossils *Monomorphichnus* and *Dimorphichnus*), walking tracks (*Diplichnites*), locomotion furrows (*Cruziana*) and resting impressions (*Rusophycus*) appear in the fossil record near the beginning of the Tommotian stage (Fig. 6) and are relatively common thereafter in rocks of suitable facies. Pellet-lined burrows like those of thalassinoid crustaceans first occur in the middle to high lower Cambrian of California (Nations and Beus, 1974) although deductions in this case would probably be misleading.

Although the ?merostome skeleton of *Gdowia* is known from the Tommotian Lontowa series of the Russian platform (Sokolov, 1976) the first mineralized trilobite skeletons appeared during the Atdabanian stage in many parts of the world (e.g. Rozanov, 1967). Reports of Tommotian trilobites in Morocco (Cowie and Glaessner, 1975) and Norway (in Martinsson, 1974, p. 200) have not been confirmed by further study. The calcareous carapace of "*Heraultia*," probably a crustacean (Müller, 1975), occurs as phosphatic internal casts in Tommotian rocks in Siberia (Matthews and Missarzhevsky, 1975). The chitinophosphatic carapaces of archaeocopid Ostracoda were present in Atdabanian times (Cobbold, 1936; Zhuravleva, 1970a) whilst the phyllocarid Crustacea were present by at least middle Cambrian times (Briggs, 1977).

(*i*) *Brachiopoda*. The Precambrian records of Brachiopoda have been dismissed in a study by Rowell (1971). Chitinophosphatic inarticulate brachiopods (e.g. *Paterina, Micromitra*) are common in lower Cambrian rocks and first appeared at the beginning of the Tommotian stage in Siberia (Zhuravleva, 1970). The phosphatic inarticulate ?*Obolella* which occurs with *Mobergella* in the upper Tommotian of England (Rushton, 1974) is probably a lingulid and all four orders of Inarticulata were present by late early Cambrian times (Rowell, 1971). Calcareous forms such as *Kutorgina* may well occur in the ?upper Tommotian to Atdabanian rocks of England (Rushton, 1974) and with early orthids such as *Nisusia* in the high lower Cambrian *Bonnia–Olenellus* zone of North America and Scotland (e.g. Palmer, 1971, p. 200; Cowie, 1974, p. 136).

(*j*) *Ectoprocta*. Missarzhevsky and Rozanov (1965) referred to the bryozoa some forms of the sponge-like *Chancelloria* (q.v.) whilst Elias (1954) regarded the alga *Cambroporella* (L. Cambrian) as a bryozoan. None of these interpretations have found favour so that Larwood *et al.* (1967) do not record the presence of any fossil Ectoprocta before the Ordovician period.

(*k*) *Phoronida*. The soft-bodied compression fossil *Odontogriphus* from the middle Cambrian Burgess Shale was a phoronid-like lophophorate according to Conway Morris (1976). Some early Cambrian trace fossils such as *Skolithos*, may also owe their origin to phoronids (Fenton and Fenton, 1934).

(*l*) *Conodontophorida*. The Conodontophorida are an extinct group known only from assemblages of small phosphatic "conodonts". These once formed part of an internal apparatus of uncertain function but were probably lophophore-bearers according to Lindström (1974). Conodont-like structures have been reported within the "lophophore" of *Odontogriphus*, a problematic organism from the middle Cambrian Burgess Shale (Conway Morris, 1976) although the similarity may be superficial. The earliest conodonts are reported from the latest Vendian Nemakit-Daldyn horizon of the Anabar Shield (Matthews and Missarzhevsky, 1975) and occur, although rarely, in many Tommotian and younger Cambrian rocks (e.g. Poulsen, 1966; Bengtson, 1976; Brasier *et al.*, 1978).

(*m*) *Mitrosagophora* and *Tommotiida*. A new order Mitrosagophora was erected by Bengtson (1970) to contain the problematica *Tommotia*, *Camenella* and *Tannuolina* on the evidence that they represent the remains of animals bearing two types of assymetrical sclerites, mitral and sellate, each with right and left symmetry forms and constructed of lamellar phosphate. Fonin and Smirnova (1967) had tentatively assigned *Tannuolina* to the cirripede Crustacea, an assignment more firmly decided on by Bischoff (1976) for *Tommotia*, *Camenella* and *Dailyatia* but not *Tannuolina*. Bengtson (1970, 1977a) however, has compared these mitrosagophorans with the Ordovician to Devonian sclerites of another problematic group, the Machaeridia, but the evidence for machaeridian affinity is barely persuasive either. The similarity between the Mitrosagophora, Machaeridia and Cirripedia is little more than that they all secreted a multi-element skeleton. But in the cirripedes and machaeridians this skeleton is usually capable of re-assembly, whereas in the Mitrosagophora this has not yet proved possible because there are no clear facets for articulation between the sclerites.

Matthews and Missarzhevsky (1975) placed the conical shells of *Lapworthella* close to those of *Tommotia* and *Camenella* in an order Tommotiida on the basis of their shared laminar phosphatic ultrastructure and lack of shell symmetry but they disregarded the evidence for multi-element arrangement of the latter two genera. Such a broad diagnosis could also admit *Sachites* and *Sunnaginia* and probably *Halkeria* and *Rhombocorniculum* as well.

It is not yet possible to indicate what the affinities of these ancient and curious fossils may be. Other contemporaneous, multi-element exoskeletons built of lamellar phosphate include the Inarticulata and the Conodontophorida. The latter may have been lophophorate like the former (q.v.). It is therefore tempting to consider whether the mitrosagophoran problematica were the exoskeletal remains of an extinct group of early lophophorates, perhaps with a life style resembling that of the Cirripedia.

(*n*) *Mobergella sp.* The phosphatic circular disc with radial "muscle scars" known as *Mobergella* occurs widely in upper Tommotian rocks (Bengtson, 1977a). This shell is a highly flattened cone closely resembling those of some tryblidiid monoplacophorans except that the inner surface may be flat or even convex and the shell is built of lamellar phosphate. Bengtson (1968, 1977a) dismissed the possibility that *Mobergella* was the operculum of the contemporaraneous phosphatic tubes of *Hyolithellus* on the grounds of their differences in size and distribution, although he favoured an opercular origin. The accumulation of *Mobergella* in conglomerates, phosphorites and nearshore facies may not be without significance. The brachial valves of the inarticulate brachiopods *Discina*, *Orbiculoidea* and *Crania* (with which morphological comparison might be made) are well adapted to attachment on rocky surfaces in littoral and shallow sublittoral regions. The possibility that *Mobergella* was a hardground dweller, perhaps a lophophorate, might therefore · deserve examination.

(*o*) *Echinodermata*. The triradial impression of *Tribrachidium* from the upper Vendian of South Australia (Glaessner and Wade, 1966) and perhaps from South Wales (Cope, 1977) may be crudely compared in morphology with early Palaeozoic echinoderms such as edrioasteroids. A lack of pentameral symmetry was not uncommon in early echinoderms, as Durham (1971) has emphasized, and the final adoption of pentamery was probably due to the pressure for efficient suspension feeding (Stephenson, 1974). Plates of eocrinoid type are the oldest known echinoderm remains, occurring with archaeócyathids and trilobites in rocks of late Atdabanian age in California (Durham, 1971). The Eocrinoidea, Helicoplacoidea, Camptostromoidea, Edrioasteroidea and Lepidocystoidea all occur at higher horizons in the same sequence, probably of Botomian age. At least seven new classes appeared during the middle Cambrian (Durham, 1971) including the calcichordates, a group referred by Jefferies (1968) to the Chordata.

(*p*) *Hemichordata*. Hemichordata in the form of Graptolithina are known from the early middle Cambrian (e.g. *Dendrograptus* of Bulman, 1955) but possible *Dictyonema*-like structures also occur in the late Vendian Valdai series of Russia according to Sokolov (1976).

(*q*) *Chaetognatha*. The soft-bodied compression fossil *Amiskwia* from the middle Cambrian Burgess Shale was once thought to represent the oldest chaetognath (Walcott, 1911) but the specimen does not have chaetognath-like jaws and, whilst actively pelagic, Conway Morris (1977c) regards it as of uncertain affinity.

(r) *Chordata*. Resser and Howell (1938) described an impression fossil from high lower Cambrian strata called *Emmonsaspis* which might be cephalochordate, although the material is insufficient for proper evaluation (Durham, 1971). Bengtson (1977b) described button-like external sclerites of phosphate called *Lenargyrion* from the upper Atdabanian of Siberia which resemble the odontodes of Palaeozoic fish, but the matter of affinities also remains open. The echinoderm-like chordates, called Calcichordata appeared first in middle Cambrian times (Jefferies, 1968) whilst the first acceptable evidence of vertebrates appears to be those odontodes of early Ordovician age reported by Bockelie and Fortey (1976).

PHYLETIC CHANGES

The fossil evidence for the time of origin of some major invertebrate animals and plants is outlined in Figs 2 and 3. Prokaryote bacteria and cyanophytes arose in the Archaean and eukaryotic algae and protists at some time during the early or middle Proterozoic. Annelid-like metazoans and problematic chitinozoans (?ciliate protists) appear to have been present in late Riphean times. The following Vendian system contains evidence of annelids, ?molluscs, coelenterates and problematic acritarchs (?flagellate protists), "protoarthropods", Porifera, Chancelloriida, "worm-tubes". Pogonophora, Conodontophorida and molluscan shells are reported from the latest Vendian Nemakit-Daldyn horizon.

The Tommotian stage at the base of the Cambrian contains the earliest remains of Archaeocyatha, inarticulate Brachiopoda, Mitrosagophora, Arthropoda (probably including crustaceans, trilobites and merostomes), Hyolitha, rhizopod protists and many problematical taxa. Echinoderms are known from the mid Atdabanian stage and priapulids, ?onycophorans, hemichordates and calcichordates from the middle Cambrian. Vertebrates and ectoproct bryozoa were not certainly present until early Ordovician times.

Although the appearance of these major invertebrate groups spans a period of about 200 million years and possibly as much as 500 million years, it is punctuated by an extraordinary biological revolution, the Cambrian radiation event. This lasted perhaps only ten million years and embraced organisms of almost every kind, including not only triploblastic but also diploblastic metazoans and the parazoans, protists, eukaryotic green and red algae and prokaryotes. The variety of body plan seen in Cambrian invertebrates may have exceeded that at any other time in earth history but a great proportion of these early Cambrian phyla and classes were geologically short-lived, excluded perhaps by the progressive race for efficiency.

How well does the fossil evidence compare with current biological ideas? Before much of the above evidence came to light, Clark (1964) had suggested that the first Bilateria were acoelomates creeping about on the seafloor or just above it, feeding on small organisms and detritus. This view is supported by the embryological researches of Jägersten (1972) which show that cnidarians, entoprocts, sipunculids, molluscs, annelids, pogonophorans, brachiopods and hemichordates may all have had ancestors with creeping organs. As the ancestral acoelomates grew in size, a need for more food was met by burrowing with the aid of a hydraulic skeleton developed in various ways. Burrows suggesting the work of early schizocoels first occur in Riphean rocks and by Vendian times we have certain evidence of eumetamerous and possibly pseudometamerous forms. The oligomerous lophophorates appear to have arisen later (supporting Clark, 1964) in either latest Vendian or Tommotian times whilst deutero-stomes followed later in Atdabanian times. Hypotheses which regard the Cnidaria and Porifera as stages in the development of the higher metazoa are not, therefore, supported by present fossil evidence.

SKELETAL CHANGES

It is convenient to subdivide the invertebrates into those which are entirely "soft" bodied (i.e. lack a hard skeleton), those which bear skeletons of chitin or other organic matter ("chitinoid") and those with mineralized skeletons of calcium phosphate, calcium carbonate, silica or agglutinated sediment grains (Fig. 4). Obviously, the structure and composition of the skeleton (if any) will affect the preservational potential of a species, as will the lithology of the enclosing rock and a range of post-mortem phenomena outlined in Fig. 5. The extent to which the early invertebrate record may have been biased by such changes in skeletal type, lithology and post-mortem processes deserves further research but a few preliminary observations can be made here.

Soft-bodied Vendian and early Cambrian faunas appear to have been most diverse in nearshore sandstones, especially in sandstone–shale alter-

---

FIG. 4. This chart shows the skeletal type and trophic-mobility-substratum niche of some invertebrates thought to have been present in early Cambrian times. Bold lines = fossil evidence; ? = questionable fossil evidence; fine lines = no fossil evidence.

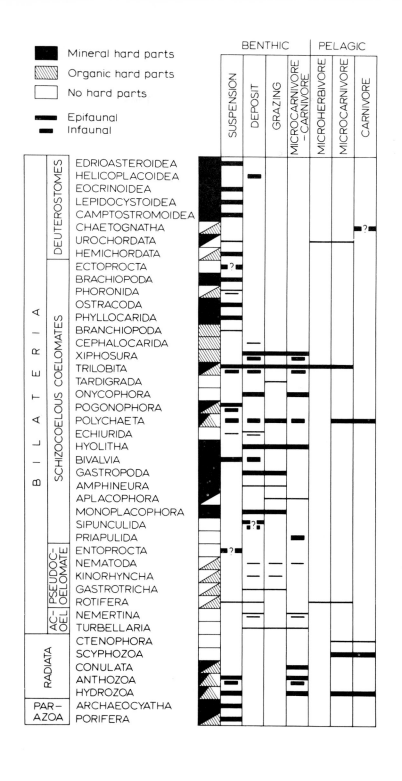

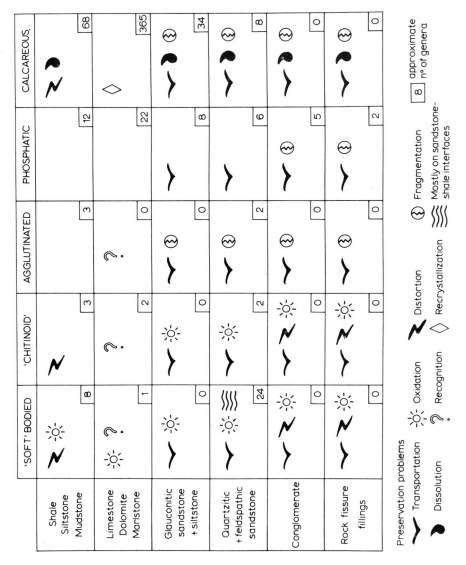

Fig. 5. For legend see opposite.

nations of Vendian age (Fig. 5). Such lithologies are favourable to preservation provided that oxidation (including scavenging) is minimal, as may have been the case. Chitinoid fossils such as *Sabellidites* are mostly known from late Vendian shales and clays, again probably for preservational reasons. Agglutinated organisms such as *Volborthella* occur predominantly in argillaceous rocks as much, perhaps, because of ease of recognition as because of better chances of preservation. Phosphatic shells, which are not prone to dissolution, are preserved in the widest range of rock types but are most diverse in carbonates (Fig. 5). Their fossil record may be quite representative but reworking and transport distort the true distribution patterns. Calcareous fossils are also most diverse in carbonate rocks (Fig. 5) their numbers partly reduced elsewhere by transport, fragmentation and dissolution.

Similar constraints apply to the preservation of trace fossils. Interbedded fine and coarse grained lithologies greatly aid their recognition whereas homogenous limestones or sandstones do not. Furthermore, most trace fossils are studied in the field where the incidence of bedding surfaces or cross sections can affect the quality of the record on a local scale.

In summary we note that preservational aspects may have biased the fossil record locally but, given a great variation in conditions across the globe, they need not have grossly distorted the story.

Riphean and Vendian organisms were predominantly soft-bodied, relying on cilia, inching or on hydrostatic skeletons for locomotion but the many novel trace fossil general that appeared in the early Cambrian are evidence for a further radiation of soft-bodied organisms, a picture enhanced by their abundance in the middle Cambrian Burgess Shale. The Cambrian radiation of soft-bodied infauna may have been a resourceful answer to the need for exoskeletons, dwelling burrows being both economical and readily modified.

As an organic skeleton provides the template for mineral deposition in most invertebrate groups it is reasonable to consider that a "chitinoid" stage may have preceeded a "mineralized" one in those taxa that have a skeleton. Protists, for example, appear to have reached this stage by late

FIG. 5. This chart illustrates the relationships between preservation lithology and skeletal type in Vendian and early Cambrian rocks. The numbers in the inset boxes give the approximate (and minimal) number of reported genera. Note that soft-bodied and chitinoid examples are mostly Vendian whilst the rest are mostly latest Vendian to early Cambrian. Data from sources in the text.

Riphean and Vendian times (i.e. acritarchs and chitinozoans), coelenterates and annelids by Vendian times (i.e. the vela of hydroid medusae, annelid setae), pogonophorans by latest Vendian times (i.e. *Sabellidites* and its relatives) and Porifera and Arthropoda by Tommotian times.

Agglutinated skeletons may be constructed with either an organic or a mineral cement or both. Currently they are favoured by polychaetes and protists in subsaline or calcium carbonate-deficient conditions. They first occur in the Tommotian genus *Platysolenites* and in the Atdabanian-Elankian *Volborthella* and *Salterella*.

Siliceous skeletons are currently typical of deep water hexactinellid sponges and oceanic plankton such as radiolarians, diatoms, ebridians and silicoflagellates. Because the oceans are relatively undersaturated with respect to silica, these siliceous skeletons are rarely massive and tend to suffer solution or replacement by calcium carbonate after death. It may therefore be that the calcareous and phosphatic hexactinellid spicules recorded from Tommotian strata were originally siliceous, as presumably were the younger hexactinellids, radiocyathans and radiolarians.

The high relative abundance of calcium phosphate skeletons in early Cambrian rocks is unusual compared with those of later times but the first examples were latest Vendian (e.g. *Hyolithellus* and conodont elements). Inarticulate brachiopods with phosphatic shells appeared in Tommotian times without prior evidence of a chitinoid stage, as did the problematic Mitrosagophora, *Morbergella*, *Lapworthella* and *Torellella*. Atdabanian examples incuded the first archaeocopid ostracods and *Lenargyrion*.

Calcarious shells were widespread in early Cambrian times but isolated Vendian occurrences are known and include *Cloudina*, latest Vendian molluscs *Anabarites* and possibly *Wyattia*. Calcareous Chancelloriida, Porifera and Mollusca appeared in the latest Vendian and Archaeocytha, Hyolitha and *Coleoloides* in early Tommotian times. Cyanophytes with calcified mucilaginous sheaths and calcareous red and green algae also appeared in early Cambrian times. Calcareous inarticulate brachiopods are known from the late Tommotian whilst articulate brachiopods, trilobites and echinoderms occurred in the Atdabanian.

Evidently, the latest Vendian–Tommotian transition was a period of intensive skeletalization almost irrespective of the physiological sophistication attained by a group (see Fig. 3). Skeletalization and the form it took, however was not unrelated to ecology and the nature of the milieu.

TROPHIC NICHE CHANGES

The value of trace fossils as guides to the story of metazoan evolution was first clearly indicated by Seilacher (1956). They have since been used to define the start of Cambrian epoch when distinctive changes in organic activity took place (e.g. Daily, 1972; Alpert, 1977). While trace fossils may give scant indication of the biological origins of their producers, they are telling indicators of their feeding and locomotion habits, of sedimentary conditions and to some extent of water depths. The early history of five major ecological niches deduced from trace and body fossils are discussed below: infaunal deposit feeding, epifaunal grazing, benthonic suspension feeding, predation and microplankton.

1. Infaunal Deposit Feeding

Infaunal feeding burrows are produced primarily by coelomates feeding directly on the sediment and detritus (i.e. deposit feeding *s.s.*), on small plants and animals within the sediment (infaunal microphagy) or on larger infaunal animals (infaunal carnivory). Unfortunately, distinctions between these are not easily made in fossil material so they are considered here together. The oldest examples occur in the 1000 m.y. old Riphean sediments of Zambia (Clemmey, 1976). Of similar age may be the *Helminthoidichnites* from the lower Vindhyans of India (Beer, 1919; Seilacher, 1956), branched cylindrical burrows up to 1300 m.y. old from Western Australia (K. Towe, personal communication, 1978) *Asterosoma? canyonensis* from the Grand Canyon series of North America (*c.* 1100–1300 m.y. B.P., Glaessner, 1969) and straight and spiral mud-eating trails of *Planolites* and *Helminthoidichnites* type, described from the upper Riphean of the Urals and Baikal region (Sokolov, 1973, 1976). Less certainly Riphean are the meandering "sediment feeding traces" called *Sabellarites* from the Brioverian (Squire, 1973).

Vendian deposit-feeding burrows are better known. On the Russian platform lower Vendian rocks contain *Planolites* and *Teichichnus*-like burrows, whilst the upper Vendian contains numerous "annelid" tracks and burrows, such as *Vendovermites*, "*Bilobites*", *Harlaniella*, *Torrowangea*, *Neonereites*, and *Planolites* (Sokolov, 1972, 1973, 1976; Fedonkin, 1977). The lower Vendian Biri limestone of central Norway also contains *Planolites*-like traces (Brasier, unpublished data) whilst Young (1972) records *Planolites* and the branched burrow *Chondrites* from probable Vendian in Canada. From California, Alpert (1975a, 1975b) and Wigett (1977) report simple horizontal *Planolites* and trails from the ?upper Vendian Wyman Formation and Reed dolomite. Banks (1970) describes simple vertical and horizontal burrows from the Stappogiedde Formation, above the Varangian tillites, whilst traces may also occur in the underlying Nyborg formation, below the upper tillite (Martinsson, 1974, p. 237). The late Vendian Pound Quartzite of South Australia contains *Cylindrichnus* (with pellet-like trails) and a range of unnamed traces of which types D and F may have been deposit feeders (see Glaessner, 1969). *Cylindrichnus* also occurs in the upper Vendian of Siberia (Sokolov, 1973). The Nama Group of South Africa contains the burrows *Taenidium* and *Archaeichnium* (Germs, 1972b).

Deposit-feeding traces become both more abundant and more "efficient" in rocks of Tommotian age so that *Phycodes* and related forms have been used as markers for the basal Cambrian. In Finmark, *Phycodes* first occurs with the feather-stich trail *Treptichnus* in strata of probable Tommotian age (Banks, 1970) whilst *Teichichnus* appear with *Planolites* in the Tommotian of England (Brasier *et al.*, 1968) and alongside *Treptichnus* in Russia (Fedonkin, 1977). Other probable deposit-feeding traces of Tommotian and younger age include *Asterichnus, Arthrophycus, Laevicyclus, Oldhamia, Palaeophycus, Taphrhelminthopsis,* | *Scolicia, Syringomorpha* and *Zoophycos* (e.g. Alpert, 1975a, 1975b, 1977; Crimes, 1976; Crimes *et al.*, 1977).

Soft-bodied and skeletal fossils of deposit feeders are less easily recognized. Adaptations that favour this mode of life include the presence of a coelom, a circular cross-section with a streamlined profile and movement in peristaltic waves, with setae, parapodia or short jointed limbs. Of the Vendian fauna, only *Vermiforma* from Carolina (Cloud *et al.*, 1976) appears tentatively to fit such a model. *Palaeoscolex* (L.-M. Cambrian) may have fed like an earthworm. The holothurian echinoderms, /which are deposit feeders are perhaps preserved as impressions in the middle Cambrian Burgess Shale (Durham, 1971).

It appears that deposit feeding was already an ancient trophic strategy before it underwent a revolution in latest Vendian to early Cambrian times (see Fig. 6). The top 10 cm of shallow marine sediments were then sufficiently rich in food matter to allow intensive and efficient exploitation. This new approach could have been the result of increased organic input, of increased competition and grazing pressure at the sediment surface, or

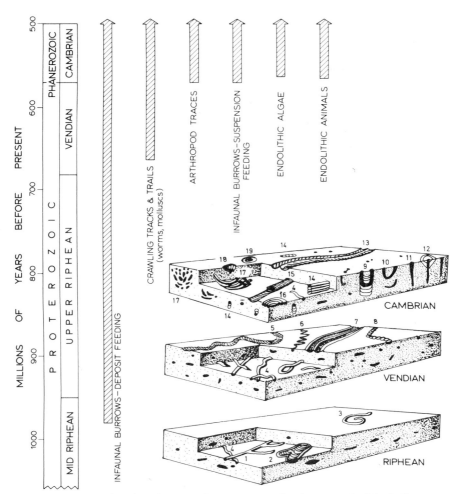

FIG. 6. Stratigraphic distribution of various trace fossil types. The block diagrams give an idea of the form of some well known examples. 1, 2, horizontal burrows of Clemmey (1976); 3, *Helminthoidichnites;* 4, *Planolites;* 5, *Nenoxites;* 6, *Cochlichnus;* 7, *Didymaulichnus;* 8, *Neonereites;* 9, *Diplocraterion;* 10, *Arenicolites;* 11, *Skolithos;* 12, *Monocraterion;* 13, *Cruziana;* 14, *Teichichnus;* 15, *Scolicia;* 16, *Syringomorpha;* 17, *Phycodes;* 18, *Astropolithon;* 19, *Bergaueria.* Data from sources in the text.

both. Skeletalization played little part in this deposit feeding radiation, presumably because rigid exoskeletons were a hindrance to movement and because the sediment itself performed many of the protective and mechanical roles of an exoskeleton. Extrapolating from the studies of Rhoads (1970) it seems likely that the deposit feeding radiation itself resulted in a great modification of the physical properties of the seafloor, with significant biological consequences.

2. Epifaunal Grazing

Grazing trails are here taken to include epifaunal or semi-infaunal trails made mainly by coelomates in their search for food, either by direct deposit feeding and microphagy or by filtering the ploughed up material (often considered as a form of suspension feeding). Locomotion trails provide no direct evidence of feeding activity (such as regular meandering or radular marks) but a distinction is not easily made so these two kinds of trail are considered alongside.

"Grazing" trails are unknown in Riphean sediments but make their first appearance soon after the Varangian glaciation, as for example in the bilobed and transversely ribbed *Bunyerichnus* (Glaessner, 1969). The tracks of creeping organisms also occur in the lower Vendian of Russia (Sokolov, 1972) and with *Plagiogmus*-like imprints in the upper Vendian (Sokolov, 1973). *Didymaulichnus* is another bilobed trail from the Vendian of Canada, South Australia and the USSR (Young, 1972; Sokolov, 1976). Fedonkin (1977) reports the grazing trails *Nenoxites, Palaeopascichnus, Suzmites, ?Cochlichnus* and "parapodial" trails from the upper Vendian of the Russian platform. Sinuous, meandering trails and feeding tracks are well known in the late Vendian Pound Quartzite of South Australia (see Glaessner, 1969, types B, C and E) and in the Manndereperelv Member of Finland (Banks, 1970).

Many tracks and "grazing" trails have been observed in the late Vendian to Tommotian strata of England including *Didymaulichnus, Cochlichnus* and *?Psammichnites* (Brasier *et al.*, 1978). Also of Tommotian age may be *Cochlichnus, Gordia, Curvolithus* and *Torrowangea* from New South Wales (Webby, 1970) and the transversely ribbed *Plagiogmus* from Australia, California, Scandinavia and Russian (e.g. Glaessner, 1969; Daily, 1972; Banks, 1970; Sokolov, 1972). The coffee-bean shaped resting (or suspension feeding?) tracks of arthropods referred to *Rusophycus* occur in Tommotian rocks on the Russian platform and possibly in England (Fedonkin, 1977; Brasier *et al.*, 1978) becoming quite common thereafter. Scratch marks made by laterally grazing or swimming, walking and furrowing arthropods also occur in many early Cambrian rocks (e.g. Banks, 1970; Cowie and Spencer, 1970; Daily, 1972; Young, 1972; Alpert, 1975a, 1975b; Crimes *et al.*, 1977).

Epifaunal grazers and deposit feeders require a broad ventral surface in contact with the substrate, hence they may be dorso-ventrally compressed when viewed in cross section. Vendian body fossils such as the annelids *Dickinsonia*, *Spriggina* and the arthropod-like problematica *Parvancorina*, *Praecambridium* and *Vendia* might have fed in this way. | Early Cambrian examples with shells may include the Monoplacophora, Gastropoda, Hyolitha, Trilobita and Merostomata. Both Garrett (1970) and Awramik (1971) suggested that the disruptive effects of invertebrate grazing and burrowing on algal mats may have caused the restriction of the latter to marginal marine and lacustrine deposits in Phanerozoic times. Probable as this is, such a grazing pressure is likely to have begun at least in the Vendian. The fossil record also suggests that the stromatolite decline was gradual rather than sudden.

Epifaunal grazing was a Proterozoic trophic strategy in which some significant changes took place near the Vendian–Cambrian boundary (see Fig. 6). Firstly, "molluscan" and worm-like tracks were joined by those of arthropods. Grazing traces became progressively more confined to offshore facies (*Nereites* type) as their place was taken in nearshore waters by other infaunal burrowers and epifauna. The grazing and bioturbation of subtidal stromatolites then caused the latter to become more confined to intertidal, supratidal and lacustrine facies. A variety of grazing organisms also developed mineralized exoskeletons, a prime function of which was probably protection from biological injury.

3. Benthonic Suspension Feeding

Infaunal suspension feeding burrows are produced by sessile or semisessile coelenterates or coelomates utilizing the sediment as a readily adaptable exoskeleton whilst feeding from the water column. Such burrows generally maintain contact with the sediment-water interface and have a prominent vertical component with modifications for changes in sedimentation rate, water level and growth. They are also typical of sediments with relatively little fine-grained detritus.

Some burrows of suspension feeding type have been reported (rarely) from Riphean and Vendian strata. Questionable examples include *Skolithos* from the Riphean of North Australia (Glaessner, 1969), *Skolithos?* from the Vendian Nama Group of South Africa (Germs, 1972) and *Bergaueria* from the upper Vendian of Russia (Sokolov, 1976). Speaking of the shallow water Vendian Pound Quartzite, however, Glaessner (1969) observes that "suspension feeders...are absent". The general lack of suspension feeding traces in Vendian strata contrasts markedly with their abundance in lower Cambrian rocks. The small *Arenicolites* burrows from the Hartshill Formation of England (Brasier *et al.*, 1978) and the small U-tubes from the upper Mannedereperelv Member of Finmark (Banks, 1970) perhaps anticipated the subsequent rapid development of *Diplocraterion, Skolithos* and *Monocraterion*. These are known from Ringsaker Quartzite of Norway, the Lontowa horizon of the Russian platform and Kalmarsund sandstone Sweden (Skjeseth, 1963; Martinsson, 1974; Fedonkin, 1977), all probably of Tommotian age. *Rhizocorallium* and vertical spiral burrows (cf. *Gyrolithes?*), which may have been suspension feeding traces, occur in the Tommotian of the Russian platform (Fedonkin, 1977) and the Atdabanian of Finmark (see Banks, 1970). The traces of benthonic actinians also appeared in the early Cambrian. By latest early Cambrian times, organisms had begun to drill into calcareous hardgrounds to form *Trypanites* borings (James *et al.*, 1977), probably suspension feeding annelids or sipunculids. Of similar origin may be the small hardground borings in the Atdabanian of Siberia mentioned by Bengtson (1977b).

It is difficult to say which of the Vendian soft-bodied fauna were benthonic suspension feeders. The attachment of the polypoid *Charnia* to the medusoid *Charniodiscus* (Ford, 1958) might be taken to indicate that the former were in reality planktonic with medusa-like umbrellas for buoyancy (cf. recent siphonophoran and chondrophoran medusoids). These "Petalonamae" and the conulariid-like *Conomedusites* were probably suspension feeding carnivores but the evidence for benthonic attachment is not compelling. The radial symmetry of *Tribrachidium* is more suggestive of a benthonic suspension feeder.

Glaessner (1976) has presented evidence to suggest that the Cribricyathida, including the ?Vendian *Cloudina*, were the skeletons of semi-infaunal, suspension feeding polychaetes. This isolated occurrence, and the finds of pogonophore-like *Hyolithellus* and *Sabellidites* and worm-like *Anabarites* and *Torellella* in the USSR appear to have heralded a great radiation of benthonic skeletonized suspension feeders in Tommotian times. New suspension feeders of Tommotian age include the Porifera, Archaeocyatha, Brachiopoda and probably the problematical *Heraultia* (?crustacean), *Chancelloria, Cambrotubulus, Coleolus, Coleolella* and *Coleoloides*. Less certainly included are the cirripede-like mitrosagophora and the ?lophophorate Conodontophorida. Atdabanian to Elankian additions to this niche perhaps included the polychaete-like *Volborthella* and *Salterella*, the early echinoderms, the archaeocopid ostracods, the early Bivalvia, the conulariid coelenterates, the stromatoporoids and the Radiocyatha.

The inference that suspension feeding from the seafloor was a novel strategy in early Cambrian times is strengthened by studies of functional morphology. For example, changes from inarticulate to articulate organization in brachiopods, from polymeral to pentameral in echinoderms, from asconoid to syconoid and leuconoid in sponges and from simple to complex porosity in archaeocyathids were all fundamental attempts to increase the efficiency of suspension feeding at these times (e.g. Gutman *et al.*, 1978; Durham, 1971; Stephenson, 1974; Finks, 1970, Rozanov and Debrenne, 1974). Many early Cambrian forms were also rather inefficient by later standards; the pogonophore-like *Sabellidites*, *Coleoloides* and *Hyolithellus*, for example, probably fed by passive collection and absorption of suspended organic matter whilst archaeocyathids may have relied on passively induced current flow (Balsam and Vogel, 1973). Whatever the methods used, skeletons and dwelling burrows assisted greatly in the formation and protection of filtration–respiration chambers and, in the case of skeletons, served to elevate the organisms upwards into the water column.

4. Predation and Cropping

If the Vendian medusoids and polyp-like Petalonamae were cnidarians then it is reasonable to infer they were carnivorous, feeding from the water column. Their importance in Vendian assemblages might be taken as indirect evidence of many other pelagic invertebrates (e.g. acritarchs, annelids and the larvae of benthonic invertebrates). If we speculate that *Chuaria* was a small *Beltanelloides*-like cnidarian, then the record of carnivory extends back to late Riphean times.

The trace fossils *Bergaueria*, *Dolopichnus*, *Anthoichnites* and *Astropolithon* are evidence that cnidarians had settled on the seafloor by early Cambrian times, perhaps joined by conulariids and corals. Of these, *Dolopichnus* was certainly carnivorous with a diet of trilobites (Alpert and Moore, 1975). Although *Teichichnus*-like traces are made today by errant, carnivorous polychaetes such as *Nereis* (Seilacher, 1957), the scolecodont jaw apparatus of such carnivores is not known until Ordovician times. Small borings into phosphatic and calcareous shells are known in Tommotian rocks (e.g.

Bengtson, 1968) but require further study. They may have been made by predators, parasites or creatures in search of a home.

Soft-bodied carnivorous priapulids were thriving in middle Cambrian times, and we may speculate on their presence in the early Cambrian, but whether the contemporary *Amiskwia*, *Aysheaia* and associated worms were carnivorous, omnivorous or microphagous is less certain. The trilobites lacked mandibles, maxillae and chelicerae and were probably only capable of eating small invertebrates, except for middle Cambrian *Olenoides* which may have grasped prey with the aid of spinose limbs (Manton, 1977). Even the early Cambrian "merostome" *Aglaspis* seems to have lacked chelicerae (Whittington, Chapter 9 this volume).

Although the radiation of skeletonized taxa in early Cambrian times has lately been ascribed to the influence of cropping pressures (e.g. Stanley, 1976), the evidence for macrocarnivory is scant. Pelagic coelenterates have, no doubt, cropped nekton and plankton since Vendian times and sedentary actinians awaited mobile benthos since early Cambrian times but early arthropods and molluscs were cropping only the smaller invertebrates. This must nonetheless have exerted great grazing pressures on benthonic juveniles. These prey could have avoided such cropping in several ways: (i) by digging into the sediment; (ii) by secreting an acid resistant or hard exoskeleton to reduce the risks of digestion or of mechanical injury; (iii) by repulsion (i.e. threatening signals, heavy armour, spines, toxins); (iv) by developing rapid locomotion for escape; (v) by size increase; (vi) by rapid reproduction. Most of these features appear to be attributes of the Cambrian radiation event.

5. Microplankton

The role played by microplankton in the evolution of ecosystems has been a popular theme in recent years. The assumption has generally been made that acritarchs were microplanktonic (probably phytoplanktonic) organisms for at least part of their life cycle (e.g. Downie, 1967; Tappan and Loeblich, 1972, 1973). Such an interpretation can follow if the following premises are accepted and met: (i) that the size range of acritarchs compares with that of known protistan plankton (the latter being 20–200 μm with an overall range of about 5–2000 μm); (ii) that their general morphology is consistent with a planktonic mode of life (i.e. sphaeroidal, discoidal, spinose. horned, streamlined, of low relative density, with buoyancy cavities or with bipolar development); (iii) that they occur in a wide range of marine deposits, including those inimicable to contemporaneous benthos; (iv) that their diversity tends to increase away from the shoreline; (v) that they are characteristic of certain water masses; (vi) that there is evidence for periodic blooms, especially those related to current upwelling. It is also reasonable to infer that ciliate and flagellate protists, by virtue of their small size and locomotory organelles, are naturally adapted to aquatic, mainly planktonic niches. Foraminiferid rhizopods, which are prominent exceptions, are planktonic for part of their life cycle (some are now wholly planktonic) and the benthonic mode of life of these and pennate diatoms may be seen as secondary adaptations, achieved variously by the secretion of unusually massive tests, sticky pseudopodia or mucilage.

The above premises can be accepted for many Palaeozoic acritarch taxa, although the evidence on iv–vi is slender. For the Proterozoic acritarchs, however, it is better to be cautious. The size range of *Chuaria* as given by Ford and Breed (1973) is 400–5000 μm but Hofmann (1977) notes bimodal size ranges with smaller forms at 100–700 μm. The larger *Chuaria* might therefore compare better with *Beltanelloides*. Some of the smaller sphaeromorph acritarchs could have been benthonic cysts, spores or zygotes of algae, as already discussed.

At least two alternative interpretations follow from the fossil record: that the first microplanktonic fossils were acanthomorph and polygono-morph acritarchs of Vendian age followed by an early Cambrian major radiation; or that many of the Riphean and Vendian sphaeromorph arcitarchs are the remains of "microplankton" but new and more complex kinds developed alongside in Vendian and Cambrian times. To the first interpretation we must question why *microplankton* (or their *cysts*) appeared in Vendian–Cambrian times and to the second we must ask why *strongly sculptured vesicles* appeared at this time? Both interpretations, nonetheless, underline the evidence for the diversification of microplankton in Vendian times with a further burgeoning in early Cambrian times.

SIZE CHANGES

In Fig. 7 are plotted the maximum recorded body diameters of the parazoan and metazoan taxa referred to previously, including the diameters of trace fossils. Although an overall trend towards increased body width with time is apparent in each "trophic group", these are not true and within-lineage trends and conclusions must therefore be cautious. Any

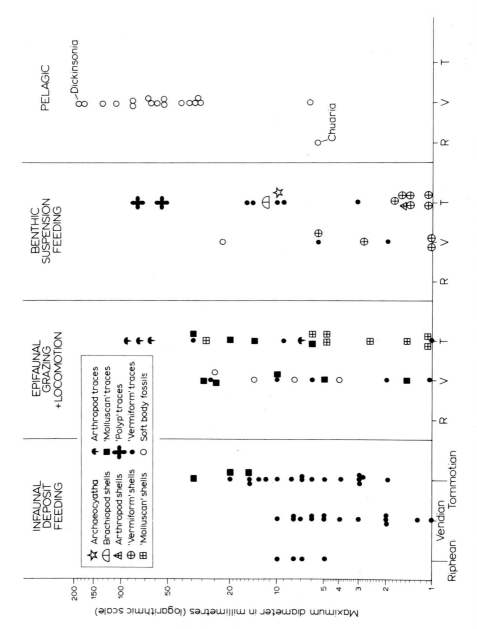

Fig. 7. For legend see opposite.

discussion of these apparent size changes must also take into account discrepancies resulting from a circular versus a flattened body cross section. In this respect, for example, surface grazers will tend to be broader than infaunal burrowers or suspension feeders of identical biovolume. The extremely large diameters of the Vendian pelagic worms and medusoids is another misleading factor if they were actually of slender body thickness, as seems likely. These problems excepted, however, several general trends can be recognized from the data.

Vendian and Tommotian organisms with shells were generally of small body diameter in comparison with contemporaneous soft-bodied and trace fossils. The contrast is so considerable that it is likely to reflect true differences in biovolume. The largest non-mineralized creatures of these times (trilobites, actinians and medusoids), were presumably less vulnerable to predatory attacks and more able to accomodate physical and chemical changes than the smaller invertebrates. The observation that it was benthonic invertebrates smaller than about 10 mm in diameter that tended to develop mineralized shells (especially the sessile suspension feeders) lends further support to the view that cropping pressures were involved in skeletalization because it was precisely these smaller, sessile forms which stood the greatest risk of being eaten.

ENVIRONMENT-RELATED CHANGES

The late Proterozoic and early Cambrian were times of considerable climatic and oceanographic change. When the ice sheets of the Varangian glaciation had melted away, a series of transgressions began to push onto the continental platforms continuing well into Ordovician times. So great and long lasting was this transgressive episode that it seems ridiculous to ascribe it to the melting of ice caps, there being no evidence for them even in Tommotian times. For the present it is as well to set aside speculations on the Varangian ice age and to concentrate on the effects of the transgressions.

Given that most rock sequences through the upper Vendian to lower Cambrian tend to reveal a trace fossil → shelly and tubular fossils → trilobite succession (e.g. Fig. 1) it is valid to wonder to what extent these

---

Fig. 7. Changes in the maximum recorded body diameter of a range of trace and body fossils of Riphean (R), Vendian (V), and Tommotian (T) age. Data from sources in the text and Brasier (unpublished).

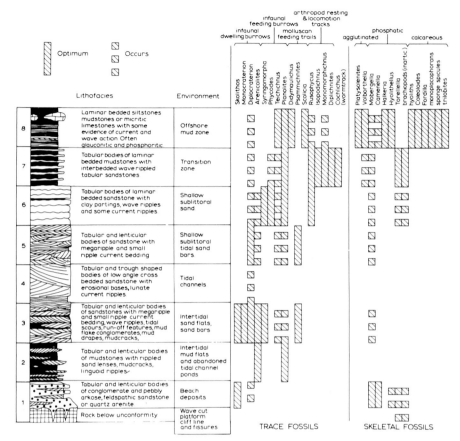

FIG. 8. A hypothetical transgressive sequence in clastic sediments of early Cambrian age. This has been put together from studies of sediments and trace
fossils on the Baltic platform of Denmark, Sweden and Central Norway
(Brasier, unpublished work) and from studies of their associated shelly
faunas (sources in Martisson, 1974 and Brasier, unpublished).

changes were merely ecological successions within neighbouring depth
communities. I have attempted to test this by arranging various early
Cambrian faunas from the same regions in hypothetical transgressive
sequences, one from predominantly clastic sediments (Fig. 8) and one from
calcareous sediments (Fig. 9). From this it seems that a typical transgressive
sequence onto the early Cambrian platforms would begin with tidal
deposits bearing low diversity burrow assemblages (e.g. *Skolithos*, *Monocraterion*, *Diplocraterion*) or stromatolites and possibly with a few inarticulate

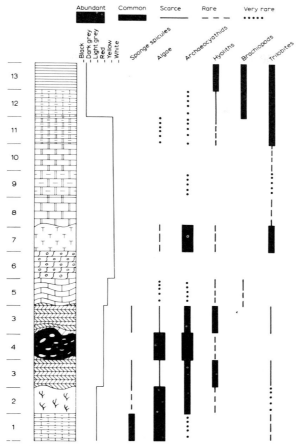

Fig. 9. A hypothetical transgressive sequence in calcareous sediments of early Cambrian age constructed from data in Zhuravleva (1972) with lithologies as follows: 1, thickly bedded argillaceous limestone; 2, massive biohermal and biostromal limestone; 3, argillaceous limestone often interbiohermal; 4, massive archaeocyathid–algal bioherms; 5, argillaceous limestomes with wavy bedding; 6, thickly bedded oolitic argillaceous and limey dolomites; 7, taphostromes of thickly bedded argillaceous limestone; 8, massive bedded pure dolomite; 9, massive bedded argillaceous limey dolomite; 10, thin bedded pure dolomite; 11, thin to medium bedded argillaceous, limey dolomites; 12, bituminus, argillaceous limestones; 13, bituminous shales.

brachiopod fragments or other phosphatic fossils. These would be followed in clastic sequences by sandstones with the above plus *Phycodes* and *Rusophycus* burrows and finally by relatively high diversity argillaceous or calcareous rocks with *Teichichnus* burrows and skeletal fossils. In calcareous sequences facies relationships can be more complex but a succession might pass from algal biostromes, through archaeocyathid bioherms into deeper water muds or limestones with hyoliths and trilobites (Fig. 9). Simple transgressive sequences are seen in Shropshire and Mjøsa, for example, but usually there is some additional evidence for regression phases within the transgression. The story of the Cambrian radiation is therefore overprinted by an element of ecological succession. This is difficult to erase because the environmental changes occurred on such a global scale.

Perhaps we should examine the problem the other way about: the succession of dominant environments through the transgression may have controlled the timing of the development and radiation of different fossil groups. In this model, littoral communities would develop during the initial stage of rising sea level. Rapid transgression allows them to spread more widely (the "radiation stage") whilst a contemporaneous assemblage develops further offshore. If the transgression proceeded in continuous pulses, the story of the Cambrian radiation would have approached the model in Fig. 10, with the evolution of different communities out of phase with each other. New stocks evolved rapidly during their development stage because they were small populations in isolated habitats. These habitats were destined to expand with time allowing global colonization, diversification and evolutionary radiation.

In general the evidence appears to fit the model fairly well. During latest Vendian and Tommotian times the platforms were, in most areas, either emergent or covered by marginal marine deposits. In terms of Palaeozoic sea level this episode was clearly "regressive" but in terms of Vendian sea level it was, of course, "transgressive". One of the most widely available habits was therefore the coastal tidal belt with dune fields and mud flats. Not surprising, therefore, is the evidence for a worldwide radiation of faunas adapted to such conditions (e.g. *Skolithos* and *Diplocraterion*). Tranquil facies of late Vendian age were scarcer but it was in such conditions that the earliest skeleton-bearers made their appearance (e.g. *Cloudina, Anabarites, Sabellidites*). Rocks bearing the Tommotian descendants of these early "experiments" are likewise quieter, offshore limestones, calcareous or glauconitic sandstones or muddy sediments, often phosphoritic. Isolation of these small populations of skeleton-bearers by intervening sand shoals and by the newly opening oceans may well have encouraged the high rate of evolution and may also account for their marked provinciality.

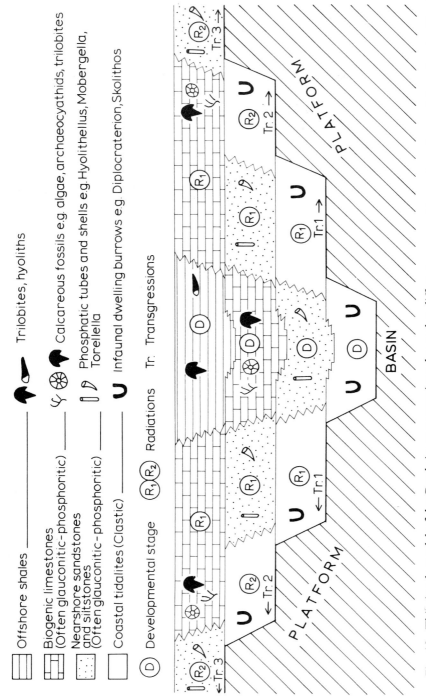

Offshore shales ⎯⎯⎯⎯

Biogenic limestones
(Often glauconitic–phosphoritic) ⎯⎯

Nearshore sandstones
and siltstones
(Often glauconitic–phosphoritic) ⎯⎯

Coastal tidalites (Clastic) ⎯⎯

Ⓓ Developmental stage   ⓇⒶ Ⓡ₂ Radiations

⎯⎯⎯⎯ Trilobites, hyoliths

ᴪ ✲ Calcareous fossils e.g. algae, archaeocyathids, trilobites

∏ ◿ Phosphatic tubes and shells e.g. Hyolithellus, Mobergella, Torellella

⊔ Infaunal dwelling burrows e.g. Diplocraterion, Skolithos

Tr. Transgressions

Fig. 10.   This simple model of the Cambrian transgressions shows that different communities need not have originated and radiated synchronously. Different stages of the transgressions preferred different opportunities.

As sea level rose and both the land and rivers diminished in influence in Atdabanian times, so the tidal deposits tended to be displaced by calcareous and argillaceous rocks. This interval witnessed a radiation of archaeocyathids, trilobites, hyoliths, molluscs and algae as their habits expanded. Correspondingly, their diversity increased, as for example in archaeocyathids which numbered about 35 genera in Tommotian times and 106 by Atdabanian times. This trend continued in Botomian times when trilobite-bearing rocks became more widespread. By the Elankian, however, regressive conditions appear to have reduced archaeocyathid diversity to 22 genera and soon afterwards to extinction. A similar rise and fall in generic diversity but without ultimate extinction took place in the trilobites, which Burrett and Richardson (1978) likewise correlate with the extent of platform flooding.

The Cambrian radiation of invertebrates and the transgressions are, for practical purposes, almost inseparable. The evolutionary changes still need to be charted within each broadly defined habitat but a continuous, unifacial and fossil-bearing sequence through the late Vendian and early Cambrian is unlikely to be found. Before an all-earth model for the Cambrian radiation can be put forward we must learn more about the significance of the abundant glauconite and phosphorite deposits and the plate tectonic and palaeomagnetic setting. Any model purporting to explain the cause of the Cambrian radiation event must then account not only for its distinctive biological nature, as outlined above, but also for its timing.

CONCLUSIONS

The fossil record suggests that the metazoan phyla originated over a period of from 200 to 500 million years. Of the several radiation events spanned by this time, the greatest was the Cambrian radiation event. Although this ranged over some 50 million years (i.e. latest Vendian and early Cambrian), evolution was particularly explosive during the relatively short, latest Vendian–Tommotian transition. Below are listed some distinctive attributes of this Cambrian radiation event.

1. Calcareous algae, rhizopod protists, parazoans (including sponges, archaeocyathids and other groups), lophophorates, conodontophorids, pogonophorans, arthropods, shelled molluscs, echinoderms and problematical hyoliths and mitrosagophorans made their first appearance during this radiation. Novel forms were also developed in the pro-

karyotes. Explanations of the radiation event that involve sexual reproduction or body organization are inappropriate here.

2. Dwelling burrows of suspension feeding invertebrates and distinctive deposit feeding and grazing trails also made their debut. These strongly imply a change in the stability and abundance of the food supply.

3. Organic-walled microplankton (acritarchs) became more diverse, with a radiation of spinose and strongly sculptured vesicles. It is possible, therefore, that the phytoplankton base of the food web was changing, partly in response to biological pressures. The abundance of phosphorite deposits on the platforms at this time is also suggestive of changing oceanic factors.

4. The soft-bodied Vendian fauna appears to have suffered worldwide extinction at this time. Perhaps increased competition and their lack of hard skeletons were responsible.

5. A Cambrian radiation of soft-bodied organisms is indicated not only by the trace fossils but also by compression and impression fossils in the Kinzer and Burgess shales. They were probably infaunal or pelagic.

6. Hard exoskeletons and endoskeletons appeared in abundance, mostly in the smaller, sessile suspension-feeding benthos but also in the vagile arthropods and molluscs. They were built by a wide range of organisms encompassing protists, algae, parazoans, cnidarians and bilaterian metazoa. The functional range of these skeletons clearly differed between groups but they shared one common denominator: protection from cropping.

7. Skeletalization varied in mode from the use of sediment (as burrows or agglutinated tubes), chitin and other organic materials to silica and a range of phosphates and carbonates. In fact, many groups "experimented" with a variety of skeletal materials at this time, of which only a limited number proved sufficiently adaptable to survive. Doubt is therefore cast on any biochemical or external chemical control over the phenomenon of skeletalization.

8. There was a general increase in the maximum body diameter of the benthos. This may have been the result of greater stability in the food supply allowing individual biomass to increase.

9. A great radiation of suspension feeders took place, each protected either by burrows or true skeletons. Many of these had relatively "inefficient" feeding methods. Stable and abundant supplies of plankton must therefore have been available over the platforms.

10. A dramatic increase in microphagous cropping pressures is also

indicated, not only within and on top of the sediment (deposit-feeding and grazing) but also in the water column (suspension feeding). Such an increase would have had self-sustaining effects on the radiation event (see above and Stanley, 1976) but the question remains as to why cropping pressures increased at that particular time?

11. These changes coincide with and were probably driven forward by a global rise in sea level that progressively flooded the continental platforms. Evolutionary patterns within the Cambrian radiation event may be linked, in part, with stages of the transgression.

One suspects there can be no final word on the Cambrian radiation event. Its dramatic qualities are untarnished by discoveries of older, more bizarre organisms. To regard this episode as the inevitable march of organic progress would at present be rather naïve. Like many extinction events it bears the heavy imprint of planetary upheaval. Had the nature, timing or degree of planetary change been different the story would not have been the same and we, perhaps, would still be earnestly sifting the ooze.

REFERENCES

AITKEN, J. D. (1967). Classification and significance of cryptalgal limestones and dolomites with illustrations from the Cambrian and Ordovician of southwestern Alberta. *J. sedim. Petrol.* **37**, 1163–1178.

ALLISON, C. W. (1975). Primitive fossil flatworm from Alaska: new evidence bearing on the ancestry of the Metazoa. *Geology*, **3**, 649–652.

ALPERT, S. P. (1973). *Bergaueria* Prantl (Cambrian and Ordovician), a probable actinian trace fossil. *J. Paleont.* **47**, 919–924.

ALPERT, S. P. (1975a). Trace fossils of the Precambrian–Cambrian succession, White-Inyo Mountains, California. *Diss. Abstr. Int.* **35**, 4072B.

ALPERT, S. P. (1975b). *Planolites* and *Skolithos* from the Upper Precambrian-Lower Cambrian White-Inyo Mountains, California. *J. Paleont.* **49**, 508–521.

ALPERT, S. P. (1977). Trace fossils and the basal Cambrian boundary. *In* "Trace Fossils 2" (T. P. Crimes and J. C. Harper, eds), pp. 1–8. Seel House Press, Liverpool.

ALPERT, S. P. and MOORE, J. N. (1975). Lower Cambrian trace fossil evidence for predation on trilobites. *Lethaia*, **8**, 223–230.

ANDERSON, M. M. and MISRA, S. B. (1968). Fossils found in the Pre-Cambrian Conception Group of south-eastern Newfoundland. *Nature, Lond.* **220**, 680–681.

AWRAMIK, S. M. (1971). Precambrian columnar stromatolite diversity, reflection of metazoan appearance. *Science, N.Y.* **174**, 825–827.

BALSAM, W. L. and VOGEL, S. (1973). Water movement in archaeocyathids: evidence and implications of passive flow in models. *J. Paleont.* **47**, 979–984.

BANKS, N. L. (1970). Trace fossils from the late Precambrian and lower Cambrian of Finmark, Norway. *In* "Trace Fossils" (T. P. Crimes and J. C. Harper, eds), pp. 19–34. Seel House Press, Liverpool.

BEER, E. J. (1919). Note on a spiral impression on Lower Vindhyan limestone. *Rec. Geol. Surv. India*, **50**, 139.

BENGTSON, S. (1968). The problematic genus *Mobergella* from the Lower Cambrian of the Baltic area. *Lethaia*, **1**, 325–351.

BENGTSON, S. (1970). The Lower Cambrian fossil *Tommotia*. *Lethaia*, **3**, 363–392.

BENGTSON, S. (1976). The structure of some Middle Cambrian conodonts and the early evolution of conodont structure and function. *Lethaia*, **9**, 185–206.

BENGTSON, S. (1977a). Aspects of problematic fossils in the early Palaeozoic. *Acta Univ. Upsaliensis*, **415**.

BENGTSON, S. (1977b). Early Cambrian button-shaped phosphatic microfossils from the Siberian Platform. *Palaeontology*, **20**, 751–762.

BERGSTRÖM, J. (1970). *Rusophycus* as an indication of early Cambrian age. *In* "Trace Fossils" (T. P. Crimes and J. C. Harper, eds), pp. 34–42. Seel House Press, Liverpool.

BINDA, P. L. (1977). Microfossils from the Lower Kundelungu (Late Precambrian) of Zambia. *Precamb. Res.* **4**, 285–306.

BISCHOFF, G. C. O. (1976). *Dailyatia*, a new genus of Tommotiidae from Cambrian strata of S.E. Australia (Crustacea, Cirripedia). *Senckenberg. leth.* **57**, 1–33.

BLOESER, B., SCHOPF, J. W., HORODYSKI, R. J. and BREED, W. J. (1977). Chitin-ozoans from the Late Precambrian Chuar Group of the Grand Canyon, Arizona. *Science, N.Y.* **195**, 675–679.

BOCKELIE, T. and FORTEY, R. A. (1976). An early Ordovician vertebrate. *Nature, Lond.* **260**, 36–38.

BONDARENKO, O. B. (1966). Lines of evolution in tabulates. *Paleont. Zh.* **4**, 8–18. (In Russian.)

BRASIER, M. D. (1976a). Early Cambrian intergrowths of archaeocyathids, *Renalcis* and pseudostromatolites from South Australia. *Palaeontology*, **19**, 223–245.

BRASIER, M. D. (1976b). An archaeocyathid-trilobite association in Sardinia and its stratigraphic significance. *Riv. ital. Paleont. Stratigr.* **82**, 267–278.

BRASIER, M. D. (1977). An early Cambrian chert biota and its implications. *Nature, Lond.*, **268**, 719–720.

BRASIER, M. D. (1979). "Microfossils." George Allen and Unwin, London. (In Press.)

BRASIER, M. D. and HEWITT, R. A. (1979). Environmental setting of some fossiliferous rocks from the uppermost Protozoic–lower Cambrian of Central England. *Palaeogeogr., Palaeoclimatol., Palaeocol.* (In Press.)

BRASIER, M. D., HEWITT, R. A. and BRASIER, C. J. (1978). On the late Precambrian-early Cambrian Hartshill Formation of Warwickshire. *Geol. Mag.* **115**, 21–36.

BRIGGS, D. E. G. (1977). Bivalved arthropods from the Cambrian Burgess Shale of British Columbia. *Palaeontology*, **20**, 595–621.

BULMAN, O. M. B. (1955). Graptolithina, with sections on Enteropneusta and Pterobranchia. *In* "Treatise on Invertebrate Paleontology, Part V" (R. C. Moore, ed.), pp. 1–101. *Geol. Soc. Amer. and Univ. Kansas Press.*

BURRETT, C. F. and RICHARDSON, R. G. (1978). Cambrian trilobite diversity related to cratonic flooding. *Nature, Lond.* **272**, 717–719.

CLARK, R. B. (1964). "The Dynamics of Metazoan Evolution." Clarendon Press, Oxford.

CLEMMEY, H. I. (1976). World's oldest animal traces. *Nature, Lond.* **261**, 576–578.

CLOUD, P. (1976). Beginnings of biospheric evolution and their biochemical consequences. *Paleobiology*, **3**, 351–382.

CLOUD, P. and GERMS, A. (1971). New Pre-Palaeozoic Nannofossils from the Stoer Formation (Torridonian) Northwest Scotland. *Bull. geol. Soc. Am.* **82**, 3469–3474.

CLOUD, P., WRIGHT, J. and GLOVE, L. (1976). Traces of animal life from 620-million-year-old rocks in North Carolina. *Am. Scient.* **64**, 396–406.

COBBOLD, E. S. (1936). The Conchostraca of the Cambrian area of Comley, Shropshire, with a note on a new variety of *Atops reticulatus* (Walcott). *Q. Jl geol. Soc. Lond.* **92**, 221–235.

CONKIN, J. E. and CONKIN, B. M. (1977). Paleozoic smaller foraminifera of the North American Atlantic borderlands. *In* "Stratigraphic Micropaleontology of Atlantic Basin and Borderlands" (F. M. Swain, ed.), pp. 49–59. Elsevier, Amsterdam.

CONWAY MORRIS, S. (1976). A new Cambrian lophophorate from the Burgess Shale of British Columbia. *Palaeontology*, **19**, 199–222.

CONWAY MORRIS, S. (1977a). A new entoproct-like organism from the Burgess Shale of British Columbia. *Palaeontology*, **20**, 833–845.

CONWAY MORRIS, S. (1977b). Fossil priapulid worms. *Spec. Pap. Palaeontology*, **20**, 95pp.

CONWAY MORRIS, S. (1977c). A redescription of the Middle Cambrian *Amiskwia sagittiformis* WALCOTT from the Burgess Shale of British Columbia. *Paläont. Z.* **51**, 271–287.

COPE, J. C. W. (1977). An Ediacara-type fauna from South Wales. *Nature, Lond.* **268**, 624.

COWIE, J. W. (1974). Cambrian of Spitsbergen and Scotland. *In* "Cambrian of the British Isles, Norden and Spitsbergen" (C. H. Holland, ed.), pp. 123–166. Wiley and Sons, London.

COWIE, J. W. and GLAESSNER, M. F. (1975). The Precambrian–Cambrian boundary: a symposium. *Earth-Sci. Revs* **11**, 209–251.

COWIE, J. W. and ROZANOV, A. Yu. (1974). I.U.G.S. Precambrian/Cambrian Boundary Working group in Siberia, 1973. *Geol. Mag.* **111**, 237–252.

COWIE, J. W. and SPENCER, A. M. (1970). Trace fossils from the late Precambrian/lower Cambrian of East Greenland. *In* "Trace Fossils" (T. P. Crimes and J. C. Harper, eds), pp. 91–100. Seel House Press, Liverpool.

COWIE, J. W., RUSHTON, A. W. A. and STUBBLEFIELD, C. J. (1972). Cambrian. *Spec. Rep. geol. Soc.* **2**, 42pp.

CRIMES, T. P. (1976). Trace fossils from the Bray Group (Cambrian) at Howth, Co. Dublin. *Bull. geol. Surv. Ireland*, **2**, 53–67.

CRIMES, T. P., LEGG, I., MARCOS, A. and ARBOLEYA, M. (1977). ?Late Precambrian-low Lower Cambrian trace fossils from Spain. *In* "Trace Fossils 2" (T. P. Crimes and J. C. Harper, eds), pp. 91–138. Seel House Press, Liverpool.

DAILY, B. (1972). The base of the Cambrian and the first Cambrian faunas. *Univ. Adelaide Centre for Precambrian Research, Spec. Pap.* **1**, 13–42.

DALE, B. (1976). Cyst formation, sedimentation and preservation: factors affecting dinoflagellate assemblages in Recent sediments from Trondheimsfjord, Norway. *Rev. Palaeobot. Palynol.* **22**, 39–60.

DEBRENNE, F., TERMIER, H. and TERMIER, G. (1970). Radiocyatha. Une nouvelle classe d'organismes primitifs du Cambrien inférieur. *Bull. Soc. géol. Fr.* **12**, 120–125.

DEBRENNE, F., TERMIER, H. and TERMIER, G. (1971). Sur de nouveaux représentants de la classe Radiocyatha. Essai sur l'evolution des Metazoaires primitifs. *Bull. Soc. géol. Fr.* **13**, 439–444.

DEFLANDRE, G. (1950). Le soi-disant Radiolaires du Précambrien de Bretagne et la question de l'existence de Radiolaires embryonnaires fossils. *Bull. Soc. zool. Fr.* **74**, 351–352.

DEFLANDRE, G. (1968). Sur l'existence des le Precambrien d'Acritarches du type Acanthomorphitae: *Eomicrhystridium* nov. gen. Typification du genre *Palaeocryptidium* Defl. 1955. *C. r. hebd. Séanc. Acad. Sci., Paris, D*, **266**, 2358–2389.

DORÉ, F. and REID, R. E. (1965). *Allonia tripodophoria* n.g., n.s., nouvelle eponge du Cambrian inférieur de Carteret (Manche). *C. r. somm. Séanc. Soc. géol. Fr. 1965*, 20–21.

DOWNIE, C. (1967). The geological history of the microplankton. *Rev. Palaeobot. Palynol.* **1**, 269–281.

DOWNIE, C. (1973). Observations on the nature of the acritarchs. *Palaeontology*, **16**, 239–259.

DOWNIE, C. (1974). Acritarchs from near the Precambrian–Cambrian boundary—a preliminary account. *Rev. Palaeobot. Palynol.* **18**, 57–60.

DURHAM, J. W. (1971). The fossil record and the origin of the Deuterostomata. *Proc. N. Am. Pal. Conv. 1969, H*, 1104–1132.

DZULYŃSKI, S. and WALTON, E. K. (1965). Sedimentary features of Flysch and Greywackes. *Developments in Sedimentology* **7**. Elsevier, Amsterdam.

ELIAS, M. K. (1954). *Cambroporella* and *Coeloclema*, Lower Cambrian and Ordovician bryozoans. *J. Paleont.* **28**, 52–58.

EVITT, W. R. (1963). A discussion and proposals concerning fossil dinoflagellates, hystrichospheres and acritarchs. *Proc. natn. Acad. Sci. U.S.A.* **49**, 158–164, 298–302.

FEDONKIN, M. A. (1977). Precambrian-Cambrian ichnocoenoses of the east European Platform. *In* "Trace Fossils 2" (T. P. Crimes and J. C. Harper, eds), pp. 183–194. Seel House Press, Liverpool.

FENTON, M. A. and FENTON, C. L. (1934). *Scolithus* as a fossil phoronid. *Pan-Am. Geol.* **61**, 341–348.

FINKS, R. M. (1970). The evolution and ecologic history of sponges during Palaeozoic times. *Symp. zool. Soc. Lond.* **25**, 3–22.

FIRBY, J. B. and DURHAM, J. W. (1974). Molluscan radula from earliest Cambrian. *J. Paleont.* **48**, 1109–1119.

FLÜGEL, E. (1977). "Fossil Algae." Springer-Verlag, Berlin.

FONIN, V. D. and SMIRNOVA, T. N. (1967). A new group of problematical early Cambrian organisms and some methods for their preparation. *Paleont. Zh.* **2**, 15–27. (In Russian.)

FORD, T. D. (1958). Pre-Cambrian fossils from Charnwood Forest. *Proc. Yorks. geol. Soc.* **31**, 211–217.

FORD, T. D. and BREED, W. J. (1973). The problematical Precambrian fossil *Chuaria*. *Palaeontology*, **16**, 535–550.

FØYN, S. and GLAESSNER, M. F. (1978). *Platysolenites*, other animal fossils and the Precambrian-Cambrian transition in Norway. *Norsk. geol. Tidsskr.* (In Press.)

FRITZ, W. H. (1972). Lower Cambrian trilobites from the Sekwi Formation type section, Mackenzie Mountains, Northwestern Canada. *Bull. geol. Surv. Can.* **212**, 1–58.

GARRETT, P. (1970). Phanerozoic stromatolites: noncompetitive ecologic restriction by grazing and burrowing animals. *Science, N.Y.* **169**, 171–173.

GERMS, G. J. B. (1968). Discovery of a new fossil in the Nama System, South West Africa. *Nature, Lond.* **219**, 53–54.

GERMS, G. J. B. (1972a). New shelly fossils from the Nama Group, Southwest Africa. *Am. J. Sci.* **272**, 752–761.

GERMS, G. J. B. (1972b). Trace fossils from the Nama Group, South-west Africa. *J. Paleont.* **46**, 864–870.

GERMS, G. J. B. (1973a). A reinterpretation of *Rangea schneiderhoehni* and the discovery of a related new fossil from the Nama Group, South-west Africa. *Lethaia*, **6**, 1–10.

GERMS, G. J. B. (1973b). Possible sprigginid worm and a new trace fossil from the Nama Group, South West Africa. *Geology*, **1**, 69–70.

GLAESSNER, M. F. (1963a). Major trends in the evolution of the Foraminifera. *In* "Evolutionary Trends in Foraminifera" (G. H. R. Koenigswald, J. D. Emeis, W. L. Buning and C. W. Wagner, eds), pp. 9–24. Elsevier, Amsterdam.

GLAESSNER, M. F. (1963b). Zur Kenntnis der Nama-Fossilien Sudwest-Afrikas. *Ann. naturhist. Mus. Hofmuseums, Wien.* **66**, 113–120.

GLAESSNER, M. F. (1969). Trace fossils from the Precambrian and basal Cambrian. *Lethaia*, **2**, 369–394.

GLAESSNER, M. F. (1971a). Geographic distribution and time range of the Ediacara Precambrian fauna. *Bull. geol. Soc. Am.* **82**, 509–514.

GLAESSNER, M. F. (1971b). The genus *Conomedusites* Glaessner and Wade and the diversification of the Cnidaria. *Paläont. Z.* **45**, 7–17.

GLAESSNER, M. F. (1975). The oldest foraminifera. *Bull. Bur. Miner. Resour. Geol. Geophys. Aust.* **192**, 61–65.

GLAESSNER, M. F. (1976). Early Phanerozoic annelid worms and their geological and biological significance. *J. geol. Soc.* **132**, 259–275.

GLAESSNER, M. F. and WADE, M. (1966). The late Precambrian fossils from Ediacara, South Australia. *Palaeontology,* **9**, 599–628.

GLAESSNER, M. F. and WADE, M. (1971). *Praecambridium*—a primitive arthropod. *Lethaia,* **4**, 71–77.

GLAESSNER, M. F. and WALTER, M. R. (1975). New Precambrian fossils from the Arumbera Sandstone, Northern Territory, Australia. *Alcheringa,* **1**, 59–69.

GNILOVSKAJA, M. B. (1971). The oldest aquatic plants of the Vendian of the Russian Platform (late Precambrian). *Paleont. J.* **5**, 372–378.

GORIANSKY, W. (1973). On the necessity of excluding *Chancelloria* from the sponges. *In* "Problems of Paleontology and Biostratigraphy in the Lower Cambrian of the far-eastern part of Siberia", pp. 39–44. *Izdat. Nauka, Sib. Otdel.* (In Russian.)

GUTMAN, W. F., VOGEL, K. and ZORN, H. (1978). Brachiopods: biomechanical interdependences governing their origin and phylogeny. *Science,* **199**, 890–893.

HAMAR, G. (1967). *Platysolenites antiquissimus* Eichw. (Vermes) from the Lower Cambrian of Northern Norway. *Norg. geol. Unders.* **249**, 89–95.

HANDFIELD, R. C. and MCKINNEY, F. K. (1975). Form and function in an atypical archaeocyathid. *J. Paleont.* **49**, 799–807.

HARLAND, W. B. (1974). The Pre-Cambrian–Cambrian boundary. *In* "Cambrian of the British Isles, Norden and Spitsbergen" (C. H. Holland, ed.), pp. 15–42. Wiley and Sons, London.

HAUGHTON, S. H. (1960). An archaeocyathid from the Nama System. *Trans. R. Soc. S. Afr.* **3b**, 57–59.

HILL, D. (1972). Archaeocyatha. *In* "Treatise on Invertebrate Paleontology, Part E, volume 1" (C. Teichert, ed.). *Geol. Soc. Amer. and Univ. Kansas Press.*

HOFMANN, H. J. (1975). Stratiform Precambrian stromatolites, Belcher Islands, Canada: relations between silicified microfossils and microstructure. *Am. J. Sci.* **275**, 1121–1132.

HOFMANN, H. J. (1977). The problematic fossil *Chuaria* from the late Precambrian Uinta Mountain Group, Utah. *Precamb. Res.* **4**, 1–11.

HOLDSWORTH, B. K. (1977). Paleozoic Radiolaria: stratigraphic distribution in Atlantic borderlands. *In* "Stratigraphic Micropaleontology of Atlantic Basin and Borderlands" (F. M. Swain, ed.), pp. 167–184. Elsevier, Amsterdam.

HOWELL, B. F. (1962). Worms. *In* "Treatise on Invertebrate Paleontology, Part W" (R. C. Moore, ed.), pp. 144–177. *Geol. Soc. Amer. and Univ. Kansas Press.*

HOWELL, B. F. and DUNN, P. H. (1942). Early Cambrian "Foraminifera". *J. Paleont.* **16**, 638–639.

JAEGER, H. and MARTINSSON, A. (1967). Remarks on the problematic fossil *Xenusion auerswaldae. Geol. För. Stockh. Förh.* **88**, 435.

JÄGERSTEN, G. (1972). "Evolution of the Metazoan life Cycle." Academic Press, London and New York.

JAMES, N. P. and KOBLUK, D. R. (1978). Lower Cambrian patch reefs and associated sediments: southern Labrador, Canada. *Sedimentology,* **25**, 1–35.

JAMES, N. P., KOBLUK, D. R. and PEMBERTON, S. G. (1977). The oldest macroborers: lower Cambrian of Labrador. *Science, N.Y.* **197**, 980–983.

JANSONIUS, J. (1970). Classification and stratigraphic application of Chitinozoa. *Proc. N. Am. Paleont. Conv. 1969, G,* 789–808.

JEFFERIES, R. P. S. (1968). The subphylum Calcichordata (Jeffries, 1967), primitive fossil chordates with echinoderm affinities. *Bull. Brit. Mus. nat. Hist. A,* **16,** 243–339.

JELL, P. A. and JELL, J. S. (1976). Early Middle Cambrian corals from western New South Wales. *Alcheringa,* **1,** 181–196.

JOHNSON, J. H. (1966). A Review of the Cambrian Algae. *Colo. Sch. Mines Q.* **61,** 162 pp.

JOST, M. (1968). Microfossils of problematic systematic position from Precambrian rocks at White Pine, Michigan. *Micropaleontology,* **14,** 365–368.

KAZMIERCZAK, J. (1976). Devonian and modern relatives of the Precambrian *Eosphaera*: possible significance for the early eukaryotes. *Lethaia,* **9,** 39–50.

KARKHANIS, S. N. (1976). Fossil iron bacteria may be preserved in Precambrian ferroan carbonate. *Nature, Lond.* **261,** 406–407.

KOLOSOV, P. N. (1975). "Stratigraphy of the Upper Precambrian of Southern Yakutsk." *Akad. Nauka SSSR., Novosibirsk.* (In Russian.)

KORDE, K. B. (1963). Hydroconozoa—a new class of coelenterate animals. *Paleont. Zh.* **2,** 20–25. (In Russian.)

KRYSHTOFOVITCH, A. N. (1953). Discovery of lycopodiaceous plants in the Cambrian of eastern Siberia. *Dokl. Akad. Nauk. SSSR* **91,** 1377–1379. (In Russian.)

LARWOOD, G. P., MEDD, A. W., OWEN, D. E. and TAVENER-SMITH, R. (1967). Bryozoa. *In* "The Fossil Record" (W. B. Harland *et al.*, eds), pp. 379–395. Geol. Soc. London.

LAUBENFELS, M. W. de (1955). Porifera. *In* "Treatise on Invertebrate Paleontology, Part E" (R. C. Moore, ed.), pp. 21–112. *Geol. Soc. Amer. and Univ. Kansas Press.*

LINDSTRÖM, M. (1974). The conodont apparatus as a food-gathering mechanism. *Palaeontology,* **17,** 729–744.

LISTER, T. R. (1970). The acritarchs and Chitinozoa from the Wenlock and Ludlow Series of the Ludlow and Millichope areas, Shropshire, Part 1. *Palaeontogr. Soc. (Monogr.),* **124,** 1–100.

LOEBLICH, A. R., Jr and TAPPAN, H. (1964). Protista 2; Sarcodina, chiefly "Thecamoebians" and Foraminiferida. *In* "Treatise on Invertebrate Paleontology, Part C" (R. C. Moore, ed.), pp. 1–900. *Geol. Soc. Amer. and Univ. Kansas Press.*

MANTON, S. M. (1977). "The Arthropods. Habits, Functional Morphology and Evolution." Clarendon Press, Oxford.

MAREK, L. and YOCHELSON, E. L. (1976). Aspects of the biology of Hyolitha (Mollusca). *Lethaia,* **9,** 65–82.

MARTINSSON, A. (1974). The Cambrian of Norden. *In* "Cambrian of the British Isles, Norden and Spitsbergen" (C. H. Holland, ed.), pp. 185–284. Wiley and Sons, London.

MATTHEWS, S. C. and MISSARZHEVSKY, V. V. (1975). Small shelly fossils of late Precambrian and early Cambrian age: a review of recent work. *J. geol. Soc.* **131,** 289–304.

MELENDEZ, B. (1966). Notas estratigraficas de la Provincia de Cordoba. *Notas Commun. Inst. geol. min. Esp.* **90,** 77–84.

MILTON, D. J. (1966). Drifting organisms in the Precambrian sea. *Science, N.Y.* **153**, 293–294.

MISSARZHEVSKY, V. V. and ROZANOV, A. Yu. (1965). The organic world in the boundary layers of the Cambrian and Precambrian and the principles of establishing the lower boundary of the Cambrian and the Palaeozoic. *In* "Proceedings of the All-State Symposium on the Precambrian and Early Cambrian, 1965" (B. S. Sokolov, ed.). Novosibirsk. (In Russian.)

MISSARZHEVSKY, V. V. (1976). New data on early Cambrian monoplacophorans. *Paleont. J.* **2**, 234–236.

MÜLLER, K. J. (1975). "*Heraultia*" *varensalensis* COBBOLD (Crustacea) aus dem Unteren Kambrium, der älteste Fall von Geschlechtsdimorphismus. *Paläont. Z.* **49**, 168–180.

NATIONS, J. D. and BEUS, S. S. (1974). Pellet-lined burrows in Poleta Formation (Lower Cambrian) of White-Inyo Mountains, California. *Geol. Soc. Am., Abstracts with programs.* **6**, 225.

NAZAROV, B. B. (1975). Lower and Middle Paleozoic radiolarians of Kazakhstan. *Trudy geol. Inst. Akad. Nauk. SSSR* **275**, 202 pp. (In Russian.)

NESTOR, H. E. (1966). On ancient stromatoporoids. *Paleont. Zh.* **2**, 3–12. (In Russian.)

NORTH, F. K. (1971). The Cambrian of Canada and Alaska. *In* "Cambrian of the New World" (C. H. Holland, ed.), pp. 219–324. Wiley-Interscience, London.

PALMER, A. R. (1971). The Cambrian of the Appalachian and Eastern New England regions, Eastern United States. *In* "Cambrian of the New World" (C. H. Holland, ed.), pp. 169–217. Wiley-Interscience, London.

PFLUG, H. D. (1965). Foraminifera and ahnliche Fossilreste aus dem Kambrium und Algonkium. *Palaeontographica,* A, **125**, 46–60.

PFLUG, H. D. (1972a). The Phanerozoic-Cryptozoic boundary and the origin of the Metazoa. *24th Int. geol. Congr. Montreal,* **1**, 58–67.

PFLUG, H. D. (1972b). Systematik der jung-präkambrischen Petalonamae Pflug 1970. *Paleont. Z.* **46**, 56–67.

POJETA, J., RUNNEGAR, B. and KRIZ, J. (1973). *Fordilla troyensis* Barrande: the oldest known pelecypod. *Science, N.Y.* **180**, 866–868.

POULSEN, C. (1932). The Lower Cambrian faunas of East Greenland. *Meddr Grønland,* **87**, 29–30.

POULSON, C. (1967). Fossil from the Lower Cambrian of Bornholm. *Meddr K. danske Vidensk. Selsk. mat.-fys.* **36**.

POULSEN, V. (1963). Notes on *Hyolithellus* Billings, 1871, Class Pogonophora Johannson, 1937. *Medr. K. danske Vidensk. Selsk. biol.* **23**, 1–15.

POULSEN, V. (1966). Early Cambrian distacodontid conodonts from Bornholm. *Meddr K. danske Vidensk. Selsk. biol.* **23**, 1–9.

RAABEN, M. E. (1969). Columnar stromatolites and late Precambrian stratigraphy. *Am. J. Sci.* **267**, 1–18.

REITLINGER, E. A. (1948). The Cambrian Foraminifera of Yakutsk. *Bull. Mosk. obshch. ispyt. prir. otdel. geol.,* N.S. **53** (23), 77–81. (In Russian.)

RESSER, C. E. and HOWELL, B. F. (1938). Lower Cambrian *Olenellus* zone of the Appalachians. *Bull. geol. Soc. Am.* **49**, 195–248.

RHOADS, D. C. (1970). Mass properties, stability and ecology of marine muds related to burrowing activity. *In* "Trace Fossils" (T. P. Crimes and J. C. Harper, eds), pp. 391–406. Seel House Press, Liverpool.

RIDING, R. and BRASIER, M. D. (1975). Earliest calcareous foraminifera. *Nature, Lond.* **257**, 208–210.

ROBISON, R. A. (1969). Annelids from the Middle Cambrian Spence Shale of Utah. *J. Paleont.* **43**, 1169–1173.

ROWELL, A. J. (1971). Supposed Pre-Cambrian brachiopods. *Smithson. Contr. Paleobiol.* **3**, 71–82.

ROZANOV. A. Yu. (1967). The Cambrian lower boundary problem. *Geol. Mag.* **104**, 415–434.

ROZANOV, A. Yu. and DEBRENNE, F. (1974). Age of archaeocyathid assemblages. *Am. J. Sci.* **274**, 833–848.

ROZANOV, A. Yu., MISSARZHEVSKY, V. V., VOLKOVA, N. A., VORONOVA, L. G., KRYLOV, I. N., KELLER, B. M., KOROLYUK, I. K., LENDZION, K., MICHNIAK, R., PYCHOVA, N. G. and SIDOROV, A. D. (1969). Tommotian stage and the Cambrian lower boundary problem. *Trudy Akad. Nauk. SSSR*, **206**, 1–380. (In Russian.)

RUDAVSKAJA, V. A. (1973). New species of ancient plants and invertebrates. *Trudy VNIGRI SSSR, Leningrad*, **318**, 7. (In Russian.)

RUNNEGAR, B. and JELL, P. A. (1976). Australian Middle Cambrian molluscs and their bearing on early molluscan evolution. *Alcheringa*, **1**, 109–138.

RUNNEGAR, B. and POJETA, J. Jr. (1974). Molluscan phylogeny : the paleontological viewpoint. *Science, N.Y.* **186**, 311–317.

RUNNEGAR, B., POJETA, J. Jr., MORRIS, N. J., TAYLOR, M. E. and McCLUNG, G. (1975). Biology of the Hyolitha. *Lethaia*, **8**, 181–191.

RUSHTON, A. W. A. (1974). The Cambrian of Wales and England. *In* "Cambrian of the British Isles, Norden and Spitsbergen" (C. H. Holland, ed.), pp. 43–121. Wiley and Sons, London.

SCHINDEWOLF, O. H. (1956). Über präkambrische Fossilien. *In* "Geotekton. Symposium zu Ehren von H. Stille" (F. Lotze, ed.), pp. 455–480. Ferdinand Enke, Stuttgart.

SCHOPF, J. M. (1969). Early Paleozoic palynomorphs. *In* "Aspects of Palynology" (R. H. Tschudy and R. A. Scott, eds), pp. 163–192. Wiley-Interscience, New York.

SCHOPF, J. W. (1971). Organically preserved Precambrian microfossils. *Proc. N. Am. Pal. Conv. 1969*, **H**, 1013–1057.

SCHOPF, J. W. (1977). Biostratigraphic usefulness of stromatolitic Precambrian microbiotas; a preliminary analysis. *Precambrian Research*, **5**, 143–173.

SCHOPF, J. W., HAUGH, B. N., MOLNAR, R. E. and SATTERTHWAIT, D. F. (1973). On the development of metaphytes and metazoans. *J. Paleont.* **47**, 1–9.

SDZUY, K. (1969). Unter-und mittel kambrische Porifera (Chancelloriida and Hexactinellida). *Paläont. Z.* **43**, 115–147.

SEILACHER, A. (1956). Der Beginn des Kambriums als biologische Wende. *Neues Jb. Geol. Paläont. Abh.* **103**, 155–180.

SEILACHER, A. (1957). An-aktualistiches Wattenmeer ? *Paläont. Z.* **31**, 198–206.

SEMIKHATOV, M. A., KOMAR, V. A. and SEREBRYAKOV, S. N. (1970). The Yudomian Complex of the stratotype area. *Trudy Akad. Nauk. SSSR, Inst. Geol.* **210**, 1–207. (In Russian.)

SKJESETH, S. (1963). Contributions to the geology of the Mjøsa districts and the classical sparagmite area in southern Norway. *Norg. geol. Unders.* **220**, 1–126.

SOKOLOV, B. S. (1965). Ancient deposits of the Early Cambrian and Sabelliditids. *In* "Proceedings of the All-State Symposium on the Precambrian and Early Cambrian 1965" (B. S. Sokolov, ed.), pp. 78–91. *Akad. Nauk. SSSR, Novosibirsk.* (In Russian.)

SOKOLOV, B. S. (1972). The Vendian stage in earth history. *24th Int. geol. Congr. Montreal,* **1**, 78–84.

SOKOLOV, B. S. (1973). Vendian of Northern Eurasia. *Mem. Am. Ass. Petrol. Geol.* **19**, 204–218.

SOKOLOV, B. S. (1976). Precambrian Metazoa and the Vendian–Cambrian boundary. *Paleont. J.* **10**, 1–13.

SPJELDNAES, N. (1963). A new fossil (Papillomembrana sp.) from the Upper Pre-Cambrian of Norway. *Nature, Lond.* **200**, 63–65.

SQUIRE, A. D. (1973). Discovery of late Precambrian trace fossils in Jersey, Channel Islands. *Geol. Mag.* **110**, 223–226.

STANLEY, S. M. (1976). Ideas on the timing of metazoan diversification. *Paleobiology,* **2**, 209–219.

STASINSKA, A. (1966). *Velumbrella czarnockii* n.g., n.sp.—meduse du Cambrien inférieur des Montes de Sainte Croix. *Acta palaeont. pol.* **5**, 337–346.

STEARN, C. W. (1975). The stomatoporoid animal. *Lethaia,* **8**, 89–100.

STEPHENSON, D. G. (1974). Pentamerism and the ancestral echinoderm. *Nature, Lond.* **250**, 82–83.

TAPPAN, H. and LOEBLICH, A. R. Jr. (1968). Lorica composition of modern and fossil Tintinnida (Ciliate Protozoa), systematics, geologic distribution and some new Tertiary taxa. *J. Paleont.* **42**, 1378–1394.

TAPPAN, H. and LOEBLICH, A. R. Jr. (1972). Fluctuating rates of protistan evolution, diversification and extinction. *24th Int. geol. Congr. Montreal,* **7**, 205–213.

TAPPAN, H. and LOEBLICH, A. R. Jr. (1973). Evolution of the ocean plankton. *Earth Sci. Revs* **9**, 207–240.

TARLO, L. B. H. (1967). Xenusion—onycophoran or coelenterate? *Mercian geol.* **2**, 97–99.

TAYLOR, M. E. (1966). Precambrian mollusc-like fossils from Inyo County, California. *Science, N.Y.* **153**, 198–201.

TCHUDINOVA, N. I. (1959). On a find of Conulariids in the Lower Cambrian of Western Sayan. *Paleont. Zh.* **2**, 53–55. (In Russian.)

TERMIER, H. and TERMIER, G. (1964). Les couches a Anzalia du Cambrien inférieur du Haut Atlas. *Notes Mem. Serv. Mines Carte géol. Maroc.* **85**, 187–198.

TIMOFEEV, B. V. (1959). The ancient flora of the Baltic region and its stratigraphical significance. *Trudy Vses. neftegaz. Nauchno-issled. geol.-razved. Inst.* **129**, 350 pp. (In Russian.)

VIDAL, G. (1976a). Late Precambrian microfossils from the Visingsö Beds in southern Sweden. *Fossils and Strata*, **9**.

VIDAL, G. (1976b). Late Precambrian acritarchs from the Eleanore Bay Group and Tillite Group in East Greenland. *Grønl. geol. Unders.* **78**.

VOLKOVA, N. A. (1971). Lower Cambrian hystrichosphaerids. *J. Palynol.* **7**, 26–29.

VOLOGDIN, A. G. (1958). Lower Cambrian Foraminifera. *Dokl. Akad. Nauk. SSSR*, **120**, 405–408. (In Russian.)

WADE, M. (1969). Medusae from uppermost Precambrian or Cambrian sandstones, central Australia. *Palaeontology*, **12**, 351–365.

WADE, M. (1972a). Hydrozoa and Scyphozoa and other medusoids from the Precambrian Ediacara fauna, South Australia. *Palaeontology*, **15**, 197–225.

WADE, M. (1972b). *Dickinsonia*: polychaete worms from the late Precambrian Ediacara fauna, South Australia. *Mem. Qd Mus.* **16**, 171–190.

WALCOTT, C. D. (1911). Middle Cambrian annelids. *Smithson. misc. Collns*, **51**, 109–142.

WALCOTT, C. D. (1920). Cambrian geology and paleontology IV. No. 6. Middle Cambrian Spongiae. *Smithson. misc. Collns*, **67**, 261–364.

WALL, D. (1962). Evidence from recent plankton regarding the biological affinities of *Tasmanites* Newton 1875 and *Leiosphaeridia* Eisenack 1958. *Geol. Mag.* **99**, 353–362.

WALTER, M. R. (1972). Stromatolites and the biostratigraphy of the Australian Precambrian and Cambrian. *Spec. Pap. Palaeontology*, **11**, 190 pp.

WALTER, M. R., OEHLER, J. H. and OEHLER, D. Z. (1976). Megascopic algae 1300 million years old from the Belt Supergroup, Montana: a reinterpretation of Walcott's *Helminthoidichnites*. *J. Paleont.* **50**, 872–881.

WEBBY, B. D. (1970). Late Precambrian trace fossils from New South Wales. *Lethaia*, **3**, 79–112.

WIGGETT, G. J. (1977). Late Proterozoic—early Cambrian biostratigraphy correlation and paleoenvironments, White Inyo facies, California–Nevada. *Spec. Rep. Calif. Div. Mines Geol.* **129**, 87–92.

YAVORSKY, V. I. (1963). Stromatoporoids of the SSSR, Part 4. *Vses. Nauchno-issled. Trudy geol. Inst.* **87**, 3–160. (In Russian.)

YANKAUSKAS, T. V. (1972). Cribricyatha of the Lower Cambrian. *In* "Problems of Biostratigraphy and Paleontology in the Lower Cambrian of Siberia" (I. T. Zhuravleva, ed.), pp. 161–183. *Trudy Adad. Nauk. SSSR, Sibir. Otd., Inst. Geol. Geofis., Nauka., Moscow.* (In Russian.)

YOCHELSON, E. L. (1969). Stenothecoida, a proposed new class of Cambrian molluscs. *Lethaia*, **2**, 49–62.

YOCHELSON, E. L. (1975). Discussion of early Cambrian "molluscs". *J. geol. Soc.* **131**, 661–662.

YOCHELSON, E. L. (1977). Agmata, a proposed extinct phylum of early Cambrian age. *J. Paleont.* **51**, 437–454.

YOCHELSON, E. L., HEMINGWAY, G. and GRIFFIN, W. L. (1977). The Early Cambrian genus *Volborthella* in southern Norway. *Norsk. geol. Tidsskr.* **57**, 133–151.

YOUNG, F. G. (1972). Early Cambrian and older trace fossils from the Southern Cordillera of Canada. *Can. J. Earth Sci.* **9**, 1–17.

ZHAMOYDA, A. E. (1968). Survey of research into Radiolaria (1950–1966). *In* "Studies in Science, Geology Series", pp. 109–134. *Viniti Izdat.* (In Russian.)

ZHURAVLEVA, I. T. (1970a). Marine faunas and Lower Cambrian stratigraphy. *Am. J. Sci.* **269**, 417–445.

ZHURAVLEVA, I. T. (1970b). Porifera, Sphinctozoa, Archaeocyathi—their connections. *Symp. Zool. Soc. Lond.* **25**, 41–59.

ZHURAVLEVA, I. T. (1972). Early Cambrian facies complexes of the archaeocyathids. *In* "Problems of Biostratigraphy and Paleontology in the Lower Cambrian of Siberia" (I. T. Zhuravleva, ed.), pp. 31–109. *Akad. Nauk. SSSR, Sibir. Otd., Inst. Geol. Geofis., Nauka, Moscow.* (In Russian.)

ZHURAVLEVA, I. T. (1974). Biology of Archaeocyatha. *In* "Stratigraphical Studies, Jubilee Book in Honour of Academician B. S. Sokolov", pp. 107–129. *Izdat., Nauka, Moscow.* (In Russian.)

ZHURAVLEVA, I. T. (1975). Description of the paleontological characteristics of the Nemakit-Daldyn Horizon and its possible equivalents in the territory of the Siberian Platform. *In* "Equivalents of the Vendian Comlex in Siberia" (B. S. Sokolov and V. V. Khomentovsky, eds), pp. 62–100. *Trudy Akad. Nauk, SSSR, Sibir. Otd., Inst. Geol. Geofis., Nauka, Moscow,* **232**. (In Russian.)

ZIEGLER, B. and RIETSCHEL, S. (1970). Phylogenetic relationships of fossil calcisponges. *Symp. zool. Soc. Lond.* **25**, 23–40.

6 | Early Fossil Cnidarians

COLIN T. SCRUTTON

Department of Geology, University of Newcastle upon Tyne,
England

Abstract: The late Precambrian and early Phanerozoic record of cnidarian or supposed cnidarian fossils is reviewed. The first reliable records are considered to be among the Ediacaran fauna of Vendian age and most cnidarian groups with a significant fossil record had appeared by the Ordovician. The relationships of some important fossil groups of disputed or doubtful affinities are discussed. The Stromatoporata and Chaetetida are considered to be closer to the Porifera than the Cnidaria, in agreement with most recent work. The Conulata are of uncertain relationships but may form a separate class of cnidarians. The Hydroconozoa are also accepted as a provisional cnidarian class. A suggestion that some or all tabulate corals may be more closely related to the Porifera is disputed.

Cambrian coral records are reviewed and some are regarded as very doubtful or not of cnidarian affinities. The relationships of the remainder to the Ordovician records are discussed and possible phylogenetic schemes for the early diversification of the tabulate and rugose corals are outlined. No post-Cambrian direct phyletic link between Tabulata and Rugosa is considered likely but a common ancestor for the two orders may have existed in the late Precambrian or early Cambrian.

Various opinions on the origins and phylogeny of the Cnidaria are briefly outlined. Doubts concerning the completeness of the fossil record and uncertainties over the interpretation of Precambrian material weakens the contribution of the palaeontological evidence to the debate on the nature of the earliest cnidarians.

INTRODUCTION

The Cnidaria appear to have one of the longest fossil histories among the metazoan phyla. Precambrian records are dominantly of supposed medusoids and extend back beyond 2000 m.y. (million years ago). Many

---

Systematics Association Special Volume No. 12, "The Origin of Major Invertebrate Groups", edited by M. R. House, 1979, pp. 161–207. Academic Press, London and New York.

of these early remains, however, are of extremely doubtful validity and a number of them are probably inorganic artefacts (for example Cloud, 1968). Furthermore, the problem of separating artefacts from authentic organic remains is not restricted to the Precambrian (Cloud, 1973 and below). The first well documented and acceptable cnidarians come from rocks in now widely separated locations generally regarded as of latest Precambrian age and assigned to the Vendian (Fig. 1). This is the Ediacaran fauna of Glaessner (1971a). The only indications of metazoan life preceeding the appearance of the Cnidaria are trace fossils (Brasier, chapter 5, this volume) although many of these records are also disputed (Cloud, 1973).

Uncertainty still surrounds the dating of the various late Precambrian fossiliferous localities. Most agree in regarding them as grouped in the Vendian stage of the late Precambrian, although Cloud (1968; 1973) would prefer to see this interval as earliest Palaeozoic. Glaessner (1971a) reviewed the geographic distribution of the Ediacaran fauna and the then available evidence for its age, which indicated a range from 590–700 m.y. More recently, the limits of the Vendian *sensu stricto* (from the base of the Varangian tillites to the base of the Tommotian) have been dated as 570–680 m.y. (Sokolov, 1972b; Cowie and Glaessner, 1975). Several localities yielding an Ediacaran fauna, however, have yet to be satisfactorily dated.

Cnidarians form a high proportion of the Vendian metazoan assemblage, 63% of specimens in the Pound Sandstone of Ediacara, S. Australia according to Glaessner (1972), although some of the species he considers to be cnidarian are of doubtful affinities. Most cnidarian groups with a significant fossil record have appeared by the Ordovician. This review is therefore concerned for the most part with Vendian, Cambrian and Ordovician remains that have been considered cnidarians or of possible cnidarian affinities. Some important fossil groups at one time included in the Cnidaria are now classified elsewhere. The Stromatoporata and Chaetetida are probably poriferans, although recent suggestions that some or all tabulate corals should also be assigned to the Porifera (Flügel, 1976; Stel and de Coo, 1977) are not accepted. The affinities of the Conulata have also been disputed recently and their relationships remain uncertain. I have discussed recent work on these important questions of affinity in the appropriate sections of this review in order to clarify relationships within the phylum as far as possible. Having documented cnidarian records over the Vendian to Ordovician interval, the last part of this paper deals briefly with phylogenetic relationships, among the early tabulate and rugose corals and for the phylum as a whole.

RECORDS OF PRECAMBRIAN AND EARLY PHANEROZOIC CNIDARIANS

1. Unassigned Medusoids

The earliest supposed medusa was recorded by Houghton (1962) as *Gakarusia addisoni* from the early Proterozoic Transvaal Dolomite of South Africa, more than 2224 m.y. old. It consists of a central circular impression surrounded by open ended rays with an overall diameter of 48 mm; there is no outer margin delineated. This is a very doubtful medusa and may well be inorganic, although no alternative interpretation of the structure has been advanced. Houghton compared it to *Protolyella resseri* from the lower Cambrian of New York but this has a distinctly scalloped margin and is more acceptably organic. Edgell's (1964) medusae *incertae sedis* from the Hammersley Group, Western Australia, dated as older than 1700 m.y., are similarly doubtfully organic as are Dunnet's (1965) *Cyclomedusa*-like forms from the younger Proterozoic Ord Group of the Kimberley Region, W. Australia (Cloud, 1968, p. 27). *Beltanella*-like jellyfish have been mentioned from the McArthur Group of northern Australia (approx. 1500 m.y. old) by Dunn (1964) but are unconfirmed.

Bassler (1941) described as *Brooksella cayonensis* a medusaeform structure from the Nankoweap Group of the Grand Canyon Series of Arizona, which is correlated with approximately 1100–1300 m.y. rocks elsewhere in that state. Cloud (1968) considered the structure to be of inorganic origin, probably formed by gas-evasion from the sediments or associated with a gas blister, but Glaessner (1969) has made out a good case for it as a trace fossil of the *Asterosoma* type. Other Grand Canyon jellyfish-like structures described by Alf (1959), and considered to represent impressions of possible cyanophytan algal sheaths according to Glaessner (1966), have been compared by Cloud (1968) to rain-drop impact structures. Another late Proterozoic "Brooksellid" was described by Vologdin (1966) from the eastern Sayan, USSR as *Sajanella arshanica* but from illustrations it has little resemblance to other medusae or indeed to any cnidarian.

The earliest convincing medusae are those described from the Pound Quartzite in the Ediacara Range of South Australia and from closely equivalent horizons in other parts of the world. The remarkable Australian fauna, of uppermost Precambrian (Vendian) age, was discovered by Sprigg (1947, 1949) and has since been made well known by the work of M. F. Glaessner and M. Wade (Glaessner, 1959, etc.; Glaessner and Wade, 1966; Wade, 1968, etc.). Of the many broadly medusaeform fossils in the fauna, some have been assigned to the Hydrozoa and Scyphozoa and will be discussed later. The unassigned medusae include *Cyclomedusa plana*, *Ediacaria flindersi*, *Lorenzinites rarus*, *Mawsonites spriggi*, *Medusinites asteroides*, *Pseudorhizostomites howchini* and *Rugoconites* spp. (Glaessner and Wade, 1966; Wade, 1972). The most valuable preservation for the elucidation of structures in these soft-bodied organisms is composite moulding, showing internal and external characters on the same specimen. Lack of sufficient preservations of this sort of *Cyclomedusa* hinders interpretation of this genus which may be related to the chondrophores (Wade, 1972). *Ediacaria*, which was tentatively assigned to the Trachylinida by Harrington and Moore (1956c) was considered possibly a scyphozoan by Wade (1972): none of its internal structures is known. *Rugoconites* seems to combine hydromedusan and scyphomedusan characteristics but is also possibly scyphozoan (Wade, 1972).

Medusoids of similar age, some regarded as generically or even specifically identical with Ediacaran forms, are widely recorded. *Cyclomedusa* appears to be a commonly recognized form genus reported from the Vendian of the Russian Platform (Zaika-Novatskii *et al.*, 1968; Sokolov, 1972a), N. Sweden (Kulling, 1964 as *Madigania*) and the supposed late Precambrian of South Wales (Cope, 1977). Other Vendian medusoids from the USSR include *Beltanella* (possibly equivalent to *Ediacaria*), *Beltanelloides* (which, however, is probably related to the ?alga *Chuaria*) and *Bronicella* (Sokolov, 1972a, b). *Medusinites*, another Ediacaran genus is also recorded from South Wales (Cope, 1977). *Paramedusium africanum* is recorded from the late Precambrian Nama Group of Namibia (S.W. Africa) (Gürich, 1933) but Wade (1972) considered that the single specimen, lost during the second world war, was probably never generically identifiable. Misra (1969) has also recorded and illustrated unclassified jellyfish, some resembling *Brooksella*, from among a fauna of Ediacaran composition in the late Precambrian Conception Group of Newfoundland.

Scattered records of unassigned medusae, some more acceptable than others, occur throughout the Phanerozoic (for example Harrington and Moore, 1956a, f; Miroshnikov and Kravtsov, 1965) but will not be discussed here. It is clear that medusoids were already well diversified by the end of the Proterozoic.

2. Hydrozoa; Hydromedusae, Chondrophora and Hydroida

There appear to be no acceptable hydroid or trachyline medusae recorded from the Precambrian. Of the South Australian Ediacara fauna, Harrington and Moore (1956c) had considered marginal concentric striae or rugae in *Ediacaria flindersi* and *Beltanella gilesi* to indicate the possible presence of a velum. Sprigg (1947) had compared *B. gilesi* to the living trachyline *Rhopalonema* and Harrington and Moore placed *E. flindersi* in the Trachylinida *incertae sedis*. These two species were regarded as possible synonyms by Glaessner and Wade (1966) and Wade (1972), but they did not support the recognition of a velum and thus left them unassigned at the class level. Stasinska (1960), however, recorded well preserved impressions from the lower Cambrian (middle Cambrian according to Bednarczyk, 1970) of Poland as *Velumbrella czarnockii* which do appear to have a much clearer velum-like structure (Figs 1, 2h). This may be the earliest record of trachyline medusae (but see below). Otherwise the most convincing fossil representatives of this group appear in the later Mesozoic whilst the fossil record of possible hydroid medusae is extremely sparse and very doubtful (Harrington and Moore, 1956d; *Protodipleurosoma* is now considered a synonym of *Ediacaria* and the upper Pennsylvanian *Crucimedusina* is not convincing).

In contrast the more highly evolved Chondrophora has been claimed to be well represented in the late Precambrian Ediacaran fauna (Fig. 1). Wade (1972) has reconstructed *Eoporpita medusa* as a porpitoid chondrophore with an appropriately chambered circular float and a distribution of tentacles on the oral surface comparable to the outer ring of dactylozooids and inner ring of gonozooids of Recent chondrophores (Fig. 2i). Other chondrophoran pneumatophores, bilaterally symmetrical and of different structure, have also been identified in the Ediacaran fauna. These are *Ovatoscutum* and *Chondroplon* which Wade (1971) placed in a new family. Wade (1972) further considered it possible that *Cyclomedusa* could represent a persistent ancestral type of chondrophore, functionally intermediate between normal attached hydrozoans and pelagic forms with a float. Several Palaeozoic records of chondrophores have also been claimed, including the most convincing *Archaeonectris benderi* (Fig. 2f) from the late Ordovician or early Silurian of Jordan (Huckriede, 1967). The widely recorded lower Cambrian genus *Protolyella*, considered medusae *incertae sedis* by Harrington and Moore (1956f), has been referred to the Chondrophora by others (Glaessner, 1971b). *Plectodiscus* (upper Devonian) and

Paleonectris (lower Devonian) are regarded as velellids whereas *Discophyllum* (middle Ordovician) and *Paropsonema* (Silurian-upper Devonian) have been assigned to the Porpitidae (Harrington and Moore, 1956e; *Palaeoscia* (upper Ordovician) has been reinterpreted as a trace fossil by Osgood, 1970). Even *Velumbrella* may possibly belong in this latter family rather than among the trachyline medusae (Glaessner, 1971b). Altogether this represents an astonishingly good record for the early Chondrophora if all of these remains are correctly interpreted.

The chondrophores were considered to be a subdivision of the Siphonophora (Harrington and Moore, 1956e) but it is now more usual to separate the elaborately polymorphic siphonophores as a unit of equivalent rank to the Hydroida(-ea) (order or subclass) within the Hydrozoa, with the Chondrophora recognized as a subdivision of the Hydroida(-ea) (e.g. Bouillon, 1968). A case was made out by Madsen (1956, 1957, 1962) for the interpretation of the middle Cambrian fossil *Eldonia ludwigi* from the Burgess Shale as a siphonophore. Durham (1971), however, has reasserted the echinoderm affinities of this form, Walcott (1911) and others having considered it a holothurian. No other records of fossil siphonophores appear to have been claimed.

Attached and/or benthonic polypoid hydrozoans seem to have no Precambrian fossil record in comparison with their pelagic relatives. The middle Cambrian "calyptoblastine hydroids" of Chapman (1919) and Chapman and Thomas (1936) have been referred to the Graptolithina by Kozlowski (1959) and Sdzuy (1974). Kozlowski, however, accepted Eisenack's genera of Ordovician and Silurian age as hydroids and himself described eleven new genera of hydroids from Ordovician eratic boulders in Poland (Fig. 21). His material, and some additional specimens of the same fauna from Sweden later described by Skevington (1965), was preserved with the scleroprotein of the periderm little altered allowing extraction from calcareous rocks by acid preparation. Apart from these relatively rich Ordovician faunas the hydroid fossil record is very sparse.

3. Hydrozoa; Calcified hydrozoans: affinities of the Stromatoporata and Chaetetida

Hydrozoa with calcareous skeletons are not definitely known before the Mesozoic when millepores, stylasterines and possibly hydractinians first appear. The stromatoporoids, which are first definitely known from the early Ordovician (Fig. 1; Flügel, 1975 regarded the Cambrian records as doubtful but see Zhuravleva, 1970) have long been of disputed systematic position (Lecompte, 1956) but were regarded as hydrozoans by most authors from the 1870s until recently. They have now been shown more convincingly to be related to the sponges (Hartman and Goreau, 1966, 1970), and Stearn (1972, 1975) has proposed that they should constitute a subphylum, the Stromatoporata, of the Porifera.

Sponge affinities for the stromatoporoids have not been universally accepted, although Kazmierczak's (1976) suggestion that they are perm- ineralized colonies of coccoid cyanophytes seems most unlikely (see Riding and Kershaw, 1977). Kazmierczak (1971) had earlier supported hydrozoan affinities for the group through a comparison with skeletal secretion in *Hydractinia*, setting aside the astrorhizae as the traces of symbiotic organisms and therefore not requiring explanation as a structure of the stromatoporoid animal (Kazmierczak, 1969, 1971). Mori (1970) and Stearn (1972, 1975), however, have argued convincingly against this interpretation of astro- rhizae and Kazmierczak (1969) himself had shown them not to be homologous with the hydrorhizae of hydroids like *Hydractinia*. The comparison of astrorhizae with an excurrent canal system such as that of the Sclerosponges seems very much more acceptable (Stearn, 1975). Mori (1976, 1977), however, still regarded some features of stromatoporoids to be better explained by cnidarian affinities and continued to favour either a hydrozoan relationship for them or their separation as a distinct phylum intermediate between the sponges and cnidarians. In particular, he pointed to the presence of tabulated, possibly zooidal, vertical tubes, partitions in the astrorhizae and lack of any definite spicules in stromatoporoids as evidence against sponge affinities. Hartman and Goreau (1975), however, have described a new living sclerosponge, *Acanthochaetetes wellsi*, in which the siliceous spicules in the living tissue are not incorporated into the basal skeleton. This suggests that the absence of spicules from stromatoporoid coenostea is not a strong argument against poriferan affinities. Furthermore, *A. wellsi* has a tabulated, tubular skeleton. On balance poriferan affinities seem more likely than hydrozoan affinities although a final decision on the phylogenetic links of stromatoporoids may be premature.

Hartman and Goreau (1975) assigned their new sclerosponge to a Mesozoic chaetetid genus *Acanthochaetetes* (Fischer, 1970) on the basis of very strong macromorphological and microstructural similarity. *A. wellsi* differed in its spicule free, calcitic lamellar skeleton, from all previously described sclerosponges which have siliceous spicules embedded in arag- onitic trabecular skeletons. They therefore placed *Acanthochaetetes* in a new order of the Sclerospongiae, the Tabulospongida. Mori (1976), however, regarded the living material of *A. wellsi* as distinct from the Mesozoic acanthochaetetids specifically because the latter showed no sign of siliceous spicules. It seems clear, however, from Mori's own illustrations of another Recent tabulosponge that whilst the calcareous surface above the last formed tabula in each vertical tube of the skeleton may be liberally scattered with spicules (Mori, 1976), there appears to be no sign of spicules even lying loose at lower levels in the basal skeleton (Mori, 1977).

Sclerosponge affinities for the acanthochaetetids naturally leads one to consider poriferan connections for the rest of the Palaeozoic and Mesozoic chaetetids. The earliest members of this group previously recorded were from the upper Ordovician but I regard some supposed Cambrian corals as belonging here (see below and Fig. 1). Although the Chaetetida have been considered to be tabulate corals (Hill and Stumm, 1956), Sokolov (1962a) regarded them as hydrozoans rather closely related to the stromatoporoids. Fischer (1970) supported both a direct relationship between Palaeozoic and Mesozoic chaetetids, hitherto disputed, and their hydrozoan affinities. Hartman and Goreau (1975) pointed out that the acanthochaetetids differed from other chaetetids in having a laminar rather than a trabecular microstructure to the skeleton and in possessing spines. Most sclerosponges, however, do have trabecular microstructure and some are aspinose. In addition, and perhaps most importantly, the sclerosponge *Ceratoporella* shows division of the tubes by longitudinal fission (Hartman and Goreau, 1972). This method of "increase" is a primary characteristic of the chaetetids (although they may show intramural and basal "increase" as well in some cases (Fisher, 1970)) and was one of the main lines of evidence cited by Sokolov (1955, 1962a), for the removal of the group from the tabulate corals. It seems therefore that there is nothing in chaetetid morphology and microstructure that is not duplicated among the sclerosponges and Fisher (1977) followed Hartman and Goreau (1972, 1975) in regarding the Chaetetida as an order of the class Sclerospongiae, of equal status with the Tabulospongida and Ceratoporellida.

Alternative classifications among the sponges have been proposed for stromatoporoids and chaetetids (Termier and Termier, 1973, 1977) but it is not intended to discuss these possibilities here. For present purposes it is sufficient to conclude that these groups are unlikely to be directly related to the Cnidaria on current evidence. On the other hand, poriferan affinities are strongly indicated, particularly for the Chaetetida, although the fact that spicules have yet to be convincingly demonstrated as associated with fossils of either group remains a major source of doubt for some workers.

4. *Scyphozoa; Scyphomedusae*

A number of Precambrian medusoids have been assigned more or less emphatically to the Scyphozoa (Fig. 1). These are almost all forms from Australia, including members of the Ediacara fauna and a very slightly younger fauna from central Australia (Wade, 1969, 1970, 1972). Of these records, *Kimberella quadrata* from the Pound Quartzite most recalls modern medusae, particularly those of the order Carybdeida or Cubomedusae including the painfully stinging sea-wasps (Fig. 2j). Wade (1972) did not regard *Kimberella* as a member of this order from which it differs in the reconstruction of unpaired gonads projecting into the lumen of the bell. Assignment to the Scyphozoa, however, is based on the recognition of gastric filaments similar in form to those of modern Carybdeida. *Brachina delicata*, also from the Ediacaran fauna, is quite different from *Kimberella* and lacks any obvious tetramerous symmetry. In this it is similar to two medusoids from the latest Precambrian of central Australia, *Hallidaya brueri* and *Skinnera brooksi* (Wade, 1969). These three forms have rather different gastrovascular systems but in each case the system is interpreted as of scyphozoan grade. The gastrovascular and gonad systems of *Ediacaria flindersi* are so far unknown but this form may eventually prove to be close to *Brachina delicata* (Wade, 1972). In addition *Rugoconites*, represented by two species, *R. enigmaticus* and *R. tenuirugosus*, has been provisionally restored as a scyphozoan by Wade (1972).

Phanerozoic medusae assigned to the Scyphozoa are revised by Harrington and Moore (1956b). These are mostly rather tentatively or doubtfully classified here with the few more convincing records coming from the upper Jurassic Solenhofen Limestone. Of the claimed early Palaeozoic scyphomedusae, *Camptostroma roddyi* from the lower Cambrian of Pennsylvania has been referred to the Echinodermata (Durham, 1971) and other species assigned to the genus are medusoids of indeterminate affinities (e.g. Popov, 1967). The scyphomedusa of Van der Meer Mohr and Okulitch (1967) from the lower Cambrian of Spain is inadequately known and *Peytoia nathorsti* from the middle Cambrian Burgess Shale, described as a rhizostomatid scyphomedusa by Walcott (1911), was removed to the medusae *incertae sedis* by Harrington and Moore (1956f) (see also Conway Morris, 1978).

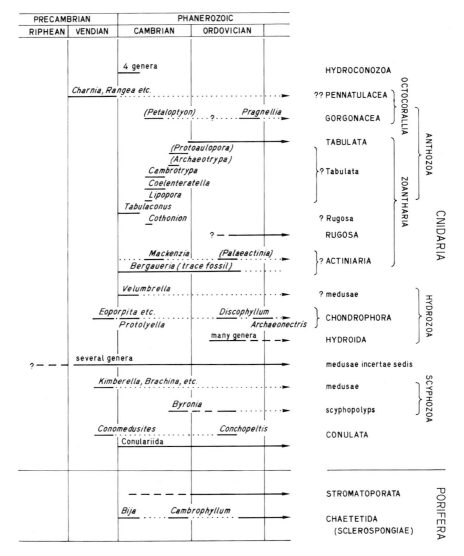

FIG. 1. Precambrian and early Phanerozoic records of fossils currently or recently assigned to the Cnidaria. For discussion see text.

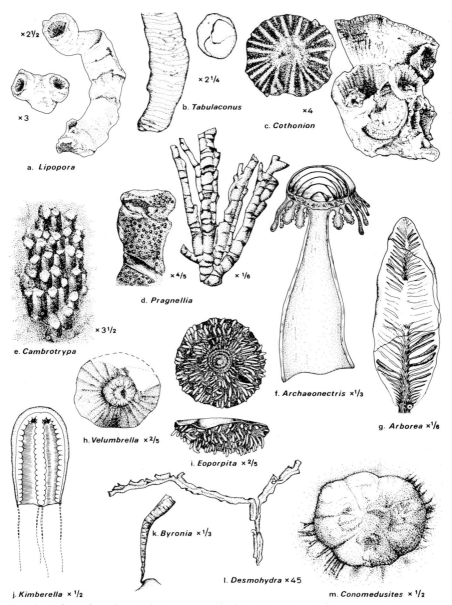

×2½

×3

a. *Lipopora*

×2¼

b. *Tabulaconus*

×4

c. *Cothonion*

×4/5   ×⅙

d. *Pragnellia*

×3½

e. *Cambrotrypa*

f. *Archaeonectris* ×⅓

g. *Arborea* ×⅙

h. *Velumbrella* ×²/₅

i. *Eoporpita* ×²/₅

j. *Kimberella* ×½

k. *Byronia* ×⅓

l. *Desmohydra* ×45

m. *Conomedusites* ×½

FIG. 2. Selected early cnidarians recorded in Fig. 1 and discussed in the text.
Redrawn from or based on the following sources:
a. *Lipopora lissa* (Jell and Jell, 1976; figs 11a, g). b. *Tabulaconus kordeae*
(Handfield, 1969; pl. 1, figs 2, 5). c. *Cothonion sympomatum* (Jell and Jell,

5. Scyphozoa; Scyphopolyps and the affinities of the Conulata

Since the work of Kiderlen (1937), the chitinophosphatic Conulata have been considered by most palaeontologists to be related to the Scyphozoa. The interpretation and affinities of this group have been discussed extensively recently (Moore and Harrington, 1956a; Chapman, 1966; Werner, 1966, 1967a, b; Kozlowski, 1968 and Glaessner, 1971b) and although the majority support scyphozoan affinities, Kozlowski has taken a more critical line. The Conulata of most authors comprises two major subdivisions, the first based on the Ordovician genus *Conchopeltis* and the second including the M. Cambrian to Trias conulariids *sensu stricto*. These were recognized as the suborders Conchopeltina and Conulariina respectively by Moore and Harrington (1956b) but were given ordinal status by Glaessner (1971b). The Ediacaran species *Conomedusites lobatus* (Fig. 2m) has now been interpreted as related to *Conchopeltis* (Glaessner and Wade, 1966; Glaessner, 1971b) thus extending the range of the Conchopeltida into the late Precambrian (Fig. 1). *Conomedusites* and *Conchopeltis* are extremely similar in form, both showing strong tetrameral symmetry and the preservation of marginal tentacles (Glaessner, 1971b; Moore and Harrington, 1956a). They differ mainly in the possession of a thin, distinctly sculptured but apparently unmineralized periderm in *Conchopeltis*. Glaessner (1971b) recorded that *Conomedusites* is invariably found with the apex of the broadly conical body orientated downwards in the rock and argued that this represented the growth position.

A major difference of opinion concerns the relationship of the Conchopeltida to the Conulariida. Kozlowski (1968) has argued that there is no link between the two groups, pointing out that the Conulariida are all narrowly conical with chitinophosphatic tests characterized by tuberculated ribs. The tubercules are perforated near the peristome but become progressively and completely sealed towards the apex of the test.

Fig. 2. (cont.)

1976; figs 2, 9g). d. *Pragnellia arborescens* (Leith, 1952; pl. 116, figs 1, 4). e. *Cambrotrypa montanensis* (Bolton and Copeland, 1963; pl. 143, fig. 3). f. *Archaeonectris benderi* (Huckriede, 1967; fig. 1). g. *Arborea arborea* (Glaessner and Wade, 1966; fig. 2). h. *Velumbrella czarnockii* (Stasinska, 1960; fig. 1). i. *Eoporpita medusa* (Wade, 1972; pl. 41, fig. 6). j. *Kimberella quadrata* (Wade, 1972; fig. 6a). k. *Byronia annulata* (Kozlowski, 1967; fig. 3b). l. *Desmohydra flexuosa* (Ordovician hydroid) (Kozlowski, 1959; fig. 8a). m. *Conomedusites lobatus* (Glaessner, 1971b; pl. 1, fig. 1).

He regarded the perforations, termed choanophymes, as probably related to the presence of sense organs and concluded that the test was secreted internally. Kozlowski considered that the tetrameral symmetry was the only strong reason in favour of scyphozoan affinities. He pointed out that no traces of tentacles have been found associated with fossil conulariids, in contrast to the conchopeltids, and dismissed the similarity between the four branching septa of the middle Silurian *Eoconularia loculata* and the endodermal infolding observed in such scyphomedusae as *Craterolophus* and *Haliclystus* as in the former, the only known case, the structure is mineralized skeleton whereas in the latter it is soft tissue only.

Kozlowski concluded that the Conulariida represented a group of organisms completely distinct from all other phyla. This is in striking contrast to the views of Chapman (1966) and Werner (1966, 1967a, b) who regarded conulariids as directly ancestral to the scyphozoan scyphistoma. Werner compared conulariids with the living coronatid scyphozoan *Stephanoscyphus*, which possesses a chitinous tube. The homology is poor, however, as the latter is circular in cross-section and is unphosphatized. Furthermore, Kozlowski (1967, 1968) has pointed out that the rare, chitinous, early Palaeozoic fossil *Byronia* (Fig. 2k) agrees much more closely in form with *Stephanoscyphus*. *Byronia*, which is currently known by two species from the upper Cambrian of British Columbia and the middle Ordovician of Poland, seems very reasonably interpreted as the earliest likely scyphopolyp and there are no known intermediates between it and contemporary conulariids, or for that matter, with the earlier conchopeltids.

Glaessner (1971b) argued against Kozlowski's conclusions to the extent of maintaining the conulariids as cnidarians although less closely related to the Scyphozoa than advocated by Chapman and Werner. He considered both the Conulariida and Conchopeltida to be essentially polypoid and not transitional to medusae despite Kozlowski's (1968) suggestion of medusoid affinities for the conchopeltids and earlier views that conulariids became free-living, floating with the oral end downwards, in adult life (Kiderlen, 1937; Moore and Harrington, 1956a). A comparison with the mode of life of the living, sessile Stauromedusae, however, seems not inappropriate for the Conulariida if these are indeed cnidarians. The more convincingly cnidarian Conchopeltida may perhaps have been little more than primitive medusae resting on a slightly modified exumbrella surface. It is difficult to dismiss any likelihood of cnidarian connections in view of the various tetrameral structures shown by the Conulariida but direct phylogenetic links between the Conchopeltida on the one hand, and the Conulariida and *Byronia* on the other as advocated by Glaessner (1971b) seem doubtful in the light of the available evidence. A pre-Ediacaran common ancestry might be more likely and I have retained the Conulata as a class *sensu* Glaessner (1971b) although conchopeltids and conulariids may ultimately prove to be less closely related.

Finally, Bischoff (1973) has suggested that the conodonts were associated with conulariids and were therefore cnidarians. Certainly he presents evidence to suggest that conodont-like simple cones may have capped spinose processes comparable to those arranged along the transverse ribs of the conulariid test. Cones have yet to be found attached to conulariids, however, or to the "plates" with the appearance of fragments of conulariid test. Furthermore, convincing confirmation of an association between the cones and conulariids would to my mind reinforce Kozlowski's (1968) arguments for considering the Conulariida (but not the Conchopeltida) as distinct from the Cnidaria, rather than drawing the conodonts into that phylum. The claim that all conodonts were related to the Conulata, however, seems even less well founded and reconstructions of the conodont-bearing animal as a lophophorate (Lindstrom, 1974; Conway Morris, 1976) are more convincing.

6. *Hydroconozoa*

Korde (1963) described a new group of organisms from the Lower Cambrian of the asiatic USSR as an independent class of cnidarians, the Hydroconozoa. They are small, up to 1·5 cm height, conical or cylindrical, solitary, attached organisms. There is a "calicular" depression in the upper part of the skeleton which presumably housed the soft-parts. The skeleton is laminar or structureless, often much thickened, and may contain radiating plates somewhat like septa, basal layers thought to resemble tabulae and frequently a series of longitudinal and radial canals in the axial area. Korde (1963) did not mention wall composition although Handfield (1969) recorded granular calcite in one of the three recognized families, the Gastroconidae.

Korde (1963) regarded the Hydroconozoa as combining rugose coral and scyphozoan characteristics but occupying a position distinct from both groups. Handfield (1969) described a new species *Tabulaconus kordeae* (Fig. 2b) from the lower Cambrian of western Canada as a member of the Gastroconidae and suggested that this family should be removed from the

Hydroconozoa and assigned to the Anthozoa. His material certainly has a strongly coralline appearance but its assignment to the Gastroconidae seems very doubtful. New material shows the species to be a colonial, fasciculate form (Françoise Debrenne, personal communication). At the moment, the Hydroconozoa is best retained intact and excluding *Tabulaconus* whose affinities are discussed further below. Zhuravleva (1970) saw no direct evidence favouring assignment of the Hydroconozoa to the Cnidaria although Jell and Jell (1976) regarded them as "extremely similar to anthozoans". Cnidarian affinities seem likely and the classification proposed by Korde (1963) seems reasonable as a provisional measure. More recently Korde (1975) claimed that hydroconozoan development could be traced up to Jurassic times although no elaboration of this statement was given and in view of a reiteration of a lower Cambrian range for the class in the same paper is of uncertain meaning.

7. *Anthozoa; Octocorallia and non-calcifying Zoantharia*

The early fossil record of non-coralline anthozoans is extremely sparse. Ceriantipatharians are not known fossil before the Miocene. Octocorallia have doubtful Precambrian and Cambrian representatives, one definite Ordovician record but are thereafter not recorded until the Mesozoic, except possibly for the curious Permian form *Trachypsammia* (Montanaro-Gallitelli, 1956; but see Hill, 1960, who regarded this as possibly an encrusted *Cladochonus*). The Zoantharia, within which subclass the bulk of the calcareous, coralline anthozoans are classified, have only two possible soft-body fossil representatives known to me.

Discounting Haughton's (1962) polyp-like mould in an algal mat from the Transvaal Dolomite which is doubtfully of animal origin (Glaessner, 1966), the earliest claimed soft fossil anthozoans are the genera *Rangea*, *Pteridinium*, *Charnia*, *Arborea* (Fig. 2g) and *Glaessnerina*, all of late Precambrian, Vendian age, one or more of which are now known from South Australia (Glaessner and Wade, 1966), Namibia (Gürich, 1930a, b; Glaessner, 1963; Pflug, 1970a and other papers; Germs, 1973), |England (Ford, 1958, 1963), N. Siberia (Glaessner, 1966) and S.E. Newfoundland (Misra, 1969). Richter (1955) had considered *Rangea* and *Pteridinium* to be Gorgonacea whilst Ford (1958), who pointed out the possible relationship between *Charnia* and the discoidal *Charniodiscus*, originally thought the former most likely to be an algal frond with the latter the basal part of the same organism. The currently popular interpretation of this group of fossils as pennatulaceans is due to Glaessner (1959) and the comparison between some of the Precambrian forms and some modern sea pens is superficially extremely good. Among recent forms *Pteroeides* and *Pennatula* show particular similarities to the fossil material (see Kükenthal and Broch, 1913) and comparisons have also been made with *Renilla*. Glaessner and Wade (1966) interpreted grooves on *Arborea arborea* as spicular impressions. No modern sea pens have basal discs but it is possible that the apparent association of *Charnia* and *Charniodiscus* is fortuitous, with the latter interpreted as a medusa. Glaessner (1959) suggested, however, that if they are related in a single organism this need not necessarily rule out pennatulacean affinities for *Charnia*.

Recently, an alternative interpretation of the *Rangea* group has been advanced by Pflug (1970a, b; 1972, 1974a, b) who described silicified material from the Nama Group of Namibia as the remains of a new primitive phylum, the Petalonamae. As well as the five genera discussed above, the phylum includes other form types which show no similarity to cnidarians. The Petalonamae were considered to be related to other "petalo-organisms" which together combine "metaphytan" as well as metazoan characteristics representing an evolutionary stage preceeding the separation of these two groups (Pflug, 1974a). A full discussion of Pflug's work is beyond the scope of this paper. As far as the supposed pennatu-laceans are concerned. Pflug's material does seem to be rather better preserved than most earlier material (see for example Pflug, 1970b) but it is very difficult to relate his sketches of internal structures to the plate illustrations and his interpretations are unconvincing. It is not clear, for example, how preservation may have affected the appearance of the reconstructed body cavities, although the arrangement of these proposed by Pflug is not totally dissimilar to that seen in pennatulaceans. The mode of life he suggested, with the "leaf" flat on the sea floor or partially buried in sediment is also found in some living sea pens. On the other hand, the lateral branches in *Rangea* and particularly *Pteridinium* are nowhere near so convincing as polyp leaves as in Glaessner and Wade's (1966) recon-struction of the much more poorly preserved *Arborea* for example. Glaessner and Walter (1975), who have also discussed Pflug's work, now regard this group of fossils provisionally as "Coelenterata of uncertain systematic position". Cnidarian affinities, indeed even pennatulacean links for some of these forms, cannot be entirely ruled out but it seems equally probable, notwithstanding criticisms of Pflug's work, that they represent an entirely separate and ultimately unsuccessful early metazoan group of similar grade with no living or Phanerozoic descendants.

Two other early octocorals, both gorgonians, have been claimed (Fig. 1). The first is *Petaloptyon danei* from the middle Cambrian of western Canada interpreted by Raymond (1931) as a probable gorgonian. Raymond described possible spicules associated with a fan-shaped network of presumed chitinous material but the interpretation is uncertain and the record needs re-examination. The second example seems much more convincing. *Pragnellia arborescens* (Fig. 2d) from the upper Ordovician of Manitoba was described by Leith (1952) as related to the heliolitid *Protaraea* but Hill (1960) has argued convincingly for its interpretation as a member of the Gorgonacea.

The two possible early actinian body fossils are also in need of re-examination. The earliest is *Mackenzia costalis* from the middle Cambrian Burgess Shale, originally described as a holothurian but now generally regarded as an actinian (Clark, 1913; Wells and Hill, 1956; Durham, 1971). *Palaeactinia halli* from the middle Ordovician of New York is also a possible actinian (Wells and Hill, 1956) but *Palaeactis*, which they doubtfully refered here, may be a synonym of *Bergaueria*, which is discussed below. The record of anemone body fossils is thus much poorer than that of medusoids although it seems reasonable to suppose from the record of skeletonized zoantharians that soft anemones may have been well diversified by the early Palaeozoic. Further evidence of the presence of anemones, however, may be provided by the trace fossil *Bergaueria*, found in the lower Cambrian to middle Ordovician of North America and Europe and interpreted as an actinian dwelling burrow (Arai and McGugan, 1968, 1969; Alpert, 1973). The structures, which show radio-bilateral symmetry and occur in clusters, have been compared with the form of the contracted physa of *Peachia hastata* and the traces in sediment left by *Phyllactis conguilega*. The burrows of modern ceriantipatharians such as *Cerianthus*, however, are quite unlike *Bergaueria* (Frey, 1970; Alpert, 1973). *Conostichus*, from the Devonian to lower Permian of various localities including North America, may also represent the dwelling burrows of animals with actinian affinities (Chamberlain, 1971; Alpert, 1973) although these structures have received a variety of different earlier interpretations. Finally, it is a possibility, but one unlikely ever to be resolved, that some supposed medusae may be the impressions of anemone pedal discs.

8. Anthozoa; Pre-Ordovician Coralline Zoantharians

There is a surprising number of skeletal organisms of pre-Ordovician age that have been assigned to the Zoantharia. These are of particular interest for the light they may throw on the early evolution of the rugose and tabulate corals that become common and widespread as fossils during the Ordovician. Some of these records are of doubtful affinities, however, and few can be accepted with any confidence as true corals (Fig. 1).

The most doubtful of all is *Aseptalia ukrainika* from a limestone pebble reputedly more than 2000 m.y. old from the Ukraine for which Vologdin and Strygin (1969) erected a new anthozoan family, the Aseptaliidae. Although the published sections appear to show a horn shaped object, only 0·3 mm maximum width and associated with single celled algae, there is no convincing reason for considering it to be a coral. Cambrian records of doubtful validity include members of Sedlak's (1977) new order Corallicyathida from lower Cambrian quartzites in Poland. He described two genera, *Heliomiria* which he regarded as resembling archaeocyathids and *Silimorpha* more similar to the Rugosa. In fact both consist of radiating silicified ridges lacking any sign of a peripheral wall and are very similar to each other. Their proposed relationships are extremely doubtful and an inorganic origin seems quite probable. Yochelson and Herrera (1974) described conical calcareous fossils from the lower Cambrian of Argentina as *Cloudina? borrelloi*. *Cloudina* was originally described by Germs (1972) as a cribricyathid, a class he doubtfully assigned to the archaeocyathids but considered possibly related to the serpulids. The latter view was supported by Yochelson and Herrera and confirmed by Glaessner (1976). Jell and Jell (1976), however, considered *C.? borrelloi* more like their middle Cambrian coral *Cothonion sympomatum* than other *Cloudina*. However, the resemblance, as they themselves noted, is superficial and *C.? borrelloi* is best regarded as of uncertain affinity at the moment. Rusconi has recorded several Cambrian and early Ordovician corals from the Argentine none of which is acceptable. *Mendoconularia lasherensis* (Rusconi, 1951; 1952), from the upper Cambrian and assigned to the Rugosa, is poorly illustrated and unconvincing as a coral, whilst two lower Ordovician species of the new genus *Challapora* (Rusconi, 1949, 1950a, b) are doubtfully organic. The late lower Ordovician *Favosites pichanensis* (Rusconi, 1955) has apparently never been figured and the supposed middle Cambrian rugose coral *Mendopora difusa* (Rusconi, 1948) seems to lack both description and illustrations and is a *nomen nudum*. *Archaeotrypa*, from the early upper

Cambrian of western Canada, was described by Fritz (1947, 1948) as bryozoan but allowing the possibility that it might be a microscopic coral. It has a cerioid form with well developed tabulae. This possibility has been subsequently restated by Fritz and Howell (1955, 1959) and Bondarenko (1966), but the very small size of the "corallites", *c.* 0·3 mm diameter, and the distinctly zigzag walls in longitudinal section tend more to support Fritz's original assignment. It is difficult to judge, however, how significant diameter is in these Cambrian forms. *Protoaulopora ramosa*, first recorded from the upper Cambrian of the Chingiz Range as *Syringopora* by Vologdin (1931), has assumed great significance in coral phylogenies proposed by Russian authors (Sokolov, 1955, 1962b; Ivanovskii, 1965; Bondarenko, 1966) who regard the auloporids as ancestral to some or all tabulate and rugose corals. Sokolov (1962b) classified the genus in the family Auloporidae, but the species seems very poorly known and from Vologdin's (1931) original description which recorded tube diameters of 0·1 mm, its identity as a coral is uncertain. Vologdin's (1939) record of *Pachypora* from the middle Cambrian of the Urals is also questionably coralline and the organism is of a size range suggesting bryozoan rather than coral affinities. *Cambrophyllum problematicum* from the upper Cambrian of Montana (and possibly the lower Cambrian of the Salair (Sokolov, 1962b)) was thought likely by Fritz and Howell (1955) to be a "schizo-coral". *Cambrophyllum* is very similar to *Bija*, recorded originally by Vologdin (1932) from the Altai and the Western Sayans and of lower Cambrian age (Bondarenko, 1966; Zhuravleva, 1970). *Bija* was tentatively referred to the tabulate coral family Lichenariidae by Sokolov (1962b) but what is known of their meandrine cross-sectional appearance and very small corallite size suggests that these two genera represent primitive chaetetids and should therefore be removed from the Cnidaria altogether. Their lack of tabulae is a feature found rarely in both chaetetids and early tabulae corals.

Although many Cambrian records are either not corals or are doubtful to varying degrees, there remains a small group of more convincingly coralline Cambrian fossils. *Tabulaconus kordeae* from the lower Cambrian of western Canada, which Handfield (1969) considered a hydroconozoan, has a very coralline appearance and is provisionally accepted as a zoantharian (see above). *Cambrotrypa montanensis* Fritz and Howell (1959) has rounded to subpolygonal tubes, averaging about 1 mm in diameter with a subcerioid arrangement (Fig. 2e). They reported no internal structures but small irregular outgrowths apparently link adjacent tubes when they are narrowly separated and the tube walls have an appearance reminiscent of epitheca. Bolton and Copeland (1963), who described further material, did not confirm the lateral processes but noted possible tube increase and suggested that traces of internal tablae were preserved. Their figures clearly show the adnate growth form. So far no information on wall structure and composition is available which makes this organism, from the middle Cambrian of Montana, Alberta and British Columbia, somewhat difficult to evaluate. Fritz and Howell (1959) dismissed bryozoan affinities and considered tubicolar worms only a remote possibility. Of possible coral or algal relationships they favoured the former as did Bolton and Copeland (1963) and if the skeleton is, or was, calcitic, *Cambrotrypa* would seem to be a likely Cambrian tabulate coral. Even more convincing is the solitary and colonial operculate species *Cothonion sympomatum* (Fig. 2c) described by Jell and Jell (1976) from the early middle Cambrian of New South Wales, Australia. The small conical corallites, up to 10 mm diameter, lack horizontal elements but are weakly septate, whereas the undersurface of the operculum is always strongly septate. These features suggest rugosan affinities although the plan of septal insertion cannot be demonstrated as rugosan. Also the radiobilateral septal distribution on the operculum is unlike that of other operculate rugosans and presents problems in the reconstruction of the soft tissues. Jell and Jell placed *Cothonion* in a new family, the Cothoniidae tentatively referred to the Rugosa. They also described two species of a new genus and family of possible tabulate corals from the same locality. *Lipopora lissa* and *L. daseia* are small colonial corals with long cylindrical corallites 1–3 mm diameter and internal structures restricted to 8–16 short ridge-like septa (Fig. 2a). In addition they assigned *Coelenteratella antiqua*, from the middle Cambrian of the Siberian Platform (Korde, 1959), to their new family, the Lipoporidae, and suggested a general similarity between their material and the auloporids among the Tabulata.

It is interesting to note how recent are many of these discoveries of putative Cambrian corals. An unpublished find is also known to me although it has yet to be properly described and thus no further details will be given here. Clearly there is still scope for further discoveries and a more extensive and well founded Cambrian coral record may be realized in the future.

9. Anthozoa; Tabulate corals

(*a*) *The affinities of tabulate corals,* The Tabulata have had a chequered history as a systematic entity (see for example Hill and Stumm, 1956; Sokolov, 1962b). A full discussion of their classification is outside the scope of this paper but it is necessary to establish the sense in which the order is used and its cnidarian status.

The Chaetetida, classified here by Hill and Stumm (1956), have already been discussed and concluded to have poriferan affinities. The interest generated by the recent rediscovery and further research on the sclerosponges, however, has led to a much broader consideration of tabulate coral affinities. Flügel (1976) has revived Kirkpatrick's (1912) original suggestion, made after the discovery of the sclerosponge *Merlia*, that some tabulate corals, including *Favosites*, might be sponges. The possibility was also discussed by Hartman and Goreau (1975) but considered unlikely. Stel and de Coo (1977) on the other hand considered all tabulate corals and possible also the Heliolitida to be sponges.

In his discussion, Flügel suggested that the Favositidae show little positive evidence for their classification as zoantharians. Their non-trabeculate septa, lack of solitary growth-forms and possession of mural pores are all unmatched among rugose and scleractinian corals. On the other hand he demonstrated the general similarity between the morpho-logy of the favositids and, in particular, the tabulosponges but otherwise was able to show no stronger evidence for poriferan than for zoantharian affinities. Of the various lines of argument, that of microstructure has to be treated with reserve. There is still considerable debate as to which are primary and which are secondary structures as concluded by both Hartman and Goreau (1975) and Flügel (1976). Fibrous "septal" spines, however, are known in some Rugosa, such as the cystimorphs. Mural pores, despite Flügel's (1976) ingenious analogy between them and the sclerosponge astrorhizae, are equally unknown among the sponges as among other corals. Lack otherwise of astrorhizal structures in favositids, and lack of spicules, cannot be seen as pointing either way (see discussion of the chaetetids). The Silurian genus *Nodulipora*, however, which has been considered a favositid in the past but possesses astrorhizae-like structures and differs from typical favositids in other details, should be removed from that group and assigned to the Porifera (Hartman and Goreau, 1975). On balance, I regard the evidence as favouring retention of the favositids as corals. The size and variation in size range of favositid corallites are both much greater than those in known sclerosponges but comparable to those in other colonial corals (although see Hartman and Goreau, 1975, for theoretical size range in sclerosponges). True lateral or peripheral increase occurs in Favositidae (Swann, 1947), and most cases of supposed intermural increase without apparent parental connection prove to be lateral increase with careful study (Oliver, 1968; Fedorowski and Jull, 1976; unpublished serial sectioning by J. H. Powell and myself). This method of increase is common in corals but unknown among sclerosponges. Finally the presence of intermural spaces (Swann, 1947; Ross, 1953) and subcerioid growth forms (Philip, 1960) argues strongly for calicular deposition by more or less individualized polyps rather than colony formation by an enveloping soft body of poriferan type. This evidence strengthens that suggested by the presence of commensals within and particularly between corallites of some favositids (Oekentorp, 1969; Stel, 1976) which is not of itself overwhelmingly strong evidence for individuality (Flügel, 1977). An alternative to either close sponge or coral affinities is a position for the favositids as a separate entity. The only structures that really distinguish this group strongly from other corals are the mural pores. Most workers, however, regard these as evolved by favositids from non–porous tabulate coral ancestors, representing perhaps a more highly integrated colonial state and analogous to the wall spaces of syringophyllids and the interconnecting tubules of the syringoporids. There seems to be no compelling reason at the moment therefore to remove this group from the Zoantharia. Sponge affinities seem even less likely for most other tabulates and heliolitids as these include growth forms and colonial organization completely unknown among the sclerosponges.

I therefore retain the Tabulata, including the favositids, as an order of zoantharians and, as will be clear from the discussion of their early phylogeny, I regard the group as a homogeneous entity, incorporating the heliolitids as the suborder Heliolitina.

(*a*) *Ordovician diversification of tabulate corals.* The earliest of the generally accepted tabulate coral genera is *Lichenaria*, represented first by two species from the Canadian (Tremadoc and Arenig) of the USA (Fig. 3). These cerioid colonies have very small corallites, down to 0·5 mm diameter, and tabulae are weakly developed or absent. These species have been tentatively referred to *Cryptolichenaria* by Sokolov (1955), a genus later assigned to the Tetradiina (Sokolov, 1962b, as Tetradiida), but Bassler's (1950) figures and descriptions give no indication of septal development and pending redescription the species are accepted as the earliest members of the Lichenariina (unless *Cambrotrypa* belongs here). *Cryptolichenaria* is, however, recorded from approximately contemporary horizons in Siberia (Kaljo and Klaaman, 1973). Sokolov (1962b) also recorded *Aulopora* from the lower Ordovician of Siberia. The first main diversification of tabulate corals, however, occurred in the middle Ordovician, in the American Chazy and Blackriverian faunas and their equivalents elsewhere (Fig. 3). The pattern of first appearances depends upon the currently uncertain subdivision of the American middle Ordovician and the accuracy of intercontinental correlation (Sweet and Bergström, 1976; Cooper, 1976; Williams *et al.*, 1976) and may require modification in the light of revision and new records. The scheme of Sweet and Bergström (1976) is followed here. New lichenariid and several syringophyllid genera appear in the Chazy Stage and *Tetradium* is also prominent. Blackriverian faunas of early Caradoc age yield the first halysitid *Quepora* and the first syringoporid *Labyrinthites*. Later in the Caradoc many more genera appear for the first time so that by the end of that series representatives of all eight suborders of tabulates (orders of Sokolov, 1962b) including the Heliolitina (subclass of Sokolov, 1962c) are known among some 30 genera. Kaljo and Klaamann (1973) listed 60 (and recorded about 70) genera of tabulates and heliolitids in the late Ordovician, among which the lichenariids, syringophyllids, tetradiids and heliolitids are particularly strongly diversified.

10. *Anthozoa; Rugosa*

The Ordovician diversification of the Rugosa occurred slightly later and was less spectacular than that of the Tabulata (Fig. 5). The earliest record appears to be of *Lambeophyllum* cf. *profundum* from the Chazyan (Welby, 1961) which needs verification (Webby, 1971). The lower Caradoc (Blackriverian fauna and correlatives), however, contains a small suite of rugosans representing the three major suborders of Hill's (1956) classification, *Lambeophyllum* and *Streptelasma* (Streptelasmatina), *Hillophyllum* and *Primitophyllum* (Cystiphyllina), *Favistina* and *Paleophyllum* (Columnariina) (Fig. 5). Slightly later, in the middle Caradoc, a further 13 genera are recorded, most of them streptelasmatids and including several additional representatives of a proposed new superfamily of this group, the Calostylaceae (Weyer, 1973) and the only Ordovician representative of the Cyathaxoniaceae, *Protozaphrentis*. Webby (1971), Weyer (1973) and

Neuman (1977) have attempted to show the order of generic appearances with considerable precision. Two new genera from the higher middle Ordovician of China, *Yohophyllum* and *Ningnanophyllum* (Lin, 1965) are unevaluated and are not included here. By the end of the Ordovician only some 30 genera of rugosans had appeared, dominated by solitary members of the Streptelasmataceae.

The relationships of early tabulate coral genera have been considered by a number of authors (Bassler, 1950; Sokolov, 1955, 1962b; Flower, 1961; Flower and Duncan, 1975; Bondarenko, 1966, for example) but little consensus of opinion exists.

Links between the possible Cambrian corals and the tabulates of the early and middle Ordovician are obscure. Sokolov (1962b) regarded *Cambrophyllum*, *Bija* and *Protoaulopora* as probable predecessors of the tabulates because of their structural similarity to the early Ordovician *Lichenaria*, *Cryptolichenaria* and *Aulopora*. *Bija* and *Cambrophyllum* I regard as chaetetids. *Protoaulopora* is problematical and early Ordovician records of *Aulopora* need verification. The presence of septal ridges in *Lipopora* suggests that it is less likely to be directly ancestral to the aseptate early Ordovician corals, but *Coelenteratella* may be aseptate (vague longitudinal ridges in Korde's (1959) figures are possibly septal). *Cambrotrypa* is insufficiently well known but could prove to be an important link in early tabulate evolution. If a relationship exists, then an aseptate lipoporid could be envisaged as ancestral to the auloporids on the one hand and to the lichenarids through a *Cambrotrypta*-like form on the other. *Lipopora* itself may be a specialized dead-end branch. However, this sequence is based on extremely poor evidence and is highly speculative. The Cambrian species could be early attempts to evolve supporting skeletons by side stocks with no descendants and Ordovician tabulates may represent skeleton acquisition by previously soft-bodied zoantharians.

The relationships first discussed above would agree with Sokolov's (1962b) idea of a pre-Ordovician separation of the lichenariids and auloporids. Flower and Duncan (1975), however, presupposed the antiquity of *Lichenaria* (Fig. 4e) and derive all Ordovician tabulates (excepting the tetradiids) from this basic stock. They considered the Auloporidae and incidentally the Syringoporidae to be descended from *Lichenaria* through *Eofletcheria*, a possibility also mentioned by Hill (1953). *Eofletcheria* (Fig.

4j), however, has a wall composed of thick, monacanthine trabeculae in better preserved material, although this is not clear in the type (Sinclair, 1961), suggesting a relationship with *Lyopora* and the sarcinulids. This was also suggested in part by Hill (1953). Hill and Stumm (1956) classified *Eofletcheria* as a syringoporid but Sokolov (1962b) placed it in the Lichenariina. I broadly agree with Sokolov (1962b) in terms of the related genera but the group including *Eofletcheria, Lyopora, Baikitolites, Reuschia, Nyctopora* and *Billingsaria*, I would separate from *Lichenaria* and place with the sarcinulids as the Syringophyllina (similar but not exactly equivalent to the Syringophyllidae of Hill and Stumm, 1956) thus uniting the group that develop thick septal trabeculae tending to coalesce in a peripheral stereozone (see also below). Returning to the Auloporina, I see no evidence for their descent from a lichenariid ancestor and I favour the pre-Ordovician separation of these two groups (Fig. 3). On the other hand, although it may seem superficially logical to derive the Syringoporina from an auloporid ancestor or a close common ancestor, the earliest member of this suborder appears to be *Labyrinthites* (= *Tetraporella*, see Bolton, 1965) and I agree with Sokolov (1962b) that in this case a lichenariid ancestor does seem likely (Figs 3, 4h).

Lichenariids are undoubtedly the starting point for much tabulate coral diversification (Fig. 3) as most workers agree. Bassler (1950), Flower (1961) and Flower and Duncan (1975) regarded *Saffordophyllum* (Fig. 4c) with its crenulate walls and rudimentary septal ridges or spines as the next most advanced genus. All species of this genus excepting *S. tabulatum* which is possibly a *Trabeculites*, have mural pores (Flower, 1961; the *S. franklini* of Troedsson (1928) is a *Saffordophyllum* with mural pores as shown by Scrutton (1975)) and *Saffordophyllum* is unlikely therefore to be ancestral to *Foerstephyllum* and *Trabeculites* which lack pores. On the other hand, *Saffordophyllum* is rejected as the ancestor of the Favositina by Flower (1961) because of differences in wall structure but this still seems to be a reasonable possibility as on all other grounds the two genera are very close. Flower (1961) is prepared to accept the evolution of corallite walls with an axial plate within *Foerstephyllum* coinciding with the appearance of mural pores, thus suggesting this genus was ancestral to favositids. It would seem more logical, however, to assign species of *Foerstephyllum* with these features to *Paleofavosites* from which genus they are rather artificially excluded by Flower (1961), although Jull (1976a), alternatively suggested their possible assignment to *Nyctopora*. In either case *Foerstephyllum* can be regarded as an aporous septate lichenariid with simple fibrous walls not

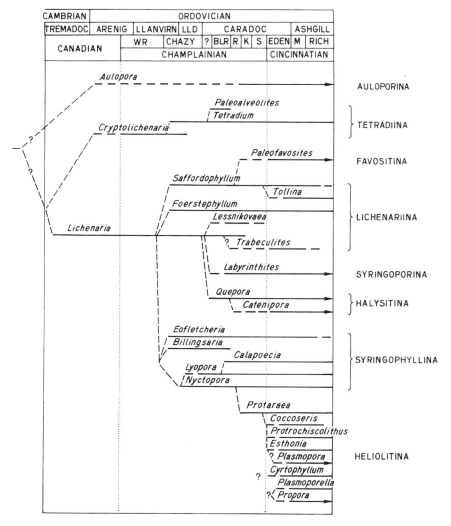

FIG. 3. Stratigraphic ranges and suggested phylogenetic links for early and middle Ordovician tabulate corals. Abbreviations: LLD—Llandeilo; WR— Whiterock; BLR—Black River; R—Rockland; K—Kirkfield; S—Sherman; M—Maysville; RICH—Richmond.

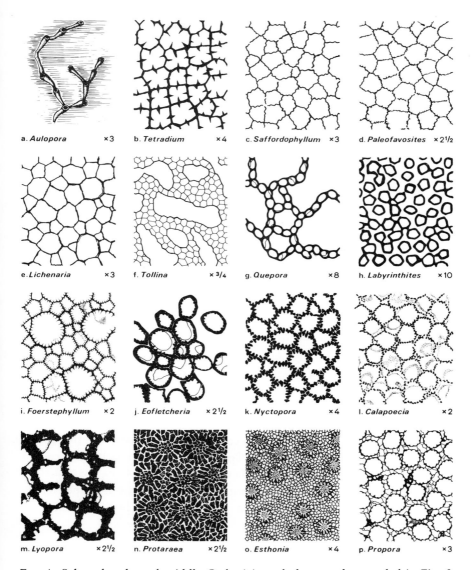

a. *Aulopora* ×3 b. *Tetradium* ×4 c. *Saffordophyllum* ×3 d. *Paleofavosites* ×2½

e. *Lichenaria* ×3 f. *Tollina* ×³/₄ g. *Quepora* ×8 h. *Labyrinthites* ×10

i. *Foerstephyllum* ×2 j. *Eofletcheria* ×2½ k. *Nyctopora* ×4 l. *Calapoecia* ×2

m. *Lyopora* ×2½ n. *Protaraea* ×2½ o. *Esthonia* ×4 p. *Propora* ×3

FIG. 4. Selected early and middle Ordovician tabulate corals recorded in Fig. 3 and discussed in the text. Redrawn from or based on the following sources:

a. *Aulopora schelonica* (Sokolov, 1955; pl. 49, fig. 1). b. *Tetradium fibratum* (Bassler, 1950; pl. 3, fig. 1). c. *Saffordophyllum undulatum* (Bassler, 1950; pl. 13, fig. 1). d. *Paleofavosites corrugatus* (Sokolov, 1962b; fig. 14a). e. *Lichenaria major* (Bassler, 1950; pl. 11, fig. 10). f. *Tollina manitoba* directly related to the Favositina, whilst *Saffordophyllum*, with slight modification of its wall structure, gave rise to *Paleofavosites* (Figs 3, 4c, d). I have retained *Saffordophyllum* in the Lichenariina here although it could be argued that the development of mural pores qualifies this genus for assignment to the Favositina. Both Webby (1971) and Jull (1976a) have discussed the proposed rugosan connections of species of *Foerstephyllum* and whilst the latter reported no characteristically rugosan pattern of septal insertion in the genus, Webby (1971) suggested that members of the *Foerstephyllum halli* group could be ancestral to the cystiphyllid *Hillophyllum*. This is more fully discussed below.

Another evolutionary trend from *Lichenaria* appears to be through *Nyctopora* and *Lyopora* to the sarcinulids, with the development of trabecular and porous walls (Figs 3, 4k–m). It is this group that I propose to call the Syringophyllina. The "pores" are formed as gaps between the trabeculae, rather irregularly developed in early genera (see Jull, 1976b) but becoming well ordered in, for example, *Calapoecia*. They do not appear to be comparable in their origin to the pores in the Favositina. Flower (1961) suggested *Trabeculites* as ancestral to *Nyctopora* but this is not supported by the ranges of these two genera as now known. *Trabeculites* is aporous and its relationships are uncertain. It could have evolved directly from *Lichenaria*, or perhaps (less likely) from *Lyopora* by the thinning and lateral closing of the trabeculae in the walls to eliminate the pores. *Eofletcheria* and *Billingsaria* would appear to be closely related to *Nyctopora* and are considered to have a common origin.

The earliest halysitid is *Quepora* (Fig. 4g) in the early Caradoc, considered by Sinclair (1955) to be a cateniform lichenariid and by Flower (1961) to be derived directly from *Lichenaria*. Bondarenko (1966) interposed *Tollina* (= *Manipora*) (Fig. 4f) in this evolutionary sequence as its structure, intermediate between cerioid and strictly cateniform, would superficially suggest. *Tollina* postdates *Quepora* in appearance, however, and its

---

Fig. 4. (cont.)
(Sokolov, 1962b; fig. 71). g. *Quepora quebecensis* (Sinclair, 1955; pl. 1, fig. 3). h. *Labyrinthites chidlensis* (Bolton, 1965; pl. 6, fig. 1). i. *Foerstephyllum halli* (Jull, 1976; pl. 1, fig. 2). j. *Eofletcheria orvikui* (Sokolov, 1962b; fig. 65a). k. *Nyctopora billingsi* (Bassler, 1950; pl. 14, fig. 1). l. *Calapoecia huronensis* (Jull, 1976; pl. 3, fig. 3). m. *Lyopora favosa* (Klaamann, 1975; pl. 4, fig. 1). n. *Protaraea ungerni* (Sokolov, 1962c; pl. 1, fig. 1b). o. *Esthonia astericus* (Sokolov, 1962c; pl. 2, fig. 1a). p. *Propora conferta* (Sokolov, 1962c; fig. 4b).

interpretation as a "cateniform" *Saffordophyllum* (Sinclair, 1955; Flower, 1961) seems more reasonable, although mural pores are either rare or apparently not present in species assigned to *Tollina*. This raises the question of how strictly cohesive the Halysitina is. Preobrazhensky (1977) has discussed this problem but I consider his suggestions for the dismemberment of this suborder to be extreme. A full discussion is beyond the scope of this paper but I regard the series *Quepora–Catenipora–Halysites* as a reasonable evolutionary lineage within the Halysitina, whilst *Tollina*, although possibly polyphyletic should be removed to the Lichenariina.

The affinities of the Tetradiina have been a subject of some controversy and separation of the group from the rest of the tabulates has been advocated (Okulitch, 1936; Flower and Duncan, 1975). I agree with Bassler (1950) and Sokolov (1962b), however, in regarding the group as a rather specialized branch of the tabulate corals. The earliest member of the suborder appears to be *Cryptolichenaria*, more or less contemporary with the earliest species of *Lichenaria* and possibly derived from one of them or possessing a close common ancestor. *Cryptolichenaria* is reported as showing corallite increase by the fusion of two opposite septa. Whether this is transitional to the standard pattern in tetradiids as Fig. 3 suggests or represents a blind side branch of *Lichenaria* from which *Tetradium* (Fig. 4b) later independently arose is unclear. All other members of the group are characterized by a distinctive tetrameral septal increase but may also show more conventionally tabulate modes of corallite increase.

Heliolitid corals have also given rise to considerable differences of opinion as to their affinities, and were assigned by some to the octocorals until quite recently on the basis of a superficial resemblance to the extant coenothecalian octocoral *Heliopora* (see Jones and Hill, 1940; Hill, 1960 for discussion). The present controversy concerns whether the heliolitids are merely a division of the Tabulata (Hill and Stumm, 1956; Hill, 1960) or a separate unit of equal rank (Sokolov, 1962c). Since Lindström's work (e.g. 1899) there has been a body of opinion against close relationships between heliolitids and tabulate corals, reinforced by Jones and Hill (1940) and followed by most Russian workers. Hill (1960), however, gave her reasons for reverting to a classification of the group within the Tabulata, pointing out that coenenchyme, so typical of the heliolitids, is also developed in the syringophyllids although it has a different structure there in which the tabularia are not aporous. Altogether the heliolitids appear to posses no features not found among other tabulate suborders. The implied monophyletic character of Sokolov's (1962c) subclass Heliolitoidea has also been questioned by Flower and Duncan (1975) who, after Flower (1961), suggested that the Protaraeida of Sokolov (1962c) evolved from *Nyctopora* separately from the rest of the heliolitids. The protaraeids are characterized by a dense, trabecular (bacular) skeleton which could quite logically be derived from the development of the trabeculae in a nyctoporid ancestor (Figs 3, 4k, n). Flower and Duncan (1975) also discussed structures in non-protaraeid heliolitids, demonstrating in a new species they assigned to *Cyrtophyllum* that the main colonial structure including the walls of the tabularium are formed by modified tabulae which are occasionally imperfectly fused to leave gaps in the tabularium wall. They suggested that all true heliolitids have hard parts completely derived from tabulae. This is certainly the case in some forms such as their *Propora lambei* but they themselves noted a clearly distinct tabularium wall in *Mcleodea loisae*, and most later heliolitids have a similar appearance. Trabeculae appear to be sparse in the earliest non-protaraeid heliolitids (Sokolov's (1962c) Heliolitida and Proporida) and ancestors for these forms are not obvious. Sokolov (1962c) showed his Proporida and *Cyrtophyllum* to have no direct links with the contemporary Protaraeida but derived his Heliolitida from the latter via *Esthonia* (Fig. 4o) in which the trabeculae are thinner and a more characteristic heliolitid corallite and coenenchymal structure is apparent. The views of Sokolov and Flower and Duncan are thus not wholly incompatible. Even so, I see no more likely signs of a proporid or cyrtophyllid ancestor outside the protareids and I suspect that these must have arisen along similar lines to that suggested for the heliolitids *sensu stricto*. There is nothing to gain from splitting up Sokolov's Heliolitoidea at this stage and I am inclined to regard this group as a single suborder of the Tabulata in view of the current evidence.

Apart from some uncertainty as to the common origin of the Auloporina and Lichenariina, and the problem of the ancestry of some heliolitids, I conclude that the tabulates are a monophyletic group.

PHYLOGENY OF THE EARLY RUGOSE CORALS

Cothonion sympomatum, which is probably more securely assigned to a zoantharian order than any of the other Cambrian corals, is a very unlikely ancestor for the first Ordovician Rugosa. If, as seems likely, it is a "rugosan", it must represent an early skeleton-evolving side branch to the main line of rugosan descent, separated from the Ordovician forms by some 70 m.y. Similarly, I am unable to relate *Tabulaconus kordeae* to an early phylogeny for either the rugose corals or the tabulates, although it has a superficial resemblance to an aseptate *Hillophyllum*. Again it seems to be an early skeletonized zoantharian, in this case not on the main line of descent of either order.

The earliest of the Ordovician Rugosa may be *Lambeophyllum* (Fig. 6b) if Welby's (1961) record is confirmed. Otherwise, about six genera appear very close together chronologically and opinions differ as to which is the more primitive (Fig. 5). Webby (1971) regarded *Hillophyllum* (Fig. 6c) as slightly predating *Lambeophyllum* (excluding Welby's record) and *Primitophyllum* (Fig. 6a), and therefore took *Hillophyllum* as ancestral to the other two genera. *Primitophyllum*, however, with its small conical corallum containing irregularly distributed septal spines only, appears to be the simplest of the early rugose corals and Ivanovskii (1968) and Weyer (1973) for example considered this genus as ancestral to all other Rugosans. Webby (1971), however, regarded *Primitophyllum* as a degenerate form and Neuman (1977) pointed out that *Primitophyllum* is not certainly older than either of the other genera, nor indeed older than *Streptelasma*, *Paleophyllum* or *Favistina* (Figs 6 d, i, j). It is not difficult to see the simple solitary genera as interrelated and having one of their number as ancestral to the others or perhaps all three with an earlier common ancestor.

What is perhaps more important is to consider firstly whether this group of solitary Ordovician rugosans is more likely to have arisen from a solitary soft zoantharian ancestor perhaps with a common progenitor with *Cothonion*, or whether they may have evolved from a more immediate colonial tabulate coral lineage. Several different views on the origin of the Ordovician Rugosa have been put forward. Most Russian authors in recent years have favoured descent from an auloporid stock (Soshkina in Soshkina *et al.*, 1962; Ivanovskii, 1965, 1966, 1968, 1972) although Sokolov (1955) suggested *Lichenaria* as an ancestor for the colonial columnariids and *Aulopora* for the solitary rugosans (Sytova, 1977). A lichenariid ancestor for the columnarids was first proposed by Bassler (1950) and developed by Flower (1961) who suggested the possibility that the solitary streptelasmatids may ultimately have evolved from such a lineage by loss of the ability to bud asexually. Flower and Duncan (1975) later tentatively added descent through the auloporids to their scheme, thus raising again the possibility of a polyphyletic origin for the Rugosa. Early Ordovician auloporids and the Cambrian *Protoaulopora* (see above) are insufficiently well known, however, for their potential as rugosan ancestors to be assessed.

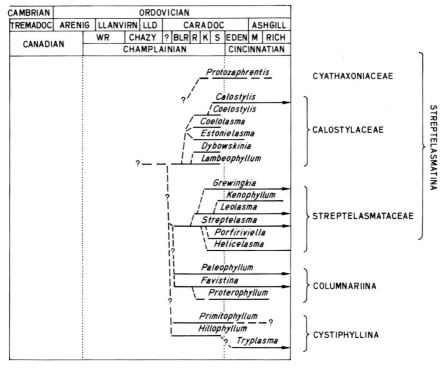

Fig. 5. Stratigraphic ranges and suggested phylogenetic links for middle Ordovician rugose corals. Abbreviations: LLD—Llandeilo; WR—Whiterock; BLR—Black River; R—Rockland; K—Kirkfield; S—Sherman; M—Maysville; RICH—Richmond.

Fig. 6. Selected middle Ordovician rugose corals recorded in Fig. 5 and discussed in the text. Redrawn from or based on the following sources: a. *Primitophyllum primum* (Ivanovskii, 1965; figs 9v, g). b. *Lambeophyllum profundum* (Stumm, 1963; pl. 1, fig. 17; Lambe, 1901; pl. 6, fig. 5a). c. *Hillophyllum priscum* (Webby, 1971; figs 1b, c). d. *Streptelasma primum* (Neuman, 1969; figs 8c, e). e. *Protozaphrentis* (Yü, 1957; pl. 2, fig. 1c). f. *Leolasma pachycolumnaris* (Neuman, 1975; fig. 4d). g. *Coelostylis toernquisti* (Weyer, 1973; pl. 12, fig. 5). h. *Grewingkia bilateralis* (Neuman, 1969; fig. 33a). i. *Paleophyllum thomi* (Flower, 1961; pl. 51, fig. 3). j. *Favistina stellata* (Flower, 1961; pl. 39, figs 2, 10).

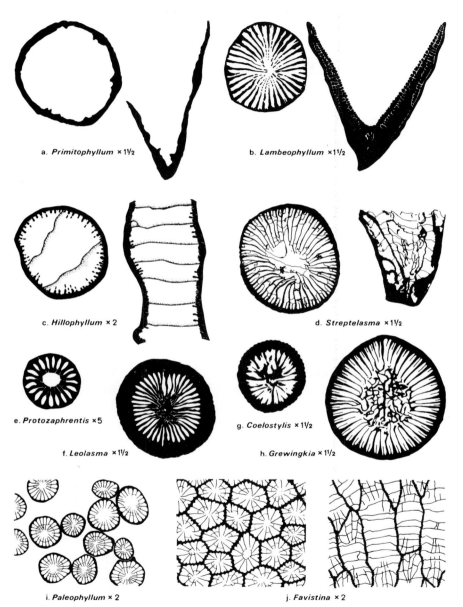

a. *Primitophyllum* ×1½   b. *Lambeophyllum* ×1½

c. *Hillophyllum* × 2   d. *Streptelasma* ×1½

e. *Protozaphrentis* ×5   g. *Coelostylis* ×1½

f. *Leolasma* ×1½   h. *Grewingkia* ×1½

i. *Paleophyllum* × 2   j. *Favistina* × 2

FIG. 6. For legend see opposite.

Hill (1960) and Webby (1971) have also considered a colonial ancestor for the cystiphyllid Rugosa in species of *Foerstephyllum* regarded as of rugosan rather than tabulate affinities. Hill regarded *Foerstephyllum* of Bassler (1950) as possibly containing unrelated rugose and tabulate corals under the one name but she did not consider this to indicate any link between the two orders. Jull (1976a) described the hystero–ontogeny of species of this genus and figured the holotype of the type species, *F. halli*. Contrary to the possibility advanced by Hill (1960), Bassler's (1950) illustrated specimens of *F. halli*, which she considered possibly rugosan, do not appear to differ significantly from the type. Jull's (1976a) evidence on septal insertion after increase is inconclusive but marginally favours tabulate affinities. If *Foerstephyllum*, either as a tabulate or rugosan, with its well developed septa, tabulae and cerioid colonial form is regarded as ancestral to such rugose coral genera as *Hillophyllum*, *Primitophyllum* and *Lambeophyllum*, the evolutionary sequence involves largely degenerate changes such as loss of asexual budding, loss of tabulae and reduction of the septa. Although coral phylogeny generally is not well known, the first two modifications are probably rare in rugose and scleractinian coral lineages. The possibility cannot be dismissed, but in the light of present evidence I regard *Foerstephyllum* as an unlikely ancestor for these early rugosans and probably a tabulate coral although it has unusually well developed septa for a tabulate. By the same token, the evolution of steptelasmatids from *Favistina* and *Paleophyllum* as proposed by Flower (1961) also seems unlikely.

The possibility that *Favistina* itself may have evolved from *Foerstephyllum*, as also advocated by Flower (1961) and Flower and Duncan (1975), is more difficult to discount. However, Fedorowski and Jull (1976) reported septal insertion during increase in *Favistina* and *Paleophyllum* to be "fundamentally dissimilar" to that in *Foerstephyllum* (Jull, 1976a) although one specimen of *Foerstephyllum vacuum* briefly shows ?rugosan features very early in septal insertion, which keeps the possibility open. An ancestry for the columnariids direct from *Lichenaria* has less to recommend it. It seems more likely that *Paleophyllum* and *Favistina* evolved from the same stock as the streptelasmatids during the first rapid diversification of the Rugosa in the middle Ordovician and that the group is indeed monophyletic as argued by Ivanovskii (1965, 1966) and Sytova (1977).

I conclude that there is no overwhelming evidence of derivation of any or all of the Rugosa from the Tabulata in the Ordovician in agreement with Hill (1960) and Sytova (1977). Acceptance of *Cothonion* as a "rugosan" implies the separation of a group of soft-bodied zoantharian precursors of the Rugosa in the early Cambrian. Any common ancestry with tabulates must date back at least as far as this. My views on the early diversification of the Rugosa in the Ordovician (Fig. 5) are close to those of Weyer (1973, fig. 2) although the ancestral position of *Primitophyllum* advocated by him is‾only morphologically suggested and not at the moment supported stratigraphically.

PHYLOGENY OF THE CNIDARIA

The origin of the Cnidaria and the phylogenetic relationships of classes within the phylum have been much discussed and are somewhat controversial. The major point at issue is whether the earliest adult cnidarians were radially symmetrical planktonic medusoids or bilaterally symmetrical benthonic polyps.

The theory that a planuloid ancestor evolved into a form close to the actinula lava of trachyline hydrozoans which then gave rise to medusae, and that the Hydrozoa represents the most primitive class, is probably the most widely held view (Hyman, 1940, 1959; Hand, 1959, 1963; Rees, 1966; Swedmark and Tessier, 1966; Hanson, 1977). In this case the polypoid form is a later development, originally as a larval stage but ultimately as the adult form in some groups. The fact that polyps can form medusae direct, but not *vice versa*, and that if both forms are present it is the medusa that produces gametes, is taken to support original medusoid form. In addition, the non-septate nature of hydrozoans, their more varied body forms and varied nematocysts tends to reinforce the likelihood that they are the more primitive (Hand, 1963). The Anthozoa are anatomically the most complex of the three classes (Hyman, 1959).

Others, however, have argued that the earliest adult cnidarians were benthonic and polypoid (Hadzi, 1963; Jägersten, 1955, 1959, 1972; Pantin, 1966; Thiel, 1966). Thiel suggested the simultaneous differentiation of the three main classes, with the independent development of medusae as dispersal agents in Hydrozoa and Scyphozoa. Hadzi's view of cnidarian evolution forms part of his scheme for metazoan phylogeny as a whole which proposed the acoel flatworms as the earliest metazoans. Cnidarians evolved from the rhabdocoel turbellarians which, being bilaterally symmetrical, gave rise first to the radiobilateral anthozoans. The Scyphozoa and Hydrozoa are derived sequentially from the Anthozoa, and the Hydrozoa are considered the most specialized of the three classes. Jägersten's (1955, 1959, 1972) bilaterogastraea theory requires the Anthozoa to be the primitive group because of the bilateral symmetry of his hypothetical planuloid ancestor leading to an enterocoelic origin of the coelom. Hardy (1953) has proposed a metaphytan origin for Cnidaria through the Anthozoa, a possibility considered by Cloud (1968) to be worth further exploration. Pflug's (1974 and other papers) rather extreme proposals, envisaging metazoan descent from a highly variable group of late Precambrian mixed metaphytan and metazoan forms termed Petalo-organisms, considerably extends this idea. Pflug's proposals have been briefly discussed above and are not considered again here.

Hand (1959, 1963) in particular has criticized Hadzi's theory, setting out the embryological and developmental evidence at variance with it, and noting important differences in basic structures and functions between the two groups. Others have also drawn attention to weaknesses in Hadzi's interpretation of rhabdocoel and anthozoan structures considered to support the linkage (Carter, 1954; Pantin, 1966) and Remane (1963) has stressed the "enormous differences" between the Anthozoa and the temnocephalid turbellarians. The case against a polypoid anthozoan ancestral form for the Cnidaria in general rests to a large extent on the arguments for primitive radial symmetry in the phylum. The maintenance of the embryological axis as the adult axis in Cnidaria, as well as radial symmetry preceeding bilateral symmetry in development, is taken as evidence in support of this. The bilateral symmetry of anthozoans is considered a necessary consequence of the partitioning of the gut linked to increase in size and the maintenance of an efficient hydrostatic skeleton. Thus bilateral symmetry is seen as a secondary, functional modification (Hyman, 1959; Hand, 1959, 1963).

The cnidarian fossil record provides rather doubtful evidence for this debate. The first reasonably well authenticated remains in the Vendian include probable representatives of two of the three major classes and both medusoid and polypoid forms. Medusae appear to be the more prominent. The levels of differentiation and organization apparently recorded, part-icularly for the Hydrozoa, suggest a preceeding phase of diversification, even if of great rapidity, which is unrepresented at the moment in the fossil record. Whether or not the Anthozoa are present in the Vendian depends on the interpretation of the supposed pennatulaceans and is thus somewhat questionable. On the other hand, the great scarcity of naked solitary polypoid cnidarians or their identifiable traces throughout the fossil record compared with the occurrence of medusoid remains, themselves scattered and sporadic, argues for caution in interpreting the fossil evidence. Although the evidence appears to favour the greater antiquity of the Hydrozoa rather than the Anthozoa, new finds and the reassessment of existing ones could considerably modify this impression.

Phylogenetic schemes based at all directly on fossil material are rare. Glaessner (1971b), however, has offered a scheme based on his work on the Conchopeltida. He argued that this group links primitive hydrozoans and scyphozoans, drawing a comparison between the chitinous wall of the chondrophoran pneumatophore which begins as a flat cone and the chitinous periderm of conchopeltids. The latter group is interpreted as giving rise on the one hand to the Conulariida and on the other, to the supposed scyphopolyp *Byronia* and the Scyphozoa. These relationships have been commented on above where, if Conularids are indeed cnidarians, common ancestry is favoured rather than direct descent.

Glaessner regarded the Anthozoa and Hydrozoa as having diverged earlier and presumably therefore takes the polypoid form as primary. In favour of their common origin, he pointed to the presence of bilateral symmetry in some hydrozoans and supposed a bilaterally symmetrical planuloid ancestor for the cnidarians. He also regarded the Conchopeltida and Conulariida as polypoid although this seems open to question. Apart from the arguments that may be advanced against bilateral symmetry in the primitive cnidarians (Hyman, 1959; Hand, 1963), this theory depends to some extent on discounting the possible late Precambrian scyphomedusae, whilst accepting the anthozoan and polypoid character of the supposed pennatulaceans. It seems to me that the latter might be the more questionable and indeed Glaessner and Walter (1975), now suggest that the *Rangea* group be provisionally classified as coelenterates of uncertain systematic position. At the moment the Hydrozoa seem the most securely established Precambrian cnidarians and the advanced stage of differentiation represented by the chondrophores tends to favour a hydrozoan as the earliest cnidarian. The zoological arguments for favouring this hypothesis, however, are somewhat stronger than the uncertain palaeontological evidence.

ACKNOWLEDGMENTS

I am grateful to Professor R. B. Clark (University of Newcastle upon Tyne) for comments on the section on the phylogeny of the Cnidaria. Several friends and colleagues have provided useful information and discussion on various other aspects of the paper. Christine Cochrane drafted Figs 1–6. I acknowledge with pleasure support from the University of Newcastle upon Tyne Research Fund towards costs involved in the preparation of this paper.

REFERENCES

ALF, R. M. (1959). Possible fossils from the early Proterozoic Bass Formation, Grand Canyon, Arizona. *Plateau*, **31**, 3, 60–63.

ALPERT, S. P. (1973). *Bergaueria* Prantl (Cambrian and Ordovician), a probable actinian trace fossil. *J. Paleont.* **47**, 919–924.

ARAI, M. N. and McGUGAN, A. (1968). A problematical coelenterate(?) from the Lower Cambrian, near Moraine Lake, Banff area, Alberta. *J. Paleont.* **42**, 205–209.

ARAI, M. N. and McGUGAN, A. (1969). A problematical Cambrian coelenterate(?). *J. Paleont.* **43**, 93–94.

BASSLER, R. S. (1941). A supposed jellyfish from the Pre-Cambrian of the Grand Canyon. *Proc. U.S. natn. Mus.* **89**, 519–522.

BASSLER, R. S. (1950). Faunal lists and descriptions of Paleozoic corals. *Mem. geol. Soc. Am.* **44**, x + 1–315.

BEDNARCZYK, W. (1970). Trilobites fauna of the Lower *Paradoxides oelandicus* Stage from the Brzechow area in the western part of the Swietokrzyskie Mts. *Bull. Acad. pol. Sci., Sér. Sci. géol. géogr.* **18**, 29–35.

BISCHOFF, G. C. O. (1973). On the nature of the conodont animal. *Geol. et Palaeont.* **7**, 147–174.

BOLTON, T. E. (1965). Ordovician and Silurian tabulate corals *Labyrinthites, Arcturia, Troedssonites, Multisolenia* and *Boreaster. Bull. geol. Surv. Can.* **134**, 15–33.

BOLTON, T. E. and COPELAND, M. J. (1963). *Cambrotrypa* and *Bradoria* from the Middle Cambrian of western Canada. *J. Paleont.* **37**, 1069–1070.

BONDARENKO, O. B. (1966). Puti razvitiya Tabulyat. *Paleont. Zh.* (1966), **4**, 8–18.

BOUILLON, J. (1968). Introduction to coelenterates. *In* "Chemical Zoology" (M. Florkin and B. T. Scheer, eds), Vol. 2, pp. 81–147. Academic Press, London and New York.

CARTER, G. S. (1954). On Hadzi's interpretation of animal phylogeny. *Syst. Zool.* **3**, 163–167.

CHAMBERLAIN, C. K. (1971). Morphology and ethology of trace fossils from the Ouachita Mountains, southeastern Oklahoma. *J. Paleont.* **45**, 212–246.

CHAPMAN, D. M. (1966). Evolution of the scyphistoma. *Symp. zool. Soc. Lond.*, **16**, 51–75.

CHAPMAN, F. (1919). On some hydroid remains of Lower Palaeozoic age from Mongetta, near Lancefield. *Proc. R. Soc. Vic.* **31**, 388–393.

CHAPMAN, F. and THOMAS, D. E. (1936). The Cambrian hydroids of the Heathcote and Monegeeta districts. *Proc. R. Soc. Vic.* **48**, 193–212.

CLARK, A. H. (1913). Cambrian holothurians. *Am. Nat.* **47**, 503.

CLOUD, P. E. (1968). Pre-Metazoan evolution and the origins of the Metazoa. *In* "Evolution and environment" (E. T. Drake, ed.), pp. 1–72. Yale Univ. Press, New Haven.

CLOUD, P. E. (1973). Pseudofossils: a plea for caution. *Geology*, **1**, 123–127.

CONWAY MORRIS, S. (1976). A new Cambrian lophophorate from the Burgess Shale of British Columbia. *Palaeontology*, **19**, 199–222.

CONWAY MORRIS, S. (1978). *Laggania cambria* Walcott: a composite fossil. *J. Paleont.* **52**, 126–131.

COOPER, G. A. (1976). Early Middle Ordovician of the United States. *In* "The Ordovician System" (M. G. Bassett, ed.), pp. 171–194. Univ. Wales Press, Cardiff.

COPE, J. C. W. (1977). An Ediacara-type fauna from South Wales. *Nature, Lond.* **268**, 624.

COWIE, J. W. and GLAESSNER, M. F. (1975). The Precambrian–Cambrian boundary—a symposium. *Earth-Sci. Rev.* **11**, 209–251.

DUNN, P. R. (1964). Triact spicules in Proterozoic rocks of the Northern Territory of Australia. *J. geol. Soc. Aust.* **11**, 195–197.

DUNNET, D. (1965). A new occurrence of Proterozoic "jellyfish" from the Kimberley region, Western Australia. *Rep. Bur. Miner. Resour. Geol. Geophys. Aust.* **134**, 1–5.

DURHAM, J. W. (1971). The fossil record and the origin of the Deuterostomata. *Proc. North Am. Paleont. Conv.* 1104–1132.

EDGELL, H. S. (1964). Precambrian fossils from the Hamersley Range, Western Australia, and their use in stratigraphic correlation. *J. geol. Soc. Austr.* **11**, 235–262.

FEDOROWSKI, J. and JULL, R. K. (1976). Review of blastogeny in Palaeozoic corals and description of lateral increase in some Upper Ordovician rugose corals. *Acta palaeont. pol.* **21**, 37–78.

FISCHER, J.-C. (1970). Révision et essai de classification des Chaetetida (Cnidaria) post-paléozoïques. *Annls Paléont. (Inv.)*, **56**, 149–220.

FISCHER, J.-C. (1977). Biogéographie des Chaetetida et des Tabulospongida post-paléozoïques. *Mém. Bur. Rech. géol. minièr.* **89**, 530–534.

FLOWER, R. H. (1961). Montoya and related colonial corals. *Mem. Inst. Min. Technol. New Mex.* **7**, 1–97.

FLOWER, R. H. and DUNCAN, H. M. (1975). Some problems in coral phylogeny and classification. *Bull. Am. Paleont.* **67**, 175–192.

FLÜGEL, E. (1975). Fossile Hydrozoen—Kenntnisstand und Probleme. *Paläont. Z.* **49**, 369–406.

FLÜGEL, H. W. (1976). Ein Spongienmodell für die Favositidae. *Lethaia*, **9**, 405–419.

FORD, T. D. (1958). Precambrian fossils from Charnwood Forest. *Proc. Yorks. geol. Soc.* **31**, 211–217.

FORD, T. D. (1963). The Precambrian fossils of Charnwood Forest. *Trans. Leicester lit. phil. Soc.* **57**, 57–62.

FREY, R. W. (1970). The lebensspuren of some common marine invertebrates near Beaufort, North Carolina, II. Anemone burrows. *J. Paleont.* **44**, 308–311.

FRITZ, M. A. (1947). Cambrian bryozoa. *J. Paleont.* **21**, 434–435.

FRITZ, M. A. (1948). Cambrian bryozoa more precisely dated. *J. Paleont.* **22**, 373.

Fritz, M. A. and Howell, B. F. (1955). An Upper Cambrian coral from Montana. *J. Paleont.* **29**, 181–183.

Fritz, M. A. and Howell, B. F. (1959). *Cambrotrypa montanensis*, a Middle Cambrian fossil of possible coral affinities. *Proc. geol. Ass. Can.* **11**, 89–93.

Germs, G. (1972). New shelly fossils from the Nama Group, South West Africa. *Am. J. Sci.* **272**, 752–761.

Germs, G. J. B. (1973). A reinterpretation of *Rangea schneiderhoehni* and the discovery of a related new fossil from the Nama Group, South West Africa. *Lethaia*, **6**, 1–10.

Glaessner, M. F. (1959). Precambrian Coelenterata from Australia, Africa and England. *Nature, Lond.* **183**, 1472–1473.

Glaessner, M. F. (1963). Zur Kenntnis der Nama-Fossilien Südwest-Afrikas. *Ann. naturh. Mus. Wien*, **66**, 113–120.

Glaessner, M. F. (1966). Precambrian palaeontology. *Earth-Sci. Rev.* **1**, 29–50.

Glaessner, M. F. (1969). Trace fossils from the Precambrian and basal Cambrian. *Lethaia*, **2**, 369–393.

Glaessner, M. F. (1971a). Geographical distribution and time range of the Ediacara Precambrian fauna. *Bull. geol. Soc. Am.* **82**, 509–513.

Glaessner, M. F. (1971b). The genus *Conomedusites* Glaessner et Wade and the diversification of the Cnidaria. *Paläont. Z.* **45**, 7–17.

Glaessner, M. F. (1972). Precambrian palaeozoology. *Spec. Pap. Univ. Adelaide Cent. Precamb. Res.* **1**, 43–52.

Glaessner, M. F. (1976). Early Phanerozoic annelid worms and their geological and biological significance. *J. geol. Soc.* **132**, 259–275.

Glaessner, M. F. and Wade, M. (1966). The late Precambrian fossils from Ediacara, South Australia. *Palaeontology*, **9**, 599–628.

Glaessner, M. F. and Walter, M. R. (1975). New Precambrian fossils from the Arumbera Sandstone, Northern Territory, Australia. *Alcheringa*, **1**, 59–69.

Gürich, G. (1930a). Die bislang ältesten Spuren von Organismen in Südafrika. *Int. geol. Congr.* **15**, 670–680.

Gürich, G. (1930b). Über den Kuibisquartzit in Südwest-afrika. *Z. dt. geol. Ges.* **82**, 637.

Gürich, G. (1933). Die Kuibis-Fossilien der Nama-Formation von Südwestafrika. *Palaeont. Z.* **15**, 137–154.

Hadzi, J. (1963). "The Evolution of the Metazoa." Pergamon Press, Oxford.

Hand, C. (1959). On the origin and phylogeny of the coelenterates. *Syst. Zool.* **8**, 191–202.

Hand, C. (1963). The early worm: a planula. *In* "The Lower Metazoa" (E. C. Dougherty, Z. N. Brown, E. D. Hanson and W. D. Hartman, eds), pp. 33–39. Univ. Calif. Press, Berkeley.

Handfield, R. S. (1969). Early Cambrian coral like fossils from the Northern Cordillera of Western Canada. *Can. J. Earth. Sci.* **6**, 782–785.

Hanson, E. D. (1977). "The Origin and Early Evolution of Animals." Pitman, London.

Hardy, A. C. (1953). On the origin of the Metazoa. *Q. Jl microsc. Sci.* **94**, 441–443.

HARRINGTON, H. J. and MOORE, R. C. (1956a). Protomedusae. *In* "Treatise on Invertebrate Paleontology, Part F" (R. C. Moore, ed.), pp. 21–23. Geol. Soc. Am. and Univ. Kansas Press.

HARRINGTON, H. J. and MOORE, R. C. (1956b). Scyphomedusae. *In* "Treatise on Invertebrate Paleontology, Part F" (R. C. Moore, ed.), pp. 38–53. Geol. Soc. Am. and Univ. Kansas Press.

HARRINGTON, H. J. and MOORE, R. C. (1956c). Trachylinida. *In* "Treatise on Invertebrate Paleontology, Part F" (R. C. Moore, ed.), pp. 68–76. Geol. Soc. Am. and Univ. Kansas Press.

HARRINGTON, H. J. and MOORE, R. C. (1956d). Medusae of the Hydroida. *In* "Treatise on Invertebrate Paleontology, Part F" (R. C. Moore, ed.), pp. 77–80. Geol. Soc. Am. and Univ. Kansas Press.

HARRINGTON, H. J. and MOORE, R. C. (1956e). Siphonophorida. *In* "Treatise on Invertebrate Paleontology, Part F" (R. C. Moore, ed.), pp. 145–152. Geol. Soc. Am. and Univ. Kansas Press.

HARRINGTON, H. J. and MOORE, R. C. (1956f). Medusae incertae sedis and unrecognizable forms. *In* "Treatise on Invertebrate Paleontology, Part F" (R. C. Moore, ed.), pp. 153–161. Geol. Soc. Am. and Univ. Kansas Press.

HARTMAN, W. D. and GOREAU, T. F. (1966). *Ceratoporella*, a living sponge with stromatoporoid affinities. *Am. Zool.* **6**, 563–564.

HARTMAN, W. D. and GOREAU, T. F. (1970). Jamaican coralline sponges: their morphology, ecology and fossil relatives. *Symp. Zool. Soc. Lond.* **25**, 205–243.

HARTMAN, W. D. and GOREAU, T. F. (1972). *Ceratoporella* (Porifera: Sclerospongiae) and the chaetetid "corals". *Trans. Conn. Acad. Arts Sci.* **44**, 133–148.

HARTMAN, W. D. and GOREAU, T. F. (1975). A Pacific tabulate sponge, living representative of a new order of sclerosponges. *Postilla*, **167**, 1–13.

HAUGHTON, S. H. (1962). Two problematic fossils from the Transvaal System. *Ann. geol. Surv. Pretoria*, **1**, 257–262.

HILL, D. (1953). The Middle Ordovician of the Oslo region, Norway. 2. Some rugose and tabulate corals. *Norsk geol. Tidsskr.* **31**, 143–168.

HILL, D. (1960). Possible intermediates between Alcyonaria Tabulata, Tabulata and Rugosa, and Rugosa and Hexacoralla. *21st Int. geol. Congr.* **22**, 51–58.

HILL, D. and STUMM, E. C. (1956). Tabulata. *In* "Treatise on Invertebrate Paleontology, Part F" (R. C. Moore, ed.), pp. 444–477. Geol. Soc. Am. and Univ. Kansas Press.

HUCKRIEDE, R. (1967). *Archaeonectris benderi* n. gen. n. sp. (Hydrozoa), eine Chondrophore von der Wende Ordovicium/Silurium aus Jordanien. *Geol. et Palaeont.* **1**, 101–109.

HYMAN, L. H. (1940). "The Invertebrates. I, Protozoa through Ctenophora." McGraw-Hill, New York.

HYMAN, L. H. (1959). "The Invertebrates. V, Smaller coelomate groups". McGraw-Hill, New York.

IVANOVSKII, A. B. (1965). "Drevneishie Rugozy". Nauka, Moskva.

IVANOVSKII, A. B. (1966). Polozheniye rugoz v sisteme korallovykh polipov. *Dokl. Acad. Nauk SSSR*, **166**, 445–458.

IVANOVSKII, A. B. (1968). Evolutsiya rugoz v Ordovike i Silure. *Mezhd. geol. Kongr.* (*dokl. sov. geol.*), **23**, 80–88.

IVANOVSKII, A. B. (1972). The evolution of the Ordovician and Silurian Rugosa. *23rd Int. geol. Congr., Proc. Int. paleont. Union*, sect 2, 69–78.

JÄGERSTEN, G. (1955). On the early phylogeny of the Metazoa. The bilatero-gastraea theory. *Zool. Bidr. Uppsala*, **30**, 321–354.

JÄGERSTEN, G. (1959). Further remarks on the early phylogeny of the Metazoa. *Zool. Bidr. Uppsala*, **33**, 79–108.

JÄGERSTEN, G. (1972). "Evolution of the Metazoan Life Cycle." Academic Press, London.

JELL, P. A. and JELL, J. S. (1976). Early Middle Cambrian corals from western New South Wales. *Alcheringa*, **1**, 181–195.

JONES, O. A. and HILL, D. (1940). The Heliolitidae of Australia, with a discussion of the morphology and systematic position of the family. *Proc. R. Soc. Qd.* **51**, 183–215.

JULL, R. K. (1976a). Septal development during hystero-ontogeny in the Ordovician tabulate coral *Foerstephyllum*. *J. Paleont.* **50**, 380–391.

JULL, R. K. (1976b). Review of some species of *Favistina*, *Nyctopora*, and *Calapoecia* (Ordovician corals from North America). *Geol. Mag.* **113**, 457–467.

KALJO, D. L. and KLAAMANN, E. (1973). Ordovician and Silurian corals. *In* "Atlas of Palaeobiogeography", (A. Hallam, ed.), pp. 37–45. Elsevier, Amsterdam.

KAZMIERCZAK, J. (1969). A new interpretation of astrorhizae in the Stromatoporoidea. *Acta palaeont. pol.* **14**, 499–535.

KAZMIERCZAK, J. (1971). Morphogenesis and systematics of the Devonian Stromatoporoidea from the Holy Cross Mountains, Poland. *Palaeont. pol.* **26**, 1–150.

KAZMIERCZAK, J. (1976). Cyanophycean nature of stromatoporoids. *Nature, Lond.* **264**, 49–51.

KIDERLEN, H. (1937). Die Conularien: über Bau und Leben der ersten Scyphozoa. *Neues Jb. Miner. Geol. Paläont. BeilBd*, **77**(B), 113–169.

KIRKPATRICK, R. (1912). On the nature of stromatoporoids. *Nature, Lond.* **89**, 607.

KLAAMANN, E. (1975). Zur Taxonomie einiger mittelordovizischer Tabulatenarten Norwegens, Schwedens und Estlands. *Izv. Akad. Nauk Est. SSR, Khim. Geol.* **24**, 219–226.

KORDE, K. B. (1959). Problematicheskie ostatki iz kembriiskikh otlozhenii yugo-vostoka sibirskoi platformy. *Dokl. Acad. Nauk SSSR*, **125**, 625–627.

KORDE, K. B. (1963). Hydroconozoa – novyi klass kishechnopolostnykh zhivotnykh. *Paleont. Zh.* (1963), 20–25.

KORDE, K. B. (1975). Kembriiskie Tselenteraty. *In* "Drevnie Cnidaria" (B. S. Sokolov, ed.), Vol. 2, pp. 53–56. Nauka, Novosibirsk.

KOZLOWSKI, R. (1959). Les hydroïdes ordoviciens à squelette chitineux. *Acta palaeont. pol.*, **4**, 209–271.

KOZLOWSKI, R. (1967). Sur certains fossiles ordoviciens à test organique. *Acta paleont. pol.* **12**, 99–132.

KOZLOWSKI, R. (1968). Nouvelles observations sur les Conulaires. *Acta palaeont. pol.* **13**, 497–529.

KÜKENTHAL, W. and BROCH, H. (1913). Pennatulacea. *Wiss. Ergebn. Deutschen Tiefsee-Exped.* **113** (2), i–vi, 113–576.

KULLING, O. (1964). Overskit över Norra Norrbottensfjällens Kaledonberggrund. *Sver. geol. Unders. Afh., Ser. Ba,* **19**, 1–166.

LAMBE, L. M. (1899, 1901). A revision of the genera and species of Canadian Palaeozoic corals. *Contr. Can. Palaeont.* **4**, Part 1, 1–96 (1899); Part 2, 97–197 (1901).

LECOMPTE, M. (1956). Stromatoporoidea. *In* "Treatise on Invertebrate Paleontology, Part F" (R. C. Moore, ed.), pp. 107–144. Geol. Soc. Am. and Univ. Kansas Press.

LEITH, E. I. (1952). Schizocoralla from the Ordovician of Manitoba. *J. Paleont.* **26**, 789–796.

LIN, B.-Y. (1965). Ordovician corals in the provinces Kweichow and Szechuan and their stratigraphic significance. *Acta palaeont. sin.* **13**, 64–93. (In Chinese, Russian summary.)

LINDSTRÖM, G. (1899). Remarks on the Heliolitidae. *K. svenska Vetensk-Akad. Handl.* **32** (1), 1–140.

LINDSTRÖM, M. (1974). The conodont apparatus as a food-gathering mechanism. *Palaeontology,* **17**, 729–744.

MADSEN, F. J. (1956). *Eldonia* a Cambrian siphonophore formerly interpreted as a Holoturian. *Vidensk. Medd. dansk. naturh. Foren.* **118**, 7–14.

MADSEN, F. J. (1957). On Walcott's supposed Cambrian holothurians. *J. Paleont.* **31**, 281–282.

MADSEN, F. J. (1962). The systematic position of the Middle Cambrian fossil *Eldonia. Medd. dansk. geol. Foren.* **15**, 87–89.

MIROSHNIKOV, L. D. and KRAVTSOV, A. G. (1965). Pozdnekembriiskie stsifomeduzy Sibirskoi platformy. *Ezheg. vses. paleont. Obshch.* **17**, 44–66.

MISRA, S. B. (1969). Late Precambrian (?) fossils from southeastern Newfoundland. *Bull geol. Soc. Am.* **80**, 2133–2140.

MONTANARO-GALLITELLI, E. (1956). Trachypsammiacea. *In* "Treatise on Invertebrate Paleontology, Part F" (R. C. Moore, ed.), pp. 190–192. Geol. Soc. Am. and Univ. Kansas Press.

MOORE, R. C. and HARRINGTON, H. J. (1956a). Scyphozoa. *In* "Treatise on Invertebrate Paleontology, Part F" (R. C. Moore, ed.), pp. 27–38. Geol. Soc. Am. and Univ. Kansas Press.

MOORE, R. C. and HARRINGTON, H. J. (1956b). Conulata. *In* "Treatise on Invertebrate Paleontology, Part F" (R. C. Moore, ed.), pp. 54–66. Geol. Soc. Am. and Univ. Kansas Press.

MORI, K. (1970). Stromatoporoids from the Silurian of Gotland, Part 2. *Stockh. Contr. Geol.* **22**, 1–152.

MORI, K. (1976). A new recent sclerosponge from Ngargol, Palau Islands and its fossil relatives. *Sci. Rep. Tohoku Univ., 2nd ser. (Geol.),* **46**, 1–9.

MORI, K. (1977). A calcitic sclerosponge from the Ishigaki-shima Coast, Ryukyu Islands, Japan. *Sci. Rep. Tohoku Univ., 2nd ser. (Geol.),* **47**, 1–5.

NEUMAN, B. (1969). Upper Ordovician streptelasmatid corals from Scandinavia. *Bull. geol. Instn Univ. Uppsala, N.S.* **1**, 1–73.

NEUMAN, B. (1975). New lower Palaeozoic streptelasmatid corals from Scandinavia. *Norsk geol. Tidsskr.* **55**, 335–359.

NEUMAN, B. E. (1978). On the taxonomy of lower Palaeozoic solitary streptelasmatids. *Mém. Bur. Rech. géol. minièr.* **89**, 69–77.

OEKENTORP, K. (1969). Kommensalismus bei Favositiden. *Münst. Forsch. Geol. Paläont.* **12**, 165–216.

OKULITCH, V. J. (1936). On the genera *Heliolites, Tetradium,* and *Chaetetes. Am. J. Sci.* **232**, 361–379.

OLIVER, W. A., Jr. (1968). Some aspects of colony development in corals. *J. Paleont.* **42**, Paleont. Soc. Mem. 2, 16–34.

OSGOOD, R. J., Jr. (1970). Trace fossils of the Cincinnati area. *Palaeontogr. Am.* **6**, 281–444.

PANTIN, C. F. A. (1966). Homology, analogy and chemical identity in the Cnidaria. *Symp. zool. Soc. Lond.* **16**, 1–17.

PFLUG, H. D. (1970a). Zur fauna der Nama-Schichten in Südwest-Afrika. I. Pteridinia, Bau und systematische Zugehörigkeit. *Palaeontographica A,* **134**, 226–262.

PFLUG, H. D. (1970b). Zur fauna der Nama-Schichten in Südwest-Afrika. II. Rangeidae, Bau und systematische Zugehörigkeit. *Palaeontographica A,* **135**, 198–231.

PFLUG, H. D. (1972). The Phanerozoic-Cryptozoic boundary and the origin of Metazoa. *24th Int. geol. Congr.,* Sect. 1, 58–67.

PFLUG, H. D. (1974a). Vor- und Frühgeschichte der Metazoen. *Neues Jb. Geol. Paläont. Abh.* **145**, 328–374.

PFLUG, H. D. (1974b). Feinstruktur und Ontogenie der jung-präkambrischen Petalo-Organismen. *Paläont. Z.* **48**, 77–109.

PHILIP, G. M. (1960). The middle Palaeozoic squamulate favositids of Victoria. *Palaeontology,* **3**, 186–207.

POPOV, Yu. N. (1967). Novaya Kembriiskaya stsifomeduza. *Paleont. Zh.* (1967), 2, 122–123.

PREOBRAZHENSKY, B. V. (1977). The structural interpretation of tabulatomorpha corals. *Mém. Bur. Rech. géol. minièr.* **89**, 97–101.

RAYMOND, P. E. (1931). Notes on invertebrate fossils, with descriptions of new species. *Bull. Mus. Comp. Zool.* **55**, 6, 165–213.

REES, W. J. (1966). The evolution of the Hydrozoa. *Symp. zool. Soc. Lond.* **16**, 199–222.

REMANE, A. (1963). The evolution of the Metazoa from colonial flagellates vs. plasmodial ciliates. *In* "The Lower Metazoa" (E. C. Dougherty, Z. N. Brown, E. D. Hanson and W. D. Hartman, eds), pp. 23–32. Univ. Calif. Press, Berkeley.

RICHTER, R. (1955). Die ältesten Fossilien Süd-Afrikas. *Senckenberg. leth.* **36**, 53–61.

RIDING, R. and KERSHAW, S. (1977). Nature of stromatoporoids. *Nature, Lond.* **268**, 178.

ROSS, M. H. (1953). The Favositidae of the Hamilton Group (Middle Devonian of New York). *Bull. Buffalo Soc. nat. Sci.* **21**, 37–89.

RUSCONI, C. (1948). Notas sobre fosiles Ordovicios y Triasicos de Mendoza. *Rev. Mus. Hist. nat. Mendoza*, **2**, 245–254.

RUSCONI, C. (1949). Nuevas especies de graptolitas Paleozoicos de Mendoza. *Rev. Mus. Hist. nat. Mendoza*, **3**, 3–8.

RUSCONI, C. (1950a). Diferentes organismos del Ordovicio y del Cambrico de Mendoza. *Rev. Mus. Hist. nat. Mendoza*, **4**, 63–70.

RUSCONI, C. (1950b). Primera contribucion al conocimiento de los graptolitas paleozoicos de Mendoza. *Rev. Mus. Hist. nat. Mendoza*, **4**, 95–164.

RUSCONI, C. (1951). Fosiles cambricos de Salagasta, Mendoza. *An. Soc. cient. argent.* **152**, 255–264.

RUSCONI, C. (1952). Los fosiles cambricos de Salagasta, Mendoza. *Rev. Mus. Hist. nat. Mendoza*, **6**, 19–62.

RUSCONI, C. (1955). Fosiles cambricos y ordovicios al Oeste de San Isidro, Mendoza. *Rev. Mus. Hist. nat. Mendoza*, **8**, 3–64.

SCRUTTON, C. T. (1975). Corals and stromatoporoids from the Ordovician and Silurian of Kronprins Christian Land, northeast Greenland. *Meddr Grønland*, **171** (4), 1–43.

SDZUY, K. (1974). Mittelkambrische Graptolithen aus NW-Spanien. *Paläont. Z.* **48**, 110–139.

SEDLAK, W. (1977). Some aspects on the stratigraphy and taxonomy of Cambrian fauna found on Lysa Góra (the Swietokrzyskie Mountains, Central Poland). *Mém. Bur. Rech. géol. minièr.* **89**, 42–48.

SINCLAIR, G. W. (1955). Some Ordovician halysitoid corals. *Proc. Trans. R. Soc. Can.* Ser. 3, **49**, Sect. 4, 95–103.

SINCLAIR, G. W. (1961). Notes on some Ordovician corals. *Bull. geol. Surv. Can.* **80**, 9–18.

SKEVINGTON, D. (1965). Chitinous hydroids from the Ontikan limestones (Ordovician) of Öland, Sweden. *Geol. Fören Stockh. Förh.* **87**, 152–161.

SOKOLOV, B. S. (1955). Tabulyaty paleozoya evropeiskoi chasti SSSR, Vvedenie. *Trudy vses. neft. nauchno-issled. geol. razv. Inst.*, N.S. **85**, 1–528.

SOKOLOV, B. S. (1962a). Gruppa Chaetetida. *In* "Osnovy paleontologii, 2. Gubki, arkheotsiaty, kishechnopolostnye, chervi" (U. A. Orlov, ed.), pp. 169–175. Izdatel'stvo Akad. Nauk, SSSR, Moskva.

SOKOLOV, B. S. (1962b). Podklass Tabulata. *In* "Osnovy paleontologii, 2. Gubki, arkheotsiaty, kishechnopolostnye, chervi" (U. A. Orlov, ed.), pp. 192–265. Izdatel'stvo Akad. Nauk, SSSR, Moskva.

SOKOLOV, B. S. (1962c). Podklass Heliolitoidea. *In* "Osnovy paleontologii, 2. Gubki, arkheotsiaty, kishechnopolostnye, chervi" (U. A. Orlov, ed.), pp. 266–285. Izdatel'stvo Akad. Nauk, SSSR, Moskva.

SOKOLOV, B. S. (1972a). The Vendian Stage in Earth History. *24th Int. geol. Congr.*, Sect. 1, 78–83.

SOKOLOV, B. S. (1972b). Vendsky etap v istorii zemli. *Mezhd. geol. Kongr. (dokl. sov. geol., Paleontologii)*, **24**, 114–124.

SOSHKINA, E. D., DOBROLUBOVA, T. A. and KABAKOVICH, N. V. (1962). Podklass Tetracoralla. *In* "Osnovi paleontologii, 2. Gubki, arkheotsiaty, kishechnopolostnye, chervi" (U. A. Orlov, ed.), pp. 286–356. Izdatel'stvo Akad. Nauk, SSSR, Moskva.

SPRIGG, R. C. (1947). Early Cambrian (?) jellyfishes from the Flinders Ranges, South Australia. *Trans. R. Soc. S. Aust.* **71**, 212–224.

SPRIGG, R. C. (1949). Early Cambrian "Jellyfishes" of Ediacara, South Australia and Mount John, Kimberley District, Western Australia. *Trans. R. Soc. S. Aust.* **73**, 72–99.

STASINSKA, A. (1960). *Velumbrella czarnockii* n. gen., n. sp.—Méduse du Cambrien inférieur des Monts de Sainte-Croix. *Acta palaeont. pol.* **5**, 339–346.

STEARN, C. (1972). The relationship of the stromatoporoids to the sclerosponges. *Lethaia*, **5**, 369–388.

STEARN, C. (1975). The stromatoporoid animal. *Lethaia*, **8**, 89–100.

STEL, J. H. (1976). The Paleozoic hard substrate trace fossils *Helicosalpinx*, *Chaetosalpinx* and *Torquaysalpinx*. *Neues Jb. Geol. Paläont., Mh.* (1976), 726–744.

STEL, J. H. and DE COO, J. C. M. (1977). The Silurian Upper Burgsvik and Lower Hamra-Sundre Beds, Gotland. *Scripta Geol.* **44**, 1–43.

STUMM, E. C. (1963). Ordovician streptelasmid rugose corals from Michigan. *Contr. Mus. Paleont. Univ. Mich.* **18**, 23–31.

SWANN, D. H. (1947). The *Favosites alpenensis* lineage in the Middle Devonian Traverse group of Michigan. *Contr. Mus. Paleont. Univ. Mich.* **6**, 235–318.

SWEDMARK, B. and TESSIER, G. (1966). The Actinulida and their evolutionary significance. *Symp. zool. Soc. Lond.* **16**, 119–133.

SWEET, W. C. and BERGSTRÖM, S. M. (1976). Conodont biostratigraphy of the Middle and Upper Ordovician of the United States midcontinent. *In* "The Ordovician System" (M. G. Bassett, ed.), pp. 121–151. Univ. Wales Press, Cardiff.

SYTOVA, V. A. (1977). On the origin of rugose corals. *Mém. Bur. Rech. géol. minièr.* **89**, 65–68.

TERMIER, H. and TERMIER, G. (1973). Stromatopores, Sclerosponges et Pharetrones: les Ischyrospongia. *Ann. Mines geol. Tunis*, **26**, 285–297.

TERMIER, H. and TERMIER, G. (1977). Ischyrosponges fossiles: paléogéographie, paléoécologie, évolution et stratigraphie. *Mém. Bur. Rech. géol. minièr.* **89**, 520–529.

THIEL, H. (1966). The evolution of the Scyphozoa. A review. *Symp. zool. Soc. Lond.* **16**, 77–117.

TROEDSSON, G. T. (1928). On the Middle and Upper Ordovician faunas of Northern Greenland, Part II. *Meddr Grønland*, **72**, 1–198.

VAN DER MEER MOHR, C. G. and OKULITCH, V. J. (1967). On the occurrence of Scyphomedusa in the Cambrian of the Cantabrian Mountains (N.W. Spain). *Geol. Mijnb.* **46**, 361–362.

VOLOGDIN, A. G. (1931). O nekotorykh okamenelostyakh khrebta Chiniz v Kazakhstane. *Yezhegodn. Russk. paleontol. o-va.* **9**, 131–147.

VOLOGDIN, A. G. (1932). Arkheotsiaty Sibiri, vypusk 2. *Trudy Vses. geol.-razved. obedin.* (1932), 1–106.

VOLOGDIN, A. G. (1939). Arkheontsiaty i vodorosli srednego kembriya Yuzhnogo Urala. *Probl. paleontologii*, **5**, 209–276.

VOLOGDIN, A. G. (1966). Ostatki protomeduz iz nizov karagasskoi svity vosto-chnogo Sayana. Dokl. Akad. Nauk SSSR, 167, 434–436.

VOLOGDIN, A. G. and STRYGIN, A. I. (1969). Otkrytiye ostatkov organizmov v verkhney svite krivorozhskoy serii dokembriya Ukrainy. Dokl. Acad. Nauk SSSR, 188, 446–449.

WADE, M. (1968). Preservation of soft-bodied animals in Precambrian sandstones at Ediacara, South Australia. Lethaia, 1, 238–267.

WADE, M. (1969). Medusae from uppermost Precambrian or Cambrian Sand-stones, Central Australia. Palaeontology, 12, 351–365.

WADE, M. (1970). The stratigraphic distribution of the Ediacara fauna in Australia. Trans. R. Soc. S. Aust. 94, 87–104.

WADE, M. (1971). Bilateral Precambrian chondrophores from the Ediacara fauna, South Australia. Proc. R. Soc. Vic. 84, 183–188.

WADE, M. (1972). Hydrozoa and scyphozoa and other medusoids from the Precambrian Ediacara fauna, South Australia. Palaeontology, 15, 197–225.

WALCOTT, C. D. (1911). Middle Cambrian holothurians and medusae. Smithsonian Misc. Coll. 57, 42–68.

WEBBY, B. D. (1971). The new Ordovician genus Hillophyllum and the early history of rugose corals with acanthine septa. Lethaia, 4, 153–168.

WELBY, C. W. (1961). Occurrence of Foerstephyllum in Chazyan rocks of Vermont. J. Paleont. 35, 391–400.

WELLS, J. W. and HILL, D. (1956). Zoanthiniaria, Corallimorpharia and Actiniaria. In "Treatise on Invertebrate Paleontology, Part F" (R. C. Moore, ed.), pp. 232–233. Geol. Soc. Am. and Univ. Kansas Press.

WERNER, B. (1966). Stephanoscyphus (Scyphozoa, Coronatae) und seine direkte Abstammung von den fossilen Conulata. Helgoländer wiss. Meeresunters. 13, 317–347.

WERNER, B. (1967a). Morphologie, Systematik und Lebensgeschichte von Steph-anoscyphus (Scyphozoa, Coronatae) sowie seine Bedeutung für die Evolution der Scyphozoa. Zool. Anz., Suppl. 30, 297–319.

WERNER, B. (1967b). Stephanoscyphus Allman (Scyphozoa Coronatae), ein rezenter Vertreter der Conulata? Paläont. Z. 41, 137–153.

WEYER, D. (1973). Über den Ursprung der Calostylidae Zittel 1879 (Anthozoa Rugosa, Ordoviz-Silur). Freiberger ForschHft C282, 23–87.

WILLIAMS, A., STRACHAN, I., BASSETT, D. A., DEAN, W. T., INGHAM, J. K., WRIGHT, A. D. and WHITTINGTON, H. B. (1976). A correlation of Ordovician rocks in the British Isles. Spec. Rep. Geol. Soc. Lond. No. 3, 1–74.

YOCHELSON, E. L. and HERRERA, H. E. (1974). Un fossil enigmatico del Cambrico Inferior de Argentina. Ameghiniana, 11, 283–294.

YÜ, C.-M. (1957). On the occurrence of a new rugose coral from the Middle Ordovician of Sinkiang Province, N.W. China. Acta palaeont. sin. 5, 307–323.

ZAIKA-NOVATSKII, V. C., VELIKANOV, V. A. and KOVAL', A. P. (1968). Pervyi predstavitel ediakarskoi fauny v vende Russkoi platformy (verkhnii Dokem-brii). Paleont. Zh. (1968), 132–134.

ZHURAVLEVA, I. T. (1970). Marine faunas and Lower Cambrian stratigraphy. Am. J. Sci. 269, 417–445.

7 | Early Structural and Ecological Diversification in the Bryozoa

G. P. LARWOOD

Department of Geological Sciences, University of Durham, England

and

P. D. TAYLOR

Department of Geology, University College of Swansea, Wales

Abstract: The Bryozoa are generally considered to have a phoronid ancestry. Unequivocal fossil bryozoans are known from the early Ordovician where they diversified rapidly and most higher taxa within the phylum became established. Known generic diversity rose from about 20 in the lower Ordovician to about 90 in the middle Ordovician. Early fossil bryozoans are dominantly stenolaematous forms which are inferred to have derived from a soft-bodied ctenostome ancestor. The transition from a ctenostome structure to that of an early stenolaemate, probably a corynotrypid cyclostome, was crucial. Tentacle protrusion in ctenostomes relies on contraction of a flexible *outer* zooid wall and consequently zooids are not usually laterally contiguous. Tentacle protrusion in stenolaemates results from deformation of flexible walls *within* the zooid. This arrangement allows for rigid calcification of the outer zooid wall and also for the evolution of colony forms with laterally contiguous zooids. The probable evolution of double-walled stenolaemates (Trepostomata, Cryptostomata, Cystoporata) from single-walled stenolaemates (Cyclostomata) may have involved paedomorphosis in the retention of hypostegal coelomic continuity between mature zooids. These structural changes during the early Ordovician resulted in a broad trend towards larger colony size and allowed the development of a wide variety of colony forms reflecting different ecological adaptations. The timing of early bryozoan diversification appears to coincide with a general increase in diversity of suspension feeding animals.

Systematics Association Special Volume No. 12, "The Origin of Major Invertebrate Groups", edited by M. R. House, 1979, pp. 209–234, Plates I and II, Academic Press, London and New York.

The Bryozoa are a phylum of colonial lophophorates with a long geological history. The majority are sessile and all are filter feeders. At the present day they are most abundant between 20 and 80 metres (Ryland, 1970) but some are found at abyssal depths. Colonies develop by asexual budding of connected zooids of various kinds. Normal tentacled feeding zooids are termed autozooids and these comprise a polypide, essentially gut and tentacles, held within a coelom which is enclosed by the body wall of the zooid (Fig. 3). Polypide eversion extrudes the tentacle crown through the orifice in the body wall of the zooid. In this position, cilia on the tentacles induce a current flow carrying food particles towards the mouth (Borg, 1926). Although the structure of the polypide is relatively uniform throughout the phylum, there is considerable variation in zooid structure, largely reflected in the form of the zooecium (calcified portion of the zooid), and an even greater variation in colony form. Many colonies have a calcified skeleton (zoarium) and zoarial diversity indicates a wide variety of ecological adaptations (Stach, 1936).

ORIGINS OF THE PHYLUM BRYOZOA

Morphological studies of extant representatives of metazoan phyla have led to two major opinions being held concerning the phylogenetic origin of the Bryozoa. Nielsen (1971) provides an extensive review of the literature relating to this topic. Features shared with the two other phyla, Phoronida and Brachiopoda, which belong to the superphylum Lophophorata (Valentine, 1973) have led many authors (e.g. Clark, 1964; Farmer et al., 1973; Farmer, 1977) to favour a phoronid ancestry for the Bryozoa (Fig. 1). A less widely held opinion considers an entoproct ancestry more likely (Nielsen, 1971, 1977a, 1977b). Forms intermediate between phoronids and bryozoans or between entoprocts and bryozoans are not known from the fossil record, nor is there any record of fossils resembling the hypothetical bryozoan ancestors envisaged by Ostroumoff (1885) or Jägersten (1972). Thus, models based on comparative zoological studies, without evidence being available from the fossil record, should take into consideration 500 m.y. or more of adaptive evolution.

THE EARLIEST RECORDED FOSSIL BRYOZOANS

The earliest known unequivocal bryozoans are recorded from the lower Ordovician. Ross (1964) has questioned the affinities of supposed Cambrian bryozoans described by Cobbold (1931) from the lower Cambrian of Comley, Shropshire, and by Fritz (1947, 1948) from the upper Cambrian

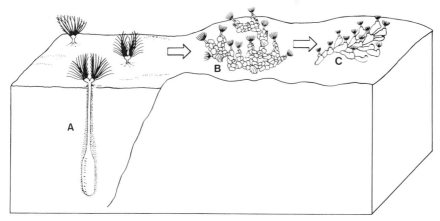

FIG. 1. Model for the evolution of the Bryozoa after Farmer *et al.* (1973). A solitary infaunal phoronid ancester (A) gives rise, via an intermediate epifaunal form (B), to an adnate bryozoan colony (C).

of Alberta, Canada. Poor preservation and a lack of diagnostic morphological characters precludes positive assignment of these Cambrian fossils. A recent discovery of possible bryozoans from the middle Cambrian has been made by Dr I. A. McIlreath (personal communication, 1978). The fossils, which are recrystallized and consist of contiguous polygonal tubes, occur in the "Thin Cathedral Formation" and the "Thick Stephen Formation" of the Canadian Rockies (McIlreath, 1977).

Heteronema priscum Bassler (1911b), the type species of *Marcusodictyon* Bassler (1952), from the Cambrian (Tremadoc) A2 horizon of Estonia has normally been regarded as the earliest known bryozoan. It was assigned by Bassler and others to the Ctenostomata. However, two topotype specimens presented by Bassler, in the collection of the British Museum (Natural History) (numbers PD 2448–9) each comprise a polygonal network of low steep-sided ridges apparently calcitic. No other external details are visible. The polygons are about 0·6 mm wide and in one of the specimens two or three such polygonal networks overlie one another without alignment. Each specimen encrusts the exterior of a valve of a small inarticulate brachiopod. The specimens are of uncertain affinities, but they do not have the characters of wholly soft-bodied ctenostomes or any other type of bryozoan.

Astrova (1965) described a cystoporate, *Profistulipora*, from the lower Ordovician of the USSR, and mentioned the occurrence of about 20 other genera in the remainder of the lower Ordovician. Ross (1966) described *Ceramopora? unapensis*, a cystoporate from the early Ordovician (early late Canadian) of Oklahoma. More recently, McLeod (1978) has described a dianulitid (?trepostome) bryozoan from the late Canadian (late Arenig) Black Rock Limestone of northeastern Akansas and southeastern Missouri. The earliest diverse bryozoan fauna to receive detailed attention is from the middle Ordovician Chazyan Series of New York State and Vermont. Ross (1964) listed 21 Chazyan lineages.

By the middle Ordovician all four stenolaemate orders (Cyclostomata, Cystoporata, Cryptostomata, Trepostomata) and probably the gymnolaematous ctenostomes had become established (Larwood et al., 1967). Highly specialized penetrant ctenostomes are preserved as trace fossils in the upper Ordovician (Pohowsky, 1974).

EARLY NUMERICAL INCREASE IN BRYOZOAN TAXA

The "standing crop" diversity of fossilizable bryozoans probable rose from no species at the onset of the Ordovician to many hundreds of species by the close of the Ordovician. Astrova (1965) gives generic longevities for lower Palaeozoic Bryozoa in North America and the USSR. Her data, summarized here in Fig. 2, indicates the occurrence of about 20 genera in the lower Ordovician, 91 genera in the middle Ordovician and 87 genera in the upper Ordovician. The detail of Astrova's analysis improves on the data available from the Treatise on Invertebrate Paleontology (Part G Bryozoa; Bassler, 1953) and is considered to be more accurate than the Ordovician portion of the diversity cladogram given by Müller (1958) and reproduced by Schopf (1977). However, the Treatise can be used to give a crude estimate of the number of supraspecific bryozoan taxa established during the Ordovician in comparison with the *total* number of supraspecific bryozoan taxa described. Bassler (1953) included four orders (to which may be added the Order Cystoporata, Astrova, 1964) with representatives in the Ordovician, 20 families and 116 genera (neglecting *Bolopora* which

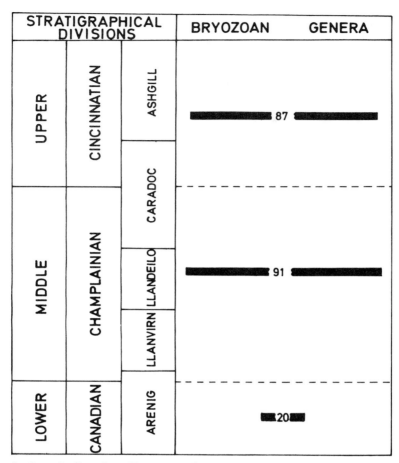

FIG. 2. Generic diversity of bryozoans for 3 intervals in the Ordovician. Numbers of genera based on Astrova (1965) and stratigraphical divisions after Williams *et al.* (1972).

has been since shown to be an oncolite, Hofmann, 1975). Thus, five of the six recognized bryozoan orders, about 17% of bryozoan families and 10% of bryozoan genera were established during the Ordovician. Raup (1976) abstracted data from the Zoological Record for 1900 to 1970 and found that although no descriptions of new Cambrian bryozoans were recorded during those years, 569 new bryozoan species were described from the Ordovician. This figure represents about 11% of the total entries of new fossil bryozoan species in the Zoological Record between 1900 and

1970. The Ordovician was therefore a time of very considerable bryozoan diversification and was unsurpassed in this respect during any other single Period of geological time with the possible exception of the Cretaceous when there was a huge diversification of newly established cheilostome taxa (Larwood *et al.*, 1967).

A MODEL FOR EARLY STRUCTURAL DIVERSIFICATION

Freshwater phylactolaematous bryozoans are often regarded by zoologists (e.g. Jebram, 1973) as the most "primitive" bryozoans, principally because they possess horseshoe-shaped tentacle crowns resembling those of phoronids. However, the phylactolaemates secrete no hard parts and have an uncertain fossil record with only one possible occurrence in the Cretaceous (Fric, 1901). Recent work by S. Mundy (personal communication, 1978) indicates that the horseshoe shape of the phylactolaemate tentacle crown may be an adaptive feature correlated with life in the freshwater environment. Therefore, the role, if any, of phylactolaemates in early bryozoan evolution cannot be assessed accurately.

Within the precision to which the fossil record can be reliably interpreted, the gymnolaemate ctenostomes and the four recognized orders of stenolaemates appeared at approximately the same time. Surprisingly, however, previous theories of bryozoan phylogeny have tended to assume that the stenolaemates were ancestral to the rather more simply organized ctenostomes (e.g. Dzik, 1975; Banta, 1976). Apparently no interpretation has considered the reciprocal possibility of a ctenostome ancestry for the stenolaemates. This is preferred here and a model for the transition from the ctenostome to stenolaemate structure is proposed along with possible subsequent evolutionary changes. The predicted consequences of these transitions compare favourably with the known early fossil record of the Bryozoa.

1. Polypide Eversion in the Bryozoa

Perhaps the most fundamental feature differentiating extant orders of marine bryozoans is their mode of polypide eversion. By everting the polypide the zooid extrudes its tentacle crown and thus is able to feed. The tentacle crown has additional roles in reproduction (Silén, 1966) and probably respiration. All bryozoans rely on a hydrostatic mechanism of polypide eversion involving muscular compression of coelomic fluid. In the majority of ctenostomes, with the exception of some carnosans (e.g.

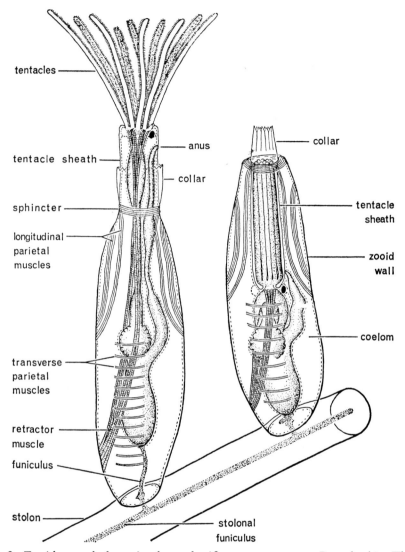

FIG. 3. Zooid morphology in the stoloniferous ctenostome *Bowerbankia*. The zooid on the left is expanded, that on the right is retracted. After Ryland (1970).

Alcyonidium and *Flustrellidra*) where eversion compares more closely with that in cheilostomes, polypide eversion is essentially brought about by the contraction of transverse parietal muscles attached to the flexible outer body wall of each zooid (Fig. 3). This muscular contraction raises coelomic

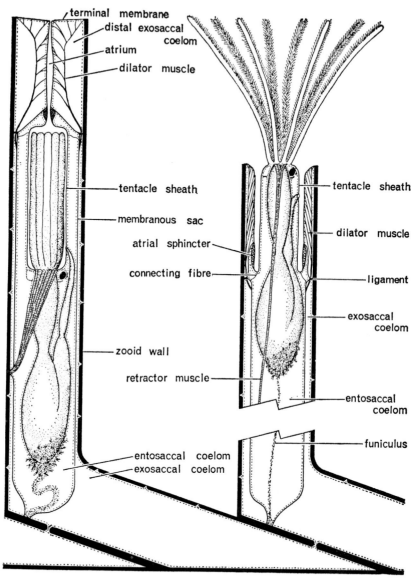

FIG. 4. Zooid morphology in a cyclostomatous stenolaemate. The zooid on the left is retracted, that on the right is expanded. After Ryland (1970).

pressure and the tentacle crown is forced out of the distal end of the zooid.

Although the cheilostomes had no part to play in early diversification it is nevertheless worthwhile considering their mode of polypide eversion. In the simplest cheilostomes (anascans), only the frontal part (the frontal membrane) of the zooid body wall is flexible and the contraction of parietal muscles attached to this frontal membrane depresses it, raising coelomic pressure and everting the polypide which extrudes the tentacle crown.

A rather different situation exists in the cyclostomes where the outer zooid wall is rigid and deformation of flexible walls within the zooid is responsible for the rise in hydrostatic coelomic pressure necessary for polypide eversion. The cyclostome zooid comprises an inner entosaccal coelom separated by the membranous sac from an exosaccal coelom which encloses the entosaccal coelom and extends distally to it forming a vestibule (Fig. 4). Until recently it was thought that polypide eversion was accomplished solely by contraction of atrial dilator muscles forcing exosaccal coelomic fluid out of the vestibule and into the proximal part of the zooid. The precise role of the membranous sac in this mechanism was unclear (Harmer, 1930). However, Nielsen and Pedersen (in prep.) have found that contraction of the atrial dilator muscles is accompanied by contraction of a series of fine circular muscle cells in the membranous sac which further raises hydrostatic pressure in the proximal portion of the zooid. The presence of a vestibulum reduces the distance by which a tentacle crown of a cyclostome zooid may protude (Fig. 5) and the mouth is rarely extended beyond the margin of the skeletal aperture (Ryland, 1976). The three extinct stenolaemate bryozoan orders are presumed to have had modes of tentacle extrusion identical to that of Recent cyclostomes. This view is supported by the recognition by McKinney (1969) of a possible fossilized membranous sac in a Palaeozoic stenolaemate.

Retraction of the polypide is identical in all three groups of extant marine bryozoans. It relies on the swift and powerful contraction of a retractor muscle which is relaxed during eversion.

2. The Relationship of Zooid Structure and Colony Form to the Mode of Polypide Eversion

The consequences of polypide eversion involving deformation of flexible walls are most important.

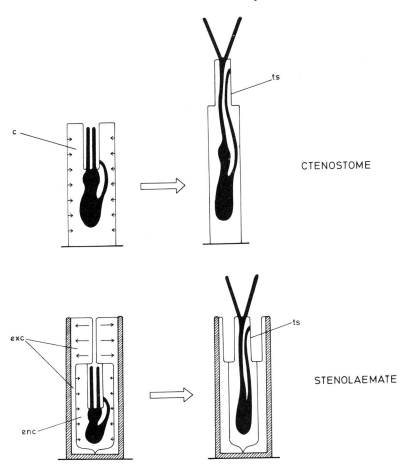

c-coelom, enc-entosaccal coelom, exc-exosaccal coelom, ts-tentacle sheath
■ polypide, ▨ rigid skeletal wall, →indicates deformation to produce polypide eversion.

FIG. 5. Modes of polypide eversion in ctenostome and in stenolaemate bryozoans.
In the ctenostome with a flexible outer zooid wall, contraction of attached
transverse parietal muscles raises coelomic hydrostatic pressure to evert the
polypide. In the stenolaemate with a rigid outer zooid wall, the polypide
is everted when flexible inner walls deform by simultaneous contraction
of atrial dilator muscles and muscles of the membranous sac.

Outer zooid walls in bryozoans commonly include a rigid calcified
layer. Rigid calcification cannot occur if it is necessary for the outer zooid
wall to be deformed during polypide eversion, as in the ctenostome
bryozoans. However, in stenolaemates the flexible wall which is deformed

to produce polypide eversion is contained within the zooid and the outer zooid wall may develop a calcified layer.

The reduction of zooid diameter (Fig. 5) necessary for polypide eversion in ctenostomes characteristically precludes lateral contiguity of zooids, except where a biserial arrangement is developed, e.g. *Amathia*, and where zooid lateral walls are thick and gelatinous and do not move during contraction of parietal muscles e.g. *Flustrellidra*. However, zooid diameter is not reduced (Fig. 5) during polypide eversion in the stenolaemates and therefore close lateral contiguity of zooids commonly develops. The limited lateral contiguity of zooids characterizing ctenostomes places constraints upon the variety of colony forms which may be developed. Most ctenostomes have uniserial colonies with autozooids adnate to the substrate (e.g. *Arachnidium*, or with creeping stolons of kenozooids connecting isolated erect autozooids (e.g. *Bowerbankia*). In contrast, colony form in stenolaemates is much more varied and includes uniserial encrusters (Pl. I, 4), multiserial encrusters (Pl. II, 2) and erect colonies with cylindrical branches (Pl. II, 1).

3. Origin of the Stenolaemates

Traditional theories explaining the rather sudden early Ordovician appearance of calcified stenolaemates exhibiting a variety of colony form have tended to invoke the abrupt appearance of a calcified skeleton in a number of independent stenolaemate stocks (e.g., Farmer, 1977). This seems improbable because no uncalcified extant stenolaemates are known and a rigid skeleton is highly beneficial if tentacle extrusion is to proceed efficiently in the manner described above. The rigidly calcified outer zooid walls of stenolaemates prevent bulging of the zooid when exosaccal coelomic fluid is forced proximally and also provide a support for the atrial dilator muscles to pull against during polypide eversion. The rapid appearance of diverse stenolaemates may be more readily accounted for by their derivation from a ctenostome stock. The structural transition involved would have conferred the potential for calcification and for the development of a wide variety of colony forms.

The postulated evolution of stenolaemates from ctenostomes may have occurred by fission of the outer zooid wall to give an additional coelomic cavity, the exosaccal coelom. If so, the circular muscle bands of the stenolaemate membranous sac may be homologous with the transverse parietal muscles of the ctenostome outer zooid wall. A less certain homology may be suggested between the longitudinal parietal muscles of ctenostomes and the atrial dilator muscles of cyclostomes. The exosaccal coelom, resulting from a split between mesoderm (the membranous sac) and ectoderm (epithelium of the outer zooid wall) may be a pseudocoel. Nielsen (1970) has suggested on embryological grounds that the cyclostome exosaccal coelom is not homologous with the visceral coelom and is not a true coelom. Following the inferred fission of the outer zooid wall, distalward extension of the exosaccal coelom to form a vestibulum, enclosing the distal portion of the zooid, would have given a zooid structure identical to that of living cyclostomes. Although the precise phylogenetic modifications in zooid structure suggested must be considered very tentative, they do indicate the ease with which a stenolaemate zooid organization could have been attained from a ctenostome ancestor.

Once the stenolaemate zooid organization had been achieved, the acquisition of a rigid outer zooid wall would have become not only a possibility but a positive advantage to the zooid. Isolated calcium carbonate crystals have been found in the zooid walls of certain living ctenostomes (Lomas, 1885) and the amalgamation of similar crystals in early Ordovician ctenostomes to form a rigid calcified skeleton seems probable.

4. The Ancestral Stenolaemate

One of the earliest known stenolaemates is the cyclostome *Corynotrypa* which first appeared in the lower Ordovician (Bassler, 1911a). *Corynotrypa* colonies consist of uniserial branching chains of adnate zooids. The structure of the zooids is extremely simple and closely resembles that of the Jurassic to Recent ctenostome *Arachnidium* (Prenant and Bobin, 1956; Voigt, 1977; Taylor, 1978). The typically pyriform zooids have a proximally tapering narrow cauda and a broad distal portion with a circular terminal aperture (Fig. 6 and Pl. I, 4). The outer zooid wall is a true exterior wall (Boardman and Cheetham, 1973) because it occurs at the zooid–environment interface. *Corynotrypa* is probably unique among cyclostomes in having no calcified walls separating contiguous zooids (interior walls) (Boardman and Cheetham, 1973; Brood, 1975). Instead, zooecial chambers are continuous through the distal end of each zooid and into the cauda of the succeeding zooid. This arrangement is reminiscent (Boardman and Cheetham, 1973) of that pertaining in some early cheilostomes with uniserial colonies e.g. *Pyripora* and *Rhammatopora*. The close evolutionary affinity between these early cheilostomes and ctenostomes is generally accepted (Banta, 1975) and a similar affinity between corynotrypid stenolaemates and ctenostomes seems very probable. Furthermore, the uniserial form of *Corynotrypa* colonies is like that predicted

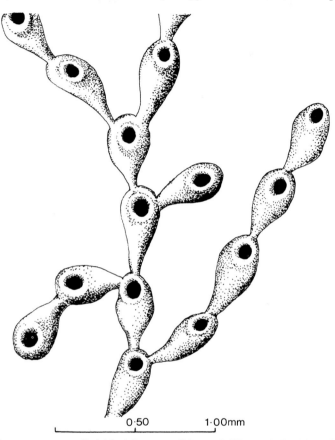

FIG. 6. *Corynotrypa* sp. British Museum (Natural History) D1253, Cincinnati Group, Ohio, USA. Group of zooids from a colony encrusting a brachiopod shell.

by Jackson (1977) for the earliest fossil representatives of colonial groups. Thus, cyclostomes of the corynotrypid type were probably the first calcified stenolaemates (and bryozoans) and arose in the early Ordovician from an *Arachnidium*-like ctenostome. Diversification from the initial corynotrypid organization was possible because of the mode of tentacle protrusion adopted by the stenolaemates.

5. Evolution of Non-corynotrypid Cyclostomes

Beginning with the basic corynotrypid cyclostome structure described above it is not difficult to derive structures corresponding to those present in other Palaeozoic cyclostomes. *Stomatopora* differs from *Corynotrypa* in the possession of tubular rather than pyriform zooids and in having interior walls separating contiguous zooids in linear series. Thus, addition of a calcified interior wall to a corynotrypid, with concommitant proximal broadening of the zooid, would give a *Stomatopora*-like organization (Brood, 1976). More complex cyclostome colonies (Brood, 1975) could have developed by modifying the patterns of zooidal budding. These include ribbon-like "*Proboscina*" colonies and discoidal *Sagenella* colonies, both encrusting with multiserially budded zooids. Erect colony forms were achieved in other genera, such as *Mitoclema*. Despite this variety in colony shape the form of the zooid in all of these genera compares closely with the *Stomatopora* zooid.

6. Evolution of Double-walled Stenolaemates

Perhaps the most fundamental difference between Palaeozoic cyclostomes and stenolaemates of the other three orders (Cryptostomata, Cystoporata and Trepostomata) is that Palaeozoic cyclostomes apparently possessed a so-called single-walled organization whereas, by analogy with certain extant cyclostomes (Boardman and Cheetham, 1973), the cystoporates, trepostomes and cryptostomes are thought to have possessed a double-walled organization. Zooecial (and zooidal) budding in stenolaemates is accomplished beneath a common or hypostegal coelom. Here interzooecial walls, secreted by epithelia lining them on both sides (i.e. they are interior body walls), may lengthen and divide to partition off new zooecia. In double-walled stenolaemates hypostegal coelom covers the whole colony surface allowing interzooecial walls separating mature zooids to be continually lengthened and hypostegal coelomic continuity to be maintained throughout colony development (Fig. 7). The outer colony wall above the hypostegal coelom invariably remains uncalcified (cf. *Stegohornera violacea*; see Brood, 1972, p. 36). In single-walled stenolaemates hypostegal coelomic continuity between mature zooids is severed by the fusion of growing interzooecial walls with the outer colony wall. Following this fusion, a calcified exterior zooecial wall is formed when the outer colony wall secretes a calcareous layer grown from within.

Zooecial (and zooidal) budding is restricted to discrete colony growth zones or common buds (Fig. 7).

Zooids of single-walled stenolaemates pass through a double-walled stage during their early ontogeny and phylogenetic separation of steno-

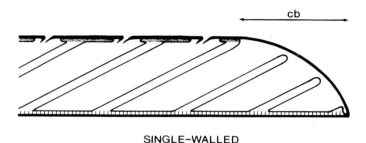

SINGLE-WALLED

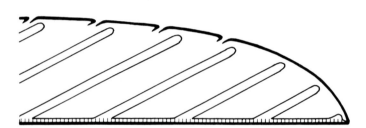

DOUBLE-WALLED

Fig. 7. Simplified diagram comparing single-walled and double-walled organization in stenolaemates. Longitudinal sections through the distal parts of adnate colonies. Zooidal epithelia and polypides are omitted. Black—cuticle; stipple—zooecial frontal walls; hatching—basal lamina; cb—common bud. The common bud of the double-walled form covers the whole colony frontal surface.

laemates into single-walled and double-walled forms is indistinct. For example, Boardman and Cheetham (1973) have described single-walled growth in an extant cyclostome, *Heteropora pacifica*, belonging to a genus otherwise composed of double-walled species. The Jurassic cyclostome *Ceriocava corymbosa* (see Walter, 1969) usually shows a double-walled organization, but some colonies (*Cyclocites primogenitum* Canu and Bassler, 1922) developed, during late zooidal ontogeny, a calcified frontal wall indicative of single-walled growth. It seems possible that single-walled

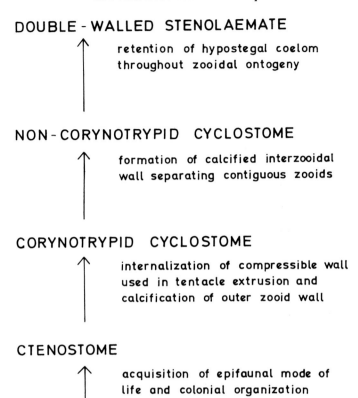

DOUBLE - WALLED STENOLAEMATE

↑   retention of hypostegal coelom
     throughout zooidal ontogeny

NON - CORYNOTRYPID CYCLOSTOME

↑   formation of calcified interzooidal
     wall separating contiguous zooids

CORYNOTRYPID CYCLOSTOME

↑   internalization of compressible wall
     used in tentacle extrusion and
     calcification of outer zooid wall

CTENOSTOME

↑   acquisition of epifaunal mode of
     life and colonial organization

PHORONID

Fig. 8. Summary of the probable evolutionary changes which occurred during
early bryozoan history.

species may have given rise to double-walled species and vice versa more
than once during bryozoan evolution. Phylogenetic transformations of
this type could be accomplished by modifying the relative timing of
developmental processes. Gould (1977) has argued that many important
evolutionary changes may have been brought about in this way. In the
case of stenolaemate bryozoans, the evolutionary change from the
postulated ancestral stenolaemates (single-walled cyclostomes) may have
been due to a relative retardation in the growth of interzooecial walls
during phylogeny so that ultimately they failed to reach the outer colony

wall above the expanding hypostegal coelom. The result of this type of change is paedomorphosis—the displacement of an ancestral feature to the later stages of the ontogeny of descendants.

By allowing lengthening of interzooecial walls throughout colony life, paedomorphosis in the Stenolaemata gave the potential for the development of larger colonies and for greater plasticity in colonial growth form. For example, thick-branched dendroid trepostomes would have been able to withstand strongly agitated environments which could not have been colonized by fragile erect cyclostomes. In addition, the hypostegal coelom linking zooids throughout the colony in double-walled stenolaemates may have had a role in coordination of colony-wide activities. Thus the acquisition during early bryozoan history of a double-walled organization expanded the range of adaptive zones available to the Bryozoa.

The first double-walled stenolaemates may well have been trepostome-like. Tavener-Smith (1975) inferred cryptostome evolution from trepostomes and the same may be true for cystoporates, although further detailed studies are necessary to confirm the relationships between stenolaemate orders.

7. Summary

The changes postulated for the early structural evolution of the Bryozoa are summarized in Fig. 8. The ctenostome–corynotrypid transition is considered to have involved the single most crucial structural innovation. This opened the way during the Ordovician for the rapid evolution of a great variety of bryozoans.

EARLY ECOLOGICAL DIVERSIFICATION

A number of authors have categorized colony form in Bryozoa (e.g. Stach, 1936; Lagaaij and Gautier, 1965; Schopf, 1969). The terminologies adopted by these authors refer especially to extant cheilostome bryozoans and have been related to named genera often implying a particular mode of life. Consequently, it seems inadvisable to use these terms for the stenolaemate dominated Ordovician bryozoan fauna and a more generally descriptive classification of colony types is given here (Table I). Most of the colony forms listed in Table I have morphological, and possibly functional parallels among extant Bryozoa. Recent bryozoans include very few additional colony types.

Encrusting colonies required a firm substrate (e.g. shell) for attachment. It seems that sheet-like colonies (developed in some cyclostomes, trepostomes and cystoporates) were in general superior competitors for substrate space. Linear branched colonies (in some cyclostomes) were, however,

TABLE I. The major categories of colony growth-form developed by Ordovician bryozoans

Relation to Substratum	Shape of Colony	Example
Encrusting	sheet-like	*Sagenella*
	linear-branched	*Stomatopora*
Erect	domed	*Orbipora*
	narrow cylindrical branched	*Mitoclema*
	massive cylindrical branched	*Stigmatella*
	frondose	*Meekopora*
	fenestrate	*Stictoporella*
	jointed	*Arthroclema*
Penetrant	linear-branched	*Vinella*

better adapted to locating spatial refuges (Buss, 1979) where resources (e.g. nutrients, substrate space) were more readily available. Erect colony forms avoided the spatial restrictions of the substrate and had the ability to over-top (cf. plants) other members of the community. Large dome-shaped or hemispherical colonies (trepostomes, cystoporates) with extensive planar bases are particularly abundant in the Ordovician. The colony base extended beyond the substrate, which provided the initial site of attachment and the development of a wide base produced a very stable colony form resistant to overturn (cf. corals, Abbott, 1974). Some small trepostomes have a distinctly concavo-convex shape comparable to that of Recent lunulitiform cheilostomes (see Lagaaij, 1963). These cheilostomes have a free living often mobile existence on unconsolidated muddy and sandy sediments (Cook and Chimonides, 1978). Narrow cylindrical branched colonies (cyclostomes, cryptostomes), with branch diameters usually less than 2 mm, were probably mechanically weak and tolerant only of low energy environments. Massive cylindrical branched colonies (trepostomes) often developed a tree-like morphology. These large and apparently often long-lived forms were certainly resistant to vigorous current action but flourished also in quieter conditions. Frondose colonies may be solid and bilamellar, or may have a central space between unilamellar expansions. Both massive cylindrical and frondose colonies often have abundant and conspicuous monticules suggesting that, in these colony forms, there was a well developed extrazooidal feeding current system over the surface of the colony (Banta *et al.*, 1974; Cook, 1977). Fenestrate colonies (crypto-stomes) have foliaceous zoaria penetrated by numerous lacunae or fenestrules which may have allowed the passage of an extrazooidal feeding current through the colony (Cowen and Rider, 1972; McKinney, 1977). Many fenestrate and frondose colonies were probably orientated at right-angles to external current flow thus maximizing the area of the colony available for food collection (cf. gorgonians, Wainwright *et al.*, 1976). This suggests adaptation to regimes of directional current flow. Jointed colonies (cryptostomes) with articulating nodes (Blake, 1979) had the ability to flex with the current and were probably capable of withstanding considerably greater agitation than their rigid counterparts. Penetrant bryozoans (ctenostomes) benefited from the protection afforded by the hard substrates into which they bored. Their borings are preserved as trace fossils.

Reefs constitute one of the important environments which were swiftly occupied by the newly diversified Bryozoa. Cambrian archaeocyathid reefs lack bryozoans (e.g. James and Kobluk, 1978) and they are also absent from a lower Ordovician reef described by Toomey (1970), but they are a very conspicuous element of the fauna in middle Ordovician reefs (Ross, 1964; Alberstadt *et al.*, 1974; Walker and Ferrigno, 1973).

Ecological studies (Ross, 1964, 1972; Anstey and Perry, 1972) of Ordovician bryozoan faunas are somewhat infrequent, but Palmer and Pamer (1977) have provided an elegant account of the bryozoans colonizing a middle Ordovician hardground.

DISCUSSION

1. Coloniality in Ordovician Bryozoa

Differing degrees of coloniality may be recognized among the Bryozoa and other colonial animals. Increasing colony dominance over zooid autonomy is manifested by higher levels of physiological integration and functional differentiation of the zooids within a colony (Beklemishev, 1970; Cowen and Rider, 1972; Boardman and Cheetham, 1973).

High levels of physiological zooidal integration were apparently attained in most Ordovician stenolaemates with the exception of the cyclostomes in which interzooidal pores were usually lacking and mature zooids were not linked by a hypostegal coelom (Brood, 1975). The trepostomes and cryptostomes also lacked interzooidal pores but evidence indicates linkage of mature zooids by hypostegal coelom (Tavener-Smith, 1975; Boardman, 1971). Cystoporates probably achieved the highest levels of integration often having both interzooidal pores and an extensive colony-wide hypostegal coelom linking zooids across areas of vesicular extrazooidal tissue (Utgaard, 1973).

Functional differentiation of zooids is indicated when more than one type of zooid is present in the colony. Feeding zooids are known as autozooids and other polymorphs may be termed heterozooids. A variety of apparent heterozooids has been found in Ordovician bryozoans e.g. trepostome mesozooecia, cystoporate exilazooecia (see Boardman and Cheetham, 1973; Utgaard, 1973). However, by comparison with the dominant group of extant bryozoans, the Cheilostomata, levels of zooidal polymorphism in the Palaeozoic were relatively low. Defensive polymorphs (notably many avicularia) are abundant in living cheilostomes but have not been recognized with certainty in any Ordovician (or Palaeozoic) bryozoans. Their development in the cheilostomes was made possible (Ryland, 1970; Schopf, 1973) by the possession of a modifiable zooidal operculum (absent in Ordovician bryozoans), and the necessity for defensive polymorphs may reflect the apparent increased level of specialized predation which evolved during the Mesozoic (Vermeij, 1977).

Another feature taken to indicate a high degree of coloniality (Beklemishev, 1970; Boardman and Cheetham, 1973) is the possession of subcolonies or cormidia composed of discrete groups of zooids. These include monticules which arose early during stenolaemate history. Anstey et al. (1976) have inferred that monticules functioned both as chimneys of exhalant extrazooidal current flow (Banta et al., 1973) and loci of zooidal budding (Anstey and Delmet, 1973) as well as releasing morphogenetic substances (?hormones) with a controlling effect on surrounding zooids.

2. Changes in Colony Size

Although data relating to individual "lineages" is lacking, it seems likely that a major trend in early bryozoan evolution was towards increased colony size following the well known Copes Rule. Increased colony size was made possible principally by two structural innovations; polypide eversion by deformation of walls within the zooid (ctenostome to corynotrypid stenolaemate transition), and retention of a hypostegal coelom throughout the colony (single-walled to double-walled stenolaemate transition). For a variety of reasons large body size in non-colonial animals tends on average to provide significant adaptive advantages over small body size (Bonner, 1965; Hallam, 1975; Stanley, 1973) and increasing body size is therefore a commonly documented phylogenetic trend. Large size in colonial animals would have had a direct selective advantage if, as Thorpe (in press) assumes for the Bryozoa, reproductive fecundity is directly proportional to the number of zooids in the colony.

3. The Timing of Bryozoan Diversification

The morphological innovations postulated above satisfactorily explain how the rapid early diversification of the Bryozoa was achieved, but they do not explain the timing of the diversification. The early Ordovician radiation of the Bryozoa coincides with a rise in numbers of other suspension feeding groups e.g. brachiopods and bivalves. Both Farmer (1977) and Cowen (1976) relate this rise in suspension feeders to environmental factors predicted by Valentine and Moores (1972) from plate tectonic evidence of land and sea configuration in the early Ordovician. Farmer stresses the overall amelioration of global marine climates and Cowen emphasizes the greater stability of planktonic food resources probably obtaining during the early Ordovician which would have favoured suspension feeding as an alternative to deposit feeding.

ACKNOWLEDGEMENTS

We wish to thank Miss P. L. Cook, Department of Zoology, British Museum (Natural History), for her valuable comments on a draft typescript of this paper and Dr H. W. Ball for authorizing the loan of material from the collections of the Department of Palaeontology, British Museum (Natural History). P.D.T. gratefully acknowledges receipt of a N.E.R.C. research fellowship.

REFERENCES

ABBOTT, B. M. (1974). Flume studies on the stability of model corals as an aid to quantitative palaeoecology. *Palaeogeogr., Palaeoclimatol., Palaeoecol.* **15**, 1–27.
ALBERSTADT, L. P., WALKER, K. R. and ZURAWSKI, R. P. (1974). Patch reefs in the Carters Limestone (Middle Ordovician) in Tennessee, and Vertical Zonation in Ordovician Reefs. *Bull. geol. Soc. Am.* **85**, 1171–1182.

ANSTEY, R. L. and DELMET, D. A. (1973). Fourier analysis of zooecial shapes in fossil tubular bryozoans. *Bull. Geol. Soc. Am.* **84**, 1753–1764.

ANSTEY, R. L. and PERRY, T. G. (1972). Eden Shale bryozoans: a numerical study (Ordovician, Ohio Valley). *Publ. Mich. St. Univ. Mus., Paleont. Ser.* **1**, 1–80.

ANSTEY, R. L., PACHUT, J. F. and PREZBINDOWSKI, D. R. (1976). Morphogenetic gradients in Paleozoic bryozoan colonies. *Paleobiology*, **2**, 131–146.

ASTROVA, G. G. (1964). O novom otryade paleozoiskikh mshanok. *Paleont. Zh.* **2** (1964), 22–31.

ASTROVA, G. G. (1965). Morfologiya, istoriya razvitiya i sistema ordovikskikh i siluriiskikh mshanok. *Trudy paleont. Inst.* **106**, 1–309.

BANTA, W. C. (1976). Origin and early evolution of cheilostome Bryozoa. "*Bryozoa 1974*" *Docum. Lab. Geol. Fac. Sci. Lyon. H.S.* **3** (2), 565–582.

BANTA, W. C., McKINNEY, F. K. and ZIMMER, R. L. (1974). Bryozoan monticules: excurrent water outlets? *Science, N.Y.* **185**, 783–784.

BASSLER, R. S. (1911a). *Corynotrypa*, a new genus of tubuloporoid Bryozoa. *Proc. U.S. natn. Mus.* **39**, 497–527.

BASSLER, R. S. (1911b). The early Paleozoic Bryozoa of the Baltic provinces. *Bull. U.S. natn. Mus.* **77**, 1–382.

BASSLER, R. S. (1952). Taxonomic notes on genera of fossil and Recent Bryozoa. *J. Wash. Acad. Sci.* **42**, 381–385.

BASSLER, R. S. (1953). Bryozoa. *In* "Treatise on Invertebrate Paleontology, Part G" (R. C. Moore, ed.). Geol. Soc. Am. and Univ. of Kansas Press.

BEKLEMISHEV, W. N. (1970). "Principles of Comparative Anatomy of Invertebrates. Volume 1: Promorphology." Oliver and Boyd, Edinburgh.

BLAKE, D. B. (1979). The Arthrostylidae and Articulated Growth Habits in Paleozoic Bryozoans. *In* "Advances in Bryozoology I" (M. B. Abbott and G. P. Larwood, eds). Academic Press, London.

BOARDMAN, R. S. (1971). Mode of growth and functional morphology of autozooids in some Recent and Paleozoic tubular Bryozoa. *Smithson. Contr. Paleobiol.* **8**, 1–28.

BOARDMAN, R. S. and CHEETHAM, A. H. (1973). Degrees of colony dominance in stenolaemate and gymnolaemate Bryozoa. *In* "Animal Colonies" (R. S. Boardman, A. H. Cheetham and W. J. Oliver, eds), pp. 121–220. Dowden, Hutchinson and Ross, Stroudsburg.

BONNER, J. T. (1965). "Size and Cycle". Princeton Univ. Press.

BORG, F. (1926). Studies on Recent cyclostomatous Bryozoa. *Zool. Bidr. Upps.* **10**, 181–507.

BROOD, K. (1972). Cyclostomatous Bryozoa from the Upper Cretaceous and Danian in Scandinavia. *Stockh. Contr. Geol.* **26**, 1–464.

BROOD, K. (1975). Cyclostomatous Bryozoa from the Silurian of Gotland. *Stockh. Contr. Geol.* **28**, 45–119.

BROOD, K. (1976). Wall structure and evolution in cyclostomate Bryozoa. *Lethaia*, **9**, 377–389.

BUSS, L. W. (1979). Habitat, Selection, Directional Growth and Spatial Refuges: Why Colonial Animals have more Hiding places. *In* "The Biology and Systematics of Colonial Organisms" (G. P. Larwood and B. R. Rosen, eds), pp. 459–497. Academic Press, London.

CANU, F. and BASSLER, R. S. (1922). Studies on the cyclostomatous Bryozoa. *Proc. U.S. natn. Mus.* **61**, 1–160.

CLARK, R. B. (1964). "Dynamics in Metazoan Evolution". Clarendon Press, Oxford.

COBBOLD, E. S. (1931). Additional fossils from the Cambrian rocks of Comley, Shropshire. *Q. Jl geol. Soc. Lond.* **87**, 459–512.

COOK, P. L. (1977). Colony-wide water currents in living Bryozoa. *Cah. Biol. mar.* **18**, 31–47.

COOK, P. L. and CHIMONIDES, P. J. (1978). Observations on living colonies of *Selenaria* (Bryozoa, Cheilostomata). Part 1. *Cah. Biol. mar.* **19**, 147–158.

COWEN, R. (1976). "History of Life". McGraw-Hill, New York.

COWEN, R. and RIDER, J. (1972). Functional analysis of fenestellid bryozoan colonies. *Lethaia*, **5**, 145–164.

DZIK, J. (1975). The Origin and Early Phylogeny of the cheilostomatous Bryozoa. *Acta palaeont. pol.* **20**, 395–423.

FARMER, J. D. (1977). An Adaptive Model for the Evolution of the Ectoproct Life Cycle. *In* "Biology of Bryozoans" (R. L. Zimmer and R. M. Woollacott, eds), pp. 487–517. Academic Press, New York.

FARMER, J. D., VALENTINE, J. W. and COWEN, R. (1973). Adaptive Strategies leading to the Ectoproct Ground-Plan. *Syst. Zool.* **22**, 233–239.

FRIC, A. (1901). Die thierischen Reste der Perucer Schichten. *Arch. naturw. LandDurchforsch. Böhm.* **11**, 163–180.

FRITZ, M. A. (1947). Cambrian Bryozoa. *J. Paleont.* **21**, 434–435.

FRITZ, M. A. (1948). Cambrian Bryozoa more precisely dated. *J. Paleont.* **22**, 373.

GOULD, S. J. (1977). "Ontogeny and Phylogeny". The Belknap Press of Harvard University Press, Cambridge, Massachusetts.

HALLAM, A. (1975). Evolutionary size increase and longevity in Jurassic bivalves and ammonites. *Nature*, **258**, 493–496.

HARMER, S. F. (1930). Polyzoa (Presidential address 1929). *Proc. Linn. Soc. Lond.* **141**, 68–114.

HOFMANN, H. J. (1975). *Bolopora* not a bryozoan, but an Ordovician phosphatic, oncolitic accretion. *Geol. Mag.* **112**, 523–526.

JACKSON, J. B. C. (1977). Growth form, zooid morphology and evolutionary rates of marine colonial animals. *J. Paleont.* **51** (pt. 2, suppl. 3. *N. Am. Paleont. Conv.* 2, *Abstr.*), 16.

JÄGERSTEN, G. (1972). "Evolution of the Metazoan Life Cycle." Academic Press, London.

JAMES, N. P. and KOBLUK, D. R. (1978). Lower Cambrian patch reefs and associated sediments: southern Labrador, Canada. *Sedimentology*, **25**, 1–35.

JEBRAM, D. (1973). The Importance of Different Growth Directions in the Phylactolaemata and Gymnolaemata for Reconstructing the Phylogeny of the Bryozoa. *In* "Living and Fossil Bryozoa" (G. P. Larwood, ed.), pp. 565–576. Academic Press, London.

LAGAAIJ, R. (1963). *Cupuladria canariensis* (Busk)—portrait of a bryozoan. *Palaeontology*, **6**, 172–217.

LAGAAIJ, R. and GAUTIER, Y. V. (1965). Bryozoan assemblages from marine sediments of the Rhone delta, France. *Micropaleontology*, **11**, 39–58.

LARWOOD, G. P., MEDD, A. W., OWEN, D. E. and TAVENER-SMITH, R. (1967). Bryozoa. *In* "The Fossil Record" (W. B. Harland *et al.*, eds), pp. 379–395. Geol. Soc., London.

LOMAS, J. (1885). On the occurrence of internal calcareous spicules in Polyzoa. *Proc. Lpool geol. Soc.* **5**, 241–243.

MCILREATH, I. A. (1977). Accumulation of a Middle Cambrian, | deep-water limestone debris apron adjacent to a vertical, submarine carbonate escarpment, Southern Rocky Mountains, Canada. *In* "Deep-water Carbonate Environments" (H. E. Cook and P. Enos, eds). *Soc. Econ. Paleontologists Mineralogists, Spec. Publ.* **25**, 113–124.

MCKINNEY, F. K. (1969). Organic structures in a late Mississippian trepostomatous ectoproct (bryozoan). *J. Paleont.* **43**, 285–288.

MCKINNEY, F. K. (1977). Functional interpretation of lyre-shaped Bryozoa. *Paleobiology*, **3**, 90–97.

MCLEOD, J. D. (1978). The Oldest Bryozoans: New Evidence from the Early Ordovician. *Science, N.Y.* **200**, 771–773.

MÜLLER, A. H. (1958). "Lehrbuch der Paläozoologie. Band II. Invertebraten. Teil 1. Protozoa-Mollusca 1". Gustav Fischer Verlag, Jena.

NIELSEN, C. (1970). On metamorphosis and ancestrula formation in cyclostomatous Bryozoa. *Ophelia*, **7**, 217–259.

NIELSEN, C. (1971). Entoproct life-cyles and the entoproct/ectoproct relationship. *Ophelia*, **9**, 209–341.

NIELSEN, C. (1977a). The Relationships of Entoprocta, Ectoprocta and Phoronida. *Am. Zool.* **17**, 149–150.

NIELSEN, C. (1977b). Phylogenetic Considerations: The Protostomian Relationships. *In* "Biology of Bryozoans" (R. L. Zimmer and R. M. Woollacott, eds), pp. 519–534. Academic Press, New York.

NIELSEN, C. and PEDERSEN, K. J. (in preparation). Cystid structure and the protrusion of the polypide in *Crisia*. (Paper read at the 4th Conf. Int. Bryozoology Ass., Woods Hole, Sept. 1977).

OSTROUMOFF, A. A. (1885). Extrait de l'oeuvre sur le morphologie des bryozoaires marines. *Zool. Anz.* **8**, 577–579.

PALMER, T. J. and PALMER, C. D. (1977). Faunal distribution and colonization strategy in a Middle Ordovician hardground community. *Lethaia*, **10**, 179–199.

POHOWSKY, R. A. (1974). Notes on the study and nomenclature of boring Bryozoa. *J. Paleont.* **48**, 556–564.

PRENANT, M. and BOBIN, G. (1956). Bryozoaires, 1re partie Entoproctes, Phylactolemes, Ctenostomes. *Faune Fr.* **60**, 1–398.

RAUP, D. M. (1976). Species diversity in the Phanerozoic: a tabulation. *Paleobiology*, **2**, 279–288.

ROSS, J. P. (1964). Morphology and phylogeny of early Ectoprocta (Bryozoa). *Bull. Geol. Soc. Am.* **75**, 927–948.

ROSS, J. P. (1966). Early Ordovician Ectoproct from Oklahoma. *Okla. Geol. Notes*, **26**, 218–224.

Ross, J. P. (1972). Paleoecology of Middle Ordovician Ectoproct Assemblages. *24th Int. geol. Congr.*, Sect. **7**, 96–102.

Ryland, J. S. (1970). "Bryozoans". Hutchinson University Library, London.

Ryland, J. S. (1976). Physiology and ecology of marine Bryozoans. *Adv. Mar. Biol.* **14**, 285–443.

Schopf, T. J. M. (1969). Paleoecology of Ectoprocts (Bryozoans). *J. Paleont.* **43**, 234–244.

Schopf, T. J. M. (1973). Ergonomics of Polymorphism. *In* "Animal Colonies". (R. S. Boardman, A. H. Cheetham and W. J. Oliver, eds), pp. 247–294. Dowden, Hutchinson and Ross, Stroudsbury.

Schopf, T. J. M. (1977). Patterns and themes of evolution among the Bryozoa. *In* "Patterns of Evolution" (A. Hallam, ed.), pp. 159–207. Elsevier, Amsterdam.

Silén, L. (1966). On the fertilization problem in gymnolaematous Bryozoa. *Ophelia*, **3**, 113–140.

Stach, L. W. (1936). Correlation of zoarial form with habitat. *J. Geol.* **44**, 60–65.

Stanley, S. M. (1973). An explanation for Cope's Rule. *Evolution*, **27**, 1–26.

Tavener-Smith, R. (1973). Some Aspects of Skeletal Organization in Bryozoa. *In* "Living and Fossil Bryozoa" (G. P. Larwood, ed.), pp. 349–359. Academic Press, London.

Tavener-Smith, R. (1975). The phylogenetic affinities of fenestelloid bryozoans. *Palaeontology*, **18**, 1–17.

Taylor, P. D. (1978). A Jurassic ctenostome bryozoan from Yorkshire. *Proc. Yorks. geol. Soc.* **42**, 211–216.

Thorpe, J. P. (1979). A model using deterministic equations to describe some possible parameters affecting growth rate and fecundity in Bryozoa. *In* "Advances in Bryozoology I" (M. B. Abbott and G. P. Larwood, eds). Academic Press, London.

Toomey, D. F. (1970). An unhurried look at a Lower Ordovician mound horizon, South Franklin Mountains, West Texas. *J. sedim. Petrol.* **40**, 1318–1334.

Utgaard, J. (1973). Mode of colony growth, autozooids and polymorphism in the bryozoan Order Cystoporata. *In* "Animal Colonies" (R. S. Boardman, A. H. Cheetham and W. J. Oliver, eds), pp. 317–360. Dowden, Hutchinson and Ross, Stroudsburg.

Valentine, J. W. (1973). Coelomate Superphyla. *Syst. Zool.* **22**, 97–102.

Valentine, J. W. and Moores, E. M. (1970). Plate-tectonic Regulation of Faunal Diversity and Sea Level: a Model. *Nature, Lond.* **228**, 657–659.

Vermeij, G. J. (1977). The Mesozoic marine revolution: evidence from snails, predators and grazers. *Paleobiology*, **3**, 245–258.

Voigt, E. (1977). *Arachnidium jurassicum* n.sp. (Bryoz. Ctenostomata) aus dem mittleren Dogger von Goslar am Harz. *Neues Jb. Geol. Paläont. Abh.* **153**, 170–179.

Wainwright, S. A., Biggs, W. D., Currey, J. D. and Gosline, J. M. (1976). "Mechanical Design in Organisms", Edward Arnold, London.

Walker, K. R. and Kerrigno, K. F. (1973). Major Middle Ordovician reef tract in East Tennessee. *Am. J. Sci.* **273**, 294–325.

WALTER, B. (1969). Les Bryozoaires Jurassiques en France. *Docum. Lab. Geol. Fac. Sci. Lyon*, **35**, 1–328.
WILLIAMS, A., STRACHAN, I., BASSETT, D. A., DEAN, W. T., INGHAM, J. K., WRIGHT, D. W. and WHITTINGTON, H. B. (1972). A Correlation of Ordovician rocks in the British Isles. *Spec. Rep. geol. Soc. Lond.* **3**, 1–74.

8 | Brachiopod Radiation

A. D. WRIGHT

Department of Geology, the Queen's University of Belfast, Northern Ireland

Abstract: The brachiopod animal, basically a sessile, epifaunal suspension feeder, is essentially protected from its external environment by the secretion of a pair of valves. Equally fundamental however is the need for the animal to maintain contact with the environment by opening these valves. As far as fossil brachiopods are concerned these two aspects, shell secretion and shell articulation, overshadow all others in importance and are strongly reflected in the radiation of the phylum. Early diversification is also evidenced by the varying mode of attachment to the substrate as revealed in the form of the pedicle opening and to a lesser extent by the development of varied lophophore supports.

The view taken here is that the brachiopods as now understood did not evolve from a single prototype at the beginning of Cambrian times, but developed from a series of infaunal lophophorates that had already undergone diversification in the late Precambrian prior to emerging as epifaunal brachiopods with mineralized valves (the ancestral lophophorates with potential to develop into brachiopods are referred to as "brachiophorates"). Thus the lower Cambrian opens with a diverse inarticulate brachiopod fauna including the orders Lingulida, Paterinida, Kutorginida, Acrotretida and an order Discinida, here separated from the Acrotretida. The Obolellida and the basic articulate stock, the billingsellaceans, appear later in the lower Cambrian. The orthide billingsellaceans with their strophic hinge lines, deltidiodont dentition and invariably calcareous shells, comprise two families which, although morphologically similar, have a fundamentally different style of shell secretion, reflected in their various descendants.

In late Cambrian times the Porambonitacea developed from the Orthacea to form a second order, the Pentamerida, with non-strophic shells associated with the deltidiodont dentition; but the major radiation of the articulate brachiopods took place in Ordovician times with the development of the Strophomenida, with essentially conservative hinge mechanisms, and the Rhynchonellida and Atrypida,

---

Systematics Association Special Volume No. 12, "The Origin of Major Invertebrate Groups", edited by M. R. House, 1979, pp. 235–252, Plate III. Academic Press, London and New York.

both non-strophic and with cyrtomatodont dentition. Although the strophic Spiriferida did not arrive until the early Silurian and the most advanced Terebratulida only with the Devonian, it is noteworthy that virtually all the morphological features known for the phylum are developed with the Ordovician radiation.

Several calcareous shelled non-"articulate" stocks, some of which are regarded as being derived from continuing brachiophorate stocks, also appeared with the Ordovician radiation, including the Craniida, here accepted as a distinct order. Except for this specialized group, attempts to develop calcareous shells met with little success apart from the two secretory regimes which, with their variations, characterize the major group of the "articulate" brachiopods.

INTRODUCTION

Studies of the last century showed that the phylum Brachiopoda could be fairly readily divided into two classes, usually referred to as the Articulata and Inarticulata. Although various morphological criteria have been used for the subdivision, the greatest importance is usually accorded to the presence or absence of a hinge mechanism (with interlocking teeth and sockets), or the absence or presence of an anal aperture. Williams and Rowell (1965) commented that "few would now dispute the validity and status of the Inarticulata and Articulata as classes within the Phylum" and concluded that only the Kutorginida could not be placed with confidence in either class.

In endeavouring to assess the diverse groups of the early brachiopod radiation, one of the striking features is the unsatisfactory nature of the twofold subdivision of the phylum. This is less apparent in the living stocks, where such articulates as *Hemithyris* and *Terebratulina* may be readily differentiated from inarticulates like *Lingula* and *Discina*, but during the Cambrian and Ordovician there are several stocks whose placement within either the articulates or inarticulates is controlled more by the belief that they must belong to one class or the other, rather than on evidence of undoubted affinity based on shell morphology. Thus my interpretation of the radiation of the phylum does not entirely accord with the usual twofold subdivision, nor with the belief of Williams and Hurst (1977) that the brachiopods are monophyletic.

ORIGINS OF THE PHYLUM

The Brachiopoda appear in the fossil record at the base of the Cambrian, where they show a diversity indicative of a substantial Precambrian ancestry, possibly as brachiopods with an organic, non-preservable skeleton or alternatively as diversified ancestral lophophorate stocks, whose establishment in the epifauna was simultaneous with the development of a mineralized exoskeleton. Clark (1964) and Valentine (1973, 1975, 1977) have focused attention on the development of different feeding habits and suggested that worm-like infaunal stocks radiated into epifaunal elements at the beginning of Cambrian times.

The group most favoured as closest to the ancestral lophophorate stock is the Phoronida (Hyman, 1959), with a vermiform body and a tentacle bearing crescentic lophophore. Cowen and Valentine (Valentine, 1973) believe that each brachiopod class arose from a phoronid-like infaunal worm along a separate adaptive pathway, with the exoskeleton primarily developing to contain feeding currents in the adaptation to an epifaunal existence. They thus regard the phylum as polyphyletic and Valentine believes that, to achieve monophyletic taxa, the distinct lineages of the early inarticulates should have phyletic status. Regardless of the merit of this proposal, the fact is that the diverse stocks of the early radiation could have had their origin in various tubicolous infaunal ancestral stocks transferring to a sessile epifaunal habitat, as much as from a single ancestral stock becoming a "brachiopod" which subsequently produced the radiation.

Accepting an origin from differentiated tubiculous lophophorates at the beginning of Cambrian times, several fundamental requirements have to be satisfied along with this change in habit. Firstly, instead of existing within the protection of a burrow, the animal became exposed on the sea bed and open to direct attack by predators. In the case of the brachiopods, the necessary protection was acquired by the secretion of a pair of mineralized valves which also acted as a container to organize the feeding currents for the lophophore. Secondly, to allow currents to enter the mantle cavity, mechanisms had to be developed for the efficient opening and closing of the valves. Thirdly, the animal must have a measure of stability on the substrate; this was primarily provided by the pedicle.

With diverse lophophorates following this general evolutionary pattern, it seems reasonable that the existing variations should also give rise to differing modes of shell secretion, patterns of articulation and methods of attachment; much diversity is undeniably present in the lower Cambrian faunas.

Although these developments took place at the beginning of the Cambrian, there seems no reason why some similar mineral skeletons could not have arisen later, so that enigmatic Ordovician forms need not necessarily be the descendants of pre-existing brachiopods with a mineralized exoskeleton; they could equally well have arisen at that time from soft-bodied "brachiophorates", i.e. infaunal lophophorate stocks with the potential to develop into epifaunal brachiopods.

As regards the skeleton in brachiopods, Williams has rightly stressed the fundamental importance of shell secretion to establishing the affinities of different groups of brachiopods in numerous studies (1956, 1968a, b, 1970, 1973) which have made an outstanding contribution to the knowledge of the brachiopod shell.

Recent important contributions on the hinge mechanisms of articulate brachiopods are those of Rudwick (1959) and Jaanusson (1971). Rudwick has recognized a fundamental difference between shells in which the axis of rotation of the valves coincides with the hinge line (strophic type) and those in which it does not (non-strophic type). Jaanusson has observed that hinge teeth may grow either by simple addition of skeletal material (deltidiodont dentition) or by a complex process of secretion and resorption (cyrtomatodont dentition). Inarticulates are supposedly without skeletal hinge mechanisms, a popular view which is not entirely correct.

The development of various styles of secretion resulting in differently structured shell substance, the establishment of varying articulation mechanisms and the varied means of attachment to the substrate are the most important features of the early brachiopod radiations. This is not to say that the form of the intestinal tract, the non-mantle fleshy tissue in general, embryology, lophophore supports, etc. are of no consequence; but at this time in the evolution of the phylum little evidence is available for the soft tissue, while lophophore supports had a relatively minor role.

The first major radiation took place in the lower Cambrian with the appearance of all inarticulate orders except the Craniida, and the first articulates. Although a further order of articulates appeared in the upper Cambrian, the main radiation of this class took place in the Ordovician. Despite the fact that two orders of articulate brachiopods (Spiriferida and Terebratulida) had not yet appeared by the end of the Ordovician, all the main morphological features seen in the phylum had, and one could say that as far as the brachiopods are concerned, nothing is new after the Ordovician.

LOWER CAMBRIAN RADIATION

The earliest of the lower Cambrian brachiopods are those recorded from the pre-trilobite Tommotian Stage of the Siberian Platform (Zhuravleva, 1970) and its equivalents; these are all inarticulate brachiopods. Inarticulates have been recorded from the lowest of the three Tommotian archaeo-cyathid zones of the Aldan River section, and cited as "numerous" in the highest zone at the Byukteleekh locality (Keller *et al.*, 1973). The Tommotian faunas are being described by Yuri Pelman and a considerable amount of important information will be available when his studies are completed. To date the obolellid *Sibiria* is the only new genus to have been described from the east Siberian lower Cambrian succession (Goryansky, 1977); this however is of post-Tommotian age.

The British "Non-Trilobite Zone" equivalent (Cowie *et al.*, 1972) has yielded *Paterina* and *Walcottina* from the lower Comley Sandstone of Comley (Cobbold, 1921) and *Acrotreta*, *Acrothele*, *Lingulella* and question-ably *Kutorgina* from equivalent horizons at Rushton (Cobbold and Pocock, 1934). Rowell (1977a) has pointed out that the commonly cited presence of *Obolella* in the Non-Trilobite Zone of Britain is in error, as the shells are phosphatic and thus cannot be ascribed to the calcareous shelled Obolellida; the species, "*Obolella*" *groomi* Matley, is possibly a lingulide.

Thus although current studies may add to the orders present, it is clear that the Acrotretida, Lingulida, Paterinida and perhaps Kutorginida are established in these earliest of Cambrian faunas.

The earliest acrotretides are represented by the highly distinctive Acrotretidae. These are characteristically minute, with a highly conical pedicle valve and a flat brachial valve. The pedicle valve has a minute apical foramen, which may continue internally as a tube. Recently Chuang (1971) suggested that the acrotretacean foramen might have functioned as an opening for an exhalant current; this radical view is not generally accepted by other workers (e.g. Rowell, 1977b) who regard the foramen more conventionally as the predicle opening.

The acrotretids undoubtedly possess a primitive hinge mechanism, as suggested by Rowell (1977b); the flattened pseudointerarae (which may be modified by a ridge or intertrough) forms an essentially straight posterior margin to the pedicle valve which is reflected in the brachial valve by a well developed pseudointerarea divided by a median groove. At the lateral edges of the groove the propareas are thickened, and a pair of processes commonly developed; in several genera (*Undiferina*, *Torynelasma*, *Spondylotreta*) these project anteroventrally to form what can only be hinge teeth (Pl. III, 1–5), fitting into recesses in the pedicle valve. With such structure it is difficult to envisage any sort of valve rotation comparable to that of modern *Lingula* or *Discina*.

Another feature of the acrotretids to receive attention recently has been the protegulum. The scanning electron microscope studies of Biernat and Williams (1970) showed the protegula to be ornamented by shallow pits which they interpreted as the moulds of a highly vesicular periostracum serving to provide extra buoyancy to the larvae before settling. On the evidence available to Biernat and Williams, no comparable development is present in the other superfamilies usually regarded as belonging to the Acrotretida i.e. the Siphonotretacea, Discinacea and Craniacea.

This distinctive protegulum would appear then to be characteristic of the Acrotretidae and the related *Curticia*, which were the acrotretaceans examined by Biernat and Williams. No Acrothelidae were studied at this time, but Henderson (1974) described an exceptionally well preserved *Orbithele* which possesses a protegular pitting of similar pattern to that of the acrotretids, although very much coarser.

I would agree with the view of Rowell (1965) that the Acrothelidae are likely ancestors for the Orbiculoidinae and the Discinacea. It may be that the protegular pitting itself is a reflection of the close relationship of the Acrotretidae and Acrothelidae at the "brachiophorate" stage, but as brachiopods the former are so distinct from all other "inarticulates" throughout their existence from Tommotian to Frasnian as to separate them as an emended order Acrotretida, leaving the remainder of the suborder Acrotretidina of *Treatise* usage to form a separate order Discinida.

The Tommotian *Lingulella* is the first of this well known and extant order in which the pedicle emerges from between the valves. Having developed its chitinophosphatic shell following the late Precambrian lophophore radiation into the epifauna, the order reverted to a secondary infaunal burrowing habit during Ordovician times. Modern *Lingula* differs markedly from the early forms in the nature of the posterior of the valves, and most obviously in the major reduction of the pseudointerareas.

Paterina and its allies form another order which in my view represents another group of "brachiophorates" which developed a skeleton independently from other stocks. The "inarticulate" aspect is the chitinophosphatic shell; but morphologically these biconvex valves with their essentially straight posterior margin, well developed pseudointerareas and medianly arched plates resemble the articulates.

The Kutorginida is the earliest brachiopod order which possesses a calcareous skeleton and, as commented by Rowell (1977a), there is still considerable uncertainty over its taxonomic placing. Recent studies of shell ultrastructure have not dealt with the Kutorginida, an omission that needs rectifying as shell ultrastructure has provided very useful evidence in the case of other enigmatic groups. The pedicle valve possesses a supra-apical foramen and an interarea (or pseudointerarea) with a triangular delthyrium partially closed by a pseudo deltidium, thus resembling the earlier articulate billingsellacean stocks. But the valve interiors are poorly known and understood, and the teeth and sockets characterizing the articulates have not been positively identified. The short-lived group does not appear closely related to other brachiopods and is perhaps best regarded as a "brachiophorate" which developed a calcareous skeleton.

The post-Tommotian Lower Cambrian saw the establishment of two other calcareous shelled brachiopod groups, the order Obolellida and the ancestral stock of the articulate brachiopods, the Billingsellacea.

The obolellides again constitute a short-lived order, apparently becoming extinct in the middle Cambrian, whose relationship to the other "inarticulates" is obscure. The group is probably best interpreted as arising from the general brachiophorate plexus and developing the ability to secrete a calcareous shell. Examination of the shell structure of *Trematobolus* by Williams and Wright (1970) revealed the presence of a coarsely crystalline primary layer succeeded by a laminar secondary layer. The laminae are interpreted as being separated by protein sheets with their growth being by a series of screw dislocations as in modern *Crania*. The obolellides have a general inarticulate external appearance and muscle scar distribution; the group is somewhat unusual in the very variable disposition of the pedicle, which may emerge from a grooved interarea, or a foramen situated either apically or anterior to the beak.

Perhaps the most important development at this time is the appearance of the first "articulate" brachiopods in the form of the billingsellacean family Nisusiidae. The articulates have long been considered to have arisen from amongst the earlier inarticulates; the presence of already well developed hinge lines and interareas in the earliest *Nisusia* makes it unlikely that the class rose directly from some brachiophorate even though the teeth, sockets and socket ridges themselves are rudimentary. In the absence of early forms with only weak interareas and hinge lines, it appears most likely that *Nisusia* arose from some biconvex obolid-like chitino-phosphatic stock; if this is the case it poses the intriguing biochemical question as to the mechanism of substitution of calcium carbonate for calcium phosphate in an already mineralized skeleton. Gutmann *et al.* (1978) have explained the development of the articulates as a gradual series of changes involving the development of the interarea and the improvement of the contact between the valves; with hinge development the comparatively bulky hydraulic valve opening mechanism could be reduced to a much smaller directly operating muscle system in the posterior part of the shell. This would give the adaptive advantage of freeing two thirds of the space within the shell cavity for the function of food gathering by the lophophore.

The secondary shell of *Nisusia* is fibrous (Williams, 1968b, 1970) with a pattern identical to that of modern terebratulides and rhynchonellides in which the calcite fibres are segregated by protein sheaths. *Billingsella*, appearing somewhat later than the nisusiids in middle Cambrian times, has a laminar secondary shell characteristic of other, quite distinct, groups of articulate brachiopods. This fundamental distinction in forms which morphologically are very similar is puzzling. Williams and Hurst (1977) point out that the laminar shell is composed of arrays of blades and not tablets as in the calcareous shelled inarticulates, and suggest that they might possibly be derived by a flattening of the inner and outer surfaces of nisusiid fibres.

The billingsellaceans as currently understood form the ancestral stock from which all the subsequent articulate brachiopods are believed to have been derived. The superfamily is conventionally placed in the Orthida, but an apparently early modification in the late lower Cambrian gave rise to the more characteristic orthaceans in which the delthyrium and notothyrium are open. This implies that a pedicle rudiment was developed as a primary segment of the larva as in modern articulates, rather than as a late outgrowth of the ventral body wall (Williams and Rowell, 1965). The orthaceans again have a fibrous secondary shell and are presumed to have arisen from some nisusiid stock, possibly by paedomorphic changes affecting pedicle development in particular.

The only major group to become established in the upper Cambrian is the order Pentamerida, represented by the early porambonitaceans which possess the single most characteristic feature of the order, the spondylium, developed either as a pseudospondylium or sessile spondylium. The earliest form ascribed to the order is the middle Cambrian *Cambrotrophia* which is a primitive form lacking a spondylium, with very rudimentary brachiophores and supporting plates and which is differentiated from contemporaneous orthacean stocks principally by its pronounced dorsal fold and ventral sulcus. The upper Cambrian *Palaeostrophia* has a sessile spondylium but is still rather primitive in the absence of cardinal process and small brachiophores, although supporting plates and fulcral plates define the sockets; the orthacean ancestry is still very apparent.

ORDOVICIAN RADIATION

The Ordovician witnessed the major radiation of the Articulata. Of the pre-existing orders, the orthides in particular displayed great morphological diversity and developed a punctate shell substance extensively. This punctation would appear to be an independent development from that inferred by Williams (1968b) as being present in the lower Cambrian nisusiid *Kotjuella*. The porambonitaceans developed a non-strophic hinge and gave rise to the cyrtomatodont Rhynchonellida which in turn produced the non-strophic spire-bearers, the Atrypida, by middle Ordovician times. Another major development was the establishment of the Strophomenida with essentially conservative strophic hinges and deltidiodont dentitions. Interesting smaller groups of inarticulate, or at least non-articulate, brachiopods which appeared in the Ordovician were the chitinophosphatic Eoconulidae and the calcareous Craniida, Craniopsidae, Trimerellacea and Dictyonellidina. It would appear that some of these last may have evolved their mineralized skeleton directly from continuing "brachiophorate" stocks at this second major phase of brachiopod evolution.

The morphological diversification of the orthides during the Ordovician radiation introduces many structures which occur or become standard in other orders. *Dicoelosia* has unusual puncta which coalesce as in *Terebratulina* (Wright, 1966); the bilobed shell resembles the Devonian strophomenide *Dicoelostrophia* and the Mesozoic terebratulide *Pygope*; the strongly concave brachial valve of some species (Wright, 1968) is typical of the strophomenides; internally the brachiophores develop blades to function in a similar fashion to rhynchonellide crura, and also show evidence of resorption typical of the more advanced articulate orders. The closely related *Epitomyonia* has, in addition to its bilobed shell, a mantle cavity further divided by a dorsal median septum, a common feature of the later enteletaceans. The open delthyrium is modified by deltidial plates in *Barbarorthis*; in *Skenidioides* a spondylium is developed; while examples of the variable external ornaments are the surface spines of the geniculate *Spinorthis*, the pitting of *Saukrodictya*, and the strong costae and granular micro-ornament of *Platystrophia*, the stock believed to have given rise to the strophic spire-bearers (Wright, 1979).

Noteworthy features of the porambonitacean radiation are the reduction of the interareas and hinge lines to produce rostrate non-strophic shells; and internally the development of elongate brachial processes comparable to crura. These features, along with the strong fold and sulcus, commonly well developed costae and the impunctate fibrous secondary shell, indicate a morphology which is very close to that of the rhynchonellides, appearing in the early Ordovician with forms like *Rostricellula*. One major morphological difference is the absence of a spondylium which Williams and Hurst (1977) suggest may have been associated with the forward shift of the ventral muscle field in the rhynchonellides; they regard the prototypic rhynchonellide as probably being descended neotonously from the porambonitacean stock by the precocious development of crura and the suppression of the tendency to form a spondylium. The other major difference in the rhynchonellides is the establishment of cyrtomatodont teeth; this was a major advance in hinge mechanisms and, as was indicated by Jaanusson (1971), required the epithelium to be capable of resorbing shell material.

From these early rhynchonellide stocks the non-strophic spire-bearing order Atrypida arose by middle Ordovician times. Apart from the presence of the calcareous spire, the ancestral rhynchonellides and atrypides are commonly nearly indistinguishable. The small size of the earlier atrypide, *Protozyga*, suggested a neotonous derivation from a rostrate impunctate rhynchonellide to Williams and Wright (1961), in a study which showed that the spire developed in ontogeny firstly from a pair of calcareous prongs, which fused medianly to form a loop before the spires themselves appeared. The implication of this is that the later Terebratulida arose neotonously from a spire-bearing stock by suppression of the growth of the spires.

The electron microscope data for the shell ultrastructure of the early strophomenide brachiopods (Williams, 1968a, 1970; Wright, 1970) supports the suggestion of other morphological data that the order is polyphyletic. The Ordovician strophomenides are characterized by the presence of pseudodeltidia and chilidia, indicating a billingsellacean ancestry; a small suprapical foramen persisted, although in many stocks this became sealed in the adults so that the animal lay free on the sea floor, and there are indications in *Eochonetes* of a potential to become attached by spines which, along with direct cementation, characterized the later evolution of the group.

A strongly concavo-convex profile (which commonly results in an extremely restricted shell space), a semi-circular outline and a pseudopunctate shell substance are typical of the strophomenides; not all however have these characters and the davidsoniaceans and triplesiaceans, appearing in the middle Ordovician, possess a biconvex shell. Further, their laminar shell substance is impunctate in the early Davidsoniace (Williams, 1970) and pseudopuncta are only occasionally developed in the Triplesiacea (Wright, 1970). Other morphological similarities (socket ridges, cardinal process, pseudodeltidia) in these two superfamilies suggest that they developed from a common ancestor close to the Billingsellidae (Williams, 1970) and are quite independent of the other strophomenides.

The fibrous shelled plectambonitacean strophomenides appearing in the Arenig are considered to have evolved directly from the nisusiids. Wide hinged, parvicostellate shells with a primitive cardinal process and socket ridges, their development from the nisusiids primarily involved the production of a concave brachial valve and a pseudopunctate shell condition. A major feature of the radiation of the superfamily is the development of pustules, septa and platforms on the valve interiors associated with the muscles and lophophore.

The Plectambonitacea are regarded as ancestral to the Strophomenacea on the basis of (i) their earlier occurrence; (ii) the primitive cardinal process; (iii) pseudopuncta; and (iv) a primary layer recorded by Williams (1970) as being partly or wholly laminar. To account for the laminar secondary shell of the Strophomenacea, Williams envisaged that the evolution from the plectambonitaceans included the paedomorphic intro-duction of lamination into the primary layer, followed by a neotonous suppression of the fibrous layer. Where known however, the stropho-menacean primary layer is composed of crystallites disposed vertically to the periostracum; this presents a problem if the primary layer has already become laminar and replaced the fibrous secondary layer during develop-ment. An alternative explanation would be that the strophomenaceans were derived directly from a stock of billingsellids which have the same primary and secondary shell as the strophomenaceans. Such an origin would entail the parallel evolution of concavo-convex, parvicostellate shells developing a generally more semi-oval outline, bilobed cardinal process and a pseudopunctate shell condition.

The possibility exists then of a three fold grouping within the Ordovician strophomenides; (i) the concavo-convex fibrous shelled plectambonitacean descendants of the nisusiids; (ii) the concavo-convex laminar shelled strophomenacean descendants of the billingsellids and (iii) the biconvex laminar shelled triplesiacean and davidsoniacean derivatives of the billingsellids.

An enigmatic non-articulate form which appeared early in the Ordovician radiation was the genus *Eoconulus*. Following Cooper (1956), the irregular form of the conical brachial valve coupled with the apparent non-existence of a pedicle valve (presumed attached by cementation) led to the genus being placed with the Craniacea. Not only is the shell chitinophosphatic and impunctate but muscle scars comparable to the anterior adductor pair of *Crania* are not known. Rowell and Krause (1973) discovered the first pedicle valves, in contact with brachial valves, in the Antelope Valley Limestone of Nevada. Although questionably an *Eoconulus*, the shells are certainly eoconulids and the pedicle valves are of interest as they show evidence of cementation, as does the associated acrotretid *Undiferina*.

The more usual absence of pedicle valves from collections of *Eoconulus* may be due to partial mineralization of the attached valves; this is also known from the Craniacea. Study of protegulum of the brachial valve revealed circular pits, but the pattern was regarded as different from that of the Acrotretacea by Biernat and Williams (1970). Thus although the Eoconulidae has been tentatively assigned to the superfamily Acrotretacea by Rowell and Krause, the precise origin of this distinct stock is still uncertain.

The irregularly conical, limpet-like craniacean brachiopods, which appear with *Philhedra* in the Llanvirn, lack an articulating hinge structure but with their calcareous shell have never fitted easily into the Inarticulata. Their anomalous position was underlined by the recent chemical work on the shell proteins of modern *Crania* by Jope (1977). She points out that the *Crania* shell has low glycine (like the inarticulates) and low alanine (like the articulates) but high serine and aspartic acid (like neither class) and concludes that *Crania* occupies an intermediate taxonomic position. This evidence could well be interpreted as indicating a *separate* taxonomic position altogether from that of the recognized classes, and I would regard the group as an order, Craniida, originating at this second major phase of brachiopod evolution from a brachiophorate stock which emerged from the infauna to become a cemented member of the epifauna.

The calcareous shell is quite distinct from that of the articulate stocks, consisting essentially of a crystalline primary layer and a laminar secondary layer growing by screw dislocations; the shell is further permeated by coalescing puncta, the detailed structure of which appears peculiar to the group (Williams and Wright, 1970). Another feature of modern *Crania*, in which it differs from the inarticulates and indeed the other lophophorates, is the position of the intestinal tract which opens in an anus in the middle of the posterior margin, and as suggested by Hyman (1959) it may be that this is the primitive condition of the ancestral lophophorate.

The Craniopsidae form another group of calcareous shelled non-articulate brachiopods that were also cemented or free-lying. These again have a laminar shell but it is impunctate.

The centrally placed raised muscle platforms distinguish the family from the Craniidae, and although in this respect there are more similarities with the lingulide Paterulidae, the calcareous shell precludes this association. The origin is uncertain; if it were to be derived from a lingulide stock it would necessitate crossing the biochemical threshold from phosphate to carbonate secretion; alternatively a "brachiophorate" origin could be envisaged.

The Trimerellacea form another group of enigmatic brachiopods apparently arising from some lingulide-like stock during the Ordovician radiation; the group differs from typical inarticulate stocks in the development of a thick calcareous shell, the possession of a rudimentary hinge-structure and in commonly attaining an unusually large size (75 mm length not being exceptional).

Although the shell substance of the Trimerellacea is of calcium carbonate, Jaanusson (1966) noted that the preservation suggested that the original shell substance was composed of aragonite, a unique development in this phylum.

The hinge structure is illustrated by the pedicle valve of *Eodinobolus* (Pl. III, 10, 11). This has well defined propareas flanking the depressed homeodeltidium; at its anterior margin, extending some way beneath it posteriorly, is the cardinal socket into which fits a transverse plate projecting from the brachial valve. The thickenings at the lateral edges of the hinge plate of this form may also have assisted in articulation, fitting into transverse furrows in the brachial valve. In addition, the scars at the lateral edges of the muscle field may be the location of a pair of muscles pulling along the line of the hinge axis to give stability to the single main calcareous pivot and socket of the hinge. This combination gives a very different articulatory arrangement to those of the articulate brachiopods.

No evidence has yet been found in any species of trimerellacean of a pedicle foramen in the ventral umbo, nor of a groove which could be interpreted as indicating that a pedicle emerged between the valves (Norford, 1960).

Eodinobolus is usually a rare element in Caradoc and Ashgill faunas, but the Silurian *Trimerella* commonly occur in a crowded setting reminiscent of the pentameride *Kirkidium*. Both are exceptionally large biconvex shells lacking pedicle attachment in the adult and evidently lived gregariously, orientated in an upright position by the weight of the thick-shelled umbonal regions. Internally the raised muscle platforms of the *Trimerella* are matched by the spondylial development of *Kirkidium*, suggesting that these structures reflect similar mechanical needs in the opening and closing of the valves.

A particularly problematic group which appeared in the Ordovician radiation is the suborder Dictyonellidina, a group of calcareous brachiopods conventionally placed with the articulates but still of uncertain origin and affinity. The ultrastructural studies of Williams (1968a) on the *Dictyonella* shell were unsatisfactory in that no tangible structure could be made out and my own current scanning electron microscope studies have so far not revealed any structure satisfactorily comparable to that of genuine articulate stocks.

The shell substance is punctate and in *Dictyonella* (Pl. III, 6–9) the puncta open into a very distinctive, but by no means unique, net-like ornament.

The valves are articulated by means of teeth and sockets in the broadest sense; these take the form of low, elongate, laterally disposed ridges and grooves, far removed from the deltidiodont and cyrtomatodont dentitions of the main groups of articulates.

A further distinctive feature of the suborder is the absence of outer shell from the pedicle valve immediately anterior to the umbo; in this region a smooth triangular umbonal plate is exposed, set just below the shell surface. Although attached to the shell laterally, it is separated anteriorly by a slot like passage opening internally. This passage is usually taken as being a modified pedicle opening following the work of Hall and Clarke (1894).

In my view the shell substance and the articulation remove the group from the articulate stocks. Some morphological similarities are seen with the chitinophosphatic siphonotretaceans, but since there is no clear evidence of origin from any pre-existing brachiopod stock, it is perhaps best to interpret the dictyonellidines as another Ordovician development from a "brachiophorate".

FIG. 1. Envisaged early radiation of the Brachiopoda. Ancestral infaunal lopho-phorate stocks with the potential to develop into brachiopod groups (brachiophorates) are variously sketched to emphasize their envisaged diversity.

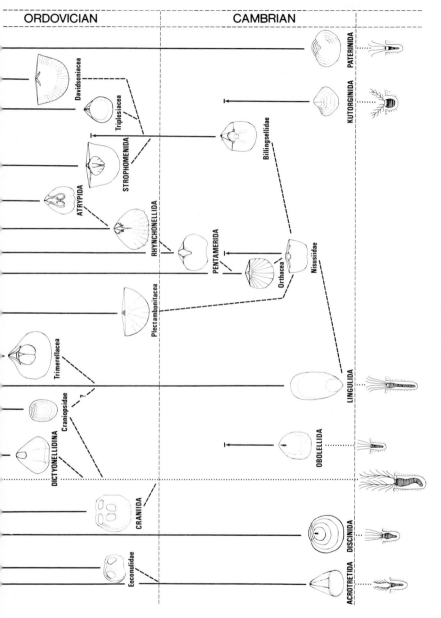

ORDOVICIAN | CAMBRIAN

Fig. 1. For legend see opposite.

CONCLUSIONS

The envisaged early brachiopod radiations (Fig. 1) may be summarized thus:

1. An already diverse series of phoronid-like infaunal brachiophorate stocks directly produced epifaunal brachiopod stocks in the lower Cambrian, the morphology of which indicates existing genetic diversity and considerable experimentation in the adaptation to the new habitats.

2. From among these stocks arose the billingsellaceans. These evidently possessed a combination of characters that was to prove a most successful formula, for these were the stem group of the major brachiopod class Articulata.

3. Some brachiophorates would appear to have retained their infaunal habit at least until the Ordovician radiation, when further stocks emerged to take up an epifaunal existence as brachiopods; the lingulides became secondarily infaunal at this time.

Several other issues become highlighted by the present study. An important aspect from the systematic viewpoint is that while the Class Articulata is monophyletic, the inarticulates (in some cases possessing mineralized skeletal articulation mechanisms) form a heterogenous collection, the grouping together of which into the single Class Inarticulata is misleading in its implied degree of unity.

From a evolutionary viewpoint, an interesting feature of the early brachiopod radiation is that practically the full gamut of morphological variation of the phylum had been tried out by the end of Ordovician times. This includes the range of modes of shell secretion, valve opening mechanisms, methods of attachment to the substrate, external shapes and internal structures.

Finally, a continuing problem is the lack of knowledge concerning the genetic and biochemical factors controlling the secretion of a particular kind of mineral exoskeleton. Associated with this is a need for substantive palaeontological evidence to support postulated mechanisms for the change from phosphate to carbonate skeletal secretion which is assumed to take place in certain stocks on morphological grounds.

REFERENCES

BIERNAT, G. and WILLIAMS, A. (1970). Ultrastructure of the protegulum of some acrotretide brachiopods. *Palaeontology*, **14**, 423–430.
CHUANG, S. H. (1971). New interpretation of the morphology of *Schizambon*

australis Ulrich and Cooper (Ordovician siphonotretid inarticulate brachiopod). *J. Paleontol.* **45**, 824–832.

CLARK, R. B. (1964). "Dynamics in Metazoan evolution". Oxford Univ. Press, London.

COBBOLD, E. S. (1921). The Cambrian Horizons of Comley (Shropshire) and their Brachiopoda, Pteropoda, Gastropoda, etc. *Q. Jl geol. Soc. Lond.* **76**, 325–386.

COBBOLD, E. S. and POCOCK, R. W. (1934). The Cambrian area of Rushton (Shropshire). *Phil. Trans. Roy. Soc. B*, **223**, 305–409.

COOPER, G. A. (1956). Chazyan and related Brachiopods. *Smithson. Misc. Coll.* **127**, 1–1245.

COWIE, J. W., RUSHTON, A. W. A. and STUBBLEFIELD, C. J. (1972). A correlation of Cambrian rocks in the British Isles. *Spec. Rep. geol. Soc. Lond.* **2**, 1–42.

GORYANSKY, V. Yu. (1977). New early Cambrian obolellids from Eastern Siberia. *In* "New species of various plants and invertebrates of the U.S.S.R. 4" (G. A. Stukalina, ed.), pp. 99–102. *Publ. Pal. Inst. U.S.S.R. Acad. Sci.*

GUTMANN, W. F., VOGEL, K. and ZORN, H. (1978). Brachiopods: Biomechanical Interdependences Governing Their Origin and Phylogeny. *Science, N.Y.* **199**, 890–893.

HALL, J. and CLARKE, J. M. (1894). An introduction to the study of the Brachiopoda. *N.Y. State Geol. Surv., Palaeontol.* **8** (2), 1–394.

HENDERSON, R. A. (1974). Shell adaptation in acrothelid brachiopods to settlement on a soft substrate. *Lethaia*, **7**, 57–61.

HYMAN, L. (1959). "The Invertebrates: Smaller Coelomate Groups. Volume 5". McGraw-Hill, New York.

JAANUSSON, V. (1966). Fossil brachiopods with probable aragonitic shell. *Geol. Fören. Förhandl.* **88**, 279–281.

JAANUSSON, V. (1971). Evolution of the brachiopod hinge. *In* "Paleozoic Perspectives: A Paleontological Tribute to G. Arthur Cooper. (J. T. Dutro Jr, ed.). *Smithson. Contr. Paleobiol.* **3**, 33–46.

JOPE, M. (1977). Brachiopod Shell Proteins: Their Functions and Taxonomic Significance. *Am. Zool.* **17**, 133–140.

KELLER, B. M., ROZANOV, A. Yu, MISSARZHEVSKY, V. V., REPINA, L. N., SHABANOV, Yu. I. and EGOROVA, L. I. (1973). Excursion Guide to the Aldan and Lena Rivers. *In* "International excursion on the problem of the Cambrian and Precambrian boundary", pp. 1–118. Acad. Nauk SSSR., Moscow.

NORFORD, B. S. (1960). A well-preserved *Dinobolus* from the Sandpile Group (Middle Silurian) of Northern British Columbia. *Palaeontology*, **3**, 242–244.

ROWELL, A. J. (1965). Inarticulata. *In* "Treatise on Invertebrate Paleontology, Part H" (R. C. Moore, ed.), pp. 260–296. Geol. Soc. Am. and Univ. Kansas Press.

ROWELL, A. J. (1977a). Early Cambrian brachiopods from the southwestern Great Basin of California and Nevada. *J. Paleontol.* **51**, 68–85.

ROWELL, A. J. (1977b). Valve orientation and functional morphology of the foramen of some siphonotretacean and acrotretacean brachiopods. *Lethaia*, **10**, 43–50.

ROWELL, A. J. and KRAUSE, F. F. (1973). Habitat diversity in the Acrotretacea (Brachiopoda, Inarticulata). *J. Paleontol.* **47**, 791–800.

RUDWICK, M. J. S. (1959). The Growth and Form of Brachiopod Shells. *Geol. Mag.* **96**, 1–24.

VALENTINE, J. W. (1973). Coelomate Superphyla. *Syst. Zool.* **22**, 97–102.

VALENTINE, J. W. (1975). Adaptive strategy and the Origin of Grades and Ground-Plans. *Am. Zool.* **15**, 391–404.

VALENTINE, J. W. (1977). General patterns of Metazoan Evolution. *In* "Patterns of Evolution as illustrated by the Fossil Record" (A. Hallam, ed.), pp. 27–57. Elsevier, Amsterdam.

WILLIAMS, A. (1956). The calcareous shell of the Brachiopoda and its importance to their classification. *Biol. Rev.* **31**, 243–287.

WILLIAMS, A. (1968a). Evolution of the shell structure of articulate brachiopods. *Spec. Pap. Palaeontology*, **2**, 1–55.

WILLIAMS, A. (1968b). Shell structure of the Billingsellacean brachiopods. *Palaeontology*, **11**, 486–90.

WILLIAMS, A. (1970). Origin of laminar-shelled articulate brachiopods. *Lethaia*, **3**, 329–342.

WILLIAMS, A. (1973). The secretion and structural evolution of the shell of thecideidine brachiopods. *Phil. Trans. Roy. Soc. B*, **264**, 439–478.

WILLIAMS, A. and HURST, J. M. (1977). Brachiopod Evolution. *In* "Patterns of Evolution as illustrated by the Fossil Record" (A. Hallam, ed.), pp. 79–121. Elsevier, Amsterdam.

WILLIAMS, A. and ROWELL, A. J. (1965). Morphology. *In* "Treatise on Invertebrate Paleontology, Part H" (R. C. Moore, ed.), pp. 57–138. Geol. Soc. Am. and Univ. Kansas Press.

WILLIAMS, A. and WRIGHT, A. D. (1961). The Origin of the loop in articulate brachiopods. *Palaeontology*, **4**, 149–176.

WILLIAMS, A. and WRIGHT, A. D. (1970). Shell structure of the Craniacea and other calcareous inarticulate Brachiopoda. *Spec. Pap. Palaeontology*, **7**, 1–51.

WRIGHT, A. D. (1966). The shell punctation of *Dicoelosia biloba* (Linnaeus). *Geol. Fören. Förhandl.* **87**, 549–557.

WRIGHT, A. D. (1968). The brachiopod *Dicoelosia biloba* (Linnaeus) and related species. *Ark. Zool.* **20**, 261–319.

WRIGHT, A. D. (1970). A note on the shell structure of the triplesiacean brachiopods. *Lethaia*, **3**, 423–426.

WRIGHT, A. D. (1979). The origin of the spiriferidine brachiopods. *Lethaia*, **12**, 29–33.

ZHURAVLEVA, I. T. (1970). Marine Faunas and Lower Cambrian Stratigraphy. *Am. J. Sci.* **269**, 417–445.

9 | Early Arthropods, Their Appendages and Relationships

H. B. WHITTINGTON

Sedgwick Museum, Cambridge, England

Abstract: New studies of the Burgess Shale are revealing some 24 genera of non-trilobite arthropods with appendages which had an unmineralized exoskeleton. They show a wider range in morphology and size than previously recognized, and were more numerous and varied than the trilobites. Assuming that such arthropods inhabited other Cambrian environments in similar variety and numbers, then these animals resulted from a major Precambrian radiation. Arthropods were the dominant Cambrian animals, many having been predators and scavengers. The Burgess Shale forms include three trilobites with appendages, early crustaceans, and a possible uniramian ancestor, the remainder represent separate lines of descent apparently unrelated to major taxa. These and other Cambrian evolutionary lines may have been largely lower Palaeozoic in age, few continuing into the upper Palaeozoic or Mesozoic. In the post-Cambrian the arthropod evolutionary pattern changed; trilobites began to decline as other phyla with hard parts radiated. Morphological advances in largely unknown ancestors led to the Siluro–Devonian continental radiation of chelicerates and uniramians, and in the upper Palaeozoic Crustacea radiated to dominate the marine environments in later eras.

INTRODUCTION

In rare circumstances arthropods may be preserved showing all or most of the appendages approximately in| their original relation to the exoskeleton. The Burgess Shale has|yielded some 24 such species, but only one other example is known from Cambrian strata. Ordovician and Silurian rocks have yielded ten further examples, and the lower Devonian Rhynie Chert and Hunsrück and Nellenköpfchen Shales 16 more (Fig. 1). The

Systematics Association Special Volume No. 12, "The Orign of Major Invertebrate Groups", edited by M. R. House, 1979, pp. 253–268. Academic Press, London and New York.

CAMBRIAN	ORDOVICIAN	SILURIAN	DEVONIAN	
	Eurypterida			CHELICERATA
Xiphosura				
			Scorpionida	
			Other Arachnida	
			Pycnogonida	
Eodiscoids	Agnostoids			TRILOBITA
Olenellids				
Redlichiids	"Polymera"			
Nektaspida				
Bradoriida	Ostracoda			CRUSTACEA
		Branchiopoda		
		Cirrepedia		
	Phyllocarida			
		Arthropleurida		UNIRAMIA
Aysheaia		Hexapoda		
		Myriapoda		
	Aglaspidida			OTHER GENERA

importance of the Burgess Shale as yielding the oldest major record of arthropod evolution is evident. The available specimens are being prepared for the first time, and studied in detail, by Drs D. E. G. Briggs, D. L. Bruton, C. P. Hughes and me, and a radically new picture of these early forms is emerging, quite different from that of Størmer (in Moore, 1959) or that in the more recent work of Simonetta (1962, 1963, 1964, 1970) and Simonetta and Cave (1975). The new finds in the Nellenköpfchen Shales (Størmer, 1970–76), and the studies of Stürmer and Bergström (1973, 1976, 1978) have thrown important new light on early Devonian arthropods. These and other palaeontological studies, and the results of zoological work (Manton, 1977; Manton and Anderson, Chapter 10 this volume), necessitate a re-assessment of views on relationships and evolution of arthropods. Phylogenetic schemes such as those of Cisne (1974) and Hessler and Newman (1975) do not take into account present knowledge. In the following pages Fig. 1 is commented upon in summarizing this knowledge, reference being made only to work published since the relevant volumes of the "Treatise on Invertebrate Paleontology" (Moore, 1955, 1959, 1961, 1969). Emphasis is placed on species in which appendages are known, because the evidence of affinity provided by the exoskeleton alone, or part of the exoskeleton, may be equivocal, as the Burgess Shale species show. Supposed arthropod tracks are not considered here to give acceptable evidence of stratigraphical ranges of groups. On the basis of this interpretation of the fossil record a possible pattern of arthropod evolution is suggested (Fig. 2).

NOTES IN EXPLANATION OF FIG. 1

1. Chelicerata

Of Ordovician eurypterids only the appendages of *Megalograptus* (Caster and Kjellesvig-Waering, 1964) are known fairly completely. I do not

---

FIG. 1. Stratigraphical range (solid line) from the early Cambrian to early Devonian of the kinds of arthropods indicated. Broken line shows supposed relationship between occurrences. Asterisks denote occurrence of examples in which the morphology of the appendages is well known. "Polymera" embraces orders Corynexochida, Ptychopariida, Phacopida, Lichida and Odontopleurida of trilobites, as given in Moore (1959). Question mark denotes uncertainty of relationship between Bradoriida and Ostracoda. Kinds of arthropods (those shown are of different taxonomic rank) are arranged in four widely recognized major groups and "other genera" of uncertain affinities.

follow Bergström (1968) in extending the range of eurypterids back to the middle Cambrian by including *Kodymirus*, because I consider the evidence inadequate. Best-preserved of all eurypterid appendages are those of the upper Silurian *Baltoeurypterus* Størmer, 1973. This genus was based on the type species *tetragonophthalmus*, which is the "*Eurypterus fischeri*" of Holm (1898) and Wills (1965). The earliest known xiphosuran is an incomplete prosoma from the lower Cambrian of Sweden (Bergström, 1968, 1975), next youngest are prosomas from the middle Ordovician (Chlupáč, 1965; Bergström, 1968, 1975). The oldest specimens from which appendages are known are of *Weinbergina*, lower Devonian (Lehmann, 1956); the presence of chelicerae is questionable, and there were six pairs of walking legs on the prosoma. Different views on merostome systematics are given by Bergström (1975) and Eldredge (1974), who describe old and new xiphosurid material; the relationships of *Chasmataspis* (Caster and Brooks, 1956), in which only an isolated appendage is known, are problematical. Størmer (1970, 1976) has described important new early Devonian arachnids with appendages and reviewed earlier work. Aquatic scorpions were present in the late Silurian and early Devonian, and air-breathing arachnids in the early Devonian of Germany and Scotland. Pycnogonida are represented in the early Devonian of Germany by *Palaeoisopus* (Lehmann, 1959) and *Palaeopantopus*.

2. *Trilobita*

In 1954 I pointed out that the major groups of trilobites appeared abruptly in the early Cambrian, and recent studies (e.g. Fritz, 1972; Sdzuy, 1978) show that the appearances are at about the same time. These groups are the eodiscoids (Jell, 1975; from which agnostoids were derived in the late lower Cambrian), the olenellids and redlichiids, and the remaining polymerid trilobites (i.e. the Corynexochida, Ptychopariida, etc. of Harrington, in Moore, 1959). A similar view of early trilobite evolution has been given by Hahn and Hahn (1975). Nothing is known of appendages of eodiscoids, agnostoids or redlichiids and only a single specimen of *Olenellus* shows antennae (Resser and Howell, 1938). Appendages of the Middle Cambrian polymerid trilobites *Olenoides* and *Kootenia* are well known (Whittington, 1975b), Cisne (1975) has re-investigated those of the upper Ordovician *Triarthrus*, and Stürmer and Bergström (1973) those of the lower Devonian phacopids. A surprising outcome of the re-study of the supposed "merostomoid" *Naraoia* (Whittington, 1977) is that the appendages and other features appear to place it as representing a distinct order of Trilobita, Nektaspida, lacking a thorax and the exoskeleton not mineralized.

3. Crustacea

Studies on the composition of the shell and of the morphology (Müller, 1964; Andres, 1969; Fleming, 1973; Kozur, 1974) of Bradoriida have shown the wide variety of carapaces presently included in this group, and raised questions of how they may be related to ostracods. Carapaces accepted as those of ostracods are abundant from the early Ordovician onwards, but appendages have not been described in species older than Carboniferous. The middle Cambrian Burgess Shale has yielded species referred to two genera of Phyllocarida, *Perspicaris* (Briggs, 1977) and *Canadaspis* (Briggs, 1978a); the genus *Plenocaris* Whittington, 1974, which I referred to Phyllocarida, is not included in this group by Briggs (1978a). The appendages of *Canadaspis* show many similarities to those of Recent leptostracans and the limbs of *Waptia* (Hughes, in preparation) show crustacean features. Thus the beginnings of one class of Crustacea, the Malacostraca, appear to be in the Cambrian, as are possibly those of a second class, the Ostracoda.

4. Uniramia

A new arthropleurid from the lower Devonian has been described by Størmer (1976), in which only portions of trunk appendages are known. The supposed lower Devonian arthropleurid *Bundenbachiellus* (Rolfe, in Moore, 1969) may be a specimen of *Cheloniellon* (Stürmer and Bergström, 1978). Following Manton (1977), arthropleurids are regarded as a taxon separate from myriapods; work on fossil myriapods by Kraus (1974) is in progress. The earliest known hexapod is the collembolan from the lower Devonian Rhynie chert (Scourfield, 1940). The middle Cambrian *Aysheaia* (Whittington, 1978) is a marine uniramian, a representative of the kind of animals from which younger uniramians may have been derived.

5. Aglaspidida

Raasch (1939) described the single specimen of *Aglaspis spinifer* in which he considered that the prosoma bore six pairs of appendages, the first being chelicerae. Re-examination of this specimen (Briggs *et al.*, 1979) shows that the anterior pair of appendages were not chelate, and that the cephalic region bore four, or perhaps five, pairs of appendages. The three or four posterior cephalic pairs and those on the anterior part of the trunk, are similar to each other, jointed walking legs, showing no evidence of a branch. Repina and Okuneva (1969) described Cambrian specimens which they attributed to aglaspidids, and which show lamellate structures beneath the trunk exoskeleton. It appears uncertain that this species is an aglaspidid and the preservation of the lamellae leaves much to be desired. Bergström's statement (1975), based on Repina and Okuneva's description, that aglaspidids have biramous appendages, is a quite unwarranted extrapolation. Størmer (1944) gave a sketch of *Aglaspella eatoni*, onto which he transferred a reconstruction of the appendages known only from the single specimen of *Aglaspis spinifer* referred to above. Appendages of *Aglaspella* are unknown, and there is no known basis for the "long-shafted chelicerae" in aglaspidids referred to by Manton (1977). Just which genera should be regarded as aglaspidids is a matter of debate (Størmer, in Moore, 1955; Bergström, 1968, 1971), but here *Paleomerus* is included. The group can no longer be referred to Chelicerata, and its affinities remain uncertain.

6. *Other Genera*

The asterisks at the level of early middle Cambrian in Fig. 1 represent genera from the Burgess Shale in which appendages are preserved; studies of some have been completed. *Branchiocaris* (Briggs, 1976) has a bivalved carapace but cannot be assigned to any group of arthropods having living members because of the unique nature of the appendages. *Burgessia* (Hughes, 1975) can no longer be considered notostracan-like, and shows a combination of characters that appears to be unique. *Marrella* (Whittington, 1971) had antennae and a setose, jointed, paired appendage on the head shield, the trunk lacked a dorsal exoskeleton, the appendages were biramous and resembled those of trilobites. It cannot be classed as a trilobite, and its nearest relative may be the early Devonian *Mimetaster* (Birenheide, 1971; Stürmer and Bergström, 1976). *Yohoia* (Whittington, 1974) stands alone, not being in the least trilobite-like; any resemblance to a crustacean is superficial. *Plenocaris* Whittington, 1974, discussed above under crustaceans, is placed here with genera of uncertain taxonomic position. Some thirteen additional genera are known, but until reinvestigation is completed relationships between them cannot be assessed.

These genera (proposed by Walcott unless otherwise indicated) are *Alalcomenaeus* Simonetta, 1970; *Anomalocaris* Whiteaves, 1892; *Emeraldella*, *Emeraldoides* Simonetta, 1964; *Habelia* Simonetta, 1964; *Helmetia*, *Leanchoilia*, *Molaria*, *Odaraia*, *Sidneyia*, *Skania*, *Tegopelte* Simonetta and Cave, 1975; and *Waptia*. Genera from the Burgess Shale in which appendages are poorly known or unknown are excluded here. In addition to the five genera from the lower Devonian Hunsrück Shale mentioned above, there are two others, *Cheloniellon* (Stürmer and Bergström, 1978) and *Vachonisia* (Stürmer and Bergström, 1976). The former is not like any Burgess Shale arthropod and the claim of Stürmer and Bergström that the trunk appendages in *Vachonisia* are biramous is not well substantiated, so that any relationship to *Mimetaster* is uncertain.

IMPLICATIONS OF PRESENT KNOWLEDGE OF APPENDAGES

Since the aglaspidids can no longer be classed as merostomes, the fossil record of Chelicerata is reduced and the earliest known chelicera are borne by the isolated appendage of *Megalograptus*. Chelicerae in place are first known in the late Silurian *Baltoeurypterus*. Knowledge of appendages of Palaeozoic xiphosurids is limited, particularly of those of the opisthosoma, so that there is no evidence for the derivation of the biramous mesosomal limb of *Limulus* (Manton, 1977).

Trilobita at their first appearance include at least three very different kinds (Fig. 1), between which no intermediates are known. Appendages have been found in only one of these groups. To consider trilobites as a single, unified stem in a phylogenetic diagram is an over-simplification. *Naraoia*, the single known nektaspid, had a non-mineralized exoskeleton, unlike known early Cambrian trilobites. Perhaps there were other such groups, indeed their existence has been suggested (Whittington, 1954) as a possible explanation for the sudden appearance of new stocks apparently unrelated to contemporaneous or older groups.

Canadaspis (Briggs, 1978a) shows that the oldest known crustacean is of phyllocarid (and therefore malacostracan) type, but there is still no evidence of how this group may be related to the other major groups which appear in the lower Palaeozoic. Burgess Shale genera other than *Canadaspis*, *Perspicaris* and possibly *Waptia*, do not appear to belong within Crustacea.

The phylum Uniramia (Manton, 1973a, b; 1977) is accepted here, and *Aysheaia* (Whittington, 1978) may represent the type of lobopodial animal from which uniramians and tardigrades may have been derived. Manton (1977) has put forward reasons for supposing that early lobopods, which had the cuticle little sclerotized, were varied; this differentiation may have taken place in the marine habitat.

The four major groups of arthropods mentioned above, three Recent and one Palaeozoic, each contains subgroups which are ranked variously as subphyla, classes or orders. In most cases there is little or no evidence of the relationship between the subgroups, for example between the classes of Crustacea (Manton, 1977). The three major groups of Chelicerata, Crustacea and Uniramia are regarded by Manton (1977) as phyla, and if this is accepted then the Trilobita may well be accorded a similar rank. To do so would be to reject the suggestion of a possible or probable trilobite–chelicerate relationship. Such a relationship was suggested by Størmer (1944) in proposing phylum Arachnomorpha to include two subphyla, Trilobitomorpha and Chelicerata. The name Trilobitomorpha was also new, a taxon proposed to include Trilobita and groups subsequently placed in Trilobitoidea Størmer, 1959, the latter a taxon almost entirely comprised of Burgess Shale genera. The taxon Arachnomorpha was not used by Størmer in 1959 (in Moore), but the acceptance of a possible or probable trilobite–chelicerate relationship has persisted (Cisne, 1974; Manton, 1977) and Bergström (e.g. Stürmer and Bergström, 1973, 1976) continues to use the term "arachnomorph". Størmer's original arguments rested to a great extent on the premises that:

1. The Burgess Shale genera (plus the Hunsrück Shale *Chelionellon*), though varied in appearance, were to be linked together in Trilobitoidea as possessing trilobite-like appendages.

2. The oldest chelicerates were aglaspidids.

3. The aglaspidids showed resemblances to certain trilobites and trilobitoids.

Arguments were also brought forward from the ontogeny of *Limulus* and the supposed homology of mesosomal appendages of *Limulus* and the trilobite limb. The first premise is no longer valid (see below) and the re-examination of the only aglaspidid specimen having appendages approximately in place (Briggs, Bruton and Whittington, 1979) removes premises (2) and (3). The dorsal exoskeleton of the olenellid trilobite has only the most superficial resemblance to that of an aglaspidid, only the antennae of one olenellid specimen are known, so that evidence of resemblances or differences in appendages is unavailable. As Manton (1977) has remarked, there is no evidence of evolutionary relationships between trilobites and merostomes, and I consider that there is no basis in fact to support the idea embraced by "Arachnomorpha".

The Class Trilobitoidea Størmer (in Moore, 1959) was proposed for animals having trilobite-like limbs but which displayed exoskeletal forms unlike those of trilobites. It contained three subclasses, the first containing only the distinctive species *Marrella splendens*. The second subclass Merostomoidea would be made even more heterogenous than originally by the addition of aglaspidid genera (Bergström, 1971) and new and old Carboniferous to Triassic genera (Schram, 1971). It was thought by Størmer (in Moore, 1959) to include *Sidneyia*, *Emeraldella*, *Leanchoilia*, *Naraoia*, and perhaps *Molaria* and *Habelia*. The first three are being studied by Bruton (1977), and display widely different limbs, none trilobite-like. *Naraoia* (Whittington, 1977) is considered a trilobite, and the remaining two genera have yet to be investigated in detail. To be of value this taxon will require restriction and re-definition. Re-study of *Burgessia* (Hughes, 1975) and *Waptia* (Hughes, 1977) is removing any bases for bringing these genera together into the third subclass Pseudonotostraca. Miscellaneous genera included in Trilobitoidea, e.g. *Yohoia* (Whittington, 1974), *Chelionellon*, seem to stand alone and *Opabinia* (Whittington, 1975b) is not an arthropod. *Anomalocaris* (Rolfe, in Moore, 1969) was not included in trilobitoids and has long been thought to be the body of a phyllocarid. However, Briggs (1978b) considers that it is the appendage of a large arthropod, perhaps 1 m long. Simonetta (1976) and Simonetta and Cave (1975) base their views on reconstructions which the new studies are showing to be erroneous in many respects. Their classification arranges the various species in 18 orders, without uniting them in a single higher taxon. This arrangement results from the wide morphological differences between species, differences which are being clarified and added to by the new studies. Not until these are completed can the relationships of the Burgess Shale arthropods be re-assessed, but it is already clear that the Class Trilobitoidea and its subdivisions, as diagnosed by Størmer (in Moore, 1959), are no longer useful.

Figure 2 is a diagram of the evolutionary pattern shown by arthropods from the beginning of the Cambrian to the Recent. The Burgess Shale arthropods are from a muddy, benthonic, oceanic environment at about 150 m depth. There were at least thirty (24 with appendages) kinds of non-trilobite arthropods in that particular middle Cambrian environment, and I suggest that many more kinds were present in other environments. Most of these arthropods would not have had a mineralized exoskeleton

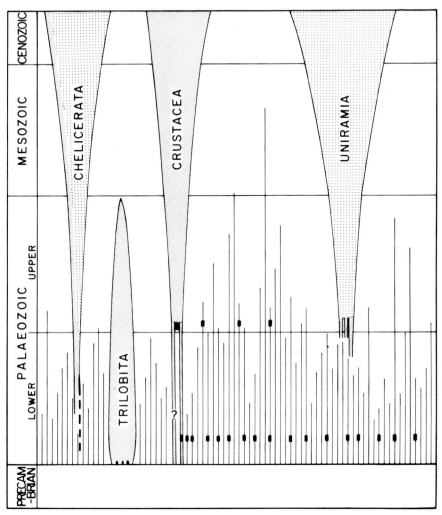

Fig. 2. Diagram of the pattern of evolution in arthropods. Shapes of areas of each major group indicate range in time and increase or decrease in numbers of kinds in time, not scaled in proportion to numbers; dense stipple indicates marine habitat, less dense the continental habitat. The many thin, parallel lines show the supposed range in time of known and unknown genera and families of other arthropods. These other genera are known mainly from the middle Cambrian and lower Devonian (indicated by thicker portion of line) and are forms which did not have a mineralized exoskeleton.

and hence only rarely have accidents of preservation revealed them. In the lower Devonian Hunsrück Shale is preserved one descendant of a Burgess Shale form, and representatives of other, previously unknown lines of descent. The many parallel lines of descent in the lower Palaeozoic, some extending into the upper Palaeozoic and even Mesozoic, represent these known and unknown forms. The number of these lines shown in Fig. 2 is arbitrary, as is the range in time, but this number may be an underestimate. Also shown are the Trilobita, the dominant *known* marine arthropods of the lower Palaeozoic, gradually reduced and becoming extinct in the upper Palaeozoic. Crustacea, from poorly known lower Palaeozoic beginnings, increase in kinds and become a recognizable major group in the marine and fresh-water environments of the upper Palaeozoic. In their subsequent radiation in the marine environment they take over the role of the Palaeozoic trilobites. Chelicerata and Uniramia, from a beginning in the lower Palaeozoic seas, show a remarkable radiation in continental environments from the early Devonian onwards. Aquatic chelicerates (eurypterids) were well defined and established in the Ordovician and Silurian, and are especially well known from marginal continental-marine environments in which xiphosurids have persisted to today. Different groups of Uniramia appear abruptly in the lower Devonian and the nature of their marine ancestors is hinted at only by the Burgess Shale *Aysheaia*. Comments on this pattern are:

(a) A major late Precambrian radiation of arthropods must have occurred, of which we know virtually nothing—disregarding the claims made for trace fossils, and noting that such fossils as *Praecambridium* (Glaessner and Wade, 1971) may be evidence of this radiation. I accept the view (Manton, 1977) that in all probability these arthropods (animals having a jointed cuticular exoskeleton on body and legs) arose independently| many times from metamerically segmented invertebrates (i.e. "worms"). Trilobites in the Cambrian, and ostracods from the Ordovician onwards, had a mineralized exoskeleton and consequently are abundant and widespread fossils. The assumption is made that they were the dominant early Palaeozoic arthropods, but in the Burgess Shale there are some 17 species of trilobites (Fritz, 1971) and some 30 kinds of other arthropods, lacking a mineralized exoskeleton. This evidence implies that the non-trilobite arthropods were both morphologically more varied and possibly more numerous than trilobites during Cambrian times. The arthropods appear to have been the dominant animals of the Cambrian, in size, variety of form, mobility and predatory and scavenging powers.

The great rise in numbers and kinds of hard-shelled animals of other phyla (coelenterates, molluscs, bryozoans, brachiopods and echinoderms) in the post-Cambrian suggests that their ancestors and those|of other |groups such as worms, both provided a food supply for, and offered little competition to, Cambrian arthropods.

(b) The rise of other invertebrate groups in the post-Cambrian may well have contributed to the decline of the Trilobita (Whittington, 1966), which was well advanced long before the marine radiation of Crustacea. The diagram is not intended to portray a direct replacement of one group by the other.

(c) The post-lower Palaeozoic radiations of Crustacea in the sea, and of chelicerates and uniramians on land, are comparable evolutionary events to the late Precambrian radiation. These radiations were possible because of the type of evolutionary change called an advance by Manton (1977). The late Precambrian radiation may have been based on the advance we call "arthropodization", the evolution of the characteristic exoskeleton. Just what structures or habits led to the successful invasions of the lands by chelicerates and uniramians is unknown.

(d) Questions of taxonomic rank are raised by the diagram. Four clearly-defined groups are evident, and if the three Recent ones are considered as phyla then the Trilobita may deserve equal rank (cf. Manton, 1977). No readily-discernible category of high rank will embrace the "other arthropods" of Fig. 1. What rank should be given to each of the many known lines of descent portrayed in the early Palaeozoic? In many of these only a single species, displaying a unique combination of characters, is known. Each may be placed in a separate genus and family, but classification into higher categories presents problems for which solutions are by no means apparent at this stage of the Burgess Shale work. These difficulties are an expression of the evolutionary pattern. In those lines in which a radiation occurred, e.g. trilobites in the marine environment or uniramians on the continents, a large group of related species—a class or phylum—is a result of the radiation and comprises a readily apparent higher taxon. In an evolutionary line such as that leading through *Marrella* and *Mimetaster* no advance in structure or habit lead to a radiation. What taxonomic rank should be given to such evolutionary lines? To group them each in a separate order, or to lump them in a class such as Trilobitoidea, that defies definition, is not useful.

REFERENCES

ANDRES, D. (1969). Ostracoden aus dem mittleren Kambrium von Öland. *Lethaia*, **2**, 165–180.

BERGSTRÖM, J. (1968). *Eolimulus*, a lower Cambrian xiphosurid from Sweden. *Geol. Fören. Förhandl.* **90**, 489–503.

BERGSTRÖM, J. (1971). *Paleomerus*—merostome or merostomoid. *Lethaia*, **4**, 393–401.

BERGSTRÖM, J. (1973). Classification of some olenellid trilobites and some Balto-Scandian species. *Norsk. Geol. Tidsskr.* **53**, 283–314.

BERGSTRÖM, J. (1975). Functional morphology and evolution of xiphosurids. *Fossils and Strata*, **4**, 291–305.

BIRENHEIDE, R. (1971). Beobachtungen am "Scheinstern" *Mimetaster* aus dem Hunsrück—Schiefer. *Senckenberg leth.* **52**, 77–91.

BRIGGS, D. E. G. (1976). The arthropod *Branchiocaris* n. gen., middle Cambrian, Burgess Shale, British Columbia. *Bull. geol. Surv. Can.* **264**, 1–29.

BRIGGS, D. E. G. (1977). Bivalved arthropods from the Cambrian Burgess Shale of British Columbia. *Palaeontology*, **20**, 595–621.

BRIGGS, D. E. G. (1978a). The morphology, mode of life, and affinities of *Canadaspis perfecta* (Crustacea : Phyllocarida), middle Cambrian, Burgess Shale, British Columbia. *Phil. Trans. R. Soc., Lond.*, B, **281**, 439–487.

BRIGGS, D. E. G. (1978b). A new trilobite—like arthropod from the lower Cambrian Kinzers Formation, Pennsylvania. *J. Paleont.* **52**, 132–140.

BRIGGS, D. E. G., BRUTON, D. L. and WHITTINGTON, H. B. (1979). Appendages of the anthropod *Aglaspis spinifer* (upper Cambrian, Wisconsin) and their significance. *Palaeontology*, **22**, 167–180.

BRUTON, D. L. (1977). Appendages of *Sidneyia*, *Emeraldella* and *Leanchoilia* and their bearing on trilobitoid classification. *J. Paleont.*, **51** (Suppl. to 2), 4–5.

CASTER, K. E. and BROOKS, H. K. (1956). New fossils from the Canadian–Chazyan (Ordovician) hiatus in Tennessee. *Bull. Am. Paleont.* **36** (157), 153–199.

CASTER, K. E. and KJELLESVIG-WAERING, E. N. (1964). Upper Ordovician eurypterids from Ohio. *Paleontogr. Am.* **IV** (32), 297–358.

CHLUPÁČ, I. (1965). Xiphosuran merostomes from the Bohemian Ordovician. *Sborn. geol. Věd., paleont.* P 5, 7–38.

CISNE, J. L. (1974). Trilobites and the origin of arthropods. *Science, N.Y.* **186**, 13–18.

CISNE, J. L. (1975). Anatomy of *Triarthrus* and the relationships of the Trilobita. *Fossils and Strata*, **4**, 45–63.

ELDREDGE, N. (1974). Revision of the Suborder Synziphosurina (Chelicerata, Merostomata), with remarks on merostome phylogeny. *Novitates*, **2543**, 1–41.

FLEMING, P. J. G. (1973). Bradoriids from the *Xystridura* zone of the Georgina Basin, Queensland. *Geol. Surv. Queensland, Pub.* **356** (Pal. Pap. 31), 1–9.

FRITZ, W. H. (1971). Geological setting of the Burgess Shale. *N. Am. Paleont. Conv., Chicago*, 1969, Proc. **I**, 1155–1170.

FRITZ, W. H. (1972). Lower Cambrian trilobites from the Sekwi Formation type section, Mackenzie Mountains, northwestern Canada. *Bull. geol. Surv. Can.* **212**, 1–58.

GLAESSNER, M. F. and WADE, M. (1971). *Praecambridium*—a primitive arthropod. *Lethaia*, **4**, 71–77.

HAHN, G. and HAHN, R. (1975). Forschungsbericht über Trilobitomorpha. *Paläont. Z.* **49**, 432–460.

HESSLER, R. R. and NEWMAN, W. A. (1975). A trilobitomorph origin for the Crustacea. *Fossils and Strata*, **4**, 437–459.

HOLM, G. (1898). Über die organisation des *Eurypterus fischeri* Eichw. *Mém. L'Acad. Imp. Sci. St. Pétersb.*, *Ser.* **VIII**, 8, 1–57.

HUGHES, C. P. (1975). Redescription of *Burgessia bella* from the middle Cambrian, Burgess Shale, British Columbia. *Fossils and Strata*, **4**, 415–435.

HUGHES, C. P. (1977). The early arthropod *Waptia fieldensis*. *J. Paleont.* **51** (Suppl. to 2), 15.

JELL, P. A. (1975). Australian middle Cambrian eodiscoids with a review of the Superfamily. *Palaeontograph.*, *A*, **150**, 1–97.

KOZUR, H. (1974). Die bedeutung der Bradoriida als vorläufer der post-kambrischen Ostracoden. *Z. geol. Wiss.*, *Berlin*, **7**, 823–830.

KRAUS, O. (1974). On the morphology of Palaeozoic diplopods. *Symp. Zool. Lond.* **32**, 13–22.

LEHMANN, W. M. (1956). Beobachtungen an *Weinbergina opitzi* (Merost., Devon.). *Senckenberg. leth.* **37**, 67–77.

LEHMANN, W. M. (1959). Neue entdeckungen an *Palaeoisopus*. *Pälont. Z.* **33**, 96–103.

MANTON, S. M. (1973a). Arthropod phylogeny—a modern synthesis. *J. Zool.*, *Lond.* **171**, 111–130.

MANTON, S. M. (1973b). The evolution of arthropodan locomotory mechanisms. Part 11. Habits, morphology and evolution of the Uniramia (Onychophora, Myriapoda, Hexapoda) and a comparison with the Arachnida, together with a functional review of uniramian musculature. *J. Linn. Soc., Zool.* **53**, 257–375.

MANTON, S. M. (1977). "The Arthropoda. Habits, Functional Morphology, and Evolution." Oxford Univ. Press, London.

MOORE, R. C. (1955). "Treatise on Invertebrate Paleontology, Part P, Arthropoda 2." Geol. Soc. Am. and Univ. Kansas Press.

MOORE, R. C. (1959). "Treatise on Invertebrate Paleontology, Part O, Arthropoda 1." Geol. Soc. Am. and Univ. Kansas Press.

MOORE, R. C. (1961). "Treatise on Invertebrate Paleontology, Part Q, Arthropoda 3." Geol. Soc. Am. and Univ. Kansas Press.

MOORE, R. C. (1969). "Treatise on Invertebrate Paleontology, Part R, Arthropoda 4." Geol. Soc. Am. and Univ. Kansas Press.

MÜLLER, K. J. (1964). Ostracoda (Bradorina) mit phosphatischen| Gehäusen aus dem Oberkambrium von Schweden. *Neues Jb. Geol. Paläont. Abh.* **121**, 1–46.

RAASCH, G. E. (1939). Cambrian Merostomata. *Spec. Pap. geol. Soc. Am.* **19**.

REPINA, L. N. and OKUNEVA, O. G. (1969). Cambrian arthropods of the Maritime Territory. *Palaeont. J.* **3**, 106–114.

RESSER, C. E. and HOWELL, B. F. (1938). Lower Cambrian *Olenellus* zone of the Appalachians. *Bull. geol. Soc. Am.* **49**, 195–248.

SCHRAM, F. R. (1971). A strange arthropod from the Mazon Creek of Illinois and the Trans Permo-Triassic Merostomoidea (Trilobitoidea). *Fieldiana*, **20**, 85–102.

SCOURFIELD, J. D. (1940). The oldest known fossil insect (*Rhyniella praecursor* Hirst and Maulik)—further details from additional specimens. *Proc. Linn. Soc., Lond.* **152**, 113–131.

SDZUY, K. (1978). The Precambrian–Cambrian boundary beds in Morocco (Preliminary Report). *Geol. Mag.* **115**, 83–94.

SIMONETTA, A. M. (1962). Note sugli artropodi non trilobiti della burgess shale, cambriano medio della Columbia Britannica (Canada), I. *Monit. Zool. Ital.* **69**, 175–185.

SIMONETTA, A. M. (1963). Osservazioni sugli artropodi non trilobiti della "Burgess Shale" (Cambriano Medio), II. *Monit. Zool. Ital.* **70–71**, 97–108.

SIMONETTA, A. M. (1964). Osservazioni sugli artropodi non trilobiti della "Burgess Shale" (Cambriano Medio), III. *Monit. Zool. Ital.* **72**, 215–231.

SIMONETTA, A. M. (1970). Studies on non Trilobite Arthropods of the Burgess Shale (middle Cambrian). *Palaeontogr. Italica*, **66** (n.s. **36**), 35–45.

SIMONETTA, A. M. (1976). Remarks on the origin of the Arthropoda. *Atti. Soc. Tosc. Sc. Nat. Mem.* B, **82**, 112–134.

SIMONETTA, A. M. and CAVE, L. D. (1975). The Cambrian non trilobite arthropods from the Burgess Shale of British Columbia. A study of their comparative morphology, taxinomy and evolutionary significance. *Palaeontogr. Ital.* **69** (n.s. **39**), 1–37.

STØRMER, L. (1944). On the relationships and phylogeny of fossil and Recent Arachnomorpha. *Skr. norske Vidensk.-Akad. Oslo, I Mat.-Nat. Kl.* **5**, 1–158.

STØRMER, L. (1970–1976). Arthropods from the lower Devonian (lower Emsian) of Alken an der Mosel, Germany. *Senckenberg. leth.* **51**, 335–369 (1, 1970); **53**, 1–29 (2, 1972); **54**, 119–205 (3, 1973); **54**, 359–451 (4, 1974); **57**, 87–183 (5, 1976).

STÜRMER, W. and BERGSTRÖM, J. (1973). New discoveries on trilobites by X-rays. *Paläont. Z.* **47**, 104–141.

STÜRMER, W. and BERGSTRÖM, J. (1976). The arthropods *Mimetaster* and *Vachonisia* from the Devonian Hunsrück Shale. *Paläont. Z.* **50**, 78–111.

STÜRMER, W. and BERGSTRÖM, J. (1978). The arthropod *Chelionellon* from the Devonian Hunsrück Shale. *Paläont. Z.* **52**, 57–81.

WHITEAVES, J. F. (1892). Description of a new genus and species of phyllocarid crustacean from the middle Cambrian of Mount Stephen, British Columbia. *Can. Rec. Sci.* **5**, 205–208.

WHITTINGTON, H. B. (1954). Status of Invertebrate Paleontology, 1953. VI. Arthropoda: Trilobita. *Bull. Mus. comp. Zool. Harv.* **112**, 193–200.

WHITTINGTON, H. B. (1966). Phylogeny and distribution of Ordovician trilobites. *J. Paleont.* **40**, 696–737.

WHITTINGTON, H. B. (1971). Redescription of *Marrella splendens* (Trilobitoidea) from the Burgess Shale, middle Cambrian, British Columbia. *Bull. geol. Surv. Can.* **209**, 1–24.

WHITTINGTON, H. B. (1974). *Yohoia* Walcott and *Plenocaris* n. gen., arthropods from the Burgess Shale, middle Cambrian, British Columbia. *Bull. geol. Surv.*

Can. **231**, 1–21 (Figs 1–6 of Pl. X should be interchanged with figs 1–8 of Pl. XII).

WHITTINGTON, H. B. (1975a). Trilobites with appendages from the middle Cambrian, Burgess Shale, British Columbia. *Phil. Trans. R. Soc. B*, **271**, 1–43.

WHITTINGTON, H. B. (1975b). The enigmatic animal *Opabina regalis*, middle Cambrian, Burgess Shale, British Columbia. *Phil. Trans. R. Soc. B*, **271**, 1–43.

WHITTINGTON, H. B. (1977). The middle Cambrian trilobite *Naraoia*, Burgess Shale, British Columbia. *Phil. Trans. R. Soc. B*, **280**, 409–433.

WHITTINGTON, H. B. (1978). The lobopod animal *Aysheaia pedunculata*, middle Cambrian, Burgess Shale, British Columbia. *Phil. Trans. R. Soc. B*, **284**, 165–197.

WILLS, L. J. (1965). A supplement to Gerhard Holm's "Uber die Organisation des *Eurypterus Fischeri* Eichw." with special reference to the organs of sight, respiration and reproduction. *Ark. f. Zool.*, *ser.* 2, **18**, 93–145.

Note added in proof:

While this chapter was in press an article by Müller (1979) has been published in which ostracods with preserved appendages from the upper Cambrian of Sweden are described. The line from Bradoriida to Ostracoda in Fig. 1 should now be continuous, with no question mark, and an asterisk added to it in upper Cambrian time.

MÜLLER, K. J. (1979). Phosphatocopine ostracodes with preserved appendages from the Upper Cambrian of Sweden. *Lethaea*, **12**, 1–27.

10 | Polyphyly and the Evolution of Arthropods

*S. M. MANTON

and

D. T. ANDERSON

School of Biological Sciences, University of Sydney, Australia

Abstract. An attempt is made to review the results from three relevant lines of investigation. There is overwhelming support for the view that the major anthropodan taxa are not monophyletic in origin. The arthropods thus become a grade of organization, independently reached a number of times. Apparently distinct phyla are the Chelicerata, the Crustacea and the Uniramia (Onychophora, Myriapoda, Hexapoda). The status of the Trilobita and of many other early arthropods cannot be assessed at present. It is considered that there is no sound evidence linking the trilobites with either crustaceans or uniramians.

The evidence for these conclusions stems firstly from direct palaeontological work, which progresses slowly and with accuracy. However so far it yields no direct fossil lineages to modern groups. We have important new evidence concerning the early armoured and the early unarmoured arthropods and the early arachnids, but this does not build up a complete picture.

Secondly we have the newer results from comparative functional morphology of living arthropods. The concept of persistent habits over long periods of geological time and the morphology correlated with these habits, is now recognized for its paramount importance in understanding the origin and diversification of the large taxa. The view that the arthropods must have each been working wholes at all stages in their evolution and that proficiencies once

---

* Dr Manton died while this volume was in press.

Systematics Association Special Volume No. 12, "The Origin of Major Invertebrate Groups", edited by M. R. House, 1979, pp. 269–321. Academic Press, London and New York.

gained cannot be replaced by lesser achievements, are principles in need of general acceptance.

The third line of advance lies with the newer comparative and functional embryology. It is perhaps the most important of all concerning the larger taxa. Fate maps, not cell lineage theories, have enabled clear comparisons to be made for the first time between the major taxa of metamerically segmented invertebrates. The ontogeny of crustaceans and uniramians are each consistently different from one another. The former could not have arisen from that of any known annelid and the latter could conceivably have been descended from some yolky-egged annelid, but the adult worms probably possessed lobopodia and a haemocoel. The embryology of arachnids and of *Limulus*, as yet insufficiently known, show no clear affinity with either of the embryonic processes of crustaceans or uniramians.

INTRODUCTION

Reassessment of the probable phylogeny of arthropods has passed in recent years from realms of personal beliefs and speculation to one in which solid evidence, although not as yet all embracing, can be assembled and suitably considered. The evidence stems from three main fields of investigation.

1. Modern Palaeontological Work

Improved techniques, mechanical, chemical, photographic and X-ray, combined with meticulous care by eminent and capable palaeontologists, have given us restorations of animals which are complete in many ways as to external features and sometimes showing a little internal structure as well. The fine detail of many of these restorations is very good and does not go beyond ascertainable facts. Knowledge of extinct arthropods is essential for comparison with extant groups, the Cambrian genera in particular being of great importance.

2. Studies on Functional Morphology

This type of work has now been executed on a sufficiently grand scale, embracing much of the characteristic structure of present-day arthropodan taxa. It has provided an understanding of how different types of arthropods work and thus an indication of the meaning of their shapes and structural detail. It has shown how habits are correlated with contrasting body form. Sure pointers are provided thereby to the paths of past evolution. Similar and equally productive work has been accomplished on gastropod and bivalve molluscs by C. M. Yonge and others. Without a knowledge of both the functional significance of the morphology and of the habits, no sure indications can be postulated as to the modes of evolution and the relationships of the animals.

3. Comparative and Functional Embryology

The "cell theory" held sway at the end of the last century and onwards. It was dependent upon the description of spiral cleavage in early ontogeny, but left an apparent impossibility of comparing one type of ontogeny with another. The conclusion that segregation of organ systems stems from spirally cleaving blastomeres does not, in fact, conform with the segregation of developmentally functional components in the blastula stage. A sufficiency of work on annelid and arthropod ontogeny has been accomplished (Anderson, 1973) which establishes the importance and usefulness of applying the fate map system to invertebrates. Herein we now have a sound basis for comparison of the ontogeny of one taxon with that of another. Reliable conclusions have been reached concerning the several different paths of ontogenetic development and consequently clear indications are given of the possible and impossible relationships between major taxa.

THE CEPHALON, PROSOMA AND HEADS OF ARTHROPODS

A simple classification of living arthropods, based on the form of the anterior tagma of the body, is not contradicted by any further considerations.

In the following survey of the fossil record we may look for enlightenment concerning the origins of the various types of anterior tagmata. Is there any evidence suggesting ancestral derivation of one type of anterior tagma from another or any grounds for supposing that any one type has passed through a stage resembling that of another? It is more probable that each type of anterior tagma arose independently from animals lacking cephalization?

1. The Cephalon of Trilobites and Early Arthropods

The major groups of living arthropods have distinctively different anterior tagmata. At its simplest the cephalon bears a small dorsal shield, often with projections; a ventral labrum hanging in front of the mouth; preoral sensory limbs; no mandibles or other feeding limbs differentiated from the following trunk limbs, see *Marrella* and *Cheloniellon* (Figs 10, 13, 14). In the trilobites (Fig. 7) the cephalic shield was large and there were three pairs of postoral limbs on the cephalon where the limbs are well known; the structure of these postoral limbs resembled that of the following trunk limbs. In *Burgessia* (Fig. 12) the cephalic shield was large and also carried three pairs of postoral limbs, but their structure differed from that of the following trunk limbs. The cephalon was not, however, obviously adapted for feeding purposes.

2. The Chelicerate Prosoma

This tagma, present in no other group, is remarkably constant in composition. The prosoma bears a united dorsal sclerite with simple or compound sessile eyes in merostomes (Xiphosura, Eurypterida). In arachnids the eyes are sessile and simple; the prosomal tergal plate, usually in one piece, is divided into separate tergites over the more posterior legs in some orders. In all groups the prosoma bears one pair of preoral limbs, usually the chelate or subchelate chelicerae; prosomal limbs 2–6 are postoral and form locomotory and feeding limbs in Xiphosura and Eurypterida. A 7th pair of legs appears to have been locomotory at one time, but forms chilaria in Xiphosura and the metastomal plate in Eurypterida. In arachnids only the first postoral pairs of limbs, the pedipalps, is concerned with chewing, and four pairs of walking legs follow, the first two of which, in scorpions, have forwardly directed gnathobases which floor the preoral cavity.

3. The Crustacean Head

This tagma is remarkably uniform throughout the immense range in general form shown by members of this group. There are two pairs of non-chelate antennae. A third preoral segment is indicated by the large preantennulary pair of somites in the embryo in some classes, giving much adult tissue. Paired mandibles are para-oral in position and are followed by paired maxillules and maxillae. Only in *Hutchinsoniella* and the Cambrian *Canadaspis* (Fig. 4) does the maxilla resemble the following trunk limbs, implying that cephalization of the maxillary segment has proceeded independently in several lines of crustacean evolution. Frequently the head is united with one or more trunk segments forming a cephalothorax.

4. The Head in Uniramian Classes

The head in hexapods and myriapods is different from that of crustaceans. There are three preoral segments, one of which carries the antennae. Embryonic preantennal and premandibular segments possess mesodermal somites and ganglionic rudiments, sometimes coelomoducts and transient limb buds. Paired mandibles, first and sometimes second maxillae follow, but there is no fusion with trunk segments, except a union with the poison claw (first trunk) segment in centipedes for stability reasons. The presence of three preoral segments in embryonic myriapods and hexapods and the same number of embryos of some crustaceans probably has a functional explanation (Manton, 1977).

Cephalization at the anterior end of the Onychophora has proceeded less far than in other extant uniramians. There are no sound indications of preantennulary somites in the embryo. The first adult segment bears the antennae, the second segment the jaws which are basically of the uniramian type (see below) and a third head segment carries limbs modified as slime papillae. No evidence suggests that this three-segmented head with jaws on the second segment was ever a stage passed through by other uniramian taxa.

CONTRIBUTIONS FROM PALAEONTOLOGY

The Cambrian arthropods were of strikingly different types. Equally striking is the absence of intermediate forms between them. A recently re-examined metamerically segmented animal is *Opabinia regalis* (Fig. 1) from the middle Cambrian Burgess Shale. This animal was a corner stone of Sharov's fantasy (1966) on arthropod evolution—replied to by Anderson (1966a) and Manton (1967)—but *Opabinia* is no arthropod at all because it has no legs (Whittington, 1975a); it has no similarity with trilobites in respiratory organs; nor was its proboscis an extension of the gut; hence this animal provides no foundation for Sharov's hypothetical Proboscidea as an arthropodan ancestral stage. We do not know where *Opabinia* belongs in the hierarchy of invertebrates. It contrasts with all known annelids and arthropods and serves to emphasize our limited comprehension of metamerically segmented invertebrates.

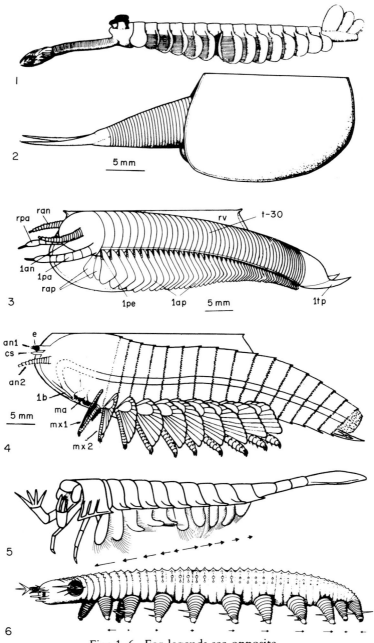

Figs 1–6. For legends see opposite.

Labels in figures:

Fig 3: ran, rpa, rv, t-30, 1an, 1pa, rap, 1pe, 1ap, 1tp, 5 mm

Fig 4: an1, e, cs, an2, 1b, ma, mx1, mx2, 5 mm

Figs 1, 2: 5 mm

1. Early Armoured Arthropods

A brief review of extinct arthropods may now be attempted, centering on the comparative compositions of the anterior tagmata, the importance of which has been noted above.

There are arthropods in plenty in the Cambrian deposits. The Phyllocarida (Cambrian to Recent) have hitherto been considered to be crustaceans and the ancestors of Malacostraca and to comprise the sure allies of modern Leptostraca, e.g. *Nebalia*. But the newly discovered detail (Figs 2–4) of Burgess Shale middle Cambrian fossils (Briggs, 1976), necessitates a revision of these ideas. The superficial appearance of a bivalved carapace, a multisegmented abdomen, a caudal furca and no knowledge of limbs, is not enough on which to base the crustacean status of *Protocaris marshi* (Fig. 2) and other genera (Briggs, 1976). *Branchiocaris pretiosa* is a well preserved arthropod (Fig. 3) and can be no crustacean (see below).

The present day ostracods are short-bodied, highly specialized crustaceans. The bivalved carapace (shell) encloses the whole body. Ostracod shells are recognized in the Cambrian and more recent deposits. Usually the shell content is unknown in fossil ostracods, but Dr K. Müller in Bonn has revealed the limbs in a Cambrian ostracod which are unlike those of extant ostracods in that all the few limbs are biramous, even the antennule.

The early merostomes appear to be close to the later members of this group in a general way and so connect with the modern *Limulus* and arachnids. There were plenty of Cambrian trilobites. Thus the origins of the phyllocarids, of many slightly crustacean-like arthropods, of ostracods, merostomes, trilobites and many other arthropods must lie in the

---

FIG. 1. *Opabinia regalis* Walcott, middle Cambrian, Burgess Shale, 70 mm, after Whittington (1975a).

FIG. 2. *Protocaris marshi* Walcott, middle Cambrian, Burgess Shale, 35 mm, after Briggs (1976).

FIG. 3 *Branchiocaris pretiosa* (Resser, 1929), middle Cambrian, Burgess Shale, 85 mm, after Briggs (1976).

FIG. 4. *Canadaspis perfecta* (Walcott), middle Cambrian, Burgess Shale, 60 mm, after Briggs (1977).

FIG. 5. *Yohoia tenuis* Walcott, middle Cambrian, Burgess Shale, 20 mm, after Whittington (1974).

FIG. 6. *Aysheaia pedunculata* Walcott, middle Cambrian, Burgess Shale, 50 mm, after Whittington (1978).

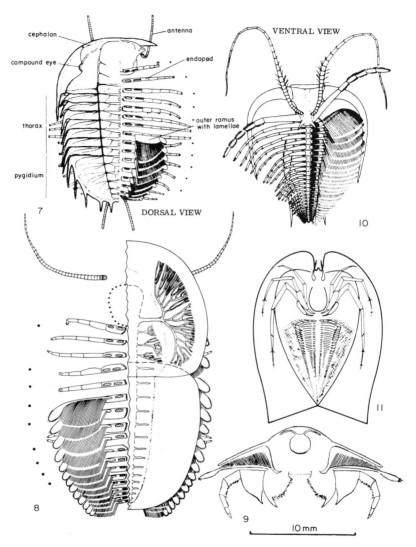

FIG. 7. *Olenoides serratus* (Rominger), middle Cambrian, Burgess Shale, 70 mm, after Whittington (1975b).

FIG. 8. *Naraoia compacta* Walcott, middle Cambrian, Burgess Shale, 30 mm, after Whittington (1977).

FIG. 9. Transverse view of same.

unknown, early Cambrian or the Precambrian. Apart from this fundamental impasse in available evidence concerning the origins of the many arthropodan groups and of the arthropods as a whole, there is much to be learned from the available fossil evidence. Do these early arthropods provide any clues as to the origin and evolution of the crustacean head, the chelicerate prosoma or the uniramian head?

Some early arthropods are shown in Figs 2–15. One pair only of preoral, sensory antennae was present in trilobites including *Naraoia*, in *Burgessia*, *Mimetaster*, *Vachonisia*, *Sidneyia*, *Emeraldella*, *Leanchoilia*, *Plenocaris*, *Yohoia* and others. In trilobites (*Naraoia*) and in *Burgessia* there were three pairs of postoral cephalic limbs, but showing no differentiation for feeding beyond that possessed by the trunk limbs. In the well preserved *Branchiocaris*, mentioned above from the Burgess Shale (Fig. 3), the head bears one pair of antenniform limbs and a second, stouter uniramous pair of limbs which could have reached the mouth. There were no further cephalic limbs and the trunk limbs formed a long, graded series.

Two pairs of antennae on a short cephalon were present in *Marrella* (Fig. 10) followed by close set trunk limbs, not differentiated for feeding purposes. The second antennae may have been used to sweep detritus towards the body. The two pairs of antennae in *Cheloniellon* (Fig. 13) are more reminiscent of crustaceans than the antennae in other early arthropods, but the resemblance stops there; no cephalic feeding limbs were present and four pairs of strongly cusped gnathobases were carried by the anterior trunk limbs. Only the bizarre *Vachonisia* (Fig. 11), with but one pair of antennae, possessed two pairs of possibly biting endites or gnathobases close to the mouth, curving laterally round the labrum. Whether these were the endites of two walking legs in the vicinity is unknown; there was a third shorter ramus directed outwards here. The correspondence of the cephalon of *Vachonisia* with the crustacean head is not exact. However the trunk of *Vachonisia* shows no resemblance to crustaceans or to any of the other extinct arthropods mentioned above.

FIG. 10. *Marrella splendens* Walcott, middle Cambrian, Burgess Shale, 15 mm. Anterior two pairs of limbs uniramous, remainder biramous, the walking endopod shown on the left and the outer ramus with lamellae on the right, after Whittington (1971).

FIG. 11. *Vachonisia rogeri* (Lehmann), Devonian, Hunsrück Shale, 60 mm, after Stürmer and Bergström (1976).

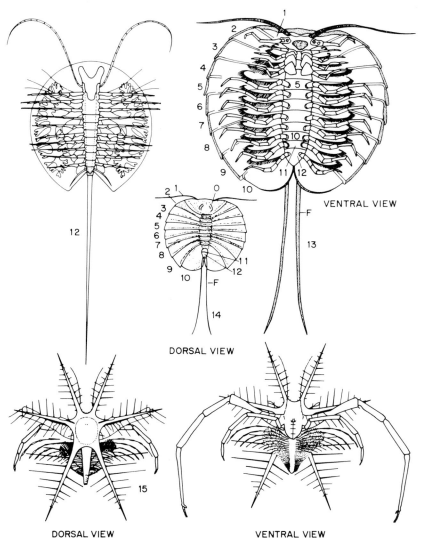

VENTRAL VIEW

DORSAL VIEW

DORSAL VIEW

VENTRAL VIEW

FIG. 12. *Burgessia bella* Walcott, middle Cambrian, Burgess Shale, 10 mm, dorsal view with carapace margin shown by dotted line, after Hughes (1975).

FIGS 13, 14. *Cheloniellon calmani* Broili, Devonian, 105 mm, from Broili (1933a, 1933b) and Tiegs and Manton (1958).

FIG. 15. *Mimetaster hexagonalis* (Gürich), Devonian, Hunsrück Shale, 25 mm, after Stürmer and Bergström (1976).

The Burgess Shale *Canadaspis* (Fig. 4), with two pairs of antennae, is more crustacean-like than any of the above. A mandible was certainly present, with enormous cusps on an incisor process. A short labrum left the cusps fully exposed, as are the cusps on a crab's mandible. Ten pairs of foliaceous, biramous limbs followed. The first was a maxillule close behind the mandible, showing strong proximal armature. The next limb was shorter than those behind and was presumably a second maxilla. The eight segments bearing the remaining limbs, all similar, appear to form the malacostracan type of thorax, while seven limbless segments and the telson must represent the leptostracan abdomen. But the telson lacked a furca and the 7th segment possessed large unarticulated spines. *Canadaspis* cannot be included within the Leptostraca because of the abdominal differences but is arguably the earliest known malacostracan crustacean (Briggs, 1977). The later *Ceratiocaris*, Ordovician to Devonian in age, is probably a leptostracan, although the head limbs are not preserved.

The simple and large cephalon of *Sidneyia* does not help us. It bore one pair of long antennae, a pair of compound eyes on short stalks and no further limbs. The trunk was differentiated into regions, bearing different kinds of limbs. The anterior trunk limbs possessed very wide, short coxae, expanded up the flanks as in *Limulus*, which might have given strength to the very strongly cusped enditic margins, suggesting much stronger biting ability than in trilobites. The remote merostome-like resemblance of *Sidneyia* is dispelled by the body terminating in a caudal fan reminiscent of swimming malacostracans.

A few other genera deserve mention. In *Emeraldella* the cephalic shield was relatively longer and the antennae very long. A second short and simple leg pair on the cephalon was situated beside the labrum. The succeeding four pairs of cephalic limbs resembled those on the trunk. No large gnathobases were present, but large endopods (inner rami), short outer rami and a huge proximal exite fringed with coarse setae carried no trilobite-like respiratory lamellae. *Leanchoilia* is well preserved and behind the elaborate complex formed by the most anterior appendage or appendages, a series of oval, flattened limbs with marginal setae followed, but no walking endopods. This animal was presumably a swimmer. The crustacean-like *Waptia* is presently under re-investigation.

The wide variety of early fossil arthropods (e.g. Figs 3–5, 12–15) were more divergent in their cephalic structure, one from another, than are any of the variously constructed extant crustaceans, all of which share the same basic plan of organization. There are, moreover, no fossil remains which are clearly intermediate between the several types mentioned above, although we need to know more about the phyllocarids before we can say what is and what is not a crustacean. *Canadaspis* comes nearest to crustaceans and the Leptostraca in the apparent direction of its cephalic and trunk evolution.

The extant Cephalocarida show less cephalization than any other living crustaceans in that the second maxilla is like the following trunk limbs, but the head and the antennae are far more advanced than that of *Canadaspis* (Fig. 4). The long abdomen of *Hutchinsoniella* without limbs and with great flexibility, cannot be a primitive feature but one suited to wriggling through certain types of bottom deposits. The earliest remains of modern crustacean taxa, other than ostracods, are much later in occurrence and their exact origins are not known.

In all the Cambrian arthropods, we see no sure indication of the origin of the crustacean head, although *Canadaspis* may be the earliest known near leptostracan showing an earlier stage of head evolution than other malacostracan crustaceans. The newer detail of many early arthropods is of great interest, but it must be remembered that the ostracods with their limbs are known from the Cambrian. It appears that we should look to the unknowable Precambrian for the origins of the trilobites, the merostomes, the phyllocarids, ostracods and other crustaceans and of the many other types of early arthropods.

With so many lines of Cambrian arthropods of decidedly different form, there seems to be nothing meaningful in grouping them into so-called higher categories, such as Trilobitomorpha, Arachnomorpha and Merostomoidea, which can have no phylogenetic significance. Even the trilobites extending through 250 million years, with some 2000 genera and a little adaptive radiation, showed a constant basic organization throughout their whole reign. Only *Naraoia* showed a marked divergence in the fusion of the trunk tergites (Whittington, 1977) a feature doubtless associated with the huge gnathobases and the muscles they must have possessed. But *Naraoia* had completed the divergence by the Cambrian, not at a later stage in the trilobite reign.

If, during 250 million years, the trilobite cephalon did not change except in minor details, it is most unlikely that just below the level at which fossils become recognizable, this cephalon metamorphosed into the crustacean or any other type of head. There must have been simpler antecedents than the known trilobites, merostomes, crustaceans and the many other types of early arthropods, but what they were like we have no clue. We do not know how the differentiation of segments and limbs into particular trunk tagmata was arrived at, but it seems probable that once a differentiation set in it could not have been transmuted into another type of tagmosis. The erection of, for example, an Ur-crustacean and then deriving other major groups from such a fictitious animal is a course fraught with more problems than it is ever likely to solve. How in the Precambrian the gene complexes were arrived at, which built up the different types of arthropods with their various potentials for success we shall never know. However the reality of the differences between the structure and potential of all these arthropods cannot be overlooked.

If we find no reasonable intermediates between many of the types of early arthropods, nor any sound reason for linking some of them together into larger taxa, we have no justification for drawing evolutionary trees representing the main lines of past history. The early arthropodan fauna can perhaps be better likened to the main branches of a bush or even to blades of grass on a lawn,* and how the branches arose below "soil" level we do not know. It is of the greatest importance to appreciate the fundamental differences between the various early arthropods and to sustain the search for further accurate detail, and it is safer to avoid hypothetical arthropods of all kinds, as has been practiced in the section on functional morphology below and in the original work which led up to it.

2. The Arachnida

This group of well diversified terrestrial arthropods is undoubtedly chelicerate in affinity. The fine work on the fossil species by Petrunkevitch (1949) classified them into 16 orders, of which half are extinct today. Recently a most important find has been described by Størmer (1974). *Alkenia mirabilis*, lower Devonian, was a thoroughly terrestrial arachnid which lived at the same time as the gilled scorpions inhabited water. Thus some arachnids must have become terrestrial millions of years before the scorpions did so. Therefore we can assume that certainly two, probably more, lines of arachnids were evolved before each became terrestrial. Functional morphology of the pycnogonids, including their general body shape and leg mechanism, indicates that they are descended from an early group of marine arachnids which never left the sea (Manton, 1978). There

---

* The latter simile was expressed to me by Profesor L. Størmer. S.M.M.

is now no necessity to try to envisage an origin of the extant orders/of arachnids from scorpions, a view which had seemed obligatory although hardly credible.

There are reasons, perhaps, for rejecting a supposed origin of arachnids from already highly advanced eurypterids (Bergström, 1975). Thus we are left with more probable early chelicerate divergencies and parallel evolution than has often been supposed (see also Van Der Hammen, 1977).

3. The Early Unarmoured Arthropods and the Uniramia

There is sound evidence for the belief that the Onychophora, Myriapoda and Hexapoda form a united group. It is probable that their diversification took place on land soon after their ancestors became terrestrial. Their ancestors were presumably soft-bodied with lobopodial limbs worked by a muscle-haemocoelic mechanism, contrasting absolutely with that of the polychaete parapodium associated with a coelomic body cavity (Manton, 1967, 1977). We know very little of the marine soft-bodied Uniramia which must have lived in the sea alongside the early armoured arthropods. The fossilization of soft-bodied animals from the land or in water is a much rarer event than the preservation of hard sclerites. But from the middle Cambrian Burgess Shale we have *Aysheaia* which probably lies on the uniramian line (Fig. 6). The limbs and texture of the cuticle are reminiscent of the Onychophora. The mouth was terminal with a pair of non-locomotory limbs behind it. The preservation makes certain that no sclerotized jaws lay within or outside the suctorial mouth region (Whittington, 1978). Sponge tissue, preserved alongside the known specimens, suggests suctorial feeding on sponges, as practised by pycnogonids and the polychaete *Spinther*. Thus the unarmoured, as well as many different armoured arthropods had, by the middle Cambrian, reached a grade or organization possessing preoral sensory limbs but no mandibles. Only *Canadaspis* and probably some other phyllocarids had progressed as far as the possession of a mandible. The incompletely known lower Cambrian *Xenusion* may perhaps be an early uniramian, but with far less certainty.

Whether the diversification of the uniramian groups had already started in the sea, as is certain for arachnids, and whether there were several uniramian landings at different times, we do not know, but both seem probable. Quiet estuaries with vegetation down to the water's edge, as seen near Newton Ferrers, Devon, today, or lagoon conditions with mangrove-like swamps in shallow water, as preserved in the Devonian Alken deposits in Germany and studied by Størmer (1970–1976), may have provided a route to dry land, but probably not the more widespread rocky or sandy beaches open to wave action. It should be remembered, however, that the first and greatest peril on land is not that of drying up but of osmotic uptake of fresh water which, if unhindered, would disrupt the whole body. Animals such as land planarians (Pantin, 1950) and onychophorans are not found under the dampest conditions, but in dryer, often open, country under the protection of stones during the day, rather than in damp forest. One example may be given in illustration.

The surface of onychophorans is extremely hydrofuge, as is that of the soft-bodied *Polyxenus* (Diplopoda), the unwettability protecting the animal from osmotic uptake of dew or rain, or water from wet surroundings (Manton, 1956a). But onychophorans have little control of water loss, drying up in room atmosphere more rapidly than earthworms, in spite of water conserving features such as uric acid excretion (Manton and Heatley, 1937). This poor control of water loss is due to the cuticle being pierced by innumerable openings leading into groups of unbranched tracheae 2 μm in diameter. Narrow tracheae are less easily deformed by outside pressure changes than are wider ones. There is a higher priority to an onychophoran than control of water loss, because such tracheae are not deformed by the extensive shape changes of the body which are of particular service in gaining shelter and are inevitably associated with changing haemocoelic forces acting on the tracheae. The centipede *Craterostigmus* uses similar minute tracheae in large numbers in association with hydrostatic forces generated by muscles at a distance which it employs in protracting the head capsule during feeding (Manton, 1965), a most unusual accomplishment. The cuticle of most myriapods and hexapods possesses an outer wax layer which functions in keeping fresh water out of the body.

There were doubtless many extinct uniramian lines. Possibly one was the myriapod-like Arthropleurida (Fig. 16). They were represented in the lower Devonian by *Eoarthropleura*, probably a few cm in length (Størmer, 1976), and by *Arthropleura* 1·5 m long, some 60–70 m.y. later in the Carboniferous. Our detailed knowledge of the structural correlations between habits and abilities of extant uniramians, some of which are summarized below, does not help us to interpret *Arthropleura*, except to stress its unsuitability as a potential ancestor of five contrasting types of leg-base mechanisms in the hexapod classes and those of the myriapods.

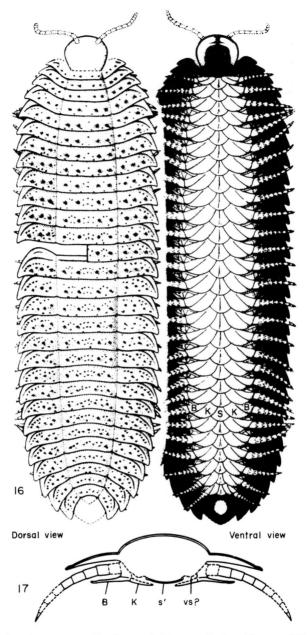

Dorsal view   Ventral view

B   K   s'   vs?

FIG. 16. *Arthropleura armata* Jordan and Meyer, Carboniferous, 800 mm. From Rolfe (1969).

FIG. 17. *Eoarthropleura devoniia* Størmer, lower Devonian, transverse view, after Størmer (1976).

The arthropleuridan sternal region is unique (Figs 16, 17). Large paired plates K overlapped the sternites in each segment and were possibly partly free from the ventral body surface, but supported laterally by the large plate B, apparently attached to the coxa on either side. Størmer (1976) suggests that plate K may have covered a "ventral sac" on the body wall above, which served a respiratory function. No spiracles were present, even in the giant *Arthropleura*. The coxal sacs present in some myriapods and hexapods and the coxal organs of certain onychophorans are not respiratory, but water absorbing, in function. If the Arthropleurida possessed some respiratory arrangement covered by a plate, as are the mesosomal gills of certain eurypterids, the Arthropleurida cannot be regarded as potential ancestors of either myriapod or hexapod classes. Even the heads of *Eoarthropleura* and *Arthropleura* are insufficiently known to demonstrate myriapodan affinity. The well sclerotized Arthropleurida may have left the water in that state; the group shows no powers of diversification. The Uniramia contrast in their great powers of diversification into the modern classes, but must have been small in size initially and little sclerotized for the known functional systems to have been evolved without reversals of evolution.

True diplopods were diversified and substantial in size, but not gigantic in the Carboniferous; chilopods with more fragile bodies are first recognizable in the Cretaceous. Winged arthropods also reached large size in the Carboniferous, but they may not be close to the modern Pterygota.

Only the Permian Monura, such as *Dasyleptus*, are close to a modern group, the Thysanura, but this is not a stepping stone to the Pterygota because the leg-mechanisms in these two classes are mutually exclusive (Manton, 1972, 1977). The leg and leg-base of *Lepisma* is far removed from a primitive hexapodan state, although this animal figures as a primitive hexapod in much entomological literature. In Recent deposits there are uniramians in plenty representing most of the modern taxa, but they provide little evidence of phylogeny.

CONTRIBUTIONS FROM FUNCTIONAL MORPHOLOGY OF LIVING ARTHROPODS

The recent work on habits and the correlations between structure and function enable us to reconstruct some of the past history of arthropods with certainty. Many of the morphological figures can be found in Manton (1977) and the rest in the original accounts.

The Cambrian arthropods show many different ways of collecting, preparing and transporting food to the mouth. The actual swallowing is done by oesophageal muscles in extant arthropods. The anterior limbs in extant arthropods usually do not push food into the mouth and there is no reason for supposing that they might have done so in extinct species. Probably the simplest method of transporting food to the mouth was by close-set limbs, as in *Marrella* (Fig. 10), which presumably moved in metachronal rhythm. A modern iuliform diplopod has very firmly articulated legs which give an apparently simple antero–posterior, promotor–remotor, swing. But the legs are not inserted exactly in the transverse plane, although this is not obvious. The median faces of the coxae come together during the backstroke and part a little during the forward movement. In a Cambrian arthropod with close-set limbs and similar movements, any food material swept towards the middle line would be alternately gripped during the backstroke, and shifted forwards and released during the forward stroke. Metachronal waves of limb movement, passing as usual from behind forwards, would effectively transport food forwards to the mouth.

The spacing of limbs further apart transversely and longitudinally, as in trilobites, would permit the enlargement of a basal endite to form a gnathobase. Food transport would continue as before but two new movements would be possible, direct transverse biting or gripping and a twisting movement of the leg-base, where the coxa-body junction lacked articulations and was narrow, which would enhance the basic effects.

1. Mandibles and Jaws

Mandibles were only one of the many solutions to feeding problems arrived at in the Cambrian. The biting, grinding or chewing organs situated behind the mouth are formed in mutually exclusive manners in the major taxa (Manton, 1964). A mandibular coxal endite, or gnathobase, may serve these purposes, the rest of the limb being large, with other functions, or reduced. Conversely, the tip of the limb and not the base may be used in feeding. Whole limb mandibles, biting with the top, are found only in the Uniramia, although the tops of the first cephalic limb in *Yohoia* and second cephalic limb in *Branchiocaris* (Figs 3, 5) may have reached the mouth. Gnathobasic mandibles occur in crustaceans where the mode of action differs from those of the many gnathobases occuring in trilobites, merostomes, arachnids and various extinct arthropods.

(a) *Merostome and trilobite gnathobases.* Chewing in the Xiphosura and Eurypterida probably resembled that in *Limulus* today, where five postoral limb pairs cut and chew the food. The number is reduced to a single pair on the pedipalps in arachnids. In *Limulus* and in arachnids the gnathobasic chewing movement is one of adduction and abduction in the transverse plane of the body, successive legs moving in opposite phase. The coxal base is wide in the transverse plane of the body and articulates laterally with the pleuron. Strong adductor muscles pull a pair of gnathobases together and weak abductor muscles operate from a short dorso–lateral coxal apodeme. The endopod of the limb in *Limulus* is used for walking or digging.

Walking movements in *Limulus*, as well as those of most arthropods, consist proximally of the promotor–remotor swing of the coxa on the body, taking place at right angles to the chewing movement. Naturally walking and chewing cannot be performed simultaneously, cf. trilobites. Probably the efficient chewing and cracking of substantial bivalve shells was achieved by a widening of the narrow coxal junction with the body possessed by many early arthropods (but not in *Sidneyia* which had a wide coxal base). In *Limulus* the extensive rim carries abundant strong extrinsic muscles.

Trilobite gnathobases, as indicated above, must have been used very differently, with their well spaced limbs, lack of coxal articulations with the body, narrow coxal head and thin sternal cuticle (Figs 7, 8, cf. 10) (Cisne, 1975; Whittington, 1975b, 1977). Some twisting of the coxal head on the body probably took place during stepping, not an easy movement to a merostome. The absence of coxal articulations was probably an important trilobite asset permitting a slight turning forward of the gnathobases during the forward stroke of the legs and the reverse on the backstroke, but gnathobasic movements must have been weak. The enormous gnathobases of *Naraoia* (Fig. 9) appear to be correlated with fusion of thoracic tergites (Fig. 8) so providing stronger support of extrinsic leg muscles than in articulated trilobites with relatively smaller gnathobases. No extant arthropods possess such enormous gnathobases as *Naraoia*; these animals did not persist for the long reign of the trilobites with articulated thorax. It is possible that the huge gnathobases, like the very large antlers of the extinct Irish Elk, were both advantageous characters but carried to extremes, ending not in success but in extinction.

(*b*) *Crustacean gnathobasic mandibles*. The crustacean mandibular gnathobase moves in an entirely different manner from that observed in *Limulus*. The strongest gnathobases of *Limulus* are on postoral legs 5, but the crustacean mandibles lie at the anterior end of the postoral series of feeding limbs and there are never more than one pair of mandibles, although smaller endites may be present on more posterior feeding limbs. In crustaceans a shortening of the mandibular limb, reducing the distal parts to a biramous or uniramous palp or none at all, and an enlargement of the gnathobase, leaves the mandible on a ventro-lateral axis of movement on the head (Manton, 1964, 1973).

The basic movement is not adduction and abduction in the transverse plane, as in the postoral limbs of *Limulus* and pedipalps of arachnids, but a movement derived directly from the promotor–remotor swing of an ambulatory leg.

The more primitive mandibles in branchiopods and malacostracans have a well formed crushing and grinding molar lobe, the pair being strongly rolled across each other during the remotor movement. Mandibular remotor muscles are larger and stronger than those performing the opposite promotor movement. The shape of the mandibles and their set on the head results in the promotor roll parting the mandibles slightly at the end of this movement, so making them ready for taking up food at the next grinding, remotor roll. The mandibular extrinsic muscles insert dorsally on the head capsule; from the internal faces of the mandible antero-lateral and postero-lateral muscles pass to their insertions on a transverse mandibular tendon. This is not an adductor tendon, as the literature often implies. The mandibular muscles inserting on the tendon contribute to the rolling mandibular movement. Those muscles arising more posteriorly are the larger and conspicuously contribute to the strong remotor grinding (see Manton, 1964, 1977, for details).

Feeding on larger food particles is associated with an increase in the postaxial part of the mandible and a decrease in the preaxial part, together with the formation of an incisor or cutting process on the postero-ventral corner. More bulky remotor mandibular muscles are accommodated thereby and the recovery, promotor, movement is served by muscles, doing no outside work, which operate from the small preaxial part of the mandible with little mechanical advantage. As the cutting ability of a mandible increases, so the dorsal end of the axis of movement on the head shifts posteriorly, thus the mandible lies in an oblique position (Manton, 1964, 1977). In some groups the grinding capacity of the mandible is reduced or abandoned as the cutting ability increases. In others, notably copepods and the Cambrian *Canadaspis,* initial predatory or scavenging habits are or were associated with a strongly cusped incisor process to the mandible. Only a few copepods possess filtratory feeding and good grinding ability. Strong mandibles cutting food by apparent movements of adduction and abduction in the transverse plane are a secondary condition, acquired independently, by several crustacean taxa. The backward and downward shift of the originally dorsal end of the coxal axis of movement progresses until it becomes horizontal and so produces a mandible with a wide gape and strong cutting, as in isopods. Here the grinding action is reduced to a minimum or may disappear. Strong cutting by mandibular marginal cusps is implemented by massive extrinsic muscles inserting via apodemes on to the head capsule. The mandibular cavity becomes devoid of muscles and the transverse tendon disappears. If present, it would restrict the gape. There are other ways in which some crustaceans acquire a transversely cutting mandible. The grinding mandibles suit fine food particles and strongly cutting mandibles suit predators and scavengers.

(c) *The preoral cavity.* Crustacean mandibles, unless strongly cutting and freely exposed as in crabs, operate within a preoral space housing the molar and incisor processes. This space is confined by the labrum, an antero-ventral fold from the body wall, laterally by the outer face of the mandibles, exactly as in a diplopod, and by the paired paragnaths postero-laterally; the mouth opens antero-dorsally deep in this space, as in uniramians. The preoral space, and its equivalent in chelicerates, is functionally much the same in all arthropodan taxa, except in the most specialized feeders. The preoral space is not restricted to the uniramians (for diagrams see Manton (1964, 1977)).

(d) *Mandibles of the Uniramia (Onychophora, Myriapoda and Hexapoda).* The contrasts between the mandibles of crustaceans and uniramians are profound. The uniramian mandibles differ from the gnathobasic feeding limbs of trilobites, merostomes, arachnids, crustaceans and many extinct aquatic arthropods in that they bite with the tip of the limb and not with the base. Since no arthropod can bite with both the tip and the base of a limb, the two types of jaw are mutually exclusive. The mode of action of uniramian jaws is various, depending on whether the mandible is jointed along its length, as in myriapods, or unjointed as in hexapods and onychophorans.

A basic hexapod mandible is seen in *Petrobius* (Manton, 1964, 1977). The axis of movement on the head is obliquely dorso-ventral, the movement being the promotor-remotor swing of an ordinary ambulatory leg, which gives grinding by the molar lobes and scratching by the incisor processes, used in scraping up algal cells. The movement and musculature resembles those of an anostracan crustacean in principle, but the mandibles themselves are entirely different, being formed by a whole limb biting with the tip in the one and by a gnathobase in the other. The hexapod, but not the crustacean, uses anterior tentorial apodemes for mandibular muscle insertions. Such mandibles in crustaceans and hexapods must be the result of entirely opposed evolutionary trends.

The mandibles in myriapods, although resembling those of onychophorans and hexapods in biting with the tip, use a contrasting movement because the mandibles are jointed along their length. This enables direct adductor muscles to pull the mandibles together, the distal podomere bearing cutting cusps or grinding ridges. Such mandibles use a transverse adductor tendon, giving strong biting in the transverse plane. The mandibles of diplopods and symphylans are strongly articulated proximally with the head. Alternatively, in chilopods and pauropods, protrusibility of mandibles takes place where there is a weak, very flexible mandibular union with the head, the mandibles being sunk in lateral pouches which are more extensive than the median preoral cavity. The condition is known as entognathy and examples occur in many taxa (Fig. 18), a convergent phenomenon (for details see Manton, 1964, 1977).

Direct transverse biting by ectognathous mandibles occupying the entire width of the head in diplopods and symphylans has brought acute problems because no extrinsic abductor muscles can be accommodated. A unique solution is provided by swinging anterior tentorial apodemes within the head which press on the mandibles, directly or indirectly, and cause abduction. The same abductor mechanism is employed by the entognathous myriapods, a probable legacy from their original ectognathous state. This solution of the problem of mandibular abduction is not found in any other taxon.

The hexapod groups possess their own types of mandibular advancement, always based on the whole-limb, rolling mandible seen in *Petrobius*, and utilizing the promotor–remotor swing of an ambulatory limb (Manton, 1972, 1977). A need for stronger biting is resolved by using the same principles as in crustaceans. The dorsal end of the mandibular axis of movement on the head shifts in a posterior direction, bringing this axis to a more or less horizontal position. The postaxial part of the mandible enlarges, as do the remotor muscles which now become adductor in function, while the preaxial part of the mandible becomes small and from it the weaker, now abductor, muscles pass into the head. A large cutting edge to the mandible provides a strong biting organ, the movement being secondarily in the transverse plane of the head in *Lepisma* and in pterygotes.

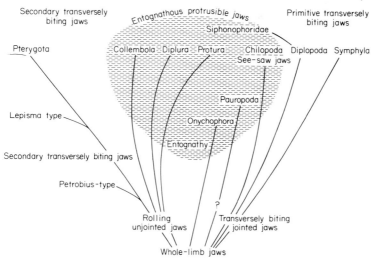

FIG. 18. Diagram showing the conclusions reached concerning the evolution of jaws or mandibles within the Uniramia from whole limbs biting with the tip. The shaded area indicates convergent evolution of entognathy and protrusible mandibles, after Manton (1964).

This biting type of mandible has formed the basis of many derived conditions suiting a variety of foods and habits.

Entognathous, whole-limb, unjointed mandibles have been acquired by Collembola, Diplura and Protura, where the advantage of protrusibility has been independently evolved by different means, but the basic promotor–remotor roll of the limb still persists.

The onychophoran entognathous whole-limb jaws are not far removed from walking legs. Each jaw is short and wide, with the two claws typical of walking legs enlarged into paired, cusped blades. The movements are antero-posterior slicing, a promotor–remotor movement, strength being conferred on the latter by massive remotor muscles originating from a long cuticular, sclerotized jaw apodeme extending posteriorly through several segments.

(e) *Tentorial apodemes.* The form and functions of the head apodemes in myriapods and hexapods until recently has been poorly understood (Manton, 1964, 1977). As noted above, myriapods possess only anterior tentorial apodemes, strong cuticular structures which are mobile, swinging within the head capsule and pushing the mandibles apart, there being no other abductor mechanism. The hexapods possess two pairs of tentorial apodemes. The anterior pair arises from a similar site to those of myriapods, but is never mobile or concerned with mandibular abduction. Anterior tentorial apodemes are absent from Diplura. The small size of the preoral cavity here makes their presence impracticable and a particular use is made by diplurans of the transverse mandibular tendon for insertion of mandibular muscles, here providing an unusual mode of action. Normally the anterior tentorial apodemes are associated with mandibular muscles but some of these may have other insertions as well.

The posterior tentorial apodemes are present in all hexapods and in no myriapods; these apodemes are associated with a well formed labium and arise from large cuticular intuckings in front of the labium. Myriapods have no true labium and the second maxillae of symphylans differ in structure and musculature from the hexapod labium and must be regarded as a partially convergent feature, lacking real homology. Where there is secondary strong transverse biting, the head apodemes fuse in pairs transversely and in pterygotes all four apodemes fuse together to form the rigid tentorium. This head endoskeleton, with four cuticular origins, primarily braces the head capsule against the tension exerted on it by the insertions of bulky extrinsic mandibular muscles. No such apodemes are needed, nor are they present, in crustaceans.

(f) *Conclusions from jaw structure and mechanisms and a note on convergence.* Full details (Manton, 1964, 1977) are available to substantiate the conclusion that the uniramian mandibles are a parallel evolution to those of crustaceans, the similarities being convergent and associated with similar needs. The crustacean gnathobasic mandibles are also different in structure and mode of action from the gnathobases of the Chelicerata. The gnathobases of the Trilobita differ again in their variety of movement, which must have been permitted by the small and unarticulated coxa-body junction; and in the lack of segmental restriction of the gnathobasic function, all postoral cephalic and trunk limbs being alike.

Thus it seems that there were at least three types of gnathobasic jaws in arthropods which contrast with the whole-limb uniramian jaws. There were many early arthropods without gnathobases or defined feeding limbs. Presumably they fed by the metachronal movement of close set limbs (see above and Fig. 10) which were not longitudinally spread out as in trilobites.

A study of the structure and mechanisms of jaws emphasizes the functional differences between the chelicerate prosoma, the trilobite cephalon and the crustacean head. The uniramian head with whole-limb jaws biting with the tips contrasts absolutely with various kinds of gnathobasic jaws present in the primarily aquatic arthropods, which appear to have been separate from one another perhaps since the Precambrian.

Few uniramians employ feeding limbs remote from the head end. The route for the food is from below upwards to the mouth region. In the basically aquatic, armoured arthropods the route is from behind forwards, close to the body, between the trunk limbs. Only in more advanced examples, such as crabs, are the feeding limbs close together so that the route does in effect pass from below upwards towards the mouth. In the Uniramia there do not appear to have been any trunk limbs associated with feeding, an exception being chilopod poison claws. Although modern annelids cannot be at all close to any arthropods, polychaete anterior parapodia are not used in feeding.

Enough has been said in outline to substantiate a supposed polyphyly between many known groups of arthropods and to indicate how deep are the clefts between them. There are parallel, independent evolutions of many structures and systems resulting in convergence. A considerable degree of convergence is implicit in any theory of arthropodan evolution and divergence; the problems attendant upon a monophyletic concept of arthropod evolution are the greater. One has to admit the convergent evolution of a mandible strongly biting in the transverse plane in certain crustaceans and in pterygotes and the independent evolution of entognathy in the groups illustrated in Fig. 18. Isopods are first found in the Triassic, much too late in time to have been involved in pterygote origin or the evolution of the land fauna. There is convergent evolution in the independent acquisition of entognathous jaws; in compound eyes, differently contrived in trilobites, crustaceans and pterygotes; in the composition of the preoral head in crustaceans and in myriapods and hexapods; and in many other features. But the fundamental differences point to the conclusion that the Uniramia, Crustacea and Chelicerata have originated independently from different pro-arthropod stocks (Fig. 19).

2. The Trunk Limbs

A consideration of limbs further substantiates the same taxonomic groupings based on heads and jaws (Fig. 19). Uniramian limbs have no branches, the onychophoran lobopodium is soft walled and sclerites on the legs of myriapods and hexapods necessitate jointing.

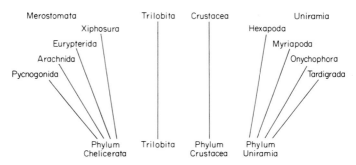

FIG. 19. The grouping of arthropodan taxa on the basis of present knowledge. The three phyla: Crustacea, Chelicerata and Uniramia appear to be quite distinct from each other. The Trilobita are distinct from the Crustacea and show no certain relationship with Merostomata. The Pycnogonida appear to have diverged from an ancestral aquatic arachnid group (Manton, 1978) and the Tardigrada probably have a distant relationship with the Onychophora, after Manton (1972, 1977).

Crustaceans possess a biramous leg, modified in a variety of ways to suit their various habits. The two rami are much the same in form in the swimming legs of copepods. The endopod and exopod are flat in branchiopods; but the endopod may be a walking leg, round in transverse section, in many groups such as malacostracans, whose endopods resemble the legs of myriapods and hexapods (see figures in Manton, 1977).

The exopod, or outer ramus, is a swimming, jointed structure richly provided with setae, which arises distally on the protopod, the proximal uniramous part of the leg in *Anaspides* and many other malacostracans. In addition, exites may be present, arising proximal to the exopod on the outer or dorso-lateral side of the protopod. Exites serve a respiratory purpose, either being gills, bearing gills, directing respiratory currents or forming valves closing and opening interlimb spaces on the outer side, as in branchiopods. *Nebalia* (Leptostraca) and crabs. Endites in variable numbers and sizes arise as lobes on the median side of the leg-base, the gnathobase being often the largest and most proximal endite. The lacinia and galea of the hexapod first maxilla are also endites.

The outer ramus of the trilobite leg may have been used for swimming. Its origin in *Triarthrus* is exactly like that of the exopod of *Anaspides*. But any homology between the trilobite outer ramus and the crustacean exite is dispelled by the close set series of lamellae, thin and presumably respiratory, carried by the trilobite outer ramus and covering the endopods and following limbs. No such structures are present in any crustacean. The outer ramus lamellae are known in some detail in many of the early arthropods and cannot be homologized with spines of either crustaceans or uniramians. The possibility of the trilobite outer ramus corresponding with a crustacean exite has been considered but the distal position of this ramus on the protopod of *Triarthrus* refutes this, although the origin of the outer ramus in *Olenoides* (Fig. 7) is more spread out and less confined to the distal end of the protopod. The non–commital term "outer ramus" is the more appropriate in trilobites and other early arthropods, since the branch in structural detail shows no crustacean resemblances.

The marine chelicerates possess a walking endopod with gnathobase on the prosoma behind the chelicerae and mouth. *Aglaspis* (Cambrian) possesses no chelicerae, and therefore can be no chelicerate. It possessed uniramous similar limbs on pro- and opisthosoma. Another aglaspid is claimed to have had an outer ramus bearing flat filaments like the pages of a book (Repina and Okuneva, 1969). In the true merostomes an outer ramus was not present on the prosoma and is restricted to the respiratory limbs on the mesosoma of *Limulus*. Arachnid limbs are simple endopods, with no trace of an outer ramus, which was probably present in their distant ancestors.

Whether the outer ramus of the biramous leg of early arthropods such as *Marrella* and *Mimetaster* (Figs 10, 15), corresponds with those of trilobites and with the claimed outer ramus of the Russian aglaspids we do not know, but meanwhile it is best not to equate them with the exopods or exites of crustaceans of very different appearance.

Thus the trunk limbs of trilobites, crustaceans and uniramians contrast in their structure and emphasize the distinction of at least these three great groups (Fig. 19).

(a) Rhythm of limb movements. The details of coordination between the limb movements are not a field which one would readily advance as having phylogenetic significance because nervous systems are capable of putting

out any rhythm, the choice in rhythm being mechanical rather than nervous. The mode of action of limbs is very plastic but there happens to be a common feature in the metachronal rhythm of uniramian limbs not shared by other arthropods or by polychaete parapodia. Long series of gaits are used in which increase in speed is obtained by progressive decrease in the duration of the backstroke relative to the forward stroke, the phase difference between successive legs being adjusted simultaneously to give mechanical efficiency. Slower progression but a greater force can be exerted by the legs using a longer relative duration of the backstroke and corresponding adjustment of the phase difference between successive legs (Manton, 1972, 1977). These series are still present in hexapods, although the smaller number of limbs gives certain restriction in the practicability of gait patterns. No long, even series of gaits are used by crustaceans or by *Limulus* or by arachnids. The latter may use three or four pairs of legs in walking or running but they do so by a quite different type of coordination from that found in hexapods (Manton, 1972, 1974, 1977). Even the polychaetes contrast in their parapodial mechanism, the uniramians flexing their legs at certain stages during the propulsive backstroke, so avoiding slipping of footholds, while the polychaete shortens its outstretched parapodium by pulling in the aciculum, no intrinsic bending being possible because of acicular rigidity. The stepping limbs of uniramians could not have evolved from the parapodium of any polychaete-like worm. It is probable that the long series of similarly coordinated gaits of the uniramians is a valid basic resemblance. The Thysanura, alone among uniramians, employ a unique type of jumping gait, used differently in the Machilidae and Lepismatidae. The thysanuran associated morphology also contrasts with other hexapods, but the series of gaits which characterizes the Uniramia still persists.

(*b*) *Conclusions concerning limbs.* The structure of the limbs of arthropods places the main taxa in the same contrasting positions as indicated by a consideration of their anterior tagmata and jaws. Trunk legs of uniramians appear to be primitively uniramous in contrast to the postoral limbs of the primarily aquatic arthropods. Here the legs are more complex, being either biramous or secondarily uniramous, the endopod or inner ramus persisting whereas the outer ramus and exites disappear, often during the life history of one crustacean.

3. Habits, Facilitating Morphology and Uniramian Evolution

The contrasting morphological features shown by onychophorans and the four myriapod classes, the Diplopoda, Chilopoda, Pauropoda and Symphyla, are intimately bound up with their very different habits, as summarized in Manton (1977). A differentiation of habits probably dated back to their early life on land when they were soft-bodied or little sclerotized, thus accounting for the evolution of head capsules of more than one type. The highly sclerotized Arthropleurida, Devonian to Carboniferous, may not be close relatives of any extant Uniramia. Until recently the functional significance of the conspicuous external features of the onychophorans and of the four myriapod and five hexapod classes has had little or no functional significance for us.

Life on land required seeking shelter and avoidance of osmotic uptake of water from dew, rain and wet surroundings, a far greater hazard than dessication (Pantin, 1950). Herbivores were likely to find food without difficulty, carnivores may have benefited by greater speed of movements for hunting prey or escaping from predators. An adoption of one or other of the following divergent habits may have led to present-day morphology: the list is not intended to be complete.

1. A habit of seeking a way, without pushing, through crevices in the surroundings, employing deformability of the body, may have led to the onychophoran type of organization including: connective tissue skeleton; unstriated muscles; slow movements; isolating hydrostatic-muscular leg mechanism as in no other known arthropod; cuticular papillae covering a sense capsule, with large apical sense seta; jaws without a transverse tendon, or any other structures which would hinder extreme distortion of the body, enabling the animal to traverse narrow crevices where sizeable predators cannot follow; minute, unbranched tracheae (see above). These and other features appear to be associated with the habits of onychophorans.

2. A contrasting habit of forcing the head end into the substratum by bulldozer-like activity, the motive force coming from the legs, appears to be associated with the diagnostic features of diplopods; the strongly armoured head and body; the pushing usually being implemented by a cap-like collum tergite or equivalent, keels, etc.; most trunk segments fused in pairs to form diplosegments, each one short, wide, with exactly cylindrical overlaps, and two leg pairs; thereby a large number of legs can

be carried by a not enormously long body, because intersegmental joints need space in which to work and space is saved when every other joint is eliminated. Some 40 other characters are recognized as associated with the diplopod pushing habit, used in penetrating soil and decaying logs for food and shelter (Manton, 1954, 1956a).

3. A habit of running faster is associated with carnivorous feeding in centipedes. The facilitating features are quite opposite from those of diplopods. At first, prey was probably stepped upon by accident in early potential centipedes, as it is in the Epimorpha today. The Anamorpha have refined their hunting techniques and associated morphology; vision is acute and they can start up in their fastest gaits, so achieving the catching of flies and spiders. But many problems had to be solved. The tendency to yaw when running fast, few legs being in contact with the ground at one moment (3 out of 40 in *Scolopendra*) has led to tergites of alternate length; long tergites over legs 7 and 8, the most stable part of the body; the formation of long sectors of many muscles traversing more than one joint; the shifting of muscle insertions off the short tergites and on to the long tergites, taking place progressively as leg length increases; very flat heads and flat poison claws, particularly in the Epimorpha, which promotes feeding in remarkably confined spaces. Entognathous mandibles of different types in the four orders of centipedes suit particular feeding techniques. The geophilomorph centipedes have exploited primitive deformity of body and have acquired an expert earthworm-like burrowing technique. The flexibility of the rolling edges of many sclerites, in contrast to those of diplopods, as well as the flexible pleuron, are associated with the ability of all centipedes to flatten themselves and hide in crevices, whether they hunt in crevices or are fleet enough to catch flies (Manton, 1965).

All centipedes (and also other myriapods) have a particular type of leg-base mechanism providing rocking on the long axis of the leg during stepping, which assists the speedy extension of extensor-less distal leg joints. Leg rocking is provided by other means in those hexapod classes which assist leg extension by this means (Manton, 1972, 1977).

The whole organization of the chilopod trunk, head, limbs, feeding mechanisms and burrowing is explicable in detail on particular habits which must have developed over long periods of time from an initial differentiation of behaviour.

4. Small arthropods with limited trunk sclerotization and a tendency to seek shelter by twisting and turning into existing crevices in leaf mould

and decaying vegetation, but without pushing, could have led to the Symphyla. Refinements of structure associated with this habit are: extra tergites; the separation of intercalary tergites from all tergites; the absence of all anti-undulation musculature, in particular the deep dorso-ventrals, deep obliques and long dorsal longitudinals; the enlargement of the trunk flexibility-promoting muscles; the superficial obliques.

The use of slow patterned gaits is obligatory because of extreme trunk flexibility, which appears to be the highest priority of symphylans. Consequently, there is no particular control of yawing and fast patterned gaits are not attempted. Instead, symphylans achieve considerable speed, in short bursts on a flat substratum, by using slow patterned gaits with many legs simultaneously on the ground, so controlling yawing, speed being achieved by the morphological features permitting a very rapid forward stroke of the leg and consequently rapid stepping. The success of symphylans appears to reside in small size and the structural and functional matters mentioned (for details see Manton, 1966).

5. Pauropod evolution is also associated with small size, but considerably smaller than symphylans. Living in the protection of draught-free, damp environments in decaying logs has probably led to little tolerance of other conditions. Pauropods, less than 1 mm in length, have achieved the greatest myriapod extremes of trunk rigidity-promoting morphology; no yawing takes place at all during their fast pattern gaits, with few legs in contact with the ground at one moment. Alternate tergites are absent; only long dorsal muscles are present, as are the deep oblique and deep dorso-ventral muscles which also promote rigidity. The flexibility pro-moting muscles, the superficial obliques, so well developed in symphylans, are absent. Exactly how the use of fast patterned gaits helps the habits of pauropods is not obvious; maybe they can travel fast for their small size in moving from one microniche to another. Only a few structural features of onychophoran and myriapodan habits have been mentioned. There are very many more (Manton, 1952–1977) but enough have been mentioned to illustrate how the detailed morphology and abilities are associated with the predominant habits. The divergencies in habits of these taxa must have set in very early in terrestrial life and must have persisted for long periods of geological time, the taxa becoming more and more different from one another structurally and more efficient at their contrasting proficiencies. Even the exact number of trunk segments is meaningful and often adjusted to achieve even loading on the legs in fast movers (Manton, 1952, 1972).

A simple plan of uniramian evolution is presented in Fig. 20 to indicate

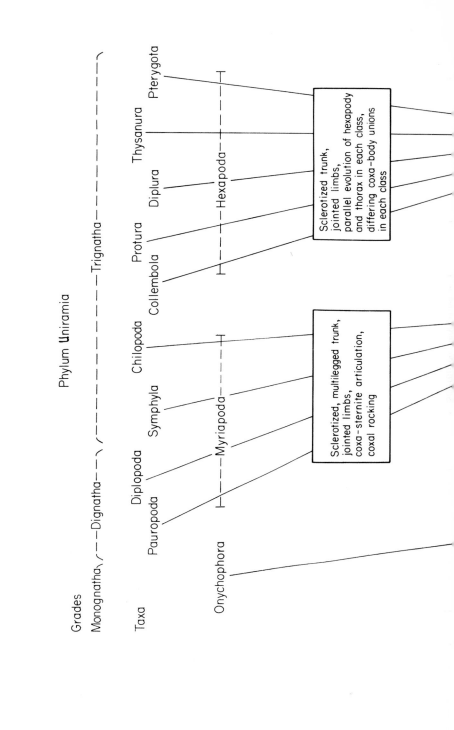

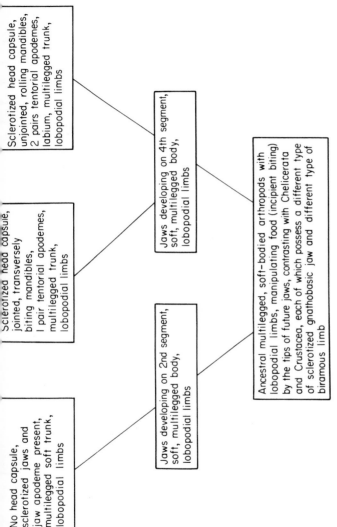

FIG. 20. A diagrammatic summary of the probable course of evolution of the several groups within the Uniramia as indicated by the evidence of functional morphology and embryology and disregarding the many taxa which must have existed but have not persisted to the present day. This diagram is not intended to indicate any support for fundamental dichotomous evolution of arthropods.

in a condensed manner the alternative paths of evolution followed by uniramian taxa. The Onychophora, the Tardigrada, which are not shown, and the nine classes of myriapods and hexapods may best be represented by independent divergent lines, as for the large groups shown in Fig. 19. Evolutionary trees do not appear to be justified and neither is there evidence of endless evolutionary dichotomies.

4. Leg-base Mechanisms and Hexapod Evolution

The evolution of hexapods as five divergent classes is bound up with locomotory habits. There are fundamental advantages in having only three pairs of legs, rather than a larger number. Fields of movement of the legs can be fanned out so that there is little or no overlapping. There is then more flexibility in stepping and ease of running in contrast to the obligation to step very exactly when legs are long and many. The habitual or facultative use of only three pairs of locomotory limbs has been evolved many times: in prawns, spider-crabs and arachnids, as well as by uniramian hexapods. But the similarities between the gaits of hexapods is due to the physical limitations imposed by the use of only six legs and is not indicative of affinity (Manton, 1972).

The hexapod classes have in common a particular type of head capsule, including two pairs of tentorial apodemes, in contrast to the myriapod head capsule with only one (Fig. 20). Both may have arisen in a soft-bodied state with evolving and contrasting mandibular mechanisms (see above). The leg mechanisms of the five hexapod classes are so different that the existence of a common basis can only be conceived at a soft-bodied stage, in which trunk sclerites were just appearing.

The leg-base mechanisms have been described in detail (Manton, 1966, 1972, summary in 1977). In outline, the dipluran and proturan coxa-body articulation is ventral and with the sternite, but the leg mechanisms are not the same. In the Diplura the axis of promotor–remotor swing of the coxa on the body lies in the transverse plane, although there is no firm articulation between the dorsal rim of the coxa and the body. A unique, intrinsic, leg-rocking arrangement is provided by particular muscles which permit and cause a twisting of the leg on its long axis and back again, facilitating the extension of distal leg joints without extensor muscles, as in myriapods which achieve leg-rocking by different means. The Protura are different again. A dorsal coxal articulation with a pleurite exists and this pleurite slides by a facet against another pleurite dorsal to it, so rocking the coxa on its ventral sternal articulation, and thus the rest of the leg.

The Collembola lack all coxa–body articulation, a condition associated with their hydrostatic jumping mechanism. The jumping organ at the end of the short abdomen is not worked by muscles, as hitherto supposed, but by trunk hydrostatic pressure; muscles put the organ away to its resting position held by the hamula. The whole body organization is geared to this production of pressure. Certain trunk muscles are enlarged whose contraction produces longitudinal body shortening. Elaborate provisions prevent arthrodial membranes from billowing outward under sudden internal increases in pressure of the haemolymph. Trunk intersegmental membranes are very short and internal struts prevent local expansions of arthrodial membrane above leg 1 and at leg joints distal to the coxa. Each coxa is suspended in an elaborate manner from thoracic tergites and segmental tendinous endoskeleton; the coxa itself possesses many unique features. A normal articulation or pair of articulations at a joint could easily slide apart under increases in internal hydrostatic pressure. Less elaborate provisions stabilize the more distal leg joints against dislocation. Leg joints are minimal in number, the leg possessing one podomere less than in other hexapods.

The many structural features of Collembola are only comprehensible on an understanding of the jumping mechanism. It is probable that these features evolved directly from a soft-bodied animal lacking the sclerotizations typical of other hexapods. The jumping organ probably corresponds with a pair of posterior limbs. A short abdomen must have been an early acquisition because the jumping organ could not function on the end of a long abdomen. There is no evidence suggesting that Collembola ever possessed the longer abdomen of other hexapods. The short abdomen must date from the time of evolution of the other trunk features associated with jumping. Even the earliest known collembolan, *Rhyniella praecursor*, Devonian, already had the pro- and mesothoracic terga fused, one of many measures promoting trunk rigidity. Collembola appear to have secondarily become small when they lead a less exposed and more subterranean life; these genera have progresssively lost the jumping organ. The structure and evolution of Collembola is clearly independant of that of other hexapods and can only be comprehended by understanding hydrostatic jumping. Many myths can be cast aside, such as the supposed existence and usefulness of subcoxal segments.

The Thysanura and Pterygota both show a coxa articulated by its dorsal rim to a pleurite, there being plenty of ventral arthrodial membrane between coxa and body in contrast to the Diplura and Protura. Again there is no further similarity. The pterygote pleurite is rigid, giving the coxa a firm base. Flight could not have originated otherwise. The pterygote coxa also articulates with a free trochantin in the pleural membrane in the antero-dorsal vicinity of the coxa. The promotor–remotor swing of the pterygote coxa takes place about the dorsal pleural articulation.

The Thysanura are quite different. The pleurites are mobile and the promotor–remotor swing of the leg takes place about the union of pleurites and body. Mobile and fixed pleurites are mutually exclusive. The one or two pleurites over the legs of the Machilidae are arranged differently from the three pleurites of the Lepismatidae, but all are associated with jumping gaits; and in the Lepismatidae with the quite extraordinary leg mechanism (Manton, 1972, 1977).

The leg-base of *Lepisma* is very far removed from a theoretically primitive state. If imaginary ring-like sub-coxae were present, they could not function. Crescentic pleurites forming a series anterior to the coxa are essential to the leg mechanism of *Lepisma* which is very different from those of other hexapods and certainly is not primitive, as has been supposed. These pleurites are associated with jumping gaits. Since there never can have been non-functional ancestral stages, many non-functional ideas about "insect" pleurites have to be abandoned.

The structure and modes of action of the leg-base mechanisms in hexapods are so different one from another, and so mutually exclusive in their actions, that independent derivation of them all from the unsclerotized or weakly sclerotized trunk of a multisegmented ancestor, with a head capsule tending towards that of hexapods, seems to be the only logical conclusion concerning the origin and evolution of hexapod classes. No evidence can be found for the generally accepted view that there once was an ancestral "insect", namely a six-legged ancestor to the five hexapod classes. On the functional basis one can only conceive of the five hexapod classes having evolved independently from multilegged ancestors (Fig. 20).

We have much information concerning flight mechanisms of pterygotes, but nothing certain concerning the origin of flying insects. There is no continuity in the fossil record between gliding and flying Carboniferous insects and modern pterygotes. We can but agree with Tiegs (1955) in supposing that the ambulatory leg and trunk muscles of walking nymphs became transformed into the flying mechanism of winged adults, as he has so ably demonstrated.

CONTRIBUTIONS FROM COMPARATIVE EMBRYOLOGY

For almost a century, comparative embryology made little impact in the debate on arthropod phylogeny, due to a fragmentary approach, a confused terminology and a lack of attention to functional considerations. A unified interpretation of arthropod embryos, based on a functional approach to the development of integrated structure, and by using fate maps to epitomize developmental patterns, was first attained by Anderson (1973). The results of this approach provide striking support for the recognition of three independently evolved phyla of modern arthropods, the Crustacea, Uniramia and Chelicerata, and for the conclusions discussed above as to relationships within these phyla. Questions concerning the Trilobita and the Pycnogonida cannot receive comment from embryology; the Trilobita for obvious reasons, the Pycnogonida because the necessary embryological studies have yet to be made.

Since 1973, a number of new embryological studies have been published on the Crustacea (Dohle, 1976; Lang, 1973), the Myriapoda (Dohle, 1974; Knoll, 1974), the Hexapoda (Ando and Kobayashi, 1975; Bernard, 1976; Bownes, 1976; Craig, 1974; Krysan, 1976; Miyakawa, 1975; Mori, 1976; Rempel, 1975; Wall, 1973) and the Chelicerata (Muñoz–Cuevas, 1973; Scholl, 1977; Weygoldt, 1975; Yoshikura, 1975), but these have not altered in any way the conclusions presented earlier.

The fate map approach to embryology allows the various basic patterns of arthropod development and the functional constraints on them to be identified. Functionally compatible, divergent patterns can be distinguished from other, functionally incompatible patterns of development. The Crustacea, for example, all share a mode of development based on modified spiral cleavage, a fate map of the blastula (Fig. 21) in which the presumptive mesoderm lies ventrally in front of the presumptive midgut, a precocious development of the anterior part of the animal as a nauplius region, and a subsequent anamorphic development of the postnaupliar region (Anderson, 1969, 1973). This pattern persists (Fig. 22) even when cleavage is centrolecithal, a blastoderm is formed and development is direct (Manton, 1928, 1934). The only feature that crustacean development shares in common with annelid development is spiral cleavage, which also occurs in platyhelminths, nemerteans, molluscs, sipunculids and echiuroids. Furthermore, crustacean development cannot be derived from annelid development. Among other things, the presumptive mesoderm of annelids (Fig. 23) lies behind the presumptive midgut in the blastula and

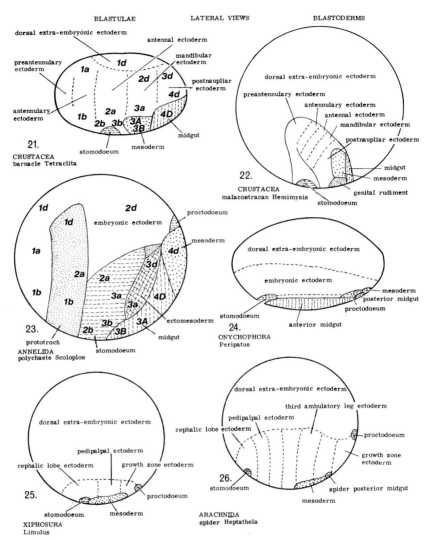

BLASTULAE LATERAL VIEWS BLASTODERMS

21.
CRUSTACEA
barnacle Tetraclita

22.
CRUSTACEA
malacostracan Hemimysis

23.
prototroch
ANNELIDA
polychaete Scoloplos

24.
ONYCHOPHORA
Peripatus

25.
XIPHOSURA
Limulus

26.
ARACHNIDA
spider Heptathela

Fig. 21. The fate map of the blastula of a cirripede crustacean *Tetraclita*, after Anderson (1969, 1973).

Fig. 22. The fate map of the blastoderm of the malacostracan crustacean *Hemimysis*, after Manton (1928) and Anderson (1973).

Fig. 23. The fate map of the blastula of the polychaete annelid *Scoloplos*, after Anderson (1959, 1966b, 1973).

Fig. 24. The fate map of the blastoderm of the onychophoran *Peripatus*, after Manton (1949) and Anderson (1966c, 1973).

is formed by the blastomere 4d. In crustaceans, 4d or its equivalent is merely a presumptive ectoderm cell behind the presumptive midgut, the mesoderm being formed by antero-ventral blastomeres 3A, 3B and 3C or their equivalent. The embryological evidence points firmly to the conclusion that the Crustacea are a phylum evolved, independently of other extant arthropods, from "worms" of the spiral cleavage assemblage that were related more closely to the Platyhelminthes than to any of the phyla with 4d mesoderm (nemerteans, annelids, molluscs, sipunculids, echiuroids) (see also Anderson, 1978.)

Similarly, the functional approach to comparative embryology adduces certain form conclusions about the Uniramia. It reveals a common basis of development in the Onychophora, Myriapoda and Hexapoda, thus supporting their inclusion in a phylum Uniramia. It shows that the Onychophora retain, in yolky-egged species, the most generalized form of uniramian embryonic development, with centrolecithal cleavage, a blastoderm and an epimorphic direct development lacking any vestige of prior larval interpolation (Anderson, 1966c, 1973; Manton, 1949). A comparison of fate maps (Fig. 24) further reveals that uniramian develop-ment is derived from spiral cleavage development of the type with 4d presumptive mesoderm behind presumptive midgut, closer to annelids and molluscs than to Crustacea. One implication of this conclusion is that metamerism had a separate origin in Uniramia and Crustacea. This is further supported by the deep-seated differences in somite development in the embryos of the two phyla (Anderson, 1973, 1978).

Within the Uniramia, embryological comparisons show that:

1. The Myriapoda and Hexapoda are each a unitary group with its own pattern of development, neither being derived from the other.

2. The scolopendromorph Chilopoda retain the most generalized type of development in the Myriapoda, with a common basis with that of the Onychophora.

3. The Symphyla, Diplopoda and Pauropoda share more develop-mental features in common than any of them does with the Chilopoda.

4. The Symphyla have developmental specializations which preclude them as ancestors of any hexapod.

Fig. 25. The fate map of the blastula of *Limulus*, after Anderson (1973).

Fig. 26. The fate map of the blastoderm of the spider *Heptathela*, after Yoshikura (1954, 1955) and Anderson (1973).

5. The Hexapoda share the development of a distinctive trignathan, labiate head structure.

6. Within the Hexapoda, the Thysanura and Pterygota, with their short embryonic primordium, their distinctive pattern of presumptive areas of the blastoderm and their development of extraembryonic membranes, may be more closely related to one another than to the other classes of hexapods.

The evidence for these views is discussed fully by Anderson (1973). In general, functional considerations of uniramian embryonic development lead to the same phylogenetic conclusions as those arising from functional morphology.

The Chelicerata are less rewarding embryologically at the present time than the other major groups of arthropods. In part this is due to lack of crucial evidence, but it also stems from the fact that no modern chelicerate retains a generalized, phylogenetically revealing mode of development equivalent to those of certain Onychophora and Crustacea. Even the embryos of *Limulus* fail to show us adequate evidence of ancestral relationships. It is not possible, for example, to establish that chelicerate embryos have any phylogenetic connection with spiral cleavage embryos (Anderson, 1973). A relationship with uniramian embryos can be ruled out on other embryological grounds, but a relationship with Crustacea cannot be unequivocally rejected embryologically. On the other hand, neither can it be supported. All chelicerate embryos share a basic mode of development with a highly distinctive fate map (Figs 25 and 26). This clearly supports the unity of the Chelicerata as a phylum. Yet, throughout chelicerate development, there is no trace of the features so distinctive of crustacean development, which might be expected if the groups were related. Somite development, in particular, is different in the Crustacea and Chelicerata. The key distinguishing feature in the fate maps of Crustacea and Uniramia, placement of presumptive mesoderm in relation to presumptive midgut, speaks in Chelicerata neither for nor against a crustacean relationship. Development is specialized (Fig. 25) and the presumptive midgut is wholly internal at the blastula stage.

Within the Chelicerata, there is clear evidence that the scorpions have a different modification of development from that of other arachnids (Anderson, 1973), supporting the idea of an independent origin. There is also some indication that the other arachnids may comprise more than one type (Anderson, 1973; Yoshikura, 1975; see also Van der Hemmen, 1977).

POINTERS TO EVOLUTIONARY PROGRESS

Many sure guides to the past history of arthropods have been touched on above. Whether the evidences are derived from palaeontology or from habits and the correlated functional morphology of living arthropods, or from comparative embryology, conclusions are valid only on a functional basis. Progressive differentiation leads towards better living in a variety of circumstances, rather than to detailed adaptations to particular niches. Also there has been plenty of adaptive radiation in the end terms of some of the larger groups, such as malacostracan crustaceans and winged insects. But such adaptations appear to be unconnected with the differentiation of major taxa.

The diversification of the principal groups of arthropods appears to be very deep-seated. The brief survey of the fossil record shows this plainly. Many parallel evolutions, without intermediates, were established by the Cambrian. The later diversification of the uniramians on land shows similar parallel lines which separated early into the several classes of extant animals and probably other taxa which are now extinct. As with many early fossil arthropods, the onychophorans and nine classes of myriapods and hexapods may be likened to the stems of a bush rather than to the branches of a tree. We have no knowledge of these arthropods below "soil level", which means at least earlier than their first divergencies on land. This type of evolution of the main lines of uniramian arthropods shows similarities with the main lines of primarily aquatic arthropods. The Xiphosura, for example, are not considered to be a simple line with few genera persisting to the present day, but rather a complex array of xiphosurans evolving in parallel with one another (Bergström, 1975) and the primarily aquatic arachnids may have been similar; but with the extant groups we can better appreciate the correlations between divergent habits and structure. One example may suffice. A symphylan ancestry of hexapods was seriously considered at one time, but the functional significance of none of the relevant morphology was understood. This view can now be set aside with certainty. The structure and mechanisms of the mandible, tentorial apodemes, leg-base, etc., are decidedly myriapodan and contrast with all hexapods. The second maxillae of symphylans are only convergently similar to a hexapod labium. The symphylan trunk organization suits particular ways of life which are unlike those of hexapods. Symphylan ontogeny is myriapodan and not hexapodan.

Symphylan gaits are unusual and are facilitated by particular limb musculature and movements which do not lead towards those of hexapods (Manton, 1966, 1972, 1977).

The categories Mono- Di- and Trignatha become grades of organization, not taxa. Evolutionary progress and the differentiation of large taxa concerns different ways of solving similar problems, the result being better living in the same or in a variety of habitats, but not in detailed adaptations to particular niches (Manton, 1977). The evolution of modern end terms of evolution, as many sets of related species usually occupying different habitats or in some other form of biological isolation, can often be represented reasonably by paper "evolutionary trees". But the main lines of arthropodan evolution, many down to the Cambrian, appear to be associated with habits and not with habitats and show little, if any, arborization. What is true for reptiles (Watson, 1949) holds also for arthropods. Many readers will look askance at the Cambrian arthropods being likened to the blades of grass in a lawn (see above), but it emphasizes the need to look again at the all too abundant drawing of imaginary evolutionary trees in zoological literature. The recognition of the impor- tance of convergence is inevitable, whatever view on arthropod phylo- geny may be favoured. Differentiated head ends have been formed several times by different means. Various degrees of sclerotization and calcification, always associated with habits and needs, are apparent. Joints of body and limbs are the inevitable consequence of cuticular stiffening. Where there is only one mode of forming a mechanically suitable joint, than it may have been evolved more than once.

Our knowledge of the origin of the arthropodan haemocoel is far less precise. Since a haemocoel is well formed in molluscs and arthropods and the latter must be far removed from the former, it is probable that the haemocoel also is a convergent phenomenon. The haemocoel in molluscs and onychophorans is used in effecting considerable shape changes of the body. The onychophoran lobopodium is dependant on the haemocoel and particular type of musculature for its functioning. The embryological evidence outlined above directs our attention to some annelid ancestor of the Uniramia. One may venture to step further by suggesting that the evolution of a lobopodium, as opposed to a parapodium, was associated with the presence of a haemocoel, the two having evolved together. Such a haemocoelic worm did not persist to the present day, its great potential having been exploited by the Uniramia. There is no evidence concerning the origin of the haemocoel in the Cambrian armoured arthropods, which must be associated with unknown Precambrian evolution.

The above pages represent a commentary upon recent published research indicating a polyphyletic evolution of arthropods. There is, however, one pointer to evolutionary progress which is very conspicuous, but appears to have passed unnoticed. A feature possessed by the Uniramia, in contrast to other arthropods, is a marked flexibility of trunk in all directions, in the horizontal plane as well as dorso-ventral. The uniramian pleuron, with its abundance of different pleurites, all serving particular needs in the several taxa (Manton, 1965, 1966, 1972), contrasts with the simple pleura★ of trilobites, chelicerates and crustaceans. In the primarily aquatic arthropods the most conspicuous and often the only trunk movement, takes place in the vertical and not the horizontal plane, except for minute bottom-wriggling crustaceans, the parallel-sided harpacticid copepods and syncarids and the greater part of the body of *Hutchinsoniella*. Lobsters and shrimps flap the tail end strongly under the thorax, so jumping backwards through the water. The isopods *Sphaeroma* and *Armadillidium* roll up into a sphere (Manton, 1977, pl. 3(d)). The mechanism is the same in all these animals; the body is shaped like a half cylinder (Fig. 27b, d, e); the axis of movement between one segment and the next being low down, as shown by the heavy line. A cylindrical body gives hindrance to dorso-ventral bending because the ventral half is difficult to compress (Fig. 27a, b). But the axis of movement in a half cylindrical body must be held in the equatorial position if the full benefit of the body shape is to be realized. In crustaceans and in oniscomorph diplopods (pill-millipedes) this is done by tergites overlapping from before backwards and sternites overlapping from behind forwards (Figs 27c, 28c).

The trilobites could also enroll (Fig. 28d, f) but their mechanism was different from that of crustaceans and oniscomorph diplopods. The paratergal extensions (pleura of palaeontologists) proximal to the paratergal (pleural) lobes were horizontal and the paratergites of successive thoracic segments articulated with one another by two peg and socket sclerotizations on either side, all four in the same horizontal plane (Fig. 29).

★ The trilobite pleuron here refers to the lateral sides of the body underneath the paratergal (pleural) expansion, just as in a crab, thysanuran, polydesmoidean diplopod or collembolan. The pleuron is large in the crustaceans and very small in extent on the collembolan thorax (with very good functional effects), but the pleuron is morphologically the same in all.

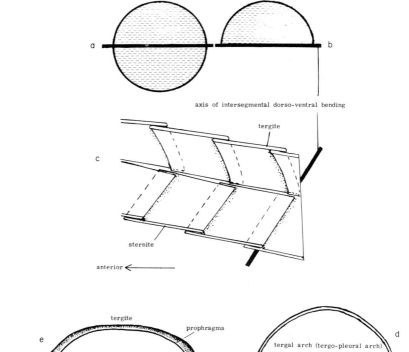

FIG. 27. One of the methods of intersegmental bending. (a), (b) Transverse
sections of a cylindrical and a half cylindrical trunk segment; the heavy
line denotes axis of swing of two successive segments on each other. (c)
Mechanism of overlap of tergites and sternites which fixes the axis of
the nodding movement to the equator, as indicated by the heavy line.
(d) Outline of a frozen section of an abdominal segment of a lobster,
showing the half cylindrical shape and the position of the axis of
intersegmental movement. (e) Anterior view of a half cylindrical
diplosegment of a pill-millipede *Sphaerotherium*; the prophragma
apodeme corresponds in position with that of trilobites (Fig. 29b, d, e),
the paratergal lobe and part of the tergo-pleural arch in (d) is the pleuron
of carcinologists, but in (e) the pleuron, morphologically ventrolateral
to the tergite, bears one large horizontal pleurite; the sternites are better
seen in Fig. 28(c).

Movement was thus restricted to exact dorso-ventral flexure, as would be done by door hinges being placed on either side. A dorsal outlet of the trilobite body on enrolment was facilitated by an antero-dorsal flange from the tergite, strengthened by a furrow. The flange tucked under the preceding tergite (Fig. 29a, c), but was no half segment, the intersegmental plane being normal. The spider's trochanter–femur joint has a similar flange from the trochanter, but distal and ventral, tucking into the femur, permitting considerable and strongly supported levator movements at this joint (Manton, 1977, Figure 10.2 row (e)).

Trilobite enrolment was not as tight as in isopods and pill-millipedes, leaving a larger central space to house the legs than in the oniscomorph, Fig. 28(a) dotted line and (e). To flank this space laterally the paratergal lobes were long in trilobites (Fig. 28d, f) but short in isopods and pill-millipedes (Fig. 28a, b). The latter have other features which enhance tight enrollment.

Since the paratergites (pleura) of trilobites take the strain of the horizontal rows of intersegmental articulations, it is not surprising to find the axial groove in a suitable position to give support by preventing buckling of the thin cuticle.

The Xiphosura either rolled or folded up; the opisthosoma of *Limulus* can fold against the prosoma. It does this by a simple dorsal hinge, very wide transversely, where little arthrodial membrane unites the stout tergite edges of pro- and opisthosoma. (The position is shown on Fig. 29d). There is no cuticular articulation. Thereby a dorsal hinge, preventing any body shortening, enables a burrowing thrust, by the depressed anterior edge of the prosoma, to be maximal. Ventral soft cuticle in front of and behind the wide genital operculum readily crushes up on dorso-ventral flexure.

Thus the trunk movements of trilobites, merostomes and crustaceans, dorso-ventral and not lateral, are dependant upon fundamentally different cuticular structures. These movements in the three taxa must have characterized them since the earliest known records but do not suggest any community in origin of the three groups.

The anterior tagmata, the coxae, the gnathobases and the limbs of crustaceans, merostomes and trilobites differ both in structure and in manner of use. These features appear to be as mutually exclusive as are the cuticular arrangements for dorso-ventral bending, again indicating divergent evolution and not common origin. The early Xiphosura with jointed opisthosoma appear to have had a series of dorsal hinges, probably of the *Limulus* type. Trilobites had their own type of gnathobasic coxa;

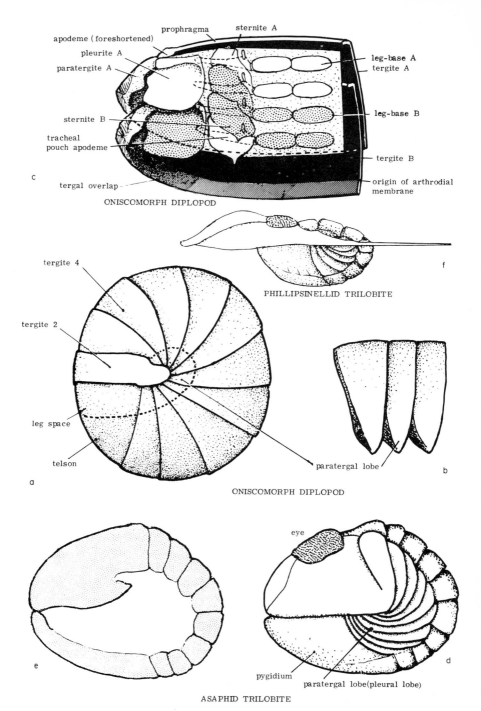

apodeme (foreshortened)
prophragma
sternite A
pleurite A
paratergite A
leg-base A
tergite A
sternite B
leg-base B
tracheal
pouch apodeme
tergite B
tergal overlap
origin of arthrodial
membrane

c

ONISCOMORPH DIPLOPOD

PHILLIPSINELLID TRILOBITE

f

tergite 4

tergite 2

leg space

telson

a

paratergal lobe

b

ONISCOMORPH DIPLOPOD

eye

e

pygidium
paratergal lobe(pleural lobe)

d

ASAPHID TRILOBITE

Fig. 28. For legend see opposite.

they can have had no use for the very different merostome coxa or for the crustacean mandible. Trilobites and merostomes each had their own proficiencies; those of the trilobites lasted essentially unchanged for 250 million years, but the stronger burrowing and stronger biting of *Limulus* and related genera has resulted in persistence to the present time. Trilobites and crustaceans with their many structural differences, do not suggest a common origin. No theoretical imaginary arthropod seems to have any relevance to the fundamental differences between the two great groups. We have real Cambrian arthropods in plenty which do not fit into any of our larger taxa. What kind of arthropod *Branchiocaris* (Fig. 3) may have been we do not know, but its trunk limbs were simpler than those of any known crustacean. We can but hunt further in the lower Cambrian for earlier members of the great groups, remembering that evolution and differentiation of the "bushes" and "lawns" may have been rapid and that the "ancestral portrait galleries" may never be found because the successful lines did not leave behind a series of imperfect stages in sufficient numbers for our inspection, cf. the explosive origin of cichlid fishes in African lakes and *Drosophila* species in Hawaii.

A proposed pointer to evolutionary progress, put forward recently by Locke and Huie (1977), cannot be accepted. They started with a description of the origin of arthropods which has long been unacceptable. They described Golgi-body phenomena in several animals and recorded similarity among the sclerite-bearing arthropods which they tested. In Cnidaria, Platyhelminthes, Nematoda, Mollusca, Annelida and Onychophora different Golgi-body phenomena were observed, contrasting with the sclerite-bearing arthropods. From this Locke and Huie proposed a monophyletic derivation of present-day arthropods and stated that the descent of the Uniramia (their Insecta) from a Peripatus-like ancestor is unlikely. Such

---

Fig. 28. Enrolment in oniscomorph diplopods and in trilobites. (a) *Sphaerotherium* (oniscomorph diplopod) enrolled in side view; the head and tergite 1 are covered by the telson and the legs pack into the small middle space demarcated by the dotted line; the paratergal lobes from the side walls of this space. (b) The same. Lateral view of three tergal arches, anterior end to the left, to show the overlap and the paratergal lobes. (c) The same. Ventral view of two diplosegments A and B with legs cut off. Each bears one pleurite, two staggered sternites and two legs on either side; stipple marks the pleurite, sternites and leg-bases of diplosegment B, after Manton (1954). (d) Asaphid trilobite, Ordovician, enrolled, after Harrington (1959). (e) The same. Median Diagrammatic section, after Harrington (1959). (f) Phillipsinellid trilobite, upper Ordovician, enrolled, after Harrington (1959).

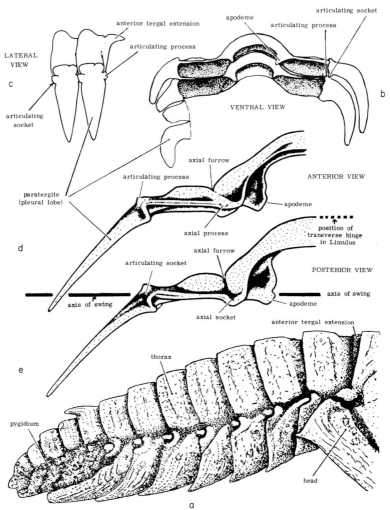

FIG. 29. Trilobite exoskeleton. (a) *Remopleurides* Ordovician, a still articulated specimen in lateral view showing thorax and pygidium; the lateral articulations between the thoracic segments are seen above the downwardly extending paratergal (pleural) lobes, after Whittington (1961). (b) *Ceraurus whittingtoni* middle Ordovician. Ventral view of the cuticles of two still articulated segments, the body cavity being stippled and the sternal cuticle absent, from Whittington (1960). (c) *Ceraurinella* middle Ordovician. Lateral view of two segments to show the position of the lateral peg and socket intersegmental articulations, seen in (a) and the forwardly extending dorsal tergal lobe, after Harrington (1959). (d) The same. Anterior view of the skeleton of one thoracic segment to show the horizontal position of the two peg-like articulating processes, after Harrington (1959). (e) The same. Posterior view of the skeleton of one thoracic segment to show the horizontal position of the two socket articulations, which take the pegs from in front, after Harrington (1959).

conclusions did not appear to have been made with reference to the relevant existing evidence, and the association of their Golgi-body phenomena with deformity of body, which includes ectoderm (the source of their material), unstriated muscle, etc. was not noted. Nothing was reported about the more rigid-bodied echinoderms, nor was there reference to the abundant solid evidence concerning the reality of the Uniramia and of polyphyly among arthropods, none of which rest upon a single feature only.

ACKNOWLEDGMENTS

We gratefully acknowledge much help from Professor H. B. Whittington F.R.S. and other palaeontologists who have generously discussed their work with us and allowed us to use their drawings of recently described genera.

REFERENCES

ANDERSON, D. T. (1959). The embryology of the polychaete *Scoloplos armiger*. *Q. Jl microsc. Sci.* **100**, 89–166.

ANDERSON, D. T. (1966a). Arthropod arboriculture. *Ann. Mag. nat. Hist.* **9**, 455–456.

ANDERSON, D. T. (1966b). The comparative embryology of the Polychaeta. *Acta zool., Stockh.* **47**, 1–47.

ANDERSON, D. T. (1966c). The comparative early embryology of the Oligochaeta, Hirudinea and Onychophora. *Proc. Linn. Soc. N.S.W.* **91**, 10–43.

ANDERSON, D. T. (1969). On the embryology of the cirripede crustaceans *Tetraclita rosea* (Krauss), *T. purpurascens* (Wood), *Chthamalus antennatus* (Darwin) and *Chamaesipho columna* (Spengler), and some considerations of crustacean phylogenetic relationships. *Phil. Trans. R. Soc. B*, **256**, 183–235.

ANDERSON, D. T. (1973). "Embryology and Phylogeny in Annelids and Arthropods". Pergamon Press, Oxford.

ANDERSON, D. T. (1978). Embryos, fate maps and the phylogeny of arthropods, Section 11, Chapter 2. In "Arthropod Phylogeny" (A. P. Gupta, ed.). Van Nostrand Reinhold Co., New York.

ANDO, H. and KOBAYASHI, H. (1975). Description of early and middle developmental stages in embryos of the firefly, *Luciola cruciata* Motchulsky (Coleoptera: Lampyridae). *Bull. Sugadaira Biol. Lab., Tokyo*, **7**, 1–11.

BERGSTRÖM, J. (1975). Functional morphology and evolution of Xiphosurids. *Fossils and Strata*, **4**, 291–305.

BERNARD, E. C. (1976). Observations on the eggs of *Eosentomon australicum* (Protura: Eosentomidae). *Trans. Am. microsc. Soc.* **95**, 129–30.

BOWNES, M. (1975). A photographic study of development in the living embryo of *Drosophila melanogaster*. *J. Embryol. exp. Morph.* **33**, 789–801.

BRIGGS, D. E. G. (1976). The arthropod *Branchiocaris* n. gen. middle Cambrian, Burgess Shale. *Bull. geol. Surv. Can.* **264**, 1–28.

BRIGGS, D. E. G. (1977). The morphology, mode of life and affinities of *Canadaspis perfecta* (Crustacea Phyllocarida), middle Cambrian, Burgess Shale, British Columbia. *Phil. Trans. R. Soc. B*, **281**, 439–487.

BROILI, F. (1933a). Ein zweites Exemplar von *Cheloniellon*. *S.B. bayer. Akad. Wiss.* Heft, **1**, 11–32.

BROILI, F. (1933b). Weitere Beobachtungen an *Palaeoisopus*. *S.B. bayer. Akad. Wiss.* Heft, **1**, 33–47.

CISNE, J. L. (1975). Anatomy of *Triarthrus* and the relationships of the Trilobita. *Fossils and Strata*, **4**, 45–63.

CRAIG, D. A. (1974). The labrum and cephalic fans of larval Simuliidae (Diptera: Nematocera). *Can. J. Zool.* **52**, 133–59.

DOHLE, W. (1974). The segmentation of the germ band of Diplopoda compared with other classes of arthropods. *Symp. zool. Soc. Lond.* **32**, 143–61.

DOHLE, W. 1976. Formation and differentiation of the postnaupliar germ band in *Diastylis rathkei* (Crustacea: Cumacea). II. Differentiation and pattern formation of the ectoderm. *Zoomorphologie* **84**, 279–300.

EVANS, M. E. G. (1975). The jump of *Petrobius* (Thysanura, Machilidae). *J. zool. Lond.* **176**, 49–65.

FRYER, G. (1969). Speciation and adaptive radiation in African Lakes. *Verh. int. Verein. Limnol.* **17**, 303–322.

FRYER, G. and ILES, T. D. (1972). "The Cichlid fishes of the great lakes of Africa: their biology and evolution". Oliver and Boyd, Edinburgh.

HAMMEN, L. VAN DER (1977). A new classification of Chelicerata. *Zool. Meded. Leiden*, **51**, 307–319.

HARRINGTON, J. H. (1959). General description of Trilobita. In "Treatise on Invertebrate Paleontology, Part O, Arthropoda 1" (R. C. Moore, ed.), pp. 038–0126. Geol. Soc. Am. and Univ. Kansas Press.

HENNIG, W. (1966). "Phylogenetic systematics". Univ. Illinois Press, Chicago and London.

HENNIG, W. (1969). Die Stammesgeschichte der Insecten. *Senkenberg-Büch*, **49**, 1–436, Frankfurt.

HUGHES, C. P. (1975). Redescription of *Burgessia bella* from the Burgess Shale, middle Cambrian, British Columbia. *Fossils and Strata*, **4**, 415–61.

HUTCHINSON, G. E. (1930). Restudy of some Burgess Shale Fossils. *Proc. U.S. nat. Mus.* no. 11.

KNOLL, H. J. (1974). Studies on the development of *Scutigera coleopterata* L. (Chilopoda). *Zool. Jb. Anat. Ont.* **91**, 47–132.

KRYSAN, J. L. (1976). The early embryology of *Diabrotica undecimpunctata howardi* (Coleoptera: Chrysomelidae). *J. Morph.* **149**, 121–37.

LANG, R. (1973). Die Ontogenese von *Maja squinado* (Crustacea, Malacostraca, Decapoda, Brachyura) under besonderer Berücksichtigung der embryonalen Ernährung under der Entwicklung der Darmtraktes. *Zool. Jb. Anat. Ont.* **90**, 389–449.

LOCKE, M. and HUIE, P. (1977). Bismuth staining of Golgi complex is a characteristic arthropod feature lacking in *Peripatus*. *Nature, Lond.* **270**, 341–343.

MANTON, S. M. (1928). On the embryology of a mysid crustacean *Hemimysis lamornae*. *Phil. Trans. R. Soc. B*, **216**, 363–463.

MANTON, S. M. (1934). On the embryology of *Nebalia bipes*. *Phil. Trans. R. Soc. B*, **223**, 168–238.

MANTON, S. M. (1937). Studies on the Onychophora II. The feeding, digestion, excretion and food storage of *Peripatopsis*. *Phil. Trans. R. Soc. B*, **227**, 411–464.

MANTON, S. M. (1949). Studies on the Onychophora VII. The early embryonic stages of *Peripatopsis* and some general considerations concerning the morphology and phylogeny of the Arthropoda. *Phil. Trans. R. Soc. B*, **233**, 483–580.

MANTON, S. M. (1952). The evolution of arthropodan locomotory mechanisms. Part 3. The locomotion of the Chilopoda and Pauropoda. *J. Linn. Soc., Zool.* **42**, 118–66.

MANTON, S. M. (1954). The evolution of arthropodan locomotory mechanisms. Part 4. The structure and evolution of the Diplopoda. *J. Linn. Soc., Zool.* **42**, 299–336.

MANTON, S. M. (1956a). The evolution of arthropodan locomotory mechanisms. Part 5. Structure, habits and evolution of the Pselaphognatha (Diplopoda). *J. Linn. Soc., Zool.* **43**, 153–187.

MANTON, S. M. (1956b). Hydrostatic pressure and leg extension in arthropods with special reference to arachnids. *Ann. Mag. nat. Hist.* ser. 13, **1**, 161–182.

MANTON, S. M. (1958). The evolution of arthropodan locomotory mechanisms. Part 6. Habits and evolution of the Lysiopetaloidea (Diplopoda), some principles of leg design in Diplopoda and Chilopoda and limb structure in Diplopoda. *J. Linn. Soc., Zool.* **43**, 487–556.

MANTON, S. M. (1961). The evolution of arthropodan locomotory mechanisms. Part 7. Functional requirements and body design in Colobognatha (Diplopoda) together with a comparative account of diplopod burrowing techniques, trunk musculature and segmentation. *J. Linn. Soc., Zool.* **44**, 383–461.

MANTON, S. M. (1964). Mandibular mechanisms and the evolution of arthropods. *Phil. Trans. R. Soc. B*, **247**, 1–183.

MANTON, S. M. (1965). The evolution of arthropodan locomotory mechanisms. Part 8. Functional requirements and body design in Chilopoda, together with a comparative account of their skeleto-muscular systems and an Appendix on a comparison between burrowing forces of annelids and chilopods and its bearing upon the evolution of the arthropodan haemocoel. *J. Linn. Soc., Zool.* **45**, 251–484.

MANTON, S. M. (1966). The evolution of arthropodan locomotory mechanisms. Part 9. Functional requirements and body design in Symphyla and Pauropoda and the relationships between Myriapoda and Plterygota. *J. Linn. Soc., Zool.* **46**, 103–41.

MANTON, S. M. (1967). The polychaete *Spinther* and the origin of the Arthropoda. *J. nat. Hist.* **1**, 1–22.

MANTON, S. M. (1972). The evolution of arthropodan locomotory mechanisms. Part 10. Locomotory habits, morphology and evolution of the hexapod classes. *Zool. J. Linn. Soc.* **51**, 203–400.

MANTON, S. M. (1973). The evolution of arthropodan locomotory mechanisms. Part 11. Habits, morphology and evolution of the Uniramia (Onychophora, Myriapoda, Hexapoda) and comparisons with Arachnida, together with a functional review of uniramian musculature. *Zool. J. Linn. Soc.* **53**, 257–375.

MANTON, S. M. (1974). Segmentation in Symphyla, Chilopoda and Pauropoda in relation to phylogeny. *Symp. zool. Soc. Lond.* **32**, 163–190.

MANTON, S. M. (1977). "The Arthropoda: Habits, Functional Morphology and Evolution". Oxford Univ. Press, Oxford.

MANTON, S. M. (1978). Habits, functional morphology and evolution of psycnogonids. *Zool. J. Linn. Soc.* **63**, 1–21.

MANTON, S. M. and HEATLEY, N. (1937). Studies on the Onychophora II. The feeding, digestion, excretion and food storage of *Peripatopsis*. *Phil. Trans. R. Soc. B*, **227**, 411–464.

MIYAKAWA, K. (1975). The embryology of the caddis fly, *Stenopsyche griseipennis* MacLachlan (Trichoptera, Stenopsychidae). V. Formation of the alimentary canal and other structures, general considerations and conclusions. *Kontyu*, **43**, 55–74.

MORI, H. (1976). Formation of the visceral musculature and origin of the midgut epithelium in the embryos of *Gerris paludum insularis* Motschulsky (Hemiptera: Gerridae). *Int. J. Insect Morph. Embryol.* **5**, 117–125.

MUÑOZ-CUEVAS, A. (1973). Embryogenèse, organogenèse et rôle des organes ventraux et neuraux de *Pachylus quinamavidensis* Muñoz (Arachnida, Opilions, Gonyleptidae). Comparaison avec les Annélides et d'autres Arthropodes. *Bull. Mus. Nat. d'Hist. Nat. Paris. ser. 3, Zool.* **128**, 1517–1537.

PANTIN, C. F. A. (1950). Locomotion in British terrestrial nemertines and planarians. *Proc. Linn. Soc.* **162**, 23–37.

PETRUNKEVITCH, A. (1949). A study of palaeozoic Arachnida. *Connecticut Acad. Arts and Sci.* **37**, 69–315.

REMPEL, J. G. (1975). The evolution of the insect head: the endless dispute. *Quaest. Entomol.* **11**, 7–25.

REPINA, L. N. and OKUNEVA, O. G. (1969). Cambrian Arthropoda of Promorye. *Paleont. Zh.* (1969), 106–114.

ROLFE, W. D. I. (1969). Phyllocarida. In "Treatise on Invertebrate Paleontology, Part R, Arthropoda 4" (R. C. Moore, ed.), R296–R331. Geol. Soc. Amer. and Univ. Kansas Press.

SCHOLL, G. (1977). Beitrage zur Embryonalentwicklung von *Limulus polyphemus*. *Zoomorphologie* **86**, 99–154.

SHAROV, A. G. (1966). "Basic Arthropodan Stock with Special Reference to Insects". Pergamon Press, Oxford.

STØRMER, L. (1970). Arthropods from the lower Devonian (lower Emsian) of Alken an der Mosel, Germany, Part 1, Arachnida. *Senckenberg. Leth.* **51**, 335–369.

STØRMER, L. (1972). Arthropods from the lower Devonian (lower Emsian) of Alken an der Mosel, Germany, Part 2, Xiphosura. *Senckenberg. leth.* **53**, 1–29.

STØRMER, L. (1973). Arthropods from the lower Devonian (lower Emsian) of Alken an der Mosel, Germany, Part 3, Eurypterida, Hughmilleriidae. *Senckenberg. leth.* **54**, 119–205.

STØRMER, L. (1974). Arthropods from the lower Devonian (lower Emsian) of Alken an der Mosel, Germany, Part 4, Eurypterida, Drepanopteridae, and other groups. *Senckenberg. leth.* **55**, 359–451.

STØRMER, L. (1976). Arthropods from the lower Devonian (lower Emsian) of

Alken an der Mosel, Germany, Part 5, Myriapoda and additional forms, with general remarks on fauna and problems regarding invasion of land by arthropods. *Senckenberg. leth.* **57**, 97–183.

STÜRMER, W. and BERGSTRÖM, J. (1976). The arthropods *Mimetaster* and *Vachonisia* from the Devonian Hunsrück Shale. *Paläont. Z.* **50**, 78–111.

TIEGS, O. W. (1955). The flight muscles of insects—their anatomy and histology; with some observations on the structure of striated muscles in general. *Phil. Trans. R. Soc. B*, **238**, 221–348.

TIEGS, O. W. and MANTON, S. M. (1958). The evolution of the Arthropoda. *Biol. Rev.* **33**, 255–337.

WALL, C. (1973). Embryonic development of two species of *Chesias* (Lepidoptera: Geometridae). *J. Zool.* **169**, 65–84.

WATSON, D. M. S. (1949). The mechanism of evolution. *Proc. Linn. Soc. Lond.* **160**, 75–84.

WEYGOLDT, P. (1975). Untersuchungen zur Embryologie und Morphologie der Geisselspinne *Tarantula marginemaculata* C. L. Koch (Arachnida, Amblypygi, Tarantulidae). *Zoomorphologie*, **82**, 137–199.

WHITTINGTON, H. B. (1960). Trilobita. *In* "McGraw-Hill Encyclopedia of Science and Technology."

WHITTINGTON, H. B. (1961). A natural history of trilobites. Natural History. *J. Am. Mus. Nat. Hist.* **70** (7), 8–17.

WHITTINGTON, H. B. (1971). Redescription of *Marrella splendens* (Trilobitoidea) from the Burgess Shale, Middle Cambrian, British Columbia. *Bull. geol. Surv. Can.* **209**, 1–24.

WHITTINGTON, H. B. (1974). *Yohoia* Walcott and *Plenocaris* n. gen., arthropods from the Burgess Shale, middle Cambrian, British Columbia. *Bull. geol. Surv. Can.* **231**, 1–26.

WHITTINGTON, H. B. (1975a). The enigmatic animal *Opabinia regalis*, middle Cambrian, Burgess Shale, British Columbia. *Phil. Trans. R. Soc. B*, **271**, 1–43.

WHITTINGTON, H. B. (1975b). Trilobites with appendages from the Burgess Shale, middle Cambrian, British Columbia. *Fossils and Strata*, **4**, 97–136.

WHITTINGTON, H. B. (1977). The middle Cambrian trilobite *Naraoia*, Burgess Shale, British Columbia. *Phil. Trans. R. Soc. B*, **280**, 409–443.

WHITTINGTON, H. B. (1978). The lobopod animal *Ayshcaia pedunculata*, Middle Cambrian, Burgess Shale, British Columbia. *Phil. Trans. R. Soc. B*, **284**, 165–197.

YOSHIKURA, M. (1954). Embryological studies on the liphistiid spider *Heptathela*. *Kumamoto J. Sci.* **3B**, 41–50.

YOSHIKURA, M. (1955). Embryological studies on the liphistiid spider *Heptathela kumurai*. II. *Kumamoto J. Sci.* **2B**, 1–86.

YOSHIKURA, M. (1975). Comparative embryology and phylogeny of Arachnida. *Kumamoto J. Sci. Biol.* **12**, 71–142.

11. | Early Radiation of Mollusca and Mollusc-like Groups

ELLIS L. YOCHELSON

US Geological Survey, Washington, USA

Abstract: To interpret history of the Mollusca, one must first ask, what is a mollusc? As no clearcut answer may be given to this question a variety of fossil organisms have been assigned to the phylum. In my view, fossil molluscs must have as a first criterion a calcium carbonate shell. In some specimens, the shell structure may assist in differentiating molluscs from other fossils whose hard parts are also composed of calcium carbonate. A second-order level of differentiation is in geometry. A fossil mollusc shell is most characteristically a closed cone, curved or coiled, and showing bilateral symmetry. Some molluscs are asymmetrical univalves, and others have more than one hard part, so again it is impossible in every specimen to differentiate mollusc from non-mollusc.

A taxonomic level is a measure of distinctiveness from other equivalent level taxa; the number of lower level taxa within a taxon is a crude measure of "success." Thus, classes are distinct, and members of some classes are greatly diverse, whereas others show limited diversity. Some early Palaeozoic fossils are judged to be morphologically so distinct that they ought to be placed in extinct classes rather than being forced into extant ones. In the scheme of Runnegar and Pojeta (1974), Cambrian strata contain representatives of the only extinct molluscan class they recognize, and all classes, save one, of extant Mollusca. I recognize several extinct classes in the Palaeozoic and am prepared to recognize more as morphologic data warrant; most classes show limited diversity.

Diversity, geologic history, or "success" are not considerations for systematic ranking; only distinctiveness concerns taxonomic rank. In my view, classes begin abruptly by major mutations of larval forms which opened new environmental niches. The basic adaptation of a new class was then shaped by Darwinian evolution.

Systematics Association Special Volume No. 12, "The Origin of Major Invertebrate Groups", edited by M. R. House, 1979, pp. 323–358, Plates IV and V. Academic Press, London and New York.

INTRODUCTION

The title of this paper is not mine! It was the theme assigned to indicate the direction I was supposed to take in organizing my thoughts. Yet it is a reasonable title. One ought to be able to determine the limits on the kinds of fossils included within a phylum before discussing relationships within that taxon, for otherwise misinterpretations on relationships could arise.

Several generations ago many forms of uncertain placement were assigned to the Mollusca. With a few exceptions these zoological puzzles were of Ordovician and middle-Palaeozoic age. Today we are faced with a diversity of forms in the early Cambrian Tommotian of the Soviet Union and the Meishucun of China (Yi, 1977). If these forms are molluscs, even extinct classes, we have some basis upon which to make speculations concerning their life habits, physiology, and evolutionary rates. If these forms are not molluscs, that basis for speculation is in error.

One colleague phrased the problem of interpretation by noting that "mollusc" is a zoological term, whereas "mollusc-like" is a palaeontological term. The lack of a unique body plan of the hard part of molluscs is what creates a problem for interpreting the word "mollusc-like." This problem is further compounded by the lack of any modern analogue of shape for some of the ancient fossils of uncertain systematic position (J. R. Derby, personal communication 1977).

One aim of the writer is to caution that we be wary. Many of these newly found forms may be unrelated to the Mollusca. Another aim is to consider the pattern of evolution within the Mollusca. Runnegar and Pojeta have vigorously espoused the view of a continuity of development within the Mollusca. I have indicated a few taxa of class rank for extinct molluscs and no close relationship among them. We disagree on most points and the controversy cannot be resolved in this essay. A summary of Pojeta and Runnegar's view of relationships within the Mollusca is given in Fig. 1. My views are given in Fig. 2.

The first task then is to outline and define what may be a mollusc. The second task is determined how one might divide up the fossil molluscs. The third task is to see what pattern is shown by these divisions and to compare the patterns envisaged by other current workers.

I guess that there are more classes of extinct molluscs than living ones and do not consider most of them to be closely related, but I do not know. Authoritarianism is dangerous, especially for scientists and the reader should approach my comments with a sceptical attitude.

THE QUESTION OF THE NUMBER OF CLASSES OF MOLLUSCS

One approach to definition of the phylum and the subdivisions within it, is historical perspective. The writer arbitrarily selected as a point of departure is Miller (1889), who published one of the early palaeontological texts in America. Miller divided organisms into the Vegetable and Animal Kingdoms, the latter containing seven subkingdoms, of which one was Mollusca. Listed in reverse of the arrangement given by Miller, the five classes he recognized were: Lamellibranchiata; Gasteropoda—which included the Families Chitonidae and Dentalidae; Cephalopoda; Pteropoda; Brachiopoda. Thus, only three scientific generations ago, at least one authority considered Brachiopoda to be molluscs. No wonder "mollusc-like" is not a clear term.

Miller is not to be simply dismissed. In his general classification he kept Molluscoida (Bryozoa) apart from Mollusca and also kept the Articulata distinct. Within Articulata were the classes Annelida, Crustacea, Arachnida, Myriapoda and Insecta. In retrospect, is it a worse interpretation to associate brachiopods with molluscs, or is it worse to place annelids with arthropods? I deliberately use *worse* for it is a comparative term; we should never forget that systematics is a comparative science. Likewise, *interpretation* is preferred, for Miller did not deal with fundamental truth or error, but with opinion; classification is subjective opinion formally codified. Unfortunately, in our thinking and writing we often indicate that the older generation had classification muddled but that we now have the placement of taxa in biologic arrangement. Few palaeontologists followed Miller; in the next two decades, his Chitonidae and Dentalidae were judged to represent groups of class-level rank, and his Pteropoda was ignored as a class-level taxon.

Thus, for the first half of this century it was "right" to place all living Mollusca in five classes. Then Knight (1952) suggested that the Polyplacophora combined with the extinct group Monoplacophora be treated as a subclass within the Gastropoda, thereby dividing Mollusca into four classes. Knight freely admitted following a concept of Miller's vintage, except it had been put forth by Sir E. Ray Lancaster, a far more impressive authority. In 1957, Lemche described *Neopilina*; abruptly Class Monoplacophora was accepted generally, presumably because a living representative of this supposed extinct group was discovered. Next, Aplacophora was split off at the class level from Polyplacophora by students of the Mollusca and that class in turn has been split in twain

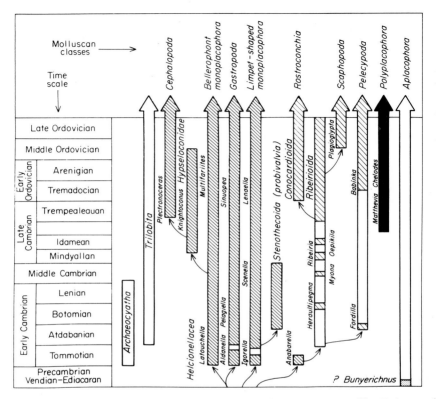

FIG. 1. The phylogenetic scheme of the early Mollusca as suggested by Pojeta and
Runnegar (1976, fig. 14). The original caption is: "Historical record of
the initial radiation of the Mollusca, scaled against time divisions based
primarily on the succession of fossil archaeocyaths and trilobites. The two
largest molluscan subphyla (Cyrtosoma, fine-shaded columns, and
Diasoma, coarse-shaded columns) separated in the early Cambrian.
Modified from Runnegar and Pojeta, 1974, figure 1". Dr Runnegar (*in
litt.*, 1978) has directed my attention to a number of points. (1) *Lamellodonta*
is now known to be a brachiopod. (2) *Heraultia* is now called *Heraultipegma*
and it first appears in the early Tommotian. (3) *Opikella* is now *Oepikella*.
(4) *Fordilla* probably has a longer stratigraphic range than shown on the
original diagram. (5) *Bunyerichnus* is now thought to be an inorganic
sedimentary structure. (6) The evolution of the Gastropoda is undoubtedly
more complex than shown on this diagram. (7) We no longer refer to the
oldest scaphopod to *Plagioglypta*.

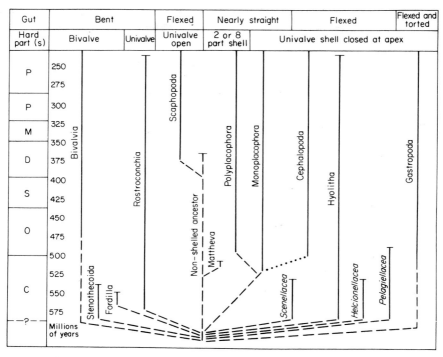

Gut	Bent		Flexed	Nearly straight	Flexed	Flexed and torted
Hard part (s)	Bivalve	Univalve	Univalve open	2 or 8 part shell	Univalve shell closed at apex	

FIG. 2. The phylogenetic scheme of the early Mollusca as suggested by Yochelson (1978, fig. 1). The original caption is: "Ranges of Mollusca through the Paleozoic. A scale of geologic periods and radioactive dates is given to the left; interpretations of the gut appear at the top just above a statement on geometry of hard parts. Classes are in full capitals and end in the Palaeozoic or continue to the Holocene; probable classes are indicated by superfamily names; possible classes are indicated by generic names."

by a few workers. Depending on which authority one follows, there are four, five, six, seven or eight classes of living molluscs (Yochelson, 1979a).

Miller was not wrong and Knight was not wrong. They simply proposed an interpretation of affinities that was not generally accepted. Which interpretation should be followed and which rejected is a murky area most persons prefer not to explore. More kinds of evidence on relationships are available to the zoologist than to the palaeontologist, yet there is no zoological concensus. As a consequence I question whether a single "palaeontological viewpoint" will ever satisfy all students of molluscan phylogeny.

A SUMMARY OF MILLER'S PALAEOZOIC PTEROPODA

Miller (1889) noted in Class Pteropoda the orders of living naked Gymnosomata and shelled Thecosomata and wrote "It may well be doubted whether or not any of the Palaeozoic fossils belong to this order." Below class rank, Miller used only species, genera and families in his text and the 15 genera of dubious Pteropoda were placed in eight families, six containing only one genus. By and large these were forms that nearly a century ago did not fit will into conventional classification of the times. Millers' assignments may give some insight into the term "mollusc-like," and the "class" is reviewed briefly.

Aspidellidae contained only *Aspidella* Billings. Today there seems general agreement that the name applies to a pseudofossil (Hofmann, 1971); Miller's younger contemporary, Walcott, had already recognized this (Yochelson, 1979b). Clathrocoeliidae contained only *Clathrocoelia* Hall. Judging from Hall's illustrations, this rare Devonian form might be an incomplete or poorly preserved hyolith, more or less what Hall suggested.

Conulariidae contained only *Conularia* Miller *in* Sowerby. My impression is that for at least two generations most palaeontologists who have studied Palaeozoic fossils exclude conulariids from Mollusca. In spite of their distinct shell, their obvious growth lines, and their symmetry, except at the earliest stages, conularids do not convey an aura of molluscan affinities. This distinctiveness is partly because the hard parts are composed of calcium phosphate rather than calcium carbonate; the "sheen" of hard parts is definitely not mulluscan. Kozłowski (1968) came within a hair's breadth of proposing an extinct phylum to include *Conularia*, though he did not take the final step of a formal systematic proposal.

Hyolithidae included seven genera and was the most heterogeneous assemblage. Miller gave no familial diagnosis; virtually the only feature members of the family have in common is a more or less slender, elongate shape. *Coleolus* Hall is quite elongate and slowly expanding. *Coleoprion* Sandberger is similar, except that its growth lines are a bit more prominent. Neither genus is much like *Hyolithes*. I consider both to be "worm tubes". *Diplotheca* Matthew can be set aside as another peculiar worm tube; the embryonic part, if correctly illustrated, is unlike that of a hyolith (Marek, 1976). *Hyolithellus* Billings, a very long slender tube, has been moved to the Pogonophora (Poulsen, 1963); the operculum or generally similar opercula (*Mobergella*) has also been removed from the molluscs (Bengtson, 1968). *Pharetrella* Hall, based on two imcomplete specimens, may be a hyolith of some sort. *Hyolithes* Eichwald and *Stenotheca* Hicks are discussed later.

Matthevidae contained only *Matthevia* Walcott. Currently the genus is considered a mollusc, but there are divergent views on the systematic level of placement (Yochelson, 1966; Runnegar and Projeta, 1974). *Pterotheca* Salter was the sole genus in Pterothecidae. For many years this wide, low, wedge-shaped form has been placed in the Bellerophontacea. Scenellidae also contained a single genus, *Scenella* Billings, variously classed today as gastropod or monoplacophoran.

Lastly, Tentaculitidae contained *Styliolina* Lesueur and *Tentaculites* Schlotheim. There is still argument as to whether these genera should be associated. Lyshenko in 1957 combined tentaculitids and hyolithids in the class Coniconchia, a systematic concept to which Yochelson (1961) took exception. In my view, Bouček (1964) made a dramatic advance with his detailed study of the styliolinids of Bohemia. He showed the great diversity within these predominately Devonian forms and indicated how different their morphology was from that of molluscs; he too stopped just short of proposing a new phylum to contain them.

Not everyone concurs. Blind (1969) has been a vigorous proponent of the notion that styliolinids are to be placed with the Cephalopoda. This placement seems unlikely, for the generally annulated and ribbed exterior of this diverse group of small forms is unlike that of most molluscs, living or fossil. The bulbous early stage is unlike that of molluscs, or at least molluscs as we know them in the Holocene.

Blind and Stürmer (1977) showed X-ray photographs of soft parts, observed in specimens from the lower Devonian Hünsrucker Shale of Germany. Subsequently, Stürmer (unpublished data) has found additional specimens with soft parts far better preserved (Pl. IV, 3). Extending from the aperture is soft tissue forming a structure much like a wrist and hand with palm open and five fingers extended upward. A dark elongate spot parallel to the wrist was independently interpreted by several specialists on Recent worms as a rod to support a feeding structure. These soft parts are not like those of cephalopods, either living ones or Devonian ones from the Hünsrucker (Stürmer, 1970; Zeiss, 1969). They demonstrate that any similarity in general shell form between orthoconic cephalopods and styliolinids is superficial; there are no homologies, only analogies.

More data are needed about Miller's Tentaculitidae, but information appears to be sufficient to demonstrate that neither *Tentaculites* nor *Styliolina* are molluscs. These two genera characterize two groups, distinctive at a high taxonomic level. In my view it would be reasonable to place each, with its allies, in two extinct classes, combined in an extinct phylum.

There may be a question in the reader's mind concerning the assignment of *Salterella* Billings, a tiny ice-cream-cone-shaped fossil from the lower Cambrian. Miller followed Billings and placed it in the Class Annelida, an eminently reasonable assignment at the time. It was not considered a mollusc until many years later. The genus is so unusual that Yochelson (1977) placed it in an extinct phylum, Agmata.

THE ISSUE OF SOME GENERAL FEATURES OF FOSSIL MOLLUSCA

It is difficult to make any assessment from Miller's classfication as to what are the salient features of molluscs. Growth lines are present between increments of hard parts in all the forms he lists, thereby demonstrating that they did not grow by moulting. Periodic growth (Lutz and Rhodes, 1977) may be a more sophisticated way of stating this attribute, but it is represented in fossils by fairly obvious growth lines and that is what palaeontologists see.

On the other hand, even though growth lines are present, one can easily accept removal from Mollusca of *Conularia* and *Hyolithellus*, for their hard parts are not composed of calcium carbonate. There is no basis for assuming that Mollusca ever used any other compound to form the shell. I dogmatically emphasize the importance of a calcium carbonate shell as the principal feature before a fossil can be considered a mollusc. Of course, it is not an exclusive criterion and fossils whose hard parts are composed of calcium carbonate with uniformly increasing increments showing growth lines are not necessarily closely related. Brachiopods, corals and molluscs all show these features. It is exceedingly difficult to define any high-level group without reference to soft parts and apart from exclusion of hard parts of calcium phosphate every attribute of molluscan shells has some exceptions.

To cite as a example a group which is not easy to assign systematically, *Tentaculites* and *Styliolina* are composed of calcium carbonate. Their shape is that of a cone (radial symmetry) which expands at a very slow rate. They do not resemble molluscs except for the shells of one or two living Pteropoda. Perhaps *Styliolina* when compared to the living *Stylio* is as close as we can come to the concept of mollusc-like. *Coeleolus* and *Coeloprion*, listed by Miller are curved tubes, but they tend toward radial symmetry.

One feature all such genera have in common is a shell wall that is either moderately thin or has an irregular thickness.

The term "moderately thin" is not easily quantified! It is a prime example of the considerations involved in framing a palaeontological oriented operational definition of Mollusca with reference only to hard parts. In spite of its vagueness the term has some utility. With the exception of the Pteropoda and Heteropoda, especially adapted to pelagic life, living molluscs tend toward a moderately thick shell, comparable to that of brachiopods. Perhaps the reason for this is that molluscs are basically benthonic animals. Among the common fossil molluscs, cephalopods have a thinner shell than gastropods, whereas pelecypods have a thicker shell. Both variations may be adaptations to particular modes of life, in that on the one hand the lighter a cephalopod is the more efficient buoyancy will be, and on the other a pelecypod requires a heavy as well as a strong shell to burrow into sediments. One may surmise that in spite of numerous exceptions an enigmatic fossil presumed to have been a benthonic mollusc ought to have a shell approximately as thick as that of a marine gastropod of about the same size.

I have not studied molluscan shell structure and my sole contribution is speculation that the primitive shell was aragonite, with calcite being a secondary development; the thought may have been original with me but others developed it far earlier. The several layers in the molluscan shell have distinct complex organic and inorganic micro- and ultrastructures (Brown, 1975). So far as I know a few of these shell structures, especially aragonitic cross-lamellar structure, are unique to the Mollusca. Unfortunately such fine details seldom survive in the oldest fossils and it is the oldest fossils which are most difficult to place in extant taxa.

Palaeozoic shells that have structure preserved are rare and none are known from the lower Palaeozoic. One may note that mollusc shells do not readily break out from limestone; characteristically the steinkern (internal filling) splits away, perhaps because typical molluscan shell structures recrystalize and bond with the matrix. This mode of breaking is not totally reliable for assignment but it suggests that the fossil was a mollusc.

On the other hand, chemical preparation may be used to dissolve matrix and shell and yet may leave a steinkern composed of insoluble material. Although the use of organic acids in digestion of rocks has been known for many years, large-scale usage is fairly new and a bewildering melange of phosphatic and phosphatized parts and steinkerns is being revealed. A steinkern from an acid residue does not provide any information on the composition or ultrastructure of the hard parts which once surrounded it, and the palaeontologist does not have available what may be the critical data for assigning a fossil to any particular phylum.

THE CONTROVERSY SURROUNDING CURRENT IDEAS ON
ZOOLOGICAL AFFINITY OF THE HYOLITHA

One cannot discuss the early history of the Mollusca and ignore the issue of placement of the Hyolitha, for it is the epitome of the problem of distinguishing and assigning an extinct form to the Mollusca by its hard parts. This group is known from very early Cambrian time until Permian, but individuals are uncommon in middle and upper Palaeozoic rocks. The principal hard part of a hyolith is a moderately thick shelled, bilaterally symmetrical logarithmic curved, tapering tube closed at one end; it is probable that all genera also had an operculum; some shells are septate.

There are two principal groups. *Hyolithes* (Pl. IV, 2) and its allies have an anterior shelf and projecting appendages like whiskers. *Orthotheca* and its allies lack the shelf and the appendages. In a broad sense the form or geometery of the shell is molluscan, but it is unlike that of the siphuncle-bearing cephalopod and the scaphopod open ended tube. Only a few very thin shelled and small living pelagic gastropods approximate the shape, though not the size of hyoliths. Currently two views are espoused on the affinities of hyoliths; either these fossils are to be placed in an extinct class of Mollusca (Marek and Yochelson, 1976) or, they are outside the Mollusca, and constitute an extinct phylum (Runnegar *et al.*, 1975).

As support and clarification of his published works, Runnegar (personal communication, 1977) would rule the hyoliths out of the phylum because his concept of a mollusc includes a dorsal integument and there is no logical reconstruction of the hyoliths which would place the hard part as dorsal. In contrast, I suggest that because hyoliths have shells of calcium carbonate, probably in the form of aragonite, an appropriate growth pattern by addition of increments, a reasonable thickness to the shell, and typical molluscan shell structure (Runnegar *et al.*, 1975, p. 185), there seems no rational basis for removing hyoliths from Mollusca. The discovery of molluscan shell microstructure by Runnegar provides a particularly strong reason for judging hyoliths to be molluscs. The operculum of hyoliths is comparable to the operculum of gastropods in growth and preservation. For those who desire a small research problem, one might note that the microstructure of the hyolith operculum is unknown.

One can argue that there may have been extinct forms which developed hard parts showing molluscan shell structure, yet were not molluscs. There seems no logical way to refute such an argument and little point in pursuing it. However, it is obvious that not all fossil molluscs show molluscan shell structure. The further one goes back in time, the poorer the record as regards details of shell microstructure. As a consequence of the vagaries in the fossil record, we are faced with forms which might be molluscs but cannot be proved to be so from their shell structure, and forms which may not be molluscs, but cannot be excluded from the phylum by objective evidence of their shell microstructure. This state of affairs is not satisfactory, but it would be naive not to recognize it. Because interpretation of fossils must necessarily include a subjective element, there is room for honest disagreement in assignment.

A FEW SPECULATIONS CONCERNING PALAEOZOIC "WORMS"

Next to the conulariids, worm shells and the opercula composed of chitinophosphatic material which go with some of these tubes seem to me to be the most biologically artificial element in Miller's classification of Pteropoda. The systematics of extinct "worms" have not all been sorted out by any matter of means. There may have been more kinds of tube-dwelling worms in the past then are extant today. Some high-level taxonomic categories are needed; use of the Linnean "Vermes" may be a fair expression of our current understanding.

The multiple-paired muscles of fossil monoplacophoran shells, for example, provide an opportunity for confusion. The operculum of *Mobergella* is tiny, circular, and shows on its inner surface multiple-paired muscle scars. It is currently interpreted as a "worm" in spite of its musculature partly because of composition of calcium phosphate. Missarzhevskiy (1976) has described a tiny oval-shaped early Cambrian fossil with paired muscle scars as the oldest monoplacophoran. I speculate that Missarzhevstiy's fossil is an operculum belonging to a worm like *Mobergella*, but one which secreted a tube that had an oval rather than a circular cross-section.

This brings us to a consideration of geometry in molluscs and mollusc-like organisms, for obviously a fossil consists of material in a particular shape. I have deliberately mixed the terms "tubes" and "shells". The few present-day worms that live in a tube do not construct this hard part by the same method that molluscs use to form a shell. Without knowing the ultrastructure of the hard parts, it is difficult to differentiate the two methods of secretion. It may be better to convey shape in geometric terms and not to attempt to use "shell" or "tube" to indicate presumed origin or biologic affinity.

Styliolina and *Tentaculites* have been mentioned in conjunction with considerations of shape and composition in judging systematic affinity, but another example may help clarify and extend this notion. *Hyolithellus* is not a mollusc because it is a straight, black, shiny tube of essentially uniform diameter; composition is of obvious help in removing this form from molluscs, but its shape is equally significant. The hard parts of a very few molluscs are straight, but they all expand at a perceptible rate.

Logarithmic curvature is characteristic of most molluscs. Gnomic growth in a logarithmic mode is not an exclusive molluscan trait, but it does help distinguish a few mollusc-like fossils from molluscs. Nevertheless some worm tubes do curve during growth, (see Pl. V, 1, 2.) The differentiation of molluscs from worm tubus is a morass of often subtle pitfalls.

However, if we wish to pursue this difference, the only class of molluscs in which the hard part is characteristically tube-shaped is the Scaphopoda. They differ from typical worm tubes in being open at both ends. Yochelson (1979a) has indicated the problem of differentiating scaphopods from extinct kinds of curved worm tubes such as the late Carboniferous *Clavulites* which may have been open at both ends, morphologically like the tube of the living *Pectenaria*, but formed of calcium carbonate rather than agglutinated grains (see Pl. II, 4). Much of our experience with fossils which do not easily fit into present-day high-level taxa, such as Miller's Pteropoda, is based on Ordovician and middle-Palaeozoic occurrences. Suddenly we are faced with a variety of far older tiny phosphatized steinkerns; we should not be too precipitous in assuming that these are molluscs.

Runnegar and Pojeta (unpublished data) are attempting to push the record of the Scaphopoda back to the middle Ordovician. I have examined this material and would assign it to a group of extinct worms like *Clavulites*. The material upon which their interpretation is based is silicified.

As noted, diagenetic change of shell and replacement seriously compound the difficulties in dealing with mollusc and mollusc-like organisms.

The writer cannot place much reliance on "typical" silicification of mollusc shells. Many undoubted Palaeozoic "worm" tubes composed of calcium carbonate when replaced by silica have a secondary texture like that of many undoubted molluscs which have also undergone silica replacement. No one has yet devised an easy method of distinguishing original phosphatic fossils from secondarily phosphatized fossils.

THE PROBLEM OF PLACEMENT OF THE APLACOPHORA

Whereas Hyolitha have no living representatives and argument surrounds their placement in the Mollusca, the Aplacophora have no fossil record and in spite of their worm-like shape, by and large zoologists consider them within this phylum. Woodward (1907), like many authors before and after, remarked on the repeated reduction or loss of shell seen throughout the living Mollusca. In the phylogenetic scheme of Runnegar and Pojeta, as in several earlier ones, a great deal of emphasis is placed on the Aplacophora, in part because these animals develop aragonitic spicules in the cuticle. Even though there is no fossil record known, Aplacophora are supposed to have begun in the Precambrian and have persisted to the present day.

Some zoologists regard aplacophorans as structurally simple. However, "simple" does not mean the same thing to all investigators. Some investigators credit the Aplacophora with some generalized or presumed primitive characters, but simple organisms may be primitive or they may be secondarily simplified. In spite of their claims, there is no independent verification that students of cladistic analysis can differentiate between the two conditions. The route to gradual development of calcified plates is not at all clear. If the record of the present-day molluscs is replete with examples of shell loss, it would seem at least as logical to consider the Aplacophora as yet another group which lost all hard parts except some spicules.

At the risk of introducing some heavy handed humour I beg to note that the conceptual archaeomollusc is a mythical animal; there is no evidence of its existence. Precisely the same argument applies here. There is no more evidence of a fossil record for the Aplacophora than there is for existence of subject matter in the field of exobiology.

A claim has been made that a rare Precambrian trace fossil *Bunyerichnus* is the trail of an early mollusc and perhaps an aplacophoran. I find it exceedingly difficult to assign any fossil trace to the action of a crawling mollusc and suggest that this line of possible evidence is particularly weak for determining the occurrence and especially the first occurrence of a particular high-level group. Even if one can reasonably assume that a trail was formed by a crawling organism, the author suggests that it might be a bit difficult to differentiate the trace of a naked organism from a shelled one. Among the living, mobile, shelled molluscs, the shell is commonly carried on the foot and does not register on the trace.

SOME FOSSIL MOLLUSCS WITH MORE THAN ONE HARD PART

Living molluscs which secrete hard parts include those with one, two, or eight principal hard parts. One may ignore internal plates, radula, spicules, and opercula as hard parts which are not particularly germane to phylogeny; of all of these rarities, opercula are the most common in the Palaeozoic. Some opercula of ancient gastropods are bizarre (Yochelson and Bridge, 1957) and may not at first show their molluscan nature. Such curiosities in the fossil record as isolated hyolith opercula occasionally have been confused with brachiopods (Bassett and Yochelson, unpublished data). Some presumed cephalopod aptychi may be better interpreted as feeding structures (Lehmann, 1976, pp. 104–106), but others well may be coverings of the aperture. An animal, such as a mollusc, which lives in a tube is well served if it can close the tube opening, particularly if it has a benthonic life habit. Obviously opercula are not confined to Mollusca and possession of an operculum may not in itself help to distinguish a mollusc from a mollusc-like fossil.

These comments, however, are beside the point because a class of molluscs with multiple hard parts, the Polyplacophora, is known today. These animals have eight parts which commonly disarticulate after death. If there were no living forms, the rare articulated fossils, best known from one species of late Carboniferous *Pterochiton* (Pl. IV, 1), would be a revelation. Zoologists place the Polyplacophora apart from all other shelled living molluscs. The oldest authentic polyplacophoron plates are latest Cambrian.

The cuticular spicules of an aplacophoran are assumed to have gradually coalesced through unknown means and to have given rise to the eight-plated polyplacophoran. If Aplacophora go back to Precambrian, it is not clear why middle or early Cambrian polyplacophorans have not yet been found. It would seem that for animals with hard parts, the fossil record is fairly reliable. In the days of Darwin one could argue that it was poorly known, but that is a weak argument today.

No evidence whatsoever exists for this sequence of Aplacophora to Polyplacophora. There is considerable difficulty in deriving them from an already calcified form by splitting of the shell but this is still easier than imagining the coalescing of isolated spicules. It seems preferable to consider that as a result of genetic mistakes, the juvenile descendants of multimuscled monoplacophorans evolved several centres of calcification and thus shell shapes different from those of the parent univalve.

The problem of the term "mollusc-like" is seldom considered in relation to polyplacophorans perhaps because so few palaeontologists are concerned in the least with fossil chitons. However, isolated plates may easily mislead one. For years the writer identified elements of Ordovician, echinoderms as "Chitons", Macuridid plates have been repeatedly confused with polyplacophorans. Even though polyplacophoran plates are bilaterally symmetrical, and macuridid plates are slightly asymmetrical, misidentification is easy.

Even if one can eliminate mollusc-like forms, the early history of polyplacophorans is not easy to interpret. Controversy surrounds the late Cambrian *Matthevia* Walcott (Pl. IV, 4), another peculiar presumed mollusc that I have assigned to the extinct Class Mattheva. This fossil has the advantage of bilateral symmetry over specimens of Macuridia, but otherwise there is no concensus as to the interpretation of its form.

In my reconstruction *Matthevia* added weight by having a strongly musculated heavy plate at each end of the body so that it could advance into a new life zone, that of rough water. Along with other mutations, polyplacophorans developed a broad foot and a low flexible body. This approach to functional morphology was far more successful than that of the high, less flexible *Matthevia*. However once having achieved a limited amount of success, essentially they did not change much. Polyplacophorans constitute a remarkably fine example of non-evolution: not changing is more important than changing, if conditions remain the same.

Runnegar and Projeta have rejected my concept of a two-part mollusc and have suggested that *Matthevia* had eight parts. They interpret it as a peculiar member of the Polyplacophora, a notion I in turn refute (Yochelson, 1978). However, they are actively engaged in further study to support their view of *Matthevia* as a chiton and their results are awaited.

ASYMMETRICAL AND SYMMETRICAL FOSSIL MOLLUSCAN BIVALVES

Some controversy, mostly generated by me, surrounds the early Cambrian genus *Fordilla*, but here, frankly, I am not on firm ground as regards the general shape of the shell. *Fordilla* is a bilaterally symmetrical bivalve. The genus appears to lack hinge teeth. It has been assigned to the Bivalvia because of presumed musculature shown on steinkerns. The musculature has been questioned (Yochelson, 1978b). This fossil might be better placed as another peculiar offshoot of Mollusca.

If one ignores the arguments on musculature this tiny form does resemble a small pelecypod, though not a protobranch, generally presumed to be the structurally most primitive group. Perhaps the arguments revolve around the question of whether a new form undergoes adaptive radiation promptly, in a geologic sense, or is delayed for a long interval. On either side of the argument, the evidence for a class assignment of *Fordilla* is inconclusive. So far as I can interpret the fossil record, Bivalvia appeared in the early Ordovician and promptly diversified.

To repeat a point concerning the principal hard parts of classes of living molluscs, more often than not, the external skeleton is bilaterally symmetrical. One obvious exception is the Class Gastropoda but even within that assemblage, some shelled forms tend toward an external symmetry. There are a variety of other exceptions; a peculiar example is *Berthelinia*, a living bivalved opisthobranch gastropod in which the two valves are not entirely symmetrical in the early growth stage. Oysters and pectens, and a few other bivalves are another kind of exception. If we ignore them, we are free to speculate that fossil mollucsc are likely to show bilateral symmetry, though once again symmetry cannot be taken as a unique or even pervasive trait of the phylum.

A cladist would probably surmise that bilateral symmetry is a shared basic character, whereas asymmetry is a derived character. It does not follow that symmetry of hard parts gives much information on relationships among molluscs. Though I judge the probability to be high that shells of molluscs are more likely bilaterally symmetrical as an original trait, this feature may not have had much significance once the phylum had evolved. Symmetry and assymmetry among molluscs is a difficult area to summarize, for assymmetry has developed repeatedly.

Symmetry seems to be related to sessile, cemented or attached life. It first appeared early in the history of molluscs, as witness the class Stenothecoida. Members of the class have been poorly understood. Some confusion surrounding the early Cambrian *Stenotheca* Hicks may have been relieved when Resser (1938) assigned it to the Arthropoda and named *Stenothecoides* for similar-appearing molluscs. On the other hand Knight and Yochelson (1960) compounded difficulties by suggesting that *Stenothecoides* and the allied early Cambrian mollusc *Cambridium* Horný were monoplacophorans.

On the basis of additional information, Yochelson (1969) suggested that these fossils were asymmetrical, inequivalved bivalves rather than univalves. An investigation of the shell composition and structure is in order, though I am confident that the fossils are molluscs. In my view the concept of Class Bivalva (= Pelecypoda) would have to have been expanded so dramatically to include the morphology (geometry) shown by these fossils that they are better treated as representatives of a distinct and extinct class. Only after such a determination on systematic placement was made did I attempt to restore soft-part morphology. Perhaps the reconstructions are wrong, but they should not be the point which is first attacked. The diversity of shape in living and undoubted fossil pelecypods is the standard that should be used to judge the distinctiveness of this assemblage.

Runnegar and Projeta attempted to remove this class-level taxon by comparing *Stenothecoides* to the limpet *Hipponyx* which secretes a calcium carbonate pad; they also suggested the concept of a bivalved monoplacophoran. Yochelson (1979) has argued the illogic of this presumed homology. Until a better interpretation of *Stenothecoides* is presented, one would like to think that the notion of a separate class is still in biological good standing. It may be worth adding that *Stenothecoides* shows no evidence of attachment or cementation. It might also be added that *Stenothecoides* is only one of several genera in the class. The *Hipponyx* comparison would be even more difficult to apply to some elongate genera.

SOME COMMONLY ACCEPTED BILATERALLY SYMMETRICAL FOSSIL MOLLUSC UNIVALVES

Let us proceed to univalve molluscs and consider form once more. Obviously a univalve has more constraints on fundamental geometry of the exoskeleton than does an organism with two or more hard parts. As noted, most hard parts of molluscan univalves and bivalves are curved in profile, following a logarithmic spiral. Orthoconic cephalopods are molluscs but are not typical because their hard parts show no logarithmic curvature and the cross-section shows external radial symmetry.

Most univalve molluscs have what is topologically the same kind of shell, a cone. Modifications of cross-section or curvature do not pose major problems in the geometry of a cone, regardless of their biological significance. However shells containing septa are geometrically quite different from cones and are topologically complex. Shells of some extinct monoplacophorans, some hyoliths and some gastropods are septate. Because of the sporadic occurrence of septa in two or perhaps three classes, it follows that the presence of septa reflects a physiological requirement of selected taxa but does not have any fundamental significance at the phylum level.

Still another peculiar shape of a molluscan shell is that of the Scaphopoda; they deserve a few words. Scaphopods are unlike other molluscs; the topology of a tube with two openings is not a cone. These peculiar molluscs show little change throughout their fossil record. An indicated earlier, problems surround determination of whether a fossil is an early scaphopod. In my view, authentic scaphopods are not known to be older than middle-Devonian.

This brings us appropriately enough to the Cephalopoda, for *Nautilus* and all fossil externally shelled cephalopods are septate. Because shells of molluscs placed in other classes are septate, it is not septa alone which are significant. The siphuncle appears to be the prime feature of Cephalopoda.

With the removal of *Salterella* and perhaps *Volbolthella* (Yochelson et al., 1977) from the molluscs, there seem to be no arguments against the late Cambrian *Plectronoceras* as the word's oldest cephalopod. Development of Cephalopoda from Monoplacophora is the only linking of classes I would envisage. Runnegar and Pojeta (1974) have added older forms to the ancestry of *Knightoconus*, a monoplacophoran assigned the role of presumed morphologic antecedent of *Pletronoceras* (Yochelson et al., 1973), but I am not sure whether these ancestors are actually related.

There may be a new notion on the origin of this class in that Runnegar (personal communication, 1977) would derive them from the middle Cambrian *Yochelcionella* which has part of the shell modified to a tube opening to the outside; by some mechanism this external tube gave rise to an internal siphuncle. This argument is not convincing and so far as I am concerned the idea of direct evolution of cephalopods from monoplacophorans (Yochelson et al., 1973) remains plausible. Cephalopods appear in the late Cambrian ocean, full blown like Botticelli's Venus, though not nearly so sexy; dimorphism did not become an important feature of the shell in this class for several hundred million years. Dimorphism is scarcely known in the gastropods and pelecypods; it would be surprising to see evidence of sexual dimorphism in early molluscs.

Arguments against *Styliolina* as a mollusc were given elsewhere but should be reconsidered here. Rugosities, or pinch and swell of the cone, are characteristic of this group, whereas cephalopods and especially orthocones are generally smooth; cephalopods commonly are much larger. Some styliolinids have multiple septa, but the aspect of their flattened and irregularly spaced septa is unlike that of the curved, more uniformly spaced cephalopod septa. Blind (1969) reported piercing of septa in a few specimens but his evidence seems weak, for only excessively rare individuals give any indication of piercement of septa; the siphuncle is a pervasive feature of Cephalopoda.

It does not follow that all molluscs in the past secreted a conical structure with or without septa. There are many possible modes of life in addition to dwelling in a tube as do the cephalopods. Pojeta and others have named and discussed in detail (Pojeta and Runnegar, 1976) the extinct class Rostroconchia. A characteristic genus is the early Ordovician *Ribeiria* which shows the features of the class. It is an elongate, laterally compressed univalve with an internal bar or thickening of shell across the presumed anterior. Its shell appears molluscan, though the microstructure is not known. Geometry of the hard part of *Ribeiria* is quite unlike that of other molluscs. I consider that the naming of this class was a major step forward in elucidating the history of the Mollusca; I am not so enthusiastic toward the linking of the younger *Conocardium* to the ribeirioids in the same class, but that is not a concern which need be pursued here.

If *Conocardium* and its allies are ignored, the group is abundant in the early Paleozoic, especially in the late Cambrian middle Ordovician interval. The early Cambrian *Heraultopegma* is reputed by Pojeta and Runnegar to be the world's earliest rostroconch mollusc, but illustrated specimens of *Heraultopegma* are steinkerns dissolved from the rock. Several workers have mentioned an alternative biologic placement for this fossil within the Arthropoda (see Yochelson, 1979a, for references). I do not have a reasonable basis for choosing between conflicting assignment to phylum and prefer not to use such a fossil as the earliest representative of the class until more is known of its integument.

COMPLEXITIES WITHIN THE MONOPLACOPHORA AND "MONOPLACOPHORA"

This discussion now brings us a morass of argument. Let us begin with the middle Cambrian *Yochelcionella*, a peculiar member of the Helcionellidae in which a bilateral tube or snorkle developed out of the shell. For purposes of methodology, the author invites comparison of two notes discussing the relationship of an enigmatic early middle Ordovician fossil, *Januspira*.

Runnegar (1977, p. 203) confidently tied *Januspira* to *Yochelcionella* because of the tube. This tie is made by suggesting that *Helcionella* (Pl. I, 2) added to three species of early middle Cambrian *Yochelcionella*, all of which occur in the same bed, represent a morphologic series showing development and migration in position of the tube. The time gap between early middle Cambrian and early middle Ordovician was bridged by a hypothetical form. Runnegar indicates no size scale for any of the fossils; the youngest is smaller than the older ones, although one of the tenets of Runnegar's phylogeny (Runnegar and Jell, 1976) is that molluscs were smaller in the older rocks. The curved cap-shaped shell became tube-like and smooth, losing the characteristic rugosities of the Helcionellidae and much of the shell thickness.

Yochelson (1977a) questioned whether *Januspira* is even a mollusc and suggested that it is a peculiar "worm". Neither of us may be correct in our interpretation, but the approach of asking the main question first may be a bit more rigorous. *Reductio ad absurdum* is a poor kind of argument, but if *Jinonicella* (Pokorný, 1978), a Silurian relative of *Januspira*, is generally accepted as a gastropod or even a mollusc, it is foolish to apply logic to classification; any univalve shape may be placed in the Mollusca.

Pojeta and Runnegar (1976) speculated on the course of water currents through the tube of *Yochelcionella* and other helcionellids which do not develop the tube, and concluded that it was inhalent so that the concave side of the organism was anterior. In the known biology of living molluscs there seems no basis to suggest that the tube was inhalent in function. Indeed, species of *Yochelcionella* with the snorkle near the apex, pointing upward rather than laterally, remove the argument that the animal crawled and waved it about, like *Busycon* testing water quality. If the tube were exhalent in function, its varying position among species makes more sense. The organism would be expected to expel spent water away from the area in which oxygenated water was taken in.

The rugosities of some helcionellids may be used to argue, admittedly in a weak way, for an organism which was sedentary for periods of time; there seems no other reason to account for the pinch and swell of the gradually widening shell. An exhalent tube is an advantage to a sedentary form.

Just as an aside, the various species of *Yochelcionella* seem to demonstrate conclusively that either the snorkle migrated anterior to posterior (or posterior to anterior) or the curvature changed from concave forward to convex forward (or its alternate). A more reasonable alternative is to suggest that a bilaterally symmetrical cap-shaped or compressed univalve does not necessarily yield information on what was anterior and what was posterior in an extinct animal.

Regardless of any arguments concerning soft parts, the helcionellids have a distinctive shape. They are bilaterally symmetrical, slightly curved, laterally compressed and commonly have prominent growth rugosities. Knight *et al.* (1960) may have made a mistake, that is, a "wrong" interpretation, in placing them as ancestral to the Bellerophontacea and hence within the Gastropoda, but that was a long time ago. If only the hard part morphology is considered, this group stands apart from other molluscs and in my view it stands sufficiently far apart to be placed in a separate class. This view has been mentioned before (Yochelson, 1979a), but a formal systematic step has not been taken here because taxonomic proposals ought to include morphologic investigation of specimens and not just theory.

It would be helpful if more information were available on the morphology of the early Cambrian *Scenella* (Pl. V, 6). It overlaps in stratigraphic range with the late early Cambrian and middle Cambrian helcionellids. It is oval, rather than laterally compressed and is a moderately high cone. I have argued that *Scenella* might be different from the members of the Helcionellidae, but perhaps the relationship is closer than was thought. Considerable attention has been focused on the muscle scars found by Rasetti (1954) in a cap-shaped mollusc from the Burgess Shale of British Columbia. It is high time to realize that regardless of what these scars may mean, they are not muscle scars of *Scenella*. The Burgess Shale form is a low, broad cone with a subcentral apex and a smooth exterior. No new genus has been named for it, but it is distinct from *Scenella*. No muscle scars are known from *Scenella*.

The compressed shell of the Helcionellidae is quite easy to differentiate from that of the low, broadly expanded Monoplacophora. Such differentiation becomes more difficult and less certain if *Scenella* is included as a helcionellacean, but it still might be possible.

Of course, some of the difficulty lies in the question of what one means by Monoplacophora. To me it implies a simple cap-shaped shell which shows multiple paired muscles on its interior. To Runnegar and Pojeta it appears to mean a bilaterally symmetrical molluscan shell combined with soft parts presumed not to have undergone torsion. Perhaps their position is overstated, perhaps not. Rostroconchs are judged by their definers not to be monoplacophorans, even though they presumably had not undergone torsion, partly because of their distinctive morphology and partly because they had a long and complex evolutionary history. This last argument is not accepted; I have argued elsewhere on the difference between distinctiveness and diversity (Yochelson, 1979a) as it applies to subdivisions of molluscs.

Basic geometry ought to be used to establish the principal distinctive categories of fossil molluscs, elaboration on that geometric plan is a function of the diversity or "success" of the category. The evidence of the distinctiveness of hard parts from those of *Proplina* (Pl. V, 5), *Plina* or *Neopilina* argues equally for removal of both the Rostroconchia and the Helcionellidae from the Monoplacophora.

THE EXTINCT MOLLUSCAN GROUP BELLEROPHONTACEA

I should now comment on another aspect of the transmogrified mono-placophorans, namely the concept that Bellerophontacea are Mono-placophora. Bellerophontaceans are common molluscan fossils in the Palaeozoic. They are commonly coiled through several whorls and characteristically bilaterally symmetrical, in contrast to the asymmetrical shell of most members of the Gastropoda; I judge them to be an extinct superfamily of gastropods.

Presumed early Cambrian bilaterally symmetrical forms may be "worms". They occur from the late Cambrian upward into the Palaeozoic: the group apparently does range up into the early Triassic before dying out. Evidence of simple paired columellar muscles in a variety of bellerophontaceans continues to mount (Peel, 1972, 1976). These scars are quite unlike the muscle scars in *Cyrtolites*, the coiled presumed mono-placophoran. Yochelson (1978) remarked that if one places much reliance on homology of muscle scars, rostroconchs then are closely related to *Cyrtolites* and hence are monoplacophorans. Further, those workers who place reliance on morphologic series of shell forms in constructing phylogeny have not yet demonstrated a missing link between the late Cambrian cap-shaped shells with discrete multiple pairs of muscles, like *Proplina* Kobayashi, and the coiled middle Ordovician *Cyrtolites*, unless it be the curved middle Devonian *Cyrtonella*!

However, it may be helpful to return to Miller who placed the middle Ordovician *Pterotheca* in the Pteropoda, even though he had Bellerophontidae as one of his 25 families in the Gastropoda. Miller knew of six genera to place in that family, one of which, *Porcellia* Levéillé, begins life as an asymmetrical pleurotomariacean and only subsequently develops a pseudo-bilateral symmetry; Miller, like everyone else who has ever tried to construct a classification, was fallible. *Pterotheca* is properly put aside at first glance for it does not resemble the coiled *Bellerophon* (Pl. V, 7). The adult shell is low, broad, and best described as wedge-shaped; it is quite unlike the coiled bellerophontacean. There is an internal platform, like that in the slipper limpet *Crepidula*, at the narrow end of the wedge. Actually a tiny coiled early shell is present and within the genus *Bucania* there are species which approach the extremely flattened form of *Pterotheca* (Pl. II, 3). This flattened shape also occurs in several Devonian genera; whether these are related or convergent is not known. What is known is that *Pterotheca* and shells of similar shape occur in geologic settings best interpreted ecologically as soft bottoms. The broad shell and broad foot seem to be a snowshoe type of adaptation.

Arguments that the Bellerophontacea are monoplacophorans call for a reconstruction which balances the coiled shell over the head in what appears to be an unstable position. If this same orientation is applied to *Pterotheca* the result is a wedge-shaped animal which moves forward across soft sediments with the point of the wedge forward and the broad end behind.

Form and function are supposed to be related, though it is easy to be fooled. The living *Polinices* has a shell which would appear to be poorly adapted for moving through sand, but the shell is covered with a large foot and mantle into a wedge shape. The shape of soft and hard parts combined cannot be deduced from the shell. If *Pterotheca* made progress by pushing the wide leading edge of its foot forward and trailed the narrower part of the wedge, the principal area of weight would be posterior. If the wedge point were forward, as those who suggest that Bellerophontacea did not undergo torsion would have us believe, the heaviest part of the shell would go first over a very soft bottom; other consequences would be little space under the shell for the head and great difficulty in moving the neck from side to side. I remain unconvinced that the Bellerophontacea are a part of the Monoplacophora and am still reactionary enough to place them in the Gastropoda.

SOME ASYMMETRICAL FOSSIL MOLLUSCS

Except in the rarest of cases, palaeontologists cannot have access to the soft parts of the fossils. The class Gastropoda is defined by torsion of the soft parts; this is a zoological concept. Gastropods are not defined by their possession of coiled shells, but some palaeontologists tend to argue that since most living gastropods have coiled shells, the coiled forms which appear in the fossil record automatically belong to dead gastropods.

A few corals have coiled early stages (Sando, 1977) and many worms form coiled tubes. Yochelson (1975) has argued that the tiny coiled steinkern found in the early Cambrian and called *Aldanella* cannot be proven to be a mollusc. This is no small point. Until we have more data on these small trochoid forms we should not call them gastropods. If they are gastropods, it is remarkable that seemingly advanced shapes appear at the dawn of the Cambrian. Before phylogeny of the molluscs is interpreted from fossils, one should be certain the included fossils are all molluscs.

If we can eliminate the worm tubes from the molluscs, four coiled shapes are known in the Cambrian. *Pelagiella* is the oldest, occurring in the early Cambrian (Pl. V, 5). Knight (1952) argued that its shape was not typical of later gastropods; partly because the whorl expands at a rapid rate and partly because, using conventional orientation, the bulk of the cross-section is below the upper whorl surface. Some exceptions within the Gastropoda approach this form but they are rare.

Unfortunately, no single feature of the *Pelagiella* shell conclusively differentiates it from the shell of gastropods. Morphologic separation of the two is far more difficult than is differentiation of helcionellids from monoplacophorans. Runnegar has found a key steinkern in the collections of the US Museum of Natural History which shows apparent muscle scars unlike those of a gastropod. We concur that *Pelagiella* need not have undergone the torsion which categorizes snails. Runnegar (personal communication, 1978) considers it an asymmetrical monoplacophoran with about 10% of torsion. I cannot understand the notion of an obviously asymmetrical monoplacophoran any more than I can understand the notion of a bivalved one. If more specimens with similar scars are found, they may provide compelling evidence for naming a new class to accommodate *Pelagiella*.

The other three shapes are first known in the late Cambrian. *Sinuopea* has a three-dimensional coil and an apertural sinus; there seems no difficulty in accepting it as an early pleurotomariacean. *Owenella* is globose and resembles a small ripe pea; it fits moderately well within the Bellerophontacea as the first representative of the superfamily. *Matherella* and its allies are not so readily placed. In conventional orientation with the spire upward they are left-handed. By making reference to the middle Ordovician *Maclurites* and working backwards, Knight (1952) judged them to be hyperstrophically coiled. This pseudo-sinistral coiling form is peculiar and these fossils ought to be in a high-level category, perhaps an order; several of my colleagues (R. M. Linsley, unpublished data; J. S. Peel, unpublished data) have at least informally discussed the concept that these constitute another class apart from the Gastropoda. Such argument would have to be based on speculation concerning the soft parts not having undergone torsion. One cannot demonstrate torsion or lack of torsion and for this group the assumption that they are true gastropods seems less speculative.

Torsion may have occurred more than once; anything is possible, but one would judge such an event most implausible. It is perhaps slightly more plausible to consider *Pelagiella* as ancestral to the Macluritacea. Soft parts for *Pelagiella* might be reconstructed without the need for torsion. One must bear in mind that it was a reasonably small animal. On the other hand, some *Maclurites* are more than 25 cm across and the same arrangement of soft parts hypothesized for *Pelagiella* might not be viable at a larger size. The whorls of macluritids expand logarithmically at a rate comparable with that of other gastropods and my guess is that the soft parts were not too dissimilar to those in present-day snail shells. However, the rapid rate of expansion of *Pelagiella* might have been able to accommodate some non-gastropod-like arrangement of its anatomy.

COMMENTS ON THE RUNNEGAR–POJETA MODEL OF PHYLOGENY

In their reconstruction of the history of the Mollusca, Runnegar and Pojeta recognized within the early Cambrian representatives of the only extinct molluscan class they accept, the Rostroconchia, and of three living classes, the Gastropoda, the Bivalvia and the Monoplacophora. In the late Cambrian they find representatives of the Polyplacophora and Cephalopoda. Finally, in the middle Ordovician, they have identified the first representatives of the final living shelled class, the Scaphopoda. Above all, they interpret a continuity of development within the Mollusca.

To summarize the relationships as they present them, they consider the multiplated Polyplacophora somewhat apart from other molluscs and submerge the matthevians within that class. They view the monoplaco-phorans as a diverse group which gave rise to two main branches. On one branch, the Rostroconchia evolved by strong lateral compression of the monoplacophoran shell. In turn, some members of this class gave rise to the Bivalvia by splitting of the shell and others, to the Scaphopoda by fusion of the ventral gape; the concept of Stenothecoida is neither accepted or clearly refuted. From the other branch of the Monoplacophora complex came the Cephalopoda and the Gastropoda. Morphologic forms presumed to be intermediate link early members of all these classes to the mono-placophorans without any gaps in the sequence of forms. Hyolitha is excluded from the phylum.

If several different molluscan forms are found in one deposit, as Runnegar and Jell (1976) have shown in their work on the middle Cambrian of Australia, this is a testimony to the variety within the molluscs, but it is not a morphologic series. A sequence may tell us of phylogeny, that is, progression of forms. A series ought to be in stratigraphic order before it is considered seriously. Morphologic series which zigzag up and down the column, as do some suggested by Runnegar and Pojeta, are not time dependent and thus are not phylogenetic sequences. Quite apart from this, a morphologic series, like any other single kind of data, cannot provide all needed information on possible relationships.

The desirability for a presumed series to be in geological order has some interesting aspects. In suggesting that early Cambrian *Heraultopegma* may not be a rostroconch, one indicates only that the evidence for its assignment to Mollusca is weak and should be reconsidered carefully. In contrast, in suggesting that early Cambrian *Fordilla* may be an independent bivalve, but not a member of Bivalvia (Yochelson, 1978b) I was strengthening the argument that Rostroconchia gave rise to Bivalvia. It seems that this might be a logical linking of the two classes. Regardless of this specific, it would seem a sounder argument for ancestor and descendant to appear in stratigraphic sequence than to be found in beds of the same age. The sequence in which fossils are found is data unlike any available to the zoologist. It is unique to the science of palaeontology and should not be ignored.

I am antipathetic toward deriving the Scaphopoda from the middle Ordovician rostroconch *Pinnocaris*. Rostroconchs grew by adding shell increments along the continuous anterior, ventral and posterior gape. There is more to the scaphopods than just a fusion of the ventral gape. Growth in *Pinnocaris* is mostly toward the posterior. In a scaphopod, the growth is entirely at the anterior end. The logarithmic growth of the scaphopod shell is an extra geometric complication. I do not know and cannot even guess as to the ancestral form of the Scaphopoda; they are most peculiar molluscs. Origin of this class from a strange abrupt mutation, rather than through an orderly progression of shapes, seems attractive. If scaphopods first appeared in the mid-Palaeozoic they have no obvious ancestor among the early molluscs. If they first appeared in the middle Ordovician, essentially contemporaneous with *Pinnocaris* as suggested by Runnegar and Pojeta, an amazing amount of morphological change must have taken place during a very short time interval.

With some difficulty one might derive the Rostroconchia from the Helcionellidae, though again it requires major abrupt changes in basic form to move from one kind of shape to the other. Virtually all the two have in common is a laterally compressed shape. Helcionellids have periodic lateral expansions of the shell, whereas the rostroconch shell is uniformly arched. The Helcionellidae have a basal aperture in one plane, whereas the Rostroconchia have an elongate narrow basal gape and narrow anterior and posterior gapes. There are no obvious intermediates between the high curved shape of *Helcionella* and the elongate low shell of *Ribeiria*. Why a mature organism should gradually change its foot shape from a broad sole which allows grater mobility to one that is narrower and of less mobility is not obvious. Perhaps the Rostroconchia lived by scuffling up the bottom to obtain food particles. Regardless of that, once again it is easier to derive a sequence of forms if the Rostroconchia first appeared in the late Cambrian after the Helcionellidae, rather than prior to them.

To move to the other branch of Runnegar and Pojeta's phylogeny, I am still quite satisfied that the Cephalopoda are best considered as specialized Monoplacophora. If the key to the cephalopods is a siphuncle, it is difficult to see how this structure could develop gradually. The origin of the Gastropoda appears even more shrouded in mystery and I am willing to appeal again to another major larval mutation as the reason for their appearance. If gastropods were not alive today, we would be exceptionally hard pressed to imagine the soft parts. Torsion of soft parts is improbable, but gradual development of such an improbable event with a continuous sequence of intermediate stages is far more difficult to imagine. In a sense, this is the problem posed by complex organs, such as the eye; intermediate stages seem ineffectual.

COMMENTS ON THE YOCHELSON MODEL OF PHYLOGENY

Earlier I summarized all the peculiar forms of early molluscs of which I was aware (Yochelson, 1963). There are a few different ways of fitting the molluscan classes into an evolutionary scheme. The Runnegar–Pojeta view is current, but it seems almost impossible to consider any arrangement which has not been considered earlier. In an abstract (Yochelson, 1963), I suggested that there were several extinct classes and that these have been replaced by the extant classes. As a result of ecologic replacement no close relationship existed among the classes; this lack of obvious arrangement is in itself the overall evolutionary pattern to be expected.

There seems to be no morphologic series, nor any gradual sequence of classes, nor even assemblages of subphyla in the fossil record of the molluscs. Each of the living molluscs seems to represent a different basic morphological adaptation. Some modes of life which result from the adaptation are quite diverse and some are quite restricted. One would then apply these data to the fossil record and use a combined geometric and palaeoecologic approach to interpret the variety of fossil molluscs. One recognizes groups of molluscs different from those which have living representatives today. If these fossils are fundamentally different, they are to be placed in extinct classes.

In the older literature Woodward (1907), expressed some pertinent concepts:

> The extreme plasticity of the Mollusca naturally renders them both peculiarly susceptible and readily responsive to the operation of the two great factors that govern the lives of all animals, namely, the influence on them of their environment, and the necessity laid on one and all of procuring food. The Molluscan mode of life is, in fact, mainly governed by the combined action of these two controlling influences, and in turn becoming itself a potent factor, completes the cycle by reacting on the animal, which is thus impelled, so long as similar conditions hold, yet further along a given line of development.
>
> Owing, however, to the paucity of stable elements to be acted on, continuous progress in any direction has, despite the antiquity of the race, been slow indeed. The total lack of anything like internal framework has militated against any such wonderful progress as exhibited in the Vertebrate kingdom; the very plasticity of the Mollusca has thwarted progressive development, as we understand the phrase, and they readily retrograde or branch off into bye-paths.
>
> Hence the study of evolution in this group is an exceedingly complex one, offering,

like a very tangled skein, so many clues to follow out that one is in doubt which thread to pursue first.

We may not believe in "progress" in evolution as the Edwardians and Victorians used it, but apart from that Woodward makes a great deal of sense. Molluscs cannot develop every life habit observed today in the animal Kingdom, but they do a great many different things remarkably well. Presumably in the past, ancient forms may have had some life habits not represented today, or may have been unable to compete with more efficient modern forms when these appeared.

A POSSIBLE MORPHOLOGICAL ROUTE TO THE MOLLUSCA

At this point it may be more useful, in the words of J. B. Knight, to try climbing up the family tree. In attempting to trace the developments that led to benthonic molluscs, I have relied entirely on the work of others; Solem (1974, pp. 41–47) has a nice summary. A satisfactory starting point is one considering the turbellarian-like flatworms as forms ancestral to molluscs. One may guess that the first step in the dramatic change from one phylum to the other began with change in digestion. A blind gut has its limitations compared with a complete digestive tract.

The ability to extract more of the food value from particles, added to the alternative of excretion of wastes by anus rather than mouth, would lead to a greater range of diet and would be viewed by us as major steps forward. Still, we have all had the experience of switching from a normal diet to an exotic one; occasionally the spice of life leads to bizarre consequences in metabolism. Of all the opinions that one reads regarding calcification the one that makes most sense to me is that an organism excreted an exoskeleton originally to remove harmful waste products from its system. From this organism, molluscs as we know them could then develop. Mollusc-like organisms may have had shells like those of molluscs but different soft parts. If we could see more of the soft parts of fossils; molluscs and mollusc-like organisms would be easy to separate.

Concurrent with loss of respiratory surface by development of the calcified integument came formation of a mantle cavity and gills. A great deal of argument has been expended on the question of whether an early mollusc had more than one set of gills. It seems easiest to start with one pair and to have them in a posterior-lateral position, definitely not anterior, but not fully posterior either. These are not original ideas, though it seems that too much emphasis has been placed on a posterior mantle cavity as primitive.

A hard part not only conveys protection but it provides an anchor for muscles. An organism living on the bottom can still use ciliary motion to get about but it can now also develop waves in the sole of the foot and crawl with a great deal more efficiency than a flatworm. This motion is quite unlike the movement of a round worm; crawling on a sole is also not like moving about on limbs. Arthropods are not necessarily better than molluscs, only different. A free-living animal, but one with no obvious locomotor adaptation, such as a hyolith, could stay in one place for considerable intervals and move with considerable difficulty, but an organism capable of moving easily may move or stay in one place as it desires. Thus is great truth revealed!

With increased locomotion also comes a need for more nutrition—this is another great truth. For some invertebrates, our categorization of food habits is still murky. Thus, a *Conus* paralysing a fish and a nudibranch chewing on a sponge are both called carnivores, but they are not earning a living in the same way. One may imagine that an early slowly crawling mollusc was a detritus eater, but this really means it was an omnivore. So long as a food item was small and did not fight back, it was eaten. For an organism crawling intermittently, a device such as palps or tentacles to collect sediment containing pieces of food would be important; less emphasis would be placed at this early stage on extensive use of a radula to rasp.

One may ask how long it took to effect these proposed changes from a turbellarian-like creature to a mollusc. We do not know and have no objective basis for making an estimate. The only fact available is the negative point that not a single authentic record exists for any Precambrian mollusc. It may be that even such dramatic changes as postulated could occur in a million years or less. We can also guess that once calcification of the organic integument began, a strong shell would have been present within a few generations, at most. Finally to make certain that I am not misunderstood, it must be stated that the concept of an early mollusc adopted here includes a calcified shell; if shell-less organisms were present in the Precambrian–Cambrian boundary time interval that had some of the morphologic features speculated about here, they were not yet molluscs.

The writer also suggests that the fossil record of the Palaeozoic harbours a great and diverse group of extinct "worms". Some evolved hard parts similar enough to those of molluscs to cause enormous confusion in interpretation of authentic Mollusca.

Did representatives of the various classes evolve gradually or abruptly? In a way, the argument between those who see gradual development and those who fall back on abrupt and dramatic changes is like the old argument between uniformitarianists and catatrophists. Geologic processes do go on continuously but most of the main events are catastrophic. Runnegar and Pojeta suggest that torsion of the snail evolved gradually because the shell became increasingly higher and fell over; others have viewed it as happening in one generation. One doubts that palaeontologists will ever have the evidence to support one or the other of these views, for the lifespan of 1000 generations or even a 100000 generations of molluscs, during which this change might have taken place, would appear as an instant in the record of the rocks.

Quite apart from this post-mortem crowding of events forced on us by the nature of the record, it seems reasonable to suggest that many changes of a morphologically fundamental nature may have occurred rapidly. On the hypothesis adopted here, some larvae of a generalized creeping form developed strong discrete multiple muscle pairs. Other larvae later developed multiple plates. Still others developed an internal siphuncle and finally, in an ecological but not a stratigraphic sense, yet other larvae developed torsion. Thus the bottom of the sea was occupied by benthonic molluscs as a result of a series of major discontinuous morphologic changes in juveniles that led to the development of fundamentally new adults. The precise changes are not in the preview of this paper.

A better case might be made for gradualism and a series of morphologic intermediates in discussing Runnegar and Pojeta's concept of evolution of the infauna. However, there seems to be no objective evidence suggesting that the early rostroconchs were infaunal. If a narrow foot developed along with the greatly folded-over shell, perhaps the animal could plow very slowly near the sediment–water interface. Most items of nutritional value are on the surface or immediately below it, so why burrow?

In the evolution of Mollusca, epifaunal life surely preceded infaunal dwelling. Again our terminology is rather vague. Is the living snail *Polinices*, for example, a member of the infauna or the epifauna? Its life habit depends on time and tide; these wait for no man and are not well understood. Clams are considered infaunal but they simply live in the sediment for protection; without contact with overlying water they die. One would not say that pelecypods are predominantly infaunal in the same sense that some worms are infaunal.

Even if the rostroconchs were infaunal, it is difficult to see how they could have evolved gradually from any monoplacophorans, even if an exceedingly broad definition of that class is used. A major modification would be required to change the shape of the shell and presumably the broad shape of the foot in *Helcionella*, to the shell and presumably the narrow elongate foot of *Ribeiria*. Further, to go from the univalved shell of a rostroconch to the bivalved shell of a pelecypod is surely a remarkable occurrence, but how can it be gradual? To move through a series of mature straight univalves which gradually become longer and longer and then to fuse ventrally and simultaneously to develop logarithmic curvature and a uniformly tapering shell open at both ends is a mind-boggling concept. The origin of the Scaphopoda is remarkable, but not that remarkable.

We may reconsider the early evolution of molluscs, using the Runnegar and Projeta model. In this, advanced pelecypods appear in the early Cambrian; after a time gap, many more pelecypods, including seemingly primitive ones, occur in the early Ordovician. Rostroconchs appear simultaneously with their possible descendant pelecypods; after a time gap, in late Cambrian time, many more rostroconchs appear. Both classes are related through morphologic series of shapes, but some of these shapes are known later in the Cambrian. Advanced gastropods appear in the early Cambrian; after a time gap more primitive gastropods appear in the late Cambrian.

If we discuss the Yochelson model of evolution of early molluscs, we have a different situation. A series of fossil entities appears in the Cambrian; the morphology of the hard parts of these fossils is so different from those of living molluscs that they deserve to be called classes. These fossils lived in different habitats at different times and, except for being related through a common form of calcified integument, there are no obvious close relationships among them. Representatives of these classes and later classes evolved abruptly.

It does seem that hopeful monsters may open up new adaptive niches and that this is followed by an exploitation of these new life habits. To make certain there is no misunderstanding, I rephrase this to fly in the face of classic assumptions and state that evolution of species, genera and perhaps families may be gradual or abrupt, but that groups of ordinal, class or phylum rank evolved abruptly. Perhaps this is the "big bang" theory of evolution (C. Roper, personal communication, 1978), but I prefer it to gradual change, especially gradual change from one adult form to another adult form.

The notion of discontinuities in evolution is not new. To cite early literature, an eminent student of molluscs, W. H. Dall (1877), was a proponent of their view, suggesting that small changes accumulated until they had a seemingly abrupt effect. Whether it is the Saltatory Evolution of Dall or the Quantum Evolution of Simpson or the Punctuated Equilibrium of Eldredge and Gould, some authors of every generation persist in noting discontinuities in the biological kingdom, in spite of all the attention that conventional wisdom devoted to slight, steady change through time.

Dall (1895, p. 485) also warned that "...systems based on a single character...are bound to prove unsatisfactory...". Although we may find forms that seem to be intermediates and that decrease a morphologic gap in one direction, they may accentuate distinctiveness in another direction. I believe that we should concentrate our efforts less on intermediate linking forms and more on the discontinuities, as perhaps being more representative of what really occurs in nature. This is also to state, once more, that just as our new data increases and our interpretation of old data changes, so classifications must change.

SUMMARY

There is no final authoritarian word in science. Dall (1895, p. 560) put it admirably, "In conclusion the reader is reminded that this summary represents not the entire truth in regard to the groups characterized, but only an approximation to our present knowledge of them. The marshalling of the characters here given will doubtless do much to call attention to discrepencies and errors hitherto unchallenged, and, by its very defects lead to an amelioration of the system."

REFERENCES

BENGTSON, S. (1968). The problematic genus *Mobergella* from the lower Cambrian of the Baltic area. *Lethaia*, **1**, 325–351.

BLIND, W. (1969). Die Systematische Stellung der tenaculiten. *Palaeontographica*, *A*, **133**, 101–145.

BLIND, W. and STÜRMER, W. (1977). *Viratellina fuchsi* Kutscher (Tentaculoidea) mit Sipho and Fangarmen. *Neues Jb. Geol. Paläont. Mh.* (1977), 513–522.

BOUČEK, B. (1964). "The tentaculites of Bohemia." Publishing House, Czech. Acad. Sci., Prague.

BROWN, C. H. (1975). "Structural materials in animals." John Wiley and sons, New York and Toronto.

DALL, W. H. (1895). Contributions to the Tertiary fauna of Florida with especial reference to the Miocene Silex beds of Tampa and the Pliocene beds of the Caloosahatchee River: Part III, a new classification of the Pelecypoda. *Trans. Wagner Free Inst. Sci. Philad.* **3**, 485–570.

DALL, W. H. (1877). On a provisional hypothesis of saltatory evolution. *Am. Nat.* **11**, 136–137.

HOFMANN, H. (1971). Precambrian fossils, pseudofossils, and problematica in Canada. *Geol. Surv. Can. Bull.* **189**, 148 pp.

KNIGHT, J. B. (1952). Primitive fossil gastropods and their bearing on gastropod classification. *Smithson. misc. Collns*, **114** (13), 55 pp.

KNIGHT, J. B. and YOCHELSON, E. L. (1960). Monoplacophora. *In* "Treatise on Invertebrate Paleontology, Part I, Mollusca 1" (R. C. Moore, ed.). Geol. Soc. Am. and Univ. Kansas Press.

KNIGHT, J. B., BATTEN and YOCHELSON, E. L. (1960). Descriptions of Paleozoic Gastropoda. *In* "Treatise on Invertebrate Paleontology, Part I, Mollusca 1" (R. C. Moore, ed.). Geol. Soc. Am. and Univ. Kansas Press.

KOZŁOWSKI, R. (1968). Nouvelles observations sur les Conulires. *Acta palaeont. pol.* **13**, 479–535.

LEHMANN, U. (1976). "Ammoniten." F. Enke, Stuttgart.

LUTZ, R. A. and RHOADS, D. C. (1977). Anaerobiosis and a theory of growth line formation. *Science, N.Y.* **198**, 1222–1227.

MAREK, L. (1976). On the ontogeny of Hyolithida. *Čas. Nár. muz., Miner., Geol.* **21**, 277–283.

MAREK, L. and YOCHELSON, E. L. (1976). Aspects of the biology of *Hyolitha* (Mollusca). *Lethaia*, **9**, 65–82.

MILLER, S. A. (1889). "North American Geology and Palaeontology for the use of Amateurs, Students, and Scientists." Western Methodist Book Concern, Cincinnati.

MISSARZHEVSKIY, V. V. (1976). New data on early Cambrian monoplacophors. *Paleont. Zh.* (1976), 129–131 (American Geological Institute translation).

PEEL, J. S. (1972). Observations on some lower Palaeozoic tremanatiform Bellerophontacea (Gastropoda) from North America. *Palaeontology*, **15**, 412–422.

PEEL, J. S. (1976). Musculature and systematic position of *Megalomphala taenia* (Bellerophontacea; Gastropoda) from the Silurian of Gotland. *Geol. Soc. Denmark Bull.* **25**, 49–55.

POJETA, J. and RUNNEGAR, B. (1976). The paleontology of rostroconch mollusks and the early history of the phylum Mollusca. *Prof. Pap. U.S. geol. Surv.* **968**, 1–88.

POJETA, J., Jr, RUNNEGAR, B., MORRIS, N. J. and NEWELL, N. D. (1972). Rostroconchia: A new class of bivalve mollusks. *Science, N.Y.* **144**, 264–267.

POKORNÝ, V. (1978). Jinonicellina, a new suborder of presumed Archaeogastropoda. *Vést. Úst. geol.* **53**, 39–42.

POULSEN, V. (1963). Notes on *Hyolithellus* Billings, 1871, class Pogonophora. Johannson, 1937. *Dan. Vid. Selsk. Biol. Medd.* **23** (12).

RASETTI, F. (1954). Internal shell structures in the middle Cambrian gastropod *Scenella* and the problematic genus *Stenothecoides*. *J. Paleont.* **28**, 59–66.

RESSER, C. E. (1938). Fourth contribution to the nomenclature of Cambrian fossils. *Smithson. misc. Collns*, **97** (10), 1–43.

RUNNEGAR, B. (1977). Found; a phylum for *Januspira. Lethaia*, **10**, 203.

RUNNEGAR, B. and JELL, P. A. (1976). Australian middle Cambrian molluscs. *Alcheringa*, **1**, 109–138.

RUNNEGAR, B. and POJETA, J., Jr (1974). Molluscan phylogeny: the palaeontological viewpoint. *Science, N.Y.* **186**, 311–317.

RUNNEGAR, B. *et al.* (1975). Biology of the Hyolitha. *Lethaia*, **8**, 181–191.

SANDO, W. J. (1977). Significance of coiled protocoralla in some Mississippian horn corals. *Palaeontology*, **20**, 47–58.

SAVAGE, T. E. (1927). Significant breaks and overlaps in the Pennsylvanian rocks of Illinois. *Am. J. Sci.*, 5th ser., **14**, 307–316.

SOLEM, G. A. (1974). "The shell makers; Introducing Mollusks." John Wiley and Sons, New York and Toronto.

STÜRMER, W. (1970). Soft parts of cephalopods and trilobites; some surprising results of X-ray examination of Devonian slates. *Science, N.Y.* **170**, 1300–1302.

WALCOTT, C. D. (1886). Second contribution to the studies on the Cambrian faunas of North America. *Bull. U.S. geol. Surv.* **30**.

WOODWARD, B. B. (1907). What evolutionary processes do the Mollusca show? *Proc. malac. Soc. Lond.* **7**, 246–256.

YI, QIAN (1977). Hyolitha and some problematica from the lower Cambrian Meishucun stage in Central and S.W. China. *Acta palaeont. sin.* **16** (2), 255–275.

YOCHELSON, E. L. (1961). Notes on the Class Coniconchia. *J. Paleont.* **35**, 162–167.

YOCHELSON, E. L. (1963). Problems of the early history of the Mollusca. *Proc. 16th Internat. Cong. Zoology, Washington, D.C., August 20–26, 1963*, **2**, 187.

YOCHELSON, E. L. (1966). Mattheva, a proposed new class of mollusks. *Prof. Pap. U.S. geol. Surv.* **532-B**, B-1–B-11.

YOCHELSON, E. L. (1969). Stenothecoida, a proposed new class of Cambrian Mollusca. *Lethaia*, **2**, 49–62.

YOCHELSON, E. L. (1975). Discussion of early Cambrian "mollusks". *J. geol. Soc. Lond.* **131**, 661–662.

YOCHELSON, E. L. (1977a). Agmata, a proposed extinct phylum of Cambrian age. *J. Paleont.* **51**, 437–454.

YOCHELSON, E. L. (1977b). Comment on *Januspira. Lethaia*, **10**, 204.

YOCHELSON, E. L. (1978b). Discussion of Pojeta, J. The origin and early diversification of pelecypods. *In* "Evolutionary systematics of bivalved molluscs" (C. M. Yonge and T. E. Thompson, eds). *Phil. Trans. R. Soc. B.* **284**, 244–245.

YOCHELSON, E. L. (1979a). An alternative approach to interpretation of phylogeny of early mollusks. *Malacologia*, **17**, 165–191.

YOCHELSON, E. L. (1979b). Charles D. Walcott, America's pioneer in Precambrian paleontology and stratigraphy. *Sp. Pap. Roy. Soc. Can.* (In Press.)

YOCHELSON, E. L. and BRIDGE, J. (1957). The Lower Ordovician gastropod *Ceratopea*. *Prof. Pap. U.S. geol. Surv.* **294–H**, 281–294.

YOCHELSON, E. L., FLOWER, R. H. and WEBERS, G. F. (1973). The bearing of the new late Cambrian monoplacophoran genus *Knightoconus* upon the origin of the Cephalopoda. *Lethaia*, **6**, 275–309.

YOCHELSON, E. L., HENNINGSMOEN, G. and GRIFFEN, W. L. (1977). The early Cambrian genus *Volborthella* in southern Norway. *Norsk geol. Tidsskr.* **57**, 133–151.

ZEISS, A. (1969). Weichteile ectocochleater Pälozoischer Cephalopoden in Röntgenaufnahmen und ihre Pälontologische Bedeutung. *Palaont. Z.* **43**, 13–27.

12 | Gastropoda

ALASTAIR GRAHAM

Department of Zoology, University of Reading, England

Abstract: Gastropods stand apart from other molluscs in exhibiting torsion, but, except at the class–subclass level, torsion is not important in their classification. It is possible, however, to give an explanation of the advantages it has conferred on both larva and adult and to connect it with two interlinked features upon which much of the radiation of the class rests.

Gastropods differ from other conchiferans and agree with aculeates (a separate evolutionary line) in that the radula is their dominant feeding organ. In other conchiferans cephalic appendages are responsible for seizure of food and the radula merely pulls it into the gut, with or without some trituration. In most of such animals only the prehensile appendages emerge from the shell and the head remains poorly developed. Cephalopods are an exception to this and have a large head with important sense organs, but their flotation devices have permitted a mode of life and adaptations totally different from those of other molluscs.

Like cephalopods, gastropods exhibit cephalization but, unlike them, usually retain a shell into which they can retract. The head has enlarged because the radula has to be manipulated by an elaborate musculature and because of the development of sense organs and its association with a penis of pedal origin. In locomotion and feeding the head is projected on a mobile neck outside an easily portable shell. The apparent irreconcilability of a large head and the power of retraction is overcome by enlargement of the mantle cavity and by the fact that it is anterior because of torsion. The evolution of a penis is a pre-adaptation for life in fresh water and on land which has proved beyond the power of all other molluscs.

Much gastropod radiation is explicable by exploitation of the radula (i) for sweeping (probably its original use); (ii) for rasping soft surfaces (vegetable and animal); (iii) with chemical help, for abrading hard surfaces; (iv) for biting; (v) for grasping.

Systematics Association Special Volume No. 12, "The Origin of Major Invertebrate Groups", edited by M. R. House, 1979, pp. 359–365. Academic Press, London and New York.

GASTROPOD EVOLUTION

There is a well-known saying about *amphioxus*, that if it had not existed someone would have had to invent it. Had gastropods not existed it is doubtful whether anyone would have been bold enough to invent animals in which torsion has made the dorsal half of the body sit back to front on the ventral half—yet gastropods form the second largest class in the animal kingdom. The effects of torsion are most obvious in prosobranchs with their forward-facing mantle cavity and their streptoneury; although pulmonates escape being streptoneurous on a technicality, their organization is fundamentally similar. It is only in opisthobranchs that torsion is not always apparent, although their anatomy can be understood only on the assumption that they derive from ancestors which had undergone torsion. Thompson (1958, 1962) denied that the two nudibranchs he studied underwent an actual untwisting during development, nevertheless he described movements which effectively undo the effects of torsion. Similar rearrangements adapt tectibranchs in general (Brace, 1977) for the burrowing mode of life which, all authorities agree, originally underlay the changes. This instance apart, neither torsion nor detorsion help much in understanding gastropod radiations: indeed, they have occurred with almost equal breadth in both torted and detorted stocks.

This is no place to discuss the mechanisms supposed to be responsible for torsion nor the numerous suggestions as to the advantages it conferred. It is important, however, to emphasize that there must be some advantage to the *adult* gastropod, even if the primary twisting takes place in the larval stage and may have been of initial importance to the larva, as claimed by Garstang (1929). The advantage he proposed was the ability to withdraw the head into the shelter of the mantle cavity and shell before the foot. Some critics countered this alleged advantage by pointing out that, as the larvae were going to be ingested whole by their most frequent predators, it hardly mattered where their head was when that happened. This, however, misses the point. Fretter (1967, 1969) has shown that the real value of withdrawing the head and velum rapidly and completely is that it enables the larva to drop quickly through the water and so avoid, not so much predators, as concentrations of algae, which interfere with locomotion, and their metabolites, which may be toxic. Notice that Garstang assumed that at this stage in their evolution gastropods had a head-foot which could be extended beyond the shell, an assumption without firm basis, and failed to ask why, if torsion was so valuable to gastropods, it did not occur in other classes.

So far as the adult is concerned Morton (1958) and Ghiselin (1966) have pointed out how the anterior position of the mantle cavity must improve its function as a respiratory chamber, both by tending to open its mouth by the backward tilting of the shell and by locating it in front of any locomotor disturbance of the substratum; it must also improve the inflow of sensory information by adding pallial to cephalic receptors, especially as the latter may not have been well developed in the initial stages of gastropod evolution. Torsion may also have introduced some sanitary complications but it seems extraordinarily unlikely that it conferred benefits solely on the larva and that adult gastropods, except for the small proportion which have undergone detorsion—some 17% of the total number of genera in the Class—have simply put up with the consequences, neither so badly affected by it as to become extinct nor taking more than modest corrective steps. On the contrary I believe that torsion offered an opportunity for a particular evolutionary trend that separates gastropods from other molluscs and explains much of their success.

The very obviousness and oddity of torsion have blinded us to the fact that there are three other ways—equally important if not so bizarre—in which gastropods differ from other molluscs.

(a) *The mantle cavity.* Gastropods differ from other molluscs (except cephalopods) in that the mantle cavity is not equal in depth at all points round the body as it is, for example, in *Neopilina* or a chiton : it is markedly deeper on the posterior face of the visceral mass in cephalopods and, because of torsion, on the anterior face in gastropods. This is tied up with the fact that the shell is spirally coiled. In a plane spiral, perhaps the original form, there is a gradient of growth along the mantle edge on each side of the body from a maximum in the mid-line posteriorly to a minimum in the mid-line anteriorly. This inevitably leads to a mantle cavity which is deep posteriorly and shallow anteriorly and to a restriction of pallial organs to the deeper part. Thus in *Neopilina* there may be a circlet of gills but they are restricted to a pair of larger posterior ones in cephalopods and gastropods. To obtain the helical shell of a gastropod there has to be superimposed on the anteroposterior growth gradient just described, another growth gradient producing a right–left gradient, maximal at the (pre-torsional) right, minimal on the left. It is inevitable that this affects internal anatomy and it may well link up with the asymmetry of cleavage described by Verdonk (1977) and of the shell muscles which Crofts (1955) showed were the direct mechanical cause of at least part of torsion. If cephalopods failed to develop this right–left asymmetry there was probably no prospect of torsion becoming part of their heritage.

(*b*) *The head.* Gastropods differ from all other molluscs—again with the exception of cephalopods—in that the head is enlarged, has become equipped with major sense organs and, when the animal is active, is extended far beyond the shelter of the shell at the end of a mobile "neck".

(*c*) *The method of feeding.* Gastropods differ from all other conchiferan molluscs but agree with Aculifera (= Polyplacophora + Aplacophora) in using the radula as the primary collector of food. (The resemblance to Aculifera may well be a parallel development. I incline to the view of Salvini-Plawen (1969) that the Aculifera represent an evolutionary line parallel to but distinct from that of the Conchifera.) A brief survey of Conchifera shows that they are primarily tentacular feeders: scaphopods, protobranchs and cephalopods all use tentacular appendages to seek and hold their food and bring it to the mouth which lies, in most, on an undeveloped snout under the shelter of the shell. In all these animals the radula is used only to break up this food and transfer it to the buccal cavity. Gastropoda, in contrast, have devoted such tentacles exclusively to sensory purposes: the search for food has now been transferred to a mobile and elongated snout under the guidance of enhanced cephalic sense organs, and the radula, worked by a complex array of muscles, has become a vastly more versatile tool. These features are together responsible for the kind of cephalization exhibited by gastropods.

These three points are interlinked. Much of the size of the head is due to the complexity of its muscles and those of the buccal mass which developed as the animals became able to position the head and use the radula with a freedom which had been impossible in the restricted space underneath the shell; the exposure of the head favoured the development of cephalic as against pallial sense organs and so contributed to further cephalization. Now the ability to protrude and withdraw this enlarged head with the foot, depends, among other things, upon the presence of a mantle cavity large enough to provide a shelter when they are withdrawn and also able to act as a compensation sac and fill with water when they are extruded. Is it just coincidence that torsion brought a part of the mantle cavity, already enlarged by the spiral growth of the shell to house the pallial complex, into the very position where it could be exploited in such a way? Is it not more likely that, once the perhaps mechanical accident of torsion brought the cavity forwards, further evolution of cavity, head, foot and feeding went on in integrated fashion? If so, then this is both a more basic and greater advantage to the adult snail to be added to the lists of Morton and Ghiselin.

Incidentally, the ability to extend the head–foot and the presence of a capacious mantle cavity have allowed gastropods to develop an efficient copulatory process, an imperative pre-adaptation for the life on land which they, alone amongst the molluscs, have been able to achieve.

RADULAR RADIATION

Once the main form of the gastropod body had been established, the radiation of the group appears to depend on factors other than torsion. Of these the most important is the buccal mass with the radula, which together form a tool permitting varied ways of feeding and so opening many ecological niches to the gastropods. I shall mention eight important ways in which the radula may be used.

1. It may be used as a brush gathering loose detrital material for ingestion. This is how the rhipidoglossan radula works. It is found in archaeogastropods, except Patellacea, and it may well have been the original type of gastropod radula. It is functionally divisible into marginal tracts of many fine teeth which brush loose or loosely-attached particles towards a central tract where fewer and more powerful teeth act, some to help loosen particles from the substratum, others, like buckets on a dredger, to pull the material into the buccal cavity. This type of radula is not powerful enough to rasp solid material unless it is friable, but it does allow the eating of sponges.

2. More power is obtained if the weak marginal tracts are lost. This gives the taenioglossan radula found in nearly all mesogastropods. It permits collection of loose material but is also capable of much more powerful rasping. It does not matter much whether the material to be grazed is vegetable (as in winkles) or animal (as in many cerithiaceans and cypraeaceans) but the radula works properly only against a solid substratum, so animal prey is limited to sponges, bryozoans and tunicates. Given chemical help, as in *Natica*, this radula can also bore shells.

3. The rachiglossan radula opens new modes of life to prosobranchs. It arises by reduction of the taenioglossan to three teeth per row. This is not an attempt to gain more power, but a narrowing of the radular ribbon so that the buccal mass may be accommodated within the limits of a

slender, retractile proboscis (an enormously elongated snout) which can be inserted into small openings. These may be natural holes within which animals (mainly moribund or dead) lie, or artificial ones made with chemical or mechanical help as when dog whelks bore shells or buckies open bivalves.

4. The most abrasive radula is the docoglossan, found in patellacean limpets. There is some reduction of the tooth row but it achieves its power primarily by changing the transverse direction of tooth action, which is characteristic of all the types so far mentioned, into a longitudinal one (Ankel, 1938). With transverse movement each row of the radula acts independently of the others and the stroke is therefore not powerful: with longitudinal action all teeth in several rows act together to give a much more powerful action.

5. The radula may be used to bite, as well as rasp and tear, when jaws are developed against which the tip of the odontophore may be brought to bear. This is frequent in pulmonates, a relatively homogeneous group from this point of view with adaptation to varied diets expressed more by changes in the shape of the teeth than in the pattern of the radula.

6. The radula may be used to grasp. This may be a simple friction grip of rather coarse nature as in Epitonacea and *Janthina* (Graham, 1965) and *Testacella* (Crampton, 1975) where pads of backwardly-directed teeth are everted to seize prey with an action comparable to that of a snake. The grasp may also be precise, as in *Philine* (Hurst, 1965), where a relatively small bunch of teeth may be manipulated to hold small prey such as forams. Incidentally, this seems the only known place where radular teeth are moved by direct muscular pull; in all other situations radular teeth move in response to changing tensions in the cuticular membrane to which they are attached.

7. The radula may be used to pierce. The sacoglossan opisthobranchs and some small rissoaceans such as *Omalogyra* and *Ammonicera* use the sharply-pointed teeth to penetrate algal cells from which they then suck the contents.

8. Radular teeth may be used, as in cones, as poison-filled darts to stun or kill prey. This allows the gastropod to capture animals (worms, fish) otherwise far too active for such a slow-moving predator.

In addition to these uses of the radula it is worth remembering that some prosobranchs (calyptraeaceans, *Turritella*, *Viviparus*) have become predominantly ciliary food-collectors, straining the current passing through the mantle cavity; others have lost the radula altogether and become dependent upon sucking fluid or cell mush from a host.

REFERENCES

ANKEL, W. E. (1938). Erwerb und Aufnahme der Nahrung bei den Gastropoden. *Verh. dt. Zool. Ges., Zool. Anz., Suppl.* **11**, 223–295.

BRACE, R. C. (1977). The functional anatomy of the mantle cavity complex and columellar muscle of tectibranch molluscs (Gastropoda: Opisthobranchia), and its bearing on the evolution of opisthobranch organization. *Phil. Trans. R. Soc. B*, **277**, 1–56.

CRAMPTON, D. M. (1975). The anatomy and method of functioning of the buccal mass of *Testacella maugei* Férussac. *Proc. malac. Soc. Lond.* **41**, 549–570.

CROFTS, D. R. (1955). Muscle morphogenesis in primitive gastropods and its relation to torsion. *Proc. Zool. Soc. Lond.* **125**, 711–750.

FRETTER, V. (1967). The prosobranch veliger. *Proc. malac. Soc. Lond.* **37**, 357–366.

FRETTER, V. (1969). Aspects of metamorphosis in prosobranch gastropods. *Proc. malac. Soc. Lond.* **38**, 375–386.

GARSTANG, W. (1929). The origin and evolution of larval forms. *Rep. Br. Ass. Advmt. Sci., Glasgow, 1928*, 77–98.

GHISELIN, M. T. (1966). The adaptive significance of gastropod torsion. *Evolution*, **20**, 337–348.

GRAHAM, A. (1965). The buccal mass of ianthinid prosobranchs. *Proc. malac. Soc. Lond.* **36**, 323–338.

HURST, A. (1965). Studies on the structure and function of the feeding apparatus of *Philine aperta* with a comparative consideration of some other opisthobranchs. *Malacologia*, **2**, 281–347.

MORTON, J. E. (1958). Torsion and the adult snail: a re-evaluation. *Proc. malac. Soc. Lond.* **33**, 89–101.

SALVINI-PLAWEN, L. V. (1969). Solenogastres und Caudofoveata (Mollusca, Aculifera): Organisation und phylogenetische Bedeutung. *Malacologia*, **9**, 191–216.

THOMPSON, T. E. (1958). The nature history, embryology, larval biology and post-larval development of *Adalaria proxima* (Alder and Hancock) (Gastropoda, Opisthobranchia). *Phil. Trans. R. Soc. B*, **242**, 1–58.

THOMPSON, T. E. (1962). Studies on the ontogeny of *Tritonia hombergi* Cuvier (Gastropoda, Opisthobranchia). *Phil. Trans. R. Soc. B*, **245**, 171–218.

VERDONK, N. H. (1977). Symmetry and asymmetry in the embryonic development of molluscs. *Abstr. 6th Eur. Malac. Congr.* 12.

13 | Early Cephalopoda

C. H. HOLLAND

Department of Geology, Trinity College, Dublin, Eire

Abstract: Systematics of the Cephalopoda within the Mollusca are briefly discussed. Origin of the Class has been sought in certain multiseptate shells of the Cambrian Monoplacophora, rather than directly from a low capulate form. The presence of a siphuncle can be regarded as diagnostic of the Class Cephalopoda, whilst the development of endogastric curvature, perhaps explicable in functional terms, has been seen as crucial to its origin. Neglecting the Volborthellidae and Salterellidae as Mollusca, the Plectronoceratina comprise the root stock of the Superorder Nautiloidea of the Subclass Ectocochlia, and hence of the Class Cephalopoda. The first known of these are rare shells from the upper Cambrian of China. After a scarce Cambrian and early Ordovician record comes evidence of a splendid radiation. Within the nautiloids as a whole, from the first known small cyrtocones to the very varied straight to coiled shells of the later Ordovician, evolution can reasonably be seen in functional terms. Widely distributed orthocone limestones of various ages, regardless of precise mode of accumulation, give a vivid impression of the abundance of nautiloids in some Palaeozoic seas.

INTRODUCTION

For some 500 million years the cephalopods have been among the more important animals in the seas. There is no particular difficulty in separating the living forms from other main molluscan groups. Nor is there any difficulty in recognizing them as molluscs: members of that remarkably successful phylum of invertebrates in which a fairly simple set of characters has proved resilient and yet plastic through an eventful history. The possession of a mantle cavity is one of the most important of molluscan characteristics and the cephalopods, which are essentially molluscs adapted

Systematics Association Special Volume No. 12, "The Origin of Major Invertebrate Groups", edited by M. R. House, 1979, pp. 367–378. Academic Press, London and New York.

for active swimming, have used this not only for its persistent respiratory function but also passively, and then more actively, with the flexible hyponome to allow that effective kind of jet propulsion which culminates in the living forms. A hyponomic sinus is already present in early Ordovician shells. The associated need for quick action has led also to the need for effective eyes and an adequate brain, and in some cephalopods to the evolution of giant nerve fibres. The failure of the cephalopods, if that is a fair statement, even when restricted to modern forms, is identified by Wells (1962) as involving restriction of physiology resulting in a limited ecological range, never involving a freshwater or terrestrial form, never a sessile form, possibly never a herbivore—there being limitations in terms of respiration, excretion and digestion.

CEPHALOPOD SYSTEMATICS

In cephalopod systematics, splitting of the living forms into coleoids and nautiloids (with a single genus) is reasonable enough. The former are definable, as Donovan (1977) has recently clearly assessed, as a group possessing a guard, a reduced body chamber, or both of these. Such characteristics indicate that soft tissue in some way covered the outside of the shell, though in fossil forms this was not necessarily internal to the extent that it is in living coleoids such as *Sepia*. Use of the gill pair number, with the tetrabranchiates versus the dibranchiates, is of course not satisfactory as we know nothing of the number of gills in the exceedingly numerous fossil cephalopods. In any case the possession of two pairs in *Nautilus* may well be a secondary feature, resulting from the inadequacy of the respiratory current in these relatively bulky forms.

Taking account of the very numerous early cephalopods, we customarily refer to a third category: the ammonoids. These forms are so common, so much discussed in evolutionary theory, and so useful in biostratigraphy that their relative uniformity compared with the diversity of the fossil nautiloids has often been disregarded in classification. Their separation from the nautiloids is seldom troublesome; we can usually recognize specimens apart. But, pressed on enumerate the differences such as the nature of the protoconch of the ammonoid, its slender marginal siphuncle, and its many whorled shell, we nevertheless tend to fall back upon the complexity of the ammonoid suture, related no doubt to their particular hydrostatic problems.

Within the nautiloids themselves three of the five superfamilies of the

Order Nautilida: the Trigonocerataceae, Clydonautilaceae, and Nautilaceae, do in fact contain members with relatively elaborate sutures. They are found especially in Triassic and early Tertiary forms (Teichert *in* Moore, 1964), of which the genus *Aturia* (Palaeocene to Miocene) of the Nautilaceae is perhaps the best known. However, the greatest elaboration of the nautiloid suture is seen in members of the Clydonautilacea where a (morphological) sutural series can be drawn up showing increasing complexity from the middle to upper Triassic genus *Styrionautilus*; through the upper Triassic genera *Proclydonautilus* and *Clydonautilus*; finally to *Siberionautilus*, also from the upper Trias (Kummel *in* Moore, 1964). The suture of *Siberionautilus* (Fig. 1a, herein) shows rather numerous saddles and lobes in the ventral region. To these records must now be added *Yakutionautilus kavalerovae* (Fig. 1b) described by Arkhipov and Barskov (1970) from the middle course of the Nel'tekhe River, north-eastern USSR. Here, uniquely within the nautiloids, some of the sutural elements show frilling and, uniquely within both the ammonoids and nautiloids, this is developed in some of the saddles rather than in the sutural lobes.

In attempting now to circumscribe the nautiloids as a taxon within a classificatory scheme (Fig. 2) which is both useful and seemingly phylogenetically sound, I follow the Soviet *Osnovy* (Ruzhentsev, 1962) in employing two Subclasses; the Ectocochlia and Endocochlia, within the Class Cephalopoda; but recognize that the name Coleoidea is likely to be maintained as an alternative to the latter. This scheme allows for the clear distinction between *Nautilus* and the remaining living cephalopods. Within the Ectochlia three Superorders can then be taken: the Nautiloidea, Bactritoidea, and Ammonoidea, but I depart from *Osnovy* in not separating the Endoceratoidea and Actinoceratoidea from the Nautiloidea. A similarly relatively high-ranking subdivision was made in the "Treatise on Invertebrate Paleontology" (Teichert and Moore *in* Moore, 1964), though here the three taxa were placed as subclasses alongside the Bactritoidea, Ammonoidea, and Coleoidea within the Class Cephalopoda. Such an arrangement does little to emphasize the diversity of the nautiloids compared with the ammonoids and I have previously followed Flower (Holland, 1967; Flower, 1964) in rejecting it. This view is strengthened by Collins's (1976) description of an endoceratoid (probably Arenig in age) from Turkey. It has thick connecting rings, suggesting that it is an early member of the group and closer therefore to the ellesmeroceratids, but has some internal structures more commonly found in the actinoceratids.

The siphuncular segments are inflated and the septal necks more curved than those of the other endoceratids.

At a lower level in taxomony, there is wide acceptance of the main nautiloid orders: Ellesmeroceratida, Discosorida, Michelinoceratida

FIG. 1. Nautiloid sutures of unusual elaboration: (A) *Siberionautilus* (after Kummel *in* Moore, 1964, Fig. 286A); (B) *Yakutionautilus* (modified from Arkhipov and Barkskov, 1970).

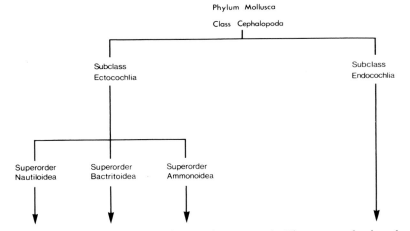

FIG. 2. Classification of the Cephalopoda (see text). The arrows lead to lower taxa.

Ascoceritida, Oncoceratida, Tarphyceratida, Barrandeoceratida, and Nautilida; and to these would here be added the Endoceratida and Actinoceratida, whose taxonomic status has already been mentioned. The commonly recognized groups of ammonoids referred to in the "Treatise" (Arkell *et al. in* Moore, 1957) as suborders would here fall into place as orders. Finally, there is no need for taxa between the Superorder Bactritoidea and its two families, the Bactritidae and Parabactritidae (Erben *in* Moore, 1964).

Recognition of the Bactritoidea as a Superorder follows easily from acceptance that these forms fall not only between the nautiloids and the coleoids, but also between the nautiloids and the ammonoids, this following the Erben's elegant and now frequently quoted demonstration of a sequence of forms leading to the latter.

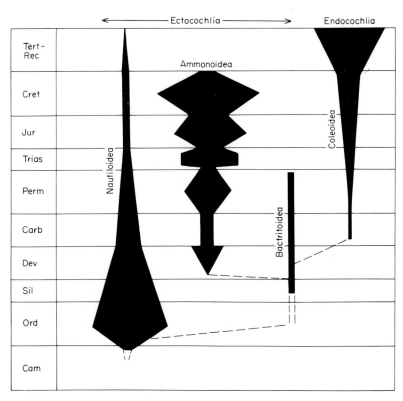

FIG. 3. Phylogeny of the cephalopods.

An acceptable phylogeny is thus that of Fig. 3, and the question of the early cephalopods is seen to be that of the early nautiloids. The cephalopods as a whole are seen to be a compact group of molluscs, within which it is not surprising to find that decisions on taxonomy are apt to depend upon niceties.

THE FIRST CEPHALOPODS: CAMBRIAN NAUTILOIDS

The first nautiloids presently known are from the upper Cambrian. Yochelson *et al.* (1973) have summarized the record of these forms, previously reviewed by Flower (1954, 1964) and Teichert (1967). The Cambrian nautiloids as a whole are small shells (the earliest only about 10 mm in length), varying in shape from endogastrically curved to straight, with one, seemingly anomalous, exogastric form. They are largely assigned to the Suborder Plectronoceratina of the Ellesmeroceratida. The first known are probably still two species from China: *Plectronoceras cambria* (Walcott 1905) and *P. liaotungense* Kobayashi 1935, respectively from the regions of Shantung (Shandong) and Manchuria. Associated trilobites have indicated a late Franconian or early Trempealeauan age, that is approximately latest Merioneth Series in British terms (Cowie *et al.*, 1972). They are endogastric shells with short, straight septal necks and have yielded one example of a curved connecting ring outlining a siphuncular bulb. Younger forms have suggested that this last feature is characteristic of the Plectronoceratina.

The straight form, *Palaeoceras mutabile* Flower 1954 |and a similar species, from high in the Trempealeauan of the Llano uplift in Texas, demonstrate lengthening of septal necks and suppression of siphuncular bulbs within individual shells. Here also is a *Plectronoceras* longer and more slender than the Chinese forms. *Balkoceras* is a slender, faintly exogastric shell. This small collection from Texas also includes the genus *Ectenolites* showing well preserved diaphragms across the tubular siphuncle (Flower, 1964). This provides the first record of the Ellesmeroceratina in the upper Cambrian, this suborder developing much more fully in the Ordovician. The Plectronoceratina lasted only until the earliest Ordovician (Tremadoc).

Although some other records of supposedly Cambrian nautiloids have been shown to be Ordovician in age, there remains the question of the genus *Olenecoceras* described by Balashov (1966) from the middle Cambrian of the Olenëk river basin in Siberia, where two fragmentary specimens were found associated with abundant agnostid trilobites. The larger is only some 6·5 mm in length. The straight shell is described as septate and with a siphuncular opening almost marginally situated and occupying 0·4 of the diameter of the shell. Septal necks and connecting rings are not known. A thin section taken longitudinally showed no structure of the siphuncle and its wall. The systematic position of the shell is said to be not quite clear and indeed the illustrations suggest that the material may be inadequate for diagnosis as a cephalopod.

The Volborthellidae and Salterellidae have been much debated as ancestral cephalopods and possibly nautiloids. Yochelson (1977) has now finally disposed of these in his "proposed extinct phylum of early Cambrian age", the Agmata. His provocative but decisive paper is beautifully illustrated with photographs of *Salterella* showing the unique morphology of a radially symmetrical, conical calcareous shell, partially filled with inclined laminae constructed of fragments and detrital grains. These shells are not septate. They lack not only cephalopod but even molluscan characteristics. Yochelson *et al.* (1977) provide similarly inform-ative illustrations of the internal structure of *Volborthella* in material from Norway.

ORIGIN OF THE CAMBRIAN NAUTILOIDS

The supposed ancestry of the nautiloids in a cap-shaped shell which could than be related back to the supposed molluscan archetype is, to say the least of it, disturbed by findings that the first Cambrian nautiloids known are not cap-shaped. Fortunately one of those rare finds of "missing links" in unlikely parts of the world has allowed Yochelson *et al.* (1973) to suggest a plausible alternative. It involves a monoplacophoran, *Knightoconus antarcticus* from the upper Cambrian, probably early Franconian, of western Antarctica. Yochelson and his fellow authors note that the late Cambrian to early Ordovician Monoplacophora are now known to be a diverse group, particularly in shell profile. *Knightoconus* is a high, curved conical shell, remarkable in having several septa across its apical part. In derivation of a first nautiloid from here it is inferred that at the time of septation the animal in question was unable to secrete calcium carbonate at the tissue strand still connecting the soft body to the apex of the shell, thus a siphuncle came into being. Some gastropods are also known to be septate. So the siphuncle must stand alone as a unique attribute and crucial to the diagnosis of the cephalopod shell. Denton (1974) reminds us that as long ago as 1696 Robert Hooke had sensed the importance of what he called "this admirable structure".

Yochelson *et al.* (1973) provide two reconstructions which are juxtaposed here in Fig. 4 in slightly modified form. *Knightoconus* would fit well between them, but detailed derivation of the cephalopod soft body from that of the monoplacophoran is a different matter. As put by Yochelson *et al.* (1973), "*Nautilus* and *Neopilina* are best viewed as somewhat distorted mirrors reflecting the past in a not completely reliable manner". These authors give detailed discussion of mode of life; but in the present context a matter of particular interest is the reason that the first nautiloids were endogastric rather than exogastric in their curvature. Monoplacophora of varied profiles existed and exogastric curvature was necessary for the eventual evolution of the highly successful and very common, more seaworthy, coiled cephalopods which were to arise after the Cambrian.

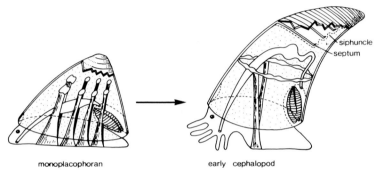

monoplacophoran   early cephalopod

Fig. 4. Reconstruction of monoplacophoran and early cephalopod (the former after Yochelson *et al.*, 1973, Fig. 5, the latter modified slightly from their Fig. 3).

It seems that forward or backward, rather than sideways, curving over of the shell was obviously to be expected. The backward version shown in Fig. 4 would not only be what intuitively seems the more "comfortable" in movement, but, facing into a current or moving against the water, was less likely to lead to upsetting of the animal from the sea bottom. Only limited lengthening with accompanying increased curvature would, however, be possible in this direction and thus perhaps straightened longer shells would arise. As these grew yet longer the familiar problems of nautiloid buoyancy and hydrostatics would appear and hence the varied solutions for bringing the long, straight shell back into a horizontal position by weighting devices in the camerae or siphuncle, or both, which resulted in the splendid evolutionary panorama of the straight-shelled nautiloids of the Palaeozoic. However, the present symposium has served especially to| emphasize that state of variety combined with relative| rarity, of plasticity, of "evolutionary experimentation", which characterizes the early history of various animal groups. So there were also exogastric shells amongst the early nautiloids. An isolated Cambrian record has been mentioned. There were certainly exogastric shells within the early Ordovician Ellesmeroceratida. As the need for greater buoyancy and mobility increased in more populated seas, some of these exogastric nautiloids gave rise to the coiled forms which eventually evolved into the modern *Nautilus*, but with very many fossil genera on that line and in divergence from it. Cowen (1973) notes that "The Cambrian was dominated by deposit-feeders, with only a few suspension-feeders low in the water column. The explosion of suspension-feeders of all types, and the roaming scavengers and carnivores dependent on them, only took place in the earliest Ordovician. So the Ordovician cephalopod radiation occurred at the first possible point in time after structural innovation *and* appropriate ecological environment had been evolved". Yochelson *et al.* (1973) refer to attainment of the "trick" of buoyancy control opening a new adaptive zone.

ORDOVICIAN CEPHALOPODS

The earliest Ordovician (Tremadoc) nautiloid faunas were dominated by the Ellesmeroceratidae, a rather uniform group with their primitively thick, layered connecting rings. They are not frequently found to be numerous as individuals, but Flower (1976) referred to over 230 species and certainly they are widespread about the world and represent the first expansion of the cephalopods. In the Arenig they were overtaken by representatives of the Tarphyceratida and Endoceratida and Flower (1976) now detects North American and Baltic–Central China nautiloid realms. Apart from the Discosorida, which, with their thinner connecting rings and expanded siphuncular bulbs, are believed to have originated directly from the Plectronoceratina, the Ellesmeroceratidae are believed to have given rise to all the other ordinal strands in Palaeozoic nautiloid evolution. The great radiation of the nautiloids was completed in the remainder of the Ordovician (Fig. 5). Nautiloids remained numerous in the remainder of the Palaeozoic but were declining in relative importance. Modest new expansions were later to affect the one order which was to outlast the Mesozoic and Tertiary; but in the end this is represented by a single genus *Nautilus*. An authoritative and exhaustive review of Ordovician cephalopods has recently been published by Flower (1976) and it would be absurd to summarize this in an extensive way here. Figure 5 illustrates the essentials of the radiation within which the main lines of Palaeozoic nautiloid evolution result from different solutions to the problems of buoyancy and mobility. The Michelinoceratida are the great generalized order of orthocones in which the weighting device is in the form of cameral deposits and there are tubular siphuncles. These are the commonest cephalopods of the Palaeozoic. The Actinoceratida and Endoceratida have

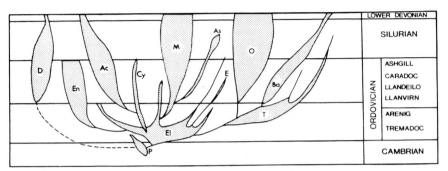

FIG. 5. The Lower Palaeozoic (and especially Ordovician) radiation of the cephalopods (from part of Flower, 1976, Fig. 1): P–Plectronoceratina; El–Ellesmeroceratina; Cy–Cyrtocerinina; D–Discosorida; En–Endoceratida; Ac–Actinoceratida; M–Michelinoceratida; As–Ascoceratida; E–Ecdyceratida; O–Oncoceratida; T–Tarphyceratida; Ba–Barrandeoceratida. The Nautilida (not shown) are thought to have arisen from the Oncoceratida in the early Devonian. The horizontal line a little above the Arenig–Llanvirn boundary represents the top of the Canadian, the lowest series of the Ordovician in North American usage and treated by Flower as separate from the Cambrian and Ordovician.

deposits also in the siphuncle, the former characterized by the familiar "beaded" siphuncle with its greatly expanded segments, the latter by the nature of the endocones in the ventrally situated siphuncle. The Discosorida with their annulosiphonate deposits, Oncoceratida with actinosiphonate deposits, and Ascoceratida with their device of shedding earlier chambers of the shell represent other solutions to the problem. The other main orders involve coiled shells and thus exploited what was to prove to be a better solution in which the buoyant shell lay largely above the soft body, but in a compact and reasonably streamlined form.

Rousseau Flower has a unique knowledge of Palaeozoic nautiloids and students of the group remain always in his debt. His style and immense grasp of detail may be illustrated by one quotation from his review of the Ordovician faunas (Flower, 1976):

The evidence [of evolution] is incomplete; cephalopods are large and hard to extract and they are usually bypassed by casual collectors. I have been able to collect from numerous exposures but many vital occurrences have eluded me, notably in the southern Appalachians. One vital Appalachian fauna is known to me only from polished stone facing the windows and doorways in the U.S. National Museum of Art; it is a Chazy fauna from Tennessee. Another important fauna is known only from polished slabs in the old railway terminal at Cincinnati...

Returning to the Ordovician seas: *Nautilus* always seems faintly improbable as a swimming animal, but how much more so were some of the huge Ordovician orthocones. Teichert and Kummel (1960), who reviewed records of the largest cephalopods, noted that in the 1890s three to five metre shells had been found in the middle Ordovician of Minnesota. There were records of five metres from Estonia and a three metre "fragment" at Harvard gives, by graphical resolution, an original length of 5·8 metres. When an estimated addition for the body chamber is made the size reaches over eight metres. Adding tentacles as well we begin to approach the size of some of the modern giant squids.

PALAEOZOIC NAUTILOID POPULATIONS

In upper Cambrian rocks, of which there is accumulated knowledge of many years in many parts of the world, it really does appear that there is an extreme rarity of cephalopods. Large populations are not of course to be expected in the early tentative phases of evolution and preliminary radiation. In the middle and upper Ordovician cephalopod shells become much more common. It is true that many genera are still known from only a few specimens but, for instance, Balashov's (1968) monograph of endoceratids from the USSR was based upon a personal collection of some 10000 specimens. From this level onwards also some of the well known Palaeozoic orthocone limestones of the world give a vivid impression that, in some places at least, the cephalopods were very numerous indeed. Concentrations of shells of modern *Nautilus* are found in the southwest Pacific and are said to be related to mass-mortality after mating. Possibly a similar explanation may hold in some of the Palaeozoic cases. Such concentrations of fossil shells are known, for instance, from the Ordovician of Scandinavia, the Silurian of Czechoslovakia, the Devonian of Western Australia and Morocco.

But to judge population size from individual finds is dangerous. Denton (1976) postulates from the known diet of sperm whales that the total mass of ammoniacal squid (forms which perhaps can be seen as showing the greatest refinement of all in buoyancy control) "is probably comparable to that of all the humans in the world taken together. Yet [he continues] these squid are animals which might from the catches in nets be thought relatively rare".

REFERENCES

ARKHIPOV, YU. V. and BARSKOV, I. S. (1970). Nautiloids with an intricately dissected suture line. *Dokl. Acad. Nauk SSSR* **195**, 464–466 (in Russian).

BALASHOV, Z. G. (1966). First discoveries of ellesmeroceratid cephalopods in the Middle Cambrian of the Olenëk River Basin. *Vop. Paleont.* **5**, 35–37, 110–111 (in Russian).

BALASHOV, Z. G. (1968). "Ordovician Endoceratids of the USSR". Leningrad University Press, pp. 278 (in Russian).

COLLINS, D. H. (1976). Origin of the Actinocerida (Cephalopoda). *25th Int. Geol. Congr. Sydney, Australia, Abst.* 290–291.

COWEN, R. (1973). Explosive radiations and early cephalopods. *Geol. Soc. Am., Abstr.* **5**, 585.

COWIE, J. W., RUSHTON, A. W. A. and STUBBLEFIELD, C. J. (1972). "A Correlation of Cambrian rocks in the British Isles." Special Report No. 2. Geol. Soc. Lond., 42 pp.

DENTON, E. J. (1974). On buoyancy and the lives of modern and fossil cephalopods. *Proc. R. Soc. Lond. B*, **185**, 273–299.

DONOVAN, D. T. (1977). Evolution of the Dibranchiate Cephalopoda. *Symp. zool. Soc. Lond.* **38**, 15–48.

FLOWER, R. H. (1954). Cambrian cephalopods. *Bull. New. Mex. St. Bur. Mines*, **40**, 1–51.

FLOWER, R. H. (1964). The Nautiloid Order Ellesmeroceratida (Cephalopoda). *Mem. Inst. Min. Technol. New Mex.* **12**, 234 pp.

FLOWER, R. H. (1976). Ordovician cephalopod faunas and their role in correlation. *In* "The Ordovician System": proceedings of a Palaeontological Association symposium, Birmingham, September 1974 (M. G. Bassett, ed.), pp. 523–552. University of Wales Press and National Museum of Wales, Cardiff.

HOLLAND, C. H. (1967). Mollusca: Cephalopoda (Nautiloidea). *In* "The Fossil Record" (W. B. Harland, C. H. Holland, M. R. House, N. F. Hughes, A. B. Reynolds, M. J. S. Rudwick, G. E. Satterthwaite, L. B. H. Tarlo and E. C. Willey, eds), pp. 431–443. Geol. Soc. Lond.

KOBAYASHI, T. (1935). On the phylogeny of the primitive nautiloids, with descriptions of *Plectronoceras liaotungense*, new species and *Iddingsia* (?) *shantungensis*, new species. *Jap. J. Geol. Georg.* **12**, 17–26.

MOORE, R. C. (ed.) (1957). "Treatise on Invertebrate Paleontology. Part L, Mollusca 4, Cephalopoda, Ammonoidea." Geol. Soc. Am. and Univ. Kansas Press.

MOORE, R. C. (ed.) (1964). "Treatise on Invertebrate Paleontology. Part K, Mollusca 3, Cephalopoda: General Features—Endoceratoidea—Actinocera-toidea—Nautiloidea—Bactritoidea." Geol. Soc. Am. and Univ. Kansas Press.

RUZHENTSEV, V. E. (ed.) (1962). *Osnovy paleontologii, Mollyuski: Golovongia,* 1. Akademii Nauk SSSR, Moscow (in Russian).

TEICHERT, C. (1967). Major features of cephalopod evolution. Essays in Paleont-ology and Stratigraphy, Raymond C. Moore Commemorative Volume, *University of Kansas, Department of Geology Special Publication,* 2, 162–210.

TEICHERT, C. and KUMMEL, B. (1960). Size of endoceroid cephalopods. *Breviora,* **128,** 1–7.

WALCOTT, C. D. (1905). Cambrian faunas of China. *Proc. U.S. National Museum,* **29,** 1–106.

WELLS, M. J. (1962). "Brain and Behaviour in Cephalopods." Heinemann, London.

YOCHELSON, E. L. (1977). Agmata, a proposed extinct phylum of early Cambrian age. *J. Paleont.* **51,** 437–454.

YOCHELSON, E. L., FLOWER, R. H. and WEBERS, G. F. (1963). The bearing of the new Late Cambrian monoplacophoran genus *Knightoconus* upon the origin of the Cephalopoda. *Lethaia,* **6,** 275–309.

YOCHELSON, E. L., HENNINGSMOEN, G. and GRIFFIN, W. L. (1977). The early Cambrian genus *Volborthella* in southern Norway. *Norsk geol. Tidsskr.* **57,** 133–151.

14 | On the origin of the Bivalvia

N. J. MORRIS

British Museum (Natural History), London, England

Abstract: The characters of early Ordovician bivalves are reviewed and compared with those of Rostroconchia and Monoplacophora. The nature of these organisms suggests that the Bivalvia share a calcareous single-shelled, monoplacophoran-like ancestor with the Rostroconchia and possibly the Scaphopoda.

This evolutionary pathway is favoured because most of the stages of evolutionary development are recognized in actual fossil organisms and a functional explanation can be given to each of the stages.

The evidence indicates that bivalves originated early in the Cambrian where relevant fossils are very scarce. The few early and mid-Cambrian bivalve-like molluscs are described and the way in which they fit into the evolutionary scheme is considered.

A modified picture is given of the early evolution and diversification of the bivalve hinge system; the common ancestor of the Ordovician bivalves probably had an opisthodetic parivincular ligament and few hinge teeth.

The lack of a radula in bivalves is interpreted as being a loss related to a change in feeding method from an ancestral epifaunal browsing or grazing type. On comparison of the two common modes of feeding in the Bivalvia, the evidence suggests that deposit feeding is just likely to be more primitive than filter feeding.

The possibility that the first Mollusca did not have larvae with a calcified shell is explored. If the veliger is a separate development for prolonged planktotrophic existence in separate molluscan classes and if the calcareous shells of the ancestor of the bivalves were actually produced after larval settling then the problem of the change from a univalved shell to a bivalved shell may not be one that affected the larva.

Systematics Association Special Volume No. 12, "The Origin of Major Invertebrate Groups", edited by M. R. House, 1979, pp. 381–413, Plate VI. Academic Press, London and New York.

INTRODUCTION

Bivalvia are a class of Mollusca which evidence suggests were primitively adapted to infaunal in soft sediment. Their life habit is facilitated by their hinge mechanism which consists of a ligament tending to hold the two calcareous valves open and countering a pair of adductor muscles

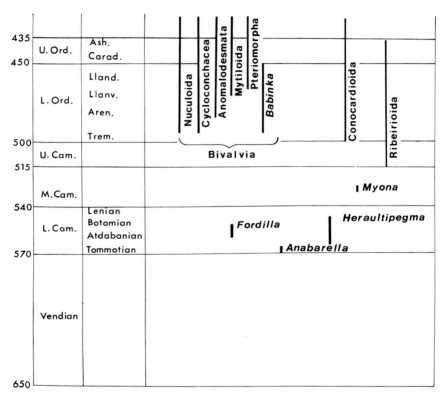

FIG. 1. Stratigraphical distribution of Cambrian and Ordovician Bivalvia and Rostroconchia (including data from Runnegar and Jell, 1976).

which contract to hold the valves closed. The ligament is often protected from shearing forces by a series of interlocking hinge teeth.

This layout has important advantages. In the burrowing process adduction pushes water out anteriorly from the mantle cavity to soften sediment before the projection of the foot (Trueman, 1966). In addition Bivalvia are able to change the water in the mantle cavity independent of ciliary current action (Morton, 1973).

Because this hinge system is a character of all Bivalvia, its evolutionary origins probably holds the key to the origin of the bivalves. Bivalvia have a substantial fossil record dating from the early Ordovician. A similar group, the Rostroconchia, which lacks the adductors and hinge apparatus is well represented from the upper Cambrian to the end of the Palaeozoic. Bivalvia are very poorly represented throughout the Cambrian, only one

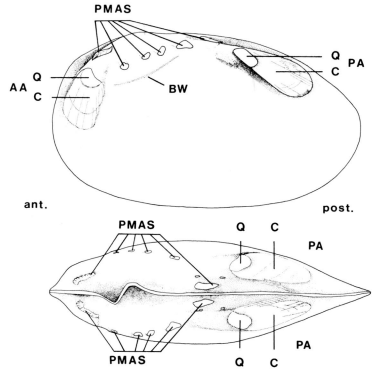

FIG. 2. Muscle attachment of *Synek* (? = *Coxiconcha*) *pellicoi* (Barrande and de Verneuil) natural internal mould: a. left lateral view, b. dorsal view, AA, anterior adductor scar, BW, body wall, C, catch muscle, PA, posterior adductor scar, PMAS, pedal muscle attachment scars, Q, quick muscle. L. Ordovician (?Llandeillo), Spain. BM(NH) L14682 (× 2·75).

named species, *Fordilla troyensis* Barrande 1881, being accepted with any certainty at present. Similarly, the Rostroconchia are represented by only a few species in the lower and middle Cambrian, and the systematic position of several of these is questionable.

This poor Cambrian record of Bivalvia and similar molluscs means that their early evolution has to be interpreted largely from our knowledge of Ordovician and later forms, with the Cambrian species being fitted into the picture as well as possible (Fig. 1).

The first well known bivalve faunas occur in the early Ordovician (Arenig) of western Europe. They are restricted to certain facies. They all appear to be infaunal and occur in inshore sands, muddy sands and muds and in offshore silts (Morris, 1979). Cycloconchacea predominate in the inshore sands whereas Nuculoida and other forms occur in addition to Cycloconchacea in the offshore facies. Bivalvia are largely replaced by Rostroconchia in the limestone facies of the early Ordovician. This distribution is taken to indicate that the primitive environment for these and all succeeding Bivalvia is among the infauna of soft sediments.

CHARACTERS OF THE EARLY BIVALVES

Early infaunal Bivalvia show many of the characters of present-day infaunal species. They are compressed, discoidal to blade-like. Many but not all lack ornament except for growth lines.

They had a simple ligament set behind the umbones (opisthodetic, parivincular) in simple furrows at the dorsal shell margins. All known pre-Llanvirn Ordovician Bivalvia have hinge teeth which, as in living species, functioned to guide the closing of the valves and prevented shearing of the ligament.

It is assumed from the known shell structure of living forms that the early bivalve shell was probably made of a complex organic matrix with calcium carbonate in the form of aragonite. Calcite is largely confined to epifaunal Bivalvia and the earliest species in which it has been recognised is the ribbed *Ambonychia radiata* (Hall, 1847) from the upper Ordovician of Scotland. The most primitive type of bivalve shell structure is thought to be an external layer of aragonite prisms with an internal layer of nacre and with a myostracum in the form of aragonite prisms (Taylor *et al.*, 1969, 1973).

The adductor muscles are of approximately equal size and are sub-circular, often truncated towards the umbones. In some genera, two distinct areas of adductor muscle attachment are visible within the same scar. These are assumed to represent separate "quick" and "catch" muscle. A pallial line is seldom obvious. The presumed byssate sub-infaunal nestler, *Modiolodon* (middle Ordovician) is the first known bivalve with a deeply inserted pallial line.

Set between the dorsal part of the adductor scars are a series of left and right pairs of body and foot attachment scars which either are well spaced and rounded in some genera (e.g. *Synek*, Fig. 2) or have a more sinuous linear arrangement (e.g. *Babinka*, Fig. 3). The arrangement of these muscles in lower and middle Ordovician Bivalvia seems to be not only of functional but also of taxonomic significance.

In *Babinka* (Fig. 3) a further line of muscle scars occurs below the main attachment scars. By analogy with recent Bivalvia this is the line where

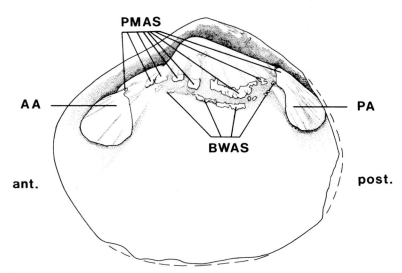

Fig. 3. Muscle attachment scars of *Babinka oelandensis* Soot-Ryen left lateral view showing cast of interior left valve, AA, anterior adductor scar, BWAS, body wall attachment scars, PA, posterior adductor scar. Ordovician (?U. Arenig) Oland, Sweden (after Soot-Ryen) (× 4).

the body wall meets the mantle. MacAlester (1964) has pointed to the possibility of a gill also being attached in this position. The join between the body wall and the mantle is probably represented in *Synek* and in a number of Cycloconchidae by an elongate low rounded ridge (Fig. 2) and in the living Nuculidae by an elongate shallow depression on the inside of the shell.

Early Nuculoida also had paired body attachment muscles but in the Praenuculidae, for example (Fig. 4), they were situated in a similar position to the accessory muscles of later Nuculidae. A downwardly convex arc of scars runs from the anterior pedal retractor scar to a scar high in the umbonal cavity, with a gap between this point and the posterior pedal retractor. In the living Nuculidae the accessory muscles occur within or just above the body wall and attach it to the shell (Heath, 1937). The body wall forms a continuous muscular sheath downwards around the foot. The gills are attached to the body wall immediately below and behind the umbonal scars close to the heart and it appears that the same arrangement

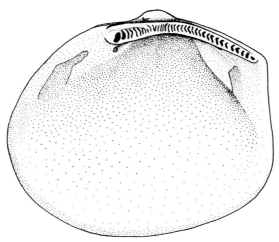

Fig. 4. *Praeleda* cf. *compar* (Barrande), interior of right valve with hinge and muscle attachment scars. L. Ordovician (U. Llanvirn) Shelve, England. BM(NH)LL31370 (× 4).

existed in the Praenuculidae. For this reason, it is assumed that the mantle cavity of the Praenuculidae was arranged in the same way as in the living Nuculidae and this in turn supports the view that early Nuculoida were deposit feeders.

Babinka and *Synek* have the shape of some later, deeper-burrowing filter feeders and may therefore belong to that trophic group.

1. The Similarities between Bivalvia and Monoplacophora

The existence of untorted univalved snails was recognized by Wenz (1943) from their fossil remains. But the eventual report of the discovery of a living representative, *Neopilina*, by Lemche and Wingstrand (1958) stimulated all those interested in the nature and relationships of the Mollusca to re-examine existing views.

Lemche and Wingstrand (1958) pointed out the great similarity between

Neopilina (Fig. 5) and the Silurian genera *Pilina* and *Tryblidium*. Like *Neopilina* these had the apex of their shell approximately above the head. Behind this, the internal surface of the shell showed a series of subequal sized pairs of body attachment scars and a number of smaller associated scars. *Neopilina* has, in addition, associated with each pair of scars, paired

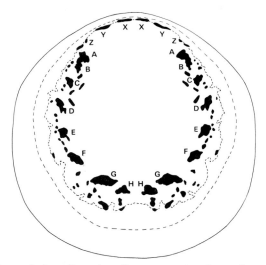

FIG. 5. *Neopilina galatheae* Lemche, diagrammatic view of muscle scars reconstructed from dorsal mantle surface, Recent (after Lemche and Wingstrand 1957) (×3.5).

nephridia, a system of nerves controlling the paired musculature and paired gills, but just two pairs of gonads. The lateral mantle cavity is very restricted. There is an immediate overall similarity to the chitons but repetition of organs in that class is limited to the shell plates and musculature. In chitons, the gills can be very numerous and are not apparently related to individual plates. They are reminiscent of the mantle gills of Patellacea.

Both Cox (1959) and MacAlester (1964) drew attention to the similarity between the multiple paired pedal attachment musculature of some early Ordovician Bivalvia and the musculature of *Neopilina*. MacAlester considered that the similarity between Monoplacophora and *Babinka*, which he recognized as the earliest lucinoid bivalve, was so great that he was led to suggest a dual origin of the Bivalvia. It is now recognized that many early Bivalvia had musculature of this type, with the result that we can accept a degree of relationship with the Monoplacophora but do not need to postulate the polyphyletic origin of the Bivalvia.

2. The Comparison of Ribeirioida and Conocardioida (Rostroconchia) With Typical Early Ordovician Bivalves

Many Ribeirioida had a similar shell shape in lateral aspect to infaunal deposit feeding Bivalvia (i.e. Nuculoida) but they lacked a hinge and any form of adductor muscles. In *Ribeiria* (Plate VI, 1) and *Technophorus* the body attachment musculature was disposed on the shell surface similarly to the body attachment musculature of typical early Ordovician infaunal Bivalvia, which have both filter feeding and deposit feeding descendants. In most species, including *Ribeiria pholadiformis* Sharpe, 1853 (Plate I, 1b), the muscles were attached to a ring-like scar although a few species of *Ribeiria* have three pairs of discrete scars and a posterior fused pair straddling the dorsum. Because separate multiple-paired muscle scars are a primitive feature for the post-Cambrian Bivalvia and occur in living and many fossil Monoplacophora, it is assumed that this feature is also primitive in the Rostroconchia.

Many Conocardioida have a crenulate margin formed by ribs on the inside of the outer shell layer which makes them easily recognizable when their shell is only partly preserved. When the steinkern is well preserved, scars of the body attachment musculature are visible (Fig. 6, Plate VI, 2). Most prominent are a posterior dorsal pair of scars in a position closely comparable with the posterior pedal retractors of infaunal bivalves; they may therefore reasonably be assumed to have the same function. Smaller muscle insertions occur anterior to and slightly below these and are interpreted as further body attachment scars.

Conocardium aliforme (Sowerby, 1815), in common with several other species, has a linear attachment scar in the same position as the bivalve pallial line (Plate VI, 3b). It turns in a dorsal direction and splits into two, half way up the flank, to form a "Y" shaped scar about half way between the umbones and the anterior margin. This is in a comparable position to body muscles that firmly fix the anterior part of the body to the shell of bivalves such as Nuculanacea and Solemyoida. These prevent adverse movement of the body cavity containing the gut when the foot is moved. Anterior to this "Y" shaped scar, several species have elongate plate-like ridges on the inner shell surface, radiating towards the anterior margin from a little in front of the umbones. These are interpreted by Pojeta and

Runnegar (1976) as supporting infoldings of the mantle that may be associated with their feeding mechanism. This agrees with my interpretation of the "Y" shaped scars, as the mantle intuckings would lead towards the mouth. A further pallial attachment line occurs concurrent with the anterior aperture a little way in from the anterior margin.

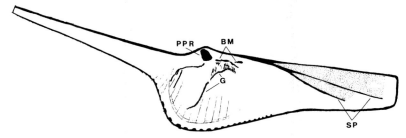

Fig. 6. *Conocardium longipennis* (Brown), composite reconstruction of interior of right side. PPR, Posterior pedal retractor, BM, Body muscle attachment, G, Possible position of gill, SP, Internal shell plates. L. Carboniferous (Viséan) Northumberland, England (× 3·5).

Conocardioida resemble the Bivalvia in more respects than do the Ribeirioida. In particular, they share the character of two calcareous halves of the shell joined along the dorsal margin (in the case, the entire dorsal margin) by what is assumed to be a more organic structure than the rest of the shell which has some flexibility. The details of this hinge structure are only clearly seen in the upper Carboniferous species, *Pseudoconocardium lanterna* (Branson, 1965) (Plate VI, 7). It is interpreted as having a high organic content as its preservation differs from the remaining structure of the shell; it shows clear evidence of bending during growth and is preserved in a similar manner to other bivalve ligaments. The similarity of this hinge structure to the bivalve ligament suggests that the two may be homologues. However, the fact that throughout the entire history of Conocardiacea the conocardiod "ligament" is nearly always preserved, whereas during that same time the bivalve ligament is preserved in only a few instances, suggests that the two may not have exactly the same composition.

Conocardioida have body attachment muscles comparable to those in the Bivalvia and Monoplacophora. Nevertheless in excellently preserved specimens, adductor muscle scars are never seen. The "ligament" in Conocardioida was constructed to work in an entirely different way to that in Bivalvia, possibly because of their convex cross-sectional shape.

Their shell opened ventrally as growth increments were added. During this growth the inner layers of the ligament were stretched while the outer layers were compressed, with the result that the whole of the ligament worked to keep the two halves of the shell firmly closed at their ventral point of contact: this is the exact opposite to its function in the Bivalvia.

Usually the hinge of the Conocardioida was linear through physical necessity, but this constraint was overcome in some species so that the posterior rostrum could be held at an angle to the hinge axis of the main part of the shell. *Pseudoconocardium lanterna* (Branson) developed shell cracks or slits behind the umbones which separate the two parts of the hinge (Pojeta *et al.*, 1972; Pojeta and Runnegar, 1976). These cracks may have been filled with periostracum as are the shell slits of some modern pandoracean bivalves, whose hinge axes are not linear. The filling of the slits is presumably analogous to the repair of damaged shell, where any part of the mantle is capable of laying down new shell tissue starting with periostracum (Beedham, 1965).

3. Bivalve Hinges

It is in the function of the hinge system that Bivalvia differ from the Conocardioida.

The ligament in Bivalvia has two separate layers below the periostracum, reflecting the two layered structure of the shell, although it is a matter of conjecture as to whether or not there is real homology. The outer organic elastic layer is stretched when the valves close, its elasticity storing energy to open the valves again, and the inner calcified layer is compressed when the valves close and is highly resilient to this stress. Both layers work together, but by opposite principles, to keep the valves open.

The ligament is delicately in balance both with the adductor muscles and also with the sediment in which the bivalve lives. Additional opening force can be applied by the foot pressing at the shell margins.

The simplest type of ligament (opisthodetic, parivincular) occurs in the Mytiloida, in the Malletiidae and many Heteroconchia. Apart from modification of the method of attachment to the shell, this is similar to the ligament of early Ordovician Bivalvia in shape, size and position. It is not known whether their ligament had a two layered structure but it does seem likely that all other forms of bivalve ligament have evolved from it. This evolutionary development can be traced in fossil material. The shape and position of this simple ligament may be considered primitive for the post-Cambrian Bivalvia.

The mechanical efficiency of a ligament seems to be directly related to the life style of the bivalve (Thomas, 1976). The simple ligament of early Ordovician Bivalvia must have been mechanically efficient and relatively powerful because it had to open the valves within the sediment in which

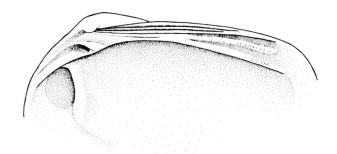

FIG. 7. *Ortonella hainesi* Miller, view of hinge area of right valve of a gerontic specimen showing repetition of the ligament groove. U. Ordovician (Richmond), Manitoulin Island, Lake Huron, Canada. BM(NH)LL31374 (× 4).

the bivalve lived. It is similar to the opisthodetic, parivincular type which fossil evidence shows to be the more primitive and not evolutionary the most advanced, as suggested by Trueman (1969).

The structure of the ligament and the method by which it joins the rest of the shell are clearly physical determinants of its efficiency. In the early Ordovician Bivalvia the structure of the ligament is unknown but its position can be determined. It is usually fixed in a narrow depression at the dorsal shell margins posterior to the umbones. Separate structures of the hinge such as nymphs or the peculiar cavernous structure of living Mytiloida had not developed at this early stage.

More complicated types of ligament, apparently with two differentiated layers, developed during the Ordovician in the Cyrtodontidae and the Ambonychiidae. A later Ordovician genus, *Ortonella*, has an area of ligament attachment intermediate between these two types (Fig. 7). With interumbonal growth, second and third ligament attachment grooves have developed in some large individuals. It is thought that the multistriate ligament area of the Cyrtodontidae and the Ambonychiidae may have developed by repetition of this process. The ligament type of these families is thought to have been ancestral to the Arcacean and the Pteriacean ligament.

Trueman's interpretation of ligament evolution can be slightly modified (Fig. 8). His type "A", exemplified by *Limopsis*, has in fact evolved by retardation of development of the Arcid type of ligament. The early

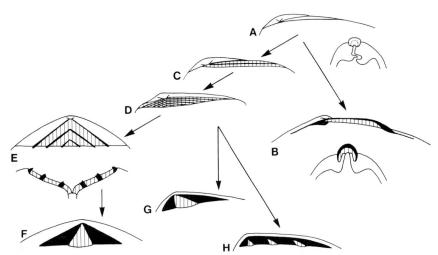

Fig. 8. Outline of the possible evolutionary relationships between some bivalve ligament types: A. Primitive parivincular opisthodetic, present in the Cycloconchacea, Praenuculidae, early Anomalodesmata and early Mytiloida. B. Advanced parivincular opisthodetic, present in many later Heteroconchia, a few later Anomalodesmata, Ctenodontacea, Lucinoida and later Mytiloida and in a modified version in the Solemyidae. C. Incipient duplivincular opisthodetic, present in *Ortonella*. D. Primitive duplivincular amphidetic, present in Ambonychiacea, Cyrtodontacea, Pteriniidae and Pterinopectinidae. E. Advanced duplivincular amphidetic, present in Arcacea. F. Secondarily parivincular amphidetic, present in Limopsacea. G. Parivincular amphidetic, present in Ostreacea, Pteriidae and most Pectinacea. H. Multivincular amphidetic or opisthodetic, present in the Isognomonidae, Bakevellidae and Inoceramidae.

ontogenetic stages of the multivincular and duplivincular types do, to some extent, reflect their evolutionary history, but the primitive type has returned in *Limopsis* and indeed in *Pteria*. Both these and the multi- and duplivincular types seem to have evolved from the type of ligament seen in the Cyrtodontacea and the Ambonychiidae and these in turn evolved from a primitive parivincular type.

These different types of ligaments have evolved in the epibyssate filter feeders. In the Limopsacea where there has been a partial reversion to a free and active life there has been also a reversion to the more primitive and simple type of ligament. The reason why the epifaunal Mytiloida have retained this parivincular ligament remains unexplained.

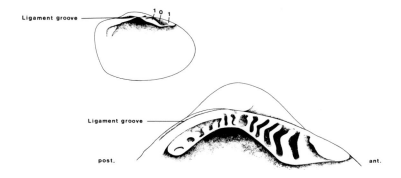

FIG. 9. Two stages in the ontogeny of the hinge of *Tironucula jugata* Morris and Fortey: a. young form with two anterior teeth present, left valve interior. SMA85185 (× 40). b. adult form with part actinodont and part taxodont teeth. SMA85182 (× 16). L. Ordovician (L. Llanvirn), Spitzbergen, Norway.

The dorsal margin of the bivalve shell, where it is joined to the ligament, is usually strengthened or modified in some way by a hinge plate. This often bears hinge teeth which interlock the valves across the dorsal margin. The hinge teeth both protect the ligament from shearing and ensure that the valves close in the correct position.

The arrangement of the hinge teeth has been used extensively in the classification of Bivalvia. This has been largely successful because it appears that different types of hinges and therefore tooth arrangements suit different basic life styles.

Although a variety of hinge teeth types are already present in the lower Ordovician Bivalvia, it is not immediately obvious which is the most primitive. However, when they are compared with the partially known ontogeny of the lower Ordovician nuculoid *Tironucula* from Spitzbergen (Morris and Fortey, 1976) it is apparent that all the different types can be derived from a common form. During development, *Tironucula* started without teeth, then one and two teeth developed on a hinge plate in front of the umbones (Fig. 9a). At this stage there was only a slight thickening below the ligament groove behind the umbones. Unfortunately intermediate growth stages are unknown. The adult has numerous teeth that are of taxodont type anterior to the umbones but actinodont to its posterior (Fig. 9b). The fact that the whole variety of lower Ordovician

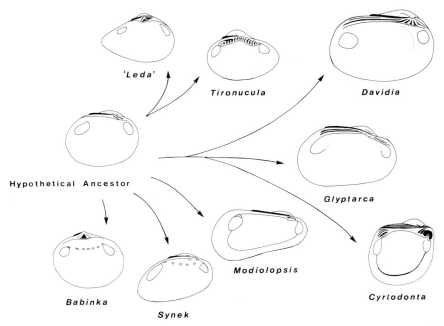

'Leda'

Tironucula   Davidia

Hypothetical Ancestor

Glyptarca

Modiolopsis

Babinka   Cyrtodonta

Synek

FIG. 10. Outline of the evolutionary relationships of early bivalve hinge teeth with a hypothetical common ancestor.

bivalve hinge types occurs within the ontogeny of a species demonstrates that they could all be derived by either the acceleration or retardation of tooth development in a common ancestor. Two facts point to evolution following the course of ontogeny in this case. First, the occurrence of an adult nuculoid "*Leda*" *languedocana* Thoral, 1935 in the earliest Arenig (L. Ord.) of southern France with three simple diverging teeth in the left valve, whereas virtually all later Nuculoida are at least partly taxodont (Fig. 15). Secondly, the ontogeny of the hinge in *Tironucula* goes from the simple condition with no teeth, building gradually to the complicated condition with many teeth in the adult. Without more knowledge it is better to assume that the likely course of evolution was from the simple to the complicated. An evolutionary model with a hypothetical ancestor with few teeth is therefore proposed (Fig. 10).

4. Bivalve Adductors

The adductor musculature counters the shell-opening property of the bivalve hinge. Adductor muscles are present in all bivalves but absent in Ribeirioida, Conocardioida and Scaphopoda. The superiority of the digging process of the Bivalvia over that of the Scaphopoda has been demonstrated by Trueman (1968). The further ability of adduction in their digging cycle enables Bivalvia to soften the sediment before insertion of the foot into the sediment. The bivalve digging process would also have been superior to that of both Conocardioida and Ribeirioida both of which lacked not only a hinge capable of opening the shell but also adductor muscles.

Yonge (1953) has suggested that bivalve adductors originated by the hypertrophy of the circular pallial musculature between the valves at both ends of the animal, indeed the adductors invariably lie at either end of the pallial line. In adult bivalves, however, the anus lies posterior and dorsal to the posterior adductor, so it seems more likely instead that the adductors developed mainly from transverse or circular muscles in the body wall, the most likely being those particular muscles that were most intimately associated with the posterior and anterior pedal retractor muscles.

5. Synthesis of the Post-Cambrian Evidence

We can now build a picture of the essentially primitive early Ordovician bivalve, which, along with their most similar contemporaries, the Conocardioida and Ribeirioida, may be compared with other Mollusca. They have been united with the Scaphopoda as the subphylum Diasoma by Runnegar and Pojeta (1974). They do not have the reduced body attachment musculature of the cyclomyan Monoplacophora and do not show the loss of bilateral symmetry typical of Gastropoda. They lack the float chambers of the primitive Cephalopoda and also the eight separate shell plates of the Polyplacophora (chitons). They do share bilateral symmetry and body attachment musculature with the tergomyan Monoplacophora and the Polyplacophora.

The Conocardioida and Bivalvia share common shell structure patterns with all other shelled Mollusca except the chitons. Unfortunately the detailed shell structure of Ribeirioida is unknown. All Mollusca possess an external dorsal integument or periostracum at some growth stage. Bivalvia are the only surviving class lacking a radula.

The Ribeirioida and the Tergomya share a single aragonitic bilaterally symmetrical shell while the Conocardioida and the Bivalvia share a similar shell which is, however, divided dorsally by a less calcareous, i.e. basically organic, structure, the ligament in the latter.

We can, therefore, arrange a group of characters, in order of their evolutionarily advanced nature, found in early molluscs that led towards the Bivalvia:

1. The dorsal integument.
2. The calcareous shell with certain varieties of shell structure.
3. Lateral compression of the shell.
4. The flexible organic dorsum of hinge.
5. The bivalve hinge system with adductor musculature.

(1) is a stage not seen in fossils, According to Salvini-Plawen (1972) the Solenogastres and Caudofoveata retain this primitive condition, although the view that these may be secondarily reduced from the Polyplacophora is held by some workers. (2) is the state of the tergomyan Monoplacophora which appear to be a stem group in Hennig's sense. (3) is the condition found in the Ribeirioida. (4) is the condition found in the Conocardioida. (5) is the primitive condition in the post–Cambrian bivalves. It has not been possible to recognize a theoretical intermediate stage where adductors are present to close the valves but where the ligament is present only for its flexibility and not to provide a positive opening force.

It remains to consider the Scaphopoda which are the only other class of primitively infaunal Mollusca that bear some resemblance to the Bivalvia. These have a shell which fuses ventrally on metamorphosis and grows mainly in an anterior direction to produce a curved tube. A significant and distinguishing difference is that Scaphopoda have a radula, which in some species is used in the trituration of the shells of foraminifers.

Various classifications have placed the Scaphopoda near the Bivalvia but Runnegar and Pojeta (1974) have suggested that they are derived from the Rostroconchia and in particular the elongate ribeiriid, *Pinnocaris*. While it is considered that ancestry from within the Rostroconchia is plausible, *Pinnocaris* does not seem to be the most likely candidate because all the most elongated part of its shell, which is suggestive of the scaphopod shape, is to the posterior, whereas the maximum shell growth rate in the scaphopoda is towards the anterior. Anterior growth is more noticeable in the Conocardioida. A conocardioid such as *Pseudotechnophorus* would resemble a possible scaphopod ancestor if it were to have undergone ventral mantle fusion. The first certain scaphopods occur in the Devonian although shells with as scaphopod shape attributed to *Plagioglypta*, are recorded from the upper Ordovician. More clearly recognizable early scaphopods will have to be found before there is any certainty as to their origin.

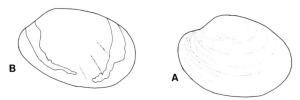

FIG. 11. *Fordilla troyensis* Barrande, view from the left side. A. exterior, B. natural internal cast. L. Cambrian (Schodak), Troy, New York, U.S.A. (after Pojeta, 1975) (× 9).

EVIDENCE FROM THE CAMBRIAN

How then does the Cambrian fossil record fit with the relationships that are suggested by the post-Cambrian forms? There are only a few laterally compressed, bivalve-like Mollusca at present known from the Cambrian. However, one well described species is known with a proper claim to acceptance as a bivalve, *Fordilla troyensis* Barrande from the early Cambrian of New York State. *Fordilla* is a laterally compressed equivalved bivalved shell with a series of muscle scars, the largest of which are interpreted as adductors (Pojeta *et al.*, 1973) (Fig. 11). It is claimed that the umbones are towards the anterior and that the ligament is posterior to them. There are growth lamellae on the shell surface which are similar to those of Mollusca. All these characters are typical of the Bivalvia. Pojeta and Runnegar (1976) are uncertain of the interpretation of the smaller muscles which may have been an early form of the pallial line. The muscle scars are not exactly those to be expected from an analysis of the post-Cambrian bivalves. Yochelson has suggested (personal communication) that although *Fordilla* is a probable mollusc it may not be a true bivalve. Some classifications of early ostracode Crustacea have considered *Fordilla* to be a common ancestor of a large group of palaeozoic ostracodes and the Conchostraca (Adamczak, 1961) but this hypothesis seems to have been superseded and does not seem to be correct.

?*Myona queenslandica* Runnegar and Jell, 1976 is a laterally compressed bivalved shell with a straight and apparently flexible hinge (Fig. 12). It is preserved as an internal mould. A little to the presumed anterior of the mid-point of the dorsal line there is a raised area which Runnegar and Jell (1976) have interpreted as a protoconch. The dorsal shell margins curve around this structure which is about the same size as the initial shells of

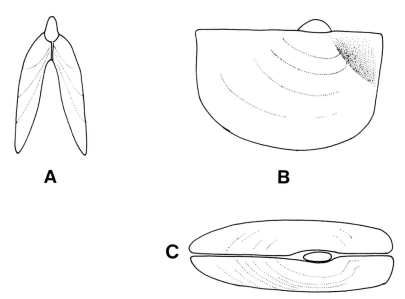

A   **B**

C

FIG. 12. ?*Myona queenslandica* Runnegar and Jell, phosphatic internal mould with structure interpreted as initial shell preserved: A. view of presumed anterior, B. view of presumed left side, C. view of dorsum. M. Cambrian (post-Templetonian), Queensland, Australia (after Runnegar and Jell, 1976) (× 50).

other Cambrian molluscs described by Runnegar and Jell (1976). Although the preservation of this structure is virtually featureless it does appear to be an undivided early shell stage. Although the inner mould of the adult shell is finely preserved, no adductor scars are visible. These feature, together with a general similarity of shape to the conocardioid family Eopteriidae, tend to support Runnegar and Jell's view that *Myona* is a rostroconch.

Heraultipegma originally described as a crustacean was recognized as the earliest ribeirioid mollusc by Morris (1967). It resembles a small and rather simple form of *Ribeiria* (Fig. 13). The specimens are internal moulds showing only growth rugae. More recently (Müller, 1975), some doubt has been case on the interpretation of *Heraultipegma* as a mollusc because of microscopic hexagonal patterns preserved within the shell structure, which are alleged to resemble the structure of the carapace of certain arthropods. However these structures may indicate the organic matrix of typical molluscan aragonite prisms such as are occasionally preserved in

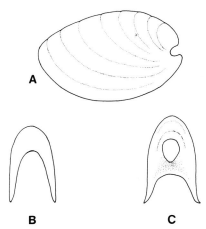

FIG. 13. *Heraultipegma varensalensis* (Cobbold), A. view of presumed right side, B. view of presumed posterior, C. view of presumed anterior. L. Cambrian, Hérault, France (after Pojeta and Runnegar, 1976) (× 23).

mesozoic Trigoniidae. *Heraultipegma* is exactly the form to be expected of an intermediate between a limpet-shaped monoplacophoran and the laterally compressed *Ribeiria*.

The order of occurrence of these few Cambrian Bivalvia and Rostroconchia does not agree with the order of appearance for the characters in the evolving bivalve hinge suggested above. This may be due to the incomplete fossil record, particularly in the Cambrian. That rich Cambrian molluscan faunas did exist is shown by Runnegar and Jell's (1976) account of the remarkable middle Cambrian of Queensland which has yielded many new species.

Myona, although younger than *Fordilla*, has a hinge structure intermediate between an early Ribeirioid such as *Heraultipegma* and the Bivalvia, even though the initial part of its shell was apparently univalved.

The early occurrence of first probable Rostroconchia and Bivalvia led Runnegar and Pojeta (1974) to suggest that the origin of the molluscan classes was in an adaptive radiation at about the Precambrian–Cambrian boundary, following the acquisition of the shell. Another aspect they have pointed out is the very small size of lower and middle Cambrian genera: this seems to have been a general feature of the earlier molluscs.

THE BIVALVE FOOT

The change from a monoplacophoran-like ancestor to the Bivalvia may represent part of an adaptive radiation in a group of animals that have recently acquired a shell. Although the living Monoplacophora are known not to be limited to a hard substrate, *Neopilina adenensis* (Tebble, 1967) was discovered clinging to rocks: this together with their limpet-like shape suggests that hard substrate dwelling may have been the primitive mode of life. As far as the Bivalvia are concerned, radiation may have been from a hard or at least firm substrate into soft substrates where complete enclosure became necessary to stop the fouling of the mantle cavity. It also would have entailed a change in locomotion. The primitive flat creeping foot is unlikely to have functioned well on unconsolidated surfaces and it could not easily have been enclosed by a shell. A more laterally compressed shell and foot would have been much more suitable for progression on and more particularly into soft sediment. The foot in the Bivalvia and Scaphopoda is an organ for digging rather than creeping. The pedal attachment scars of *Synek*, where the most anterior and posterior pairs are differently orientated and more deeply inserted than the intervening pairs, show the first recognizable signs of an advance (other than lateral compression) from the monoplacophoran-like foot, where the scars are of approximately equal size and insertion (Figs 2, 5).

FEEDING TYPES

Bivalves are the only major group of living molluscs that do not have a radula. If the assumption that the radula is a primitive character of Mollusca is correct, the Bivalvia lost the radula during their evolutionary history.

Since the radula in other molluscs is part of the feeding apparatus, its loss in Bivalvia presumably was related to a contemporaneous and profound change in feeding methods. So far the argument is simple, but in Bivalvia there are two major feeding types: deposit-feeding, the ingestion of organically rich silt and mud taken from the sediment by palp proboscides, and filter-feeding, with organic particles, usually small organisms, sieved from sea water by the gills. Both types have been postulated as the more primitive.

Yonge (1939) has maintained that deposit feeding is primitive in the Nuculoida and that the Nuculidae are the most primitive living Bivalvia. Furthermore he pointed out that deposit feeding, as opposed to filter-feeding, more closely resembles grazing which he considered to be the feeding type of the ancestral mollusc. However, Stasek (1963) pointed out that the nuculoid deposit-feeders do in fact obtain a considerable proportion of their food by filter feeding. Stanley furthermore (1975), following Pojeta and Runnegar (1974), suggested that the presumed earliest bivalve, *Fordilla*, was a filter-feeder on the grounds that its shape is typical of an infaunal filter-feeder. These points, together with the fact that filter-feeding is a more widespread character throughout the bivalves than deposit-feeding, suggest the opposite view: namely that filter-feeding is the more primitive.

Nuculoid infaunal deposit-feeders have to move fairly continuously through the sediment to obtain their food and have a distinctive variety of shell shapes which are adaptations of forward movement in fine soft sediment. They are rounded anteriorly and usually their greatest height is anterior to their mid-point. In many, but certainly not all, the umbones point to the posterior and are posterior to the mid-point. There is a strong muscular foot attached to the shell by a large number of muscles. The Malletiidae alone retain the primitive external ligament. The Nuculoida have developed a specialized larval form, the test-cell larva, in which the velar lobes have developed into a leathery cover for the developing shell, in the shape of a little barrel. This is unlikely to be a primitive bivalve character.

These deposit-feeding Nuculoida can be traced back to the earliest Ordovician both by their shape and the structure of their hinge. By the upper Ordovician, they seem to have reached a diversity of shape and form which has been maintained to the present day. However, the increase in diversity of filter-feeding bivalves since the middle Ordovician is such that the Nuculoida have been completely overtaken. Many filter-feeders are byssally fixed or cemented epibionts which apparently had not appeared at the beginning of the Ordovician.

The shapes of many Ordovician Nuculoida may be matched with those of today. In addition, however, there are orbicular shells such as *Cardiolaria* and *Inaequidens*, with anteriorly placed umbones, which cannot be

Pseudarca

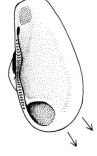

Ctenodonta

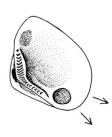

Tancrediopsis

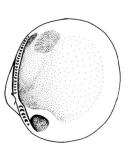

Cardiolaria

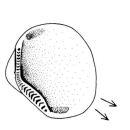

Deceptrix

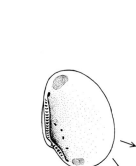

Myoplusia

Fig. 14. Shell shape of various Ordovician Nuculoida. Arrows indicate main probing direction.

compared with any living Nuculoida (Fig. 14). Their shape is more comparable, in the most general way, to some living infaunal filter-feeders. However, the arrangement of their body attachment muscle scars is clearly nuculoid and their hinge teeth are definitely of the taxodont type. It would appear that these are early Nuculoida that had not fully developed the specialized shell shape but nevertheless had typical nuculoid organization of the mantle cavity and musculature. They, therefore, presumably were of the same feeding type as living nuculoids.

 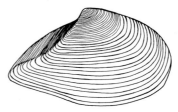

FIG. 15. "*Leda*" *languedocana* Thoral, interior view of left valve and exterior view of right valve, reconstructed from Internal and external moulds and latex casts. L. Ordovician (L. Arenig), Montagnes Noires, S. France. BM(NH)LL31371 (×6·5).

No doubt, as in the case with modern Bivalvia, the shell shape of early Ordovician bivalves was directly related to several functions including locomotion. However, it seems unlikely that trophic types can be distinguished by shell shape in every case.

The ontogeny and unusual hinge teeth pattern of *Tironucula* has already been used to support the contention that nuculoid deposit-feeders and other post-Cambrian bivalves had close common ancestry. Further evidence is provided by "*Leda*" *languedocana* Thoral (L. Arenig, S. France) (Fig. 15). Its shell shape and body attachment musculature clearly relate it to the Nuculoida, but instead of taxodont teeth, it had three small actinodont teeth set below the umbones. Its typical nuculoid shape suggests that it was a deposit-feeder in spite of its teeth.

Among those lower Ordovician genera interpreted as nuculid deposit-feeders we have *Cardiolaria* and *Inaequidens* which cannot be recognized as such by their shape and "*Leda*" *languedocana* which cannot be recognized as such by its hinge. If this atypical hinge and shape were to occur in combination in a bivalve at or before this time it would be impossible to identify its feeding type.

What happens when we approach the problem from another direction?

Is it possible to recognize early filter-feeders? If a bivalve is byssate throughout life it almost certainly cannot be a deposit-feeder and therefore has to be a filter-feeder. The first byssate families that may be recognized by their shell shape are the Cyrtodontidae, the Ambonychiidae and the Modiomorphidae.

On present evidence, they all occur later than non-byssate bivalves. Some early Ordovician actinodonts (e.g. *Glyptarca*) could either be ancestral or share close common ancestry with the Ambonychiidae and the Cyrtodontidae. The Modiomorphidae, particularly *Modiolopsis* and related genera, which either have only a very few simple teeth or none at all, must have arisen from some earlier ancestor of the actinodonts and Nuculoida, which had very few teeth.

Babinka has the shell shape of the Lucinoida and therefore we can postulate that it also had the typical sedentary infaunal filter-feeding habit. The actinodonts and possible ancestors of the Anomalodesmata such as *Synek*, however, have a similar shell shape to their filter-feeding descendants. They have a relatively smaller mantle cavity than present-day filter-feeders, and this suggests that they had smaller and less elaborate gills. Their feeding type cannot be assessed with certainty. If the actinodonts are the ancestors of both the Pteriomorpha and the Heteroconchia, as is commonly believed, and if they are more closely related to the Nuculoida than either are to the Lucinoida, the Mytiloida, or the Anomalodesmata, then filter-feeding is more widespread within the early Bivalvia than deposit-feeding. The simplest explanation for this would be that filter-feeding was primitive.

There are two other factors to be considered. First, in the living Bivalvia, there is a change in feeding habit at metamorphosis. The larva often passes through a stage of byssal fixation before burrowing. A byssal gland is present in living *Nucula* (Drew, 1901). In consequence the larva undergoes a period of development in which deposit-feeding is impossible. Retardation of the development of deposit-feeding could have given rise to the evolution of filter-feeding on any number of occasions, if a truly planktotrophic existence was taken up early in the history of bivalve larvae.

Secondly, if the evolutionary model showing the Bivalvia as sharing common ancestry with the Scaphopoda is correct, filter-feeding cannot have developed until after the epifaunal monoplacophoran-like ancestor first took to an infaunal habit. This is likely to be not until after the development of true Bivalvia had taken place. On the other hand,

Scaphopoda may not be derived, together with the Bivalvia, from an infaunal *Ribeiria*-like ancestor. This would allow for the possibility that Stasek's (1972) interpretation was correct. He implied that Bivalvia and Rostroconchia shared a common ancestry, being derived from an epifaunal, filter-feeding, monoplacophoran-like organism.

In trying to take an independent view I would suggest that the evidence does not weigh very strongly in either direction at present, although I would suggest (independent of the relationships of the Scaphopoda) that deposit-feeding is more like the assumed primitive molluscan feeding type and it could well be primitive for the Bivalvia. If this is true, then filter-feeding may have arisen by the approximation of palps and gills.

Scaphopoda, Monoplacophora and Cephalopoda have organs around the mouth that aid feeding whilst, in contrast, chitons and Gastropoda usually graze directly by using the radula within the mouth. Feeding in Bivalvia entails a pair of palps either side of the mouth which either take food filtered by the gills to the mouth, or, with the addition of distally placed proboscides, carry organically rich sediment directly to the mouth after passing the leading edge of the gills there they collect additional food. The palps and possibly the palp proboscides may be homologues of the feeding structures in Scaphopoda, Monoplacophora and Cephalopoda. If so, bivalve filter-feeding cannot have started until the palps came into contact with the gills; this type of contact does not occur in any other molluscan class. The palps may have been pre-adapted for carrying food from the gills to the mouth if their earlier function was to carry food from the exterior to the mouth. When the bivalves adopted a diagonally-downward digging posture, anteriorly directed palps may have migrated downwards and backwards to avoid damage during locomotion. Filter-feeding may have become possible when the palps overlapped the gills, and the proboscides atrophied when deposit-feeding was rejected. Even the Solemyidae, generally thought to be the most primitive of bivalve filter-feeders, would have little reason to evolve long proboscides at some distance from the potential food source. If the evolution of feeding types in the bivalves was from a filter-feeder to a deposit-feeder then the palps of the ancestral filter-feeder would have been of a different morphology to present day palps.

The loss of the radula, which, by its absence, is proved to be unnecessary to both filter-feeding and deposit-feeding bivalves may be assumed to be directly connected with the evolutionary change from a primitive grazing type to whichever of these bivalve feeding habits was first adopted.

LARVAL DEVELOPMENT

If the theory that Bivalvia have evolved from an ancestor with a undivided calcified monoplacophoran-like shell is correct, the changes in form of the shell will of course effect all post-trochophore life stages (Fig. 16).

Because the trochophore larva occurs in some Bivalvia, Gastropoda, Scaphopoda and chitons as well as a number of other possibly related phyla, it is assumed to be a primitive stage of the molluscan life history. Veliger larvae, however, occur in other Gastropoda, Bivalvia and Scaphopoda and are thought to be a prolongation of the larval stage, which may have acquired adult shell characters for different reasons in each class. The acquisition of a shelled larva must have happened at an early stage of the evolutionary history at least in the Gastropoda, since they need the larval shell for torsion. The majority of Ordovician Gastropoda are sufficiently similar to recent taxa for us to be certain that torsion took place in their larval stages.

A number of Cambrian molluscs show what appear to be initial growth stages (Runnegar and Jell, 1976). In these, as in *Helcionella*, the first growth line occurs at approximately 0.2 mm diameter. Up to that first growth line, the shell is not very similar to modern larval shells (Pl. I, 4). It is not quite a hemisphere and cannot have enclosed the animal to the same extent. In *Helcionella*, the shell resembles half an egg that has been cut through a plane on the long axis. Similar shaped larval shells of other genera are illustrated by Runnegar and Jell (1976).

The possibility remains that an additional larval shell was shed in these Cambrian genera, but the different ornament in the helcionellid, *Yochelcionella* as figured by Runnegar and Jell (1976, Fig. 11c), suggests that this is not so; two distinct stages are preserved. An explanation is that the initial calcareous shell of *Helcionella* developed when the larva settled. If *Helcionella* developed pseudometamery like *Neopilina*, its pseudometamery would have appeared before the shell was formed. Ontogenetic increase in metamery is possible in the arthropods by addition of extra segments after moults. In chitons the shell plates can develop one at a time. But in an organism with a single, encasing shell, where each metamere is attached to the shell, it is more difficult to imagine that an increase in the number of metameres could take place after the shell has formed.

In other molluscan groups, the evolutionary development of adults with fewer metameres could arise by formation of the shell at increasingly earlier stages of metameric development. Thus a *Neopilina*-like animal

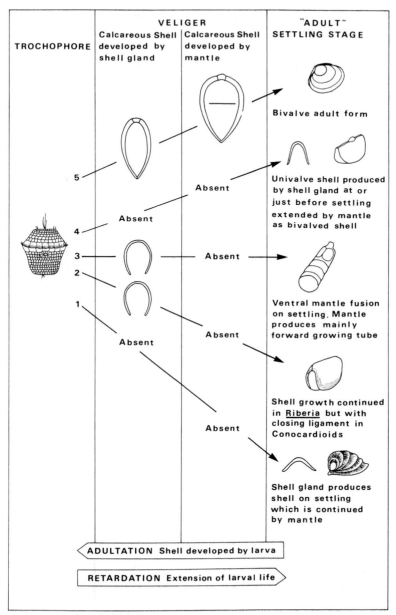

FIG. 16. Outline of the origin of the bivalved larval and adult shell in the Diasoma: 1. *Helcionella*, 2. Ribeirioid, 3. Scaphopod, 4. *Myona*, 5. Bivalve.

could have given rise first to the Cyrtolitidae with two or three metameres behind the head and, later, to a bellerophontid with only one pair of body attachment muscles. Such a reduction in numbers of metameres would have been necessary before true gastropod torsion could have taken place and would have been a combination of retardation of metameric development and also perhaps earlier development of the shell in larval life known as adultation.

In their descent from a hypothetical pseudometameric ancestor with many of the characters of *Neopilina*, bivalves did not lose the multiple paired pedal muscles in their earliest history. Loss of some of these may well take place later through atrophy related to changes in locomotion. However, loss of additional pairs of gills, nephridia and gonads may have taken place by retarded development of some pairs while those remaining underwent hypertrophy.

The initial shells of some Rostroconchia are known but their detailed features are not well preserved (Pojeta *et al.*, 1972). A very small shell with a similar shape to an adult ribeiriid occurs in a "larval fall out" deposit in the Vallhallfonna Formation, lower Llanvirn, of West Spitzbergen (Pl. VI, 5). The only rostroconch known to occur in this sequence belongs to *Euchasma* Eopteriidae, to which the larval shell is tentatively assigned. This young rostroconch is extremely similar to the larval shells of Scaphopoda (Lacaze-Duthiers, 1857). In some species of Dentaliidae the larval shell is retained on the posterior dorsum of the shell (Pl. I, 6). It can be seen that during later growth, the larval shell has opened with the two valves hinging on the dorsal axis. This results in a linear concavity resembling the dorsal line of Conocardioida.

Myona has developed a different system to this; the initial shell is apparently similar to the initial shell of *Helcionella* and may have formed after the larva settled. This shell was then incorporated within the assumed ligament and hinge movements of the shell would have taken place around it. In present-day Bivalvia, the hinge mechanism of the shell develops when the veliger shell is first calcified. This suggests that an acceleration of the development of the adult shell has taken place during evolution.

It is highly probable that many characters of the bivalve larval shell have developed as adaptations to larval life. There is no reason that they should reflect the evolution of the bivalves. At present, there is little evidence to suggest that the early formed larval hinge teeth are anything but a larval adaptation. There is no indication that the first bivalve hinges resembled those of larvae. It would seem possible, though, that larval teeth similar to those of living bivalves, were present even before the Silurian as some genera of the unusual bivalve order Praecardioida (common in the Silurian and Devonian) had this type of hinge at later growth stages. Careful consideration of this order as a whole suggests that a number of genera retained other larval characters. *Slava*, for example, retained a juvenile type of shell until it reached about half its total size. This suggests that retention of larval hinge teeth was secondary in this order.

Bivalves eventually developed a shell that was capable of complete closure at the time of larval settling. At any time that the shell was closed the animal would have totally lost its stability. The bivalve byssus may have developed to overcome this problem and then later, adult forms might have adopted the byssus to maintain their stability in an epifaunal or semi-infaunal environment (Stanley, 1975).

The occurrence of the byssus in the young stages of the Heteroconchia and the Anomalodesmata, together with the presence of a byssal gland in the young stages of Nuculoida, suggests that the byssus evolved before these groups separated. This points to their pre-early Ordovician common ancestor possessing a byssus in its larval settling stage.

Today many Bivalvia and Scaphopoda go through an initial crawling phase. The bivalve then becomes byssally fixed whereas the scaphopod starts burrowing.

A calcified univalved shell could not have been present except perhaps at the initial settling stages of the earliest true bivalves (if it had not been lost already). As the functional shell gland developed at even earlier growth stages of the larva, the adult bivalved type of shell may also have started occurring at a stage where the shell is secreted by the shell gland (i.e., prodissoconch 1 of Ocklemann, 1965). This shell stage is apparently calcified in living bivalves (Iwata and Akamatsu, 1975) and consists of two oval halves joined at the dorsum (Galtsoff, 1964).

Presumably from this shelled larva followed the development of a true planktotrophic larva with shell secreted by the mantle and with fully developed larval teeth and provinculum.

CONCLUSIONS

The pseudometamery of the body attachment musculature present in both early Bivalvia and the tergomyan Monoplacophora is evidence supporting both Salvini-Plawen's (1972) and Stasek's (1972) independant views that the Mollucsa developed from an epifaunal flatworm that developed a dorsal integument. The coelom in *Neopilina* is of a very limited type. In no mollusc is the coelom segmented as in the annelids. These considerations have led to the acceptance of the view that the metamery or pseudo-metamery of *Neopilina* is of a primitive nature, developed independently of, possibly earlier than, the coelomic segmentation of the annelids.

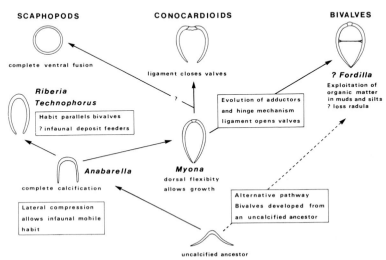

FIG. 17. Outline of the possible origin and evolutionary relationships of the Bivalvia. Names are examples of organisms with this morphology.

Locomotion in Bivalvia depends rather on the muscular control of the haemocoel than on a true coelom and evolved from a more primitive method. This depended on a ciliated, mucus-covered creeping foot, similar to that of chitons, Gastropoda and Monoplacophora as well as turbellarian flatworms.

The evolutionary development of the shell, which needs the mantle for its secretion, and also the mantle cavity, which needs the shell for its support, lead to a fundamental change in the lifestyle of the Mollusca, including the Bivalvia. This involved a complete change, not only in method of locomotion but also in the nature of many of the basic physiological processes. Moreover, many of the associated organs moved into the newly evolved mantle cavity. So fundamental was this change that some characters of a non-shelled ancestor (or "pre-mollusc") could have been completely masked. It is clearly not reasonable to base comparisons with other phyla on characters which may have developed since these sweeping changes in molluscan organization.

Bivalvia share features with the Monoplacophora to an extent that suggests they share a close common ancestry. They have characters of shell microstructure in common with other Conchifera. Their resemblance to the shell structure of the chitons is rather less. Some conocardioid and ribeirioid shells may be interpreted as intermediate stages between the low, elongate limpet-like univalve shells of the Monoplacophora and the laterally compressed, hinged bivalve shells. Although these are all best known from post-Cambrian rocks, their existence fits the theory of the origin of the Bivalvia from a calcified univalved monoplacophoran-like ancestor early in the Cambrian (Fig. 17). An alternative, which does not take account of the existence of the Rostroconchia, is an origin from a non-calcified monoplacophoran-like ancestor. If this theory were correct, the primitive shell structure types of the different Conchifera are not strictly homologues but their similarity could be related to their secretion onto and within an identical organic matrix. On balance I prefer the first view.

ACKNOWLEDGEMENTS

I would like to thank the following people for their considerable assistance; Dr J. D. Taylor for much helpful discussion of the matters considered in this paper and many related topics, Dr W. J. Kennedy for help with sectioning fossil shells, Mr C. P. Palmer for help with the names of scaphopods, Mr D. M. Shale for kindly sending specimens of bivalve larvae and information concerning their preparation and identification, Mr C. P. Nuttall for his constructive criticism of the manuscript and Solene Whybrow for preparing the illustrations.

REFERENCES

ADAMCZAK, F. (1961). Eridostraca—a new suborder of ostracods and its phylogenetic significance. *Acta palaeont. pol.* **6**, 29–104.

BEEDHAM, G. (1965). Repair of the shell in species of *Anodonta*. *Proc. zool. Soc. Lond.* **145**, 107–124.

COX, L. R. (1959). Symposium: 1. The geological history of the Protobranchia and the dual origin of taxodont Lamellibranchia. *Proc. malac. Soc. Lond.* **33**, 200–209.

DREW, G. A. (1901). The life-history of *Nucula delphimodonta* (Mighels). *Q. Jl microsc. Sci.* **44**, 313–391.

GALTSOFF, P. S. (1964). The American oyster *Crassostrea virginica* Gmelin. *Fishery Bull. Fish Wildl. Serv. U.S.* **64**, 1–480.

HEATH, H. (1937). The anatomy of some protobranch mollusks. *Mém. Mus. R. Hist. nat. Belg.* (2), **10**, 1–26.

IWATA, K. and AKAMATSU, M. (1975). A study on the larval prodissoconch of a japanese scallop (*Patinopecten yessoensis*). *Rep. Hokkaido Kaitaku Kinenkan*, **10**, 11–17.

LACAZE-DUTHIERS, H. (1856–1857). Histoire de l'organization et du dévelopment du *Dentale*. *Annls Sci. nat.*, *Ser 4, Zool.* **6**, 225–281; **7**, 5–51, 171–228; **8**, 18–44.

LEMCHE, H. and WINGSTRAND, K. G. (1958). The anatomy of *Neopilina galatheae* Lemche, 1957. *Galathea Rep.* **3**, 9–71.

MACALESTER, A. L. (1964). Transitional Ordovician bivalve with both mono-placophoran and Lucinacean affinities. *Science, N.Y.* **146** (3649), 1293–1294.

MORRIS, N. J. (1967). Mollusca: Scaphopoda and Bivalvia. *In* "The Fossil Record" (W. B. Harland), pp. 469–477. *Geol. Soc. London*.

MORRIS, N. J. (1979). The infaunal descendants of the Cycloconchidae: an outline of the evolutionary history and taxonomy of the Heteroconchia, superfamilies Cycloconchacea to Chamacea. *Phil. Trans. R. Soc. B*, **284**, 259–275.

MORRIS, N. J. and FORTEY, R. A. (1976). The significance of *Tironucula* gen. nov. to the study of bivalve evolution. *J. Paleont.* **50**, 701–709.

MORTON, B. (1973). A new theory of feeding and digestion in the filter feeding Lamellibranchia. *Malacologia*, **14**, 63–79.

MÜLLER, K. J. (1975). "*Heraultia*" *varensalensis* Cobbold (Crustacea) aus dem Unteren Kambrium, der ältests Fall von Geschlects dimorphisms. *Paläont. Z.* **49**, 168–180.

OCKLEMANN, K. W. (1965). Developmental types in marine bivalves and their distribution along the Atlantic coast of Europe. *Proc. 1st Europ. Malac. cong.* 25–35. London.

POJETA, J. Jr and RUNNEGAR, B. (1974). *Fordilla troyensis* and the early history of pelecypod mollusks. *Am. Scient.* **62**, 706–711.

POJETA, J. Jr and RUNNEGAR, B. (1976). The Paleontology of Rostroconch Mollusks and the early History of the Phylum Mollusca. *Prof. Pap. U.S. geol. Surv.* **963**, 1–78.

POJETA, J. Jr, RUNNEGAR, B., MORRIS, N. J. and NEWELL, N. D. (1972). Rostro-conchia: a new class of bivalved mollusks. *Science, N.Y.* **177**, 264–267.

POJETA, J. Jr, RUNNEGAR, B. and KRIZ, J. (1973). *Fordilla troyensis* Barrande: the oldest known pelecypod. *Science, N.Y.* **180**, 866–868.

RUNNEGAR, B. and JELL, P. A. (1976). Australian middle Cambrian molluscs and their bearing on early molluscan evolution. *Alcheringa*, **1**, 109–138.

RUNNEGAR, B. and POJETA, J. Jr (1974). Molluscan Phylogeny: the paleontological viewpoint. *Science, N.Y.* **186**, 311–317.

SALVINI PLAWEN, L. v. (1972). Zur Morphologie und Phylogenie der Mollusken: Die Beziehungen der Caudofoveata und der Solenogastres als Aculifera, als Mollusca und als Spiralia. *Z. wiss. Zool.* **184**, 205–394.

STANLEY, S. M. (1975). Adaptive themes in the evolution of the Bivalvia (Mollusca). *A. Rev. Earth Planet. Sci.* **3**, 361–385.

STASEK, C. R. (1963). Synopsis and discussion of the association of ctenidia and labial palps in the bivalved Mollusca. *Veliger*, **6**, 91–97.

STASEK, C. R. (1972). The molluscan framework. *In* "Chemical Zoology, 6, Mollusca" (Florkin, Marcel and B. T. Scheer, eds), pp. 1–43. Academic Press, New York and London.

TAYLOR, J. D., KENNEDY, W. J. and HALL, A. (1969). The shell structure and mineralogy of the Bivalvia. 1. Introduction. Nuculacea-Trigonacea. *Bull. Br. Mus. nat. Hist., Zool.*, Suppl. **3**.

TAYLOR, J. D., KENNEDY, W. J. and HALL, A. (1973). The shell structure and mineralogy of the Bivalvia. 2. Lucinacea-Clavagellacea conclusions. *Bull. Br. Mus. nat. Hist., Zool.* **22**.

THOMAS, R. D. K. (1976). Constraints of ligament growth, form and function on evolution in the Arcoida (Mollusca: Bivalvia). *Paleobiology*, **2**, 64–83.

TRUEMAN, E. R. (1966). Bivalve mollusks: Fluid dynamics of burrowing. *Science, N.Y.* **152**, 523–525.

TRUEMAN, E. R. (1968). The burrowing process of *Dentalium* (Scaph.). *J. Zool., Lond.* **154**, 19–27.

TRUEMAN, E. R. (1969). Ligament. *In* "Treatise on Invertebrate Paleontology, Part N, Mollusca 6, Bivalvia" (R. C. Moore, ed.). Geol. Soc. Amer. and Univ. Kansas Press.

WENZ, W. (1939–1943). Gastropoda: Prosobranchia. *In* "Handbuch der Paläozoologie", Vol. **6**, Part 1 (O. H. Schindewolf, ed.). Berlin.

YONGE, C. M. (1939). The protobranchiate Mollusca; a functional interpretation of their structure and evolution. *Phil. Trans. R. Soc. B*, **230**, 79–147.

YONGE, C. M. (1953). The monomyarian condition in the Lamellibranchia. *Trans. R. Soc. Edinb.* **62**, 443–478.

15 | Early Echionoderm Radiation

C. R. C. PAUL

Department of Geology, Liverpool University, England

Abstract: The fossil record of echinoderms reveals little about their origins but an early radiation at class level in the Cambrian and Ordovician. Diversity, as measured by a genera/million years ratio, increased from the Cambrian to the Carboniferous, declined to a minimum in the Trias and then rose again to a Tertiary maximum. The different evolutionary patterns at class and generic level probably result from colonization of the marine biosphere initially under conditions of low competition which allowed relatively inefficient "designs" to survive, followed by later elimination of less efficient groups as competition increased. This colonization–radiation/competition–retrenchment model implies that early, small, short-lived groups were less efficient than larger, extant classes. The model is tested against efficiency in protection, respiration and filter-feeding. Analysis of these three vital functions tends to support the model.

The classification of echinoderms should reflect their evolution. The taxonomic implications of the colonization–radiation/competition–retrenchment model are that early, small, short-lived groups with highly distinctive morphology should be given high taxonomic (class) rank. In this I am opposed to Breimer and Ubaghs (1974). A classification of echinoderms is proposed accepting five subphyla, Homalozoa, Blastozoa, Crinozoa, Asterozoa, Echinozoa. The Pelmatozoa is rejected. No characters unite all "stalked" echinoderms, rather differences in thecal and subvective morphology characterize the Blastozoa and Crinozoa as distinct subphyla. The paracrinoids are assigned to the Blastozoa on thecal characters. Their arms are not considered homologous with crinoid arms, nor is their general morphology sufficiently distinct to justify a sixth subphylum.

Systematics Association Special Volume No. 12, "The Origin of Major Invertebrate Groups", edited by M. R. House, 1979, pp. 415–434. Academic Press, London and New York.

INTRODUCTION

In the case of a well skeletized group like the echinoderms, evidence for their first appearance and subsequent radiation comes from the fossil record with all its inadequacies. Indeed the fossil record gives few clues as to the origin of echinoderms. The first lower Cambrian echinoderms are very distinctive, quite obviously echinoderms and nothing even vaguely like

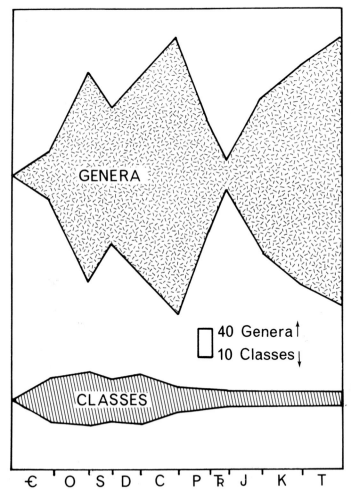

FIG. 1. Diversity of genera and classes of echinoderms through the Phanerozoic.

them occurs lower in the Cambrian or in the Precambrian, although *Tribrachidium*, from the late Precambrian Ediacara fauna, is probably an echinoderm. However, I have never seen any specimens and judge this from illustrations only. Thus, as far as the origin of this particular invertebrate group is concerned, a palaeontologist has very little to say. Nevertheless, the fossil record does provide a clear pattern of echinoderm radiation which is at least analogous to the radiation of the Metazoa and worth examining. This pattern may be an artifact of the incompleteness of the fossil record and our knowledge of it, but at least it gives a starting point from which to begin discussion.

PATTERN OF ECHINODERM EVOLUTION

The pattern of echinoderm evolution is quite different when considered at high and low taxonomic levels (Fig. 1). Echinoderm workers, unlike those of some other groups, have unashamedly created classes for the many fundamentally different "designs" of echinoderms that appear early in the fossil record of the phylum, irrespective of their size (numbers of genera or species) or longevity. At class level there is a clear early radiation, 15 classes known from the Cambrian, 19 in the Ordovician, followed by a steady decline. Only one class, the Blastoidea (Silurian–Permian) and two subclasses, the articulate crinoids and euechinoid sea urchins (Trias–Recent), appeared after the end of the Ordovician. In terms of basic "designs", and there were many weird and wonderful examples among early echinoderms (Fig. 3), virtually all echinoderm evolution was over by the end of the Ordovician. However, when the success of the phylum as a whole is considered, a completely different pattern emerges. Although the number of genera known from each geological period shows three peaks (Fig. 1), in the Ordovician, Carboniferous and Tertiary, when one allows for the different lengths of the geological periods, there is a continual rise to the Carboniferous, a decline to a minimum in the Trias following the Permo–Trias life crisis, and a second rise to a Tertiary maximum (Fig. 2). Echinoderm faunas are probably as diverse today as at any time in the past.

COLONIZATION–RADIATION/COMPETITION–RETRENCHMENT MODEL

From the divergent patterns at class and generic level, one may suggest a model, which is by no means original, to explain the radiation. When new environments or new modes of life are first colonized, competition is low and almost any "design", however bizarre or inefficient, might well survive. As more and more organisms become adapted to the new conditions, competition increases and the less efficient "designs" are eliminated while the more efficient continue to flourish. This sort of pattern characterizes the early fossil record of molluscs, brachiopods and arthropods, despite the different taxonomic treatments advocated by

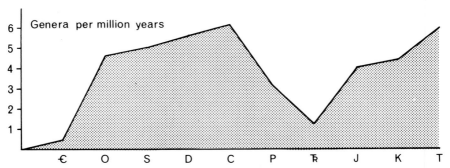

FIG. 2. Diversity of echinoderms through the Phanerozoic as estimated by number of genera per million years for each geological system.

workers in these groups. Certainly it is found in the "evolution" of motor cars, typewriters and the history of canal and railway companies in Britain. I propose to refer to it as the colonization–radiation/competition–retrenchment model.

Models are only useful insofar as they clarify our ideas or suggest tests by which they may be verified, refined or rejected. If competition eliminated the less efficient "designs" of echinoderms, there should be evidence of this in the fossil record. Cambrian echinoderms should be demonstrably less efficient at something than Ordovician echinoderms which in turn should be less efficient than Silurian and later echinoderms. All living animals have to perform a small number of vital functions to survive as did all fossils when alive. They must find a source of food (nutrition) and avoid becoming one themselves (protection). Food is converted into usable energy by metabolism which requires oxygen (respiration) and metabolic by-products must be voided (excretion), while energy goes towards ontogeny (growth and development) or the perpetuation of the species (reproduction). If we consider how fossil echinoderms performed all these vital functions, we may test the model.

Evidence for some functions is better than that for others. Excretion in particular, is a mystery in living echinoderms, let alone fossils. I wish to concentrate here on some aspects of just three functions, protection, respiration and feeding, for reasons of space. A more complete review of current knowledge and ideas will be found in Paul (1977a). Skeletal protection and respiratory gas exchange have basic requirements which are fundamentally opposed. The resolution of the resulting "design" problems has resulted in some fascinating skeletal structures. The contrast between filter-feeding in modern crinoids and in Palaeozoic "pelmatozoans" is equally instructive and rather surprising. I am aware that I am selecting the best examples, but it is the approach rather than the results, that I wish to stress.

VITAL FUNCTIONS

1. General Considerations

In most echinoderms the skeleton provides mechanical protection to the vital organs and, in effect, forms a protective envelope. The paradigm, or ideal structure, for a protective envelope would completely enclose the soft tissue with no gaps that predators or parasites could penetrate and it would be as thick and strong as possible. However, such a skeleton would isolate the animal from the surrounding sea water which contains the food and oxygen necessary for life. It might also pose problems in growth. In respiration, gases are exchanged between sea water and the internal medium, normally oxygen diffuses inwards and carbon dioxide outwards. An exchange surface is required to prevent mixing of the internal and external fluids which is inefficient at best and can be fatal. The paradigm for an exchange surface would be as thin and extensive as possible to allow maximum exchange. Thus there is a fundamental conflict between the requirements of a protective envelope, which should be thick and all embracing, and those of an exchange surface, which should be thin and extensive. In some organisms such as brachiopods, bivalves, limpets and other gastropods, this conflict is resolved by a temporal alternation of functions: the strategy of a mediaeval castle. The animals emerge to feed, breathe, reproduce, etc. but clamp the shells shut on the approach of danger, just as mediaeval peasants lived outside the castle in times of peace, but withdrew inside, raised the drawbridge and hoped to withstand the siege, when an enemy army was sighted. Echinoderms, on the other hand, have attempted to resolve these conflicting requirements by developing

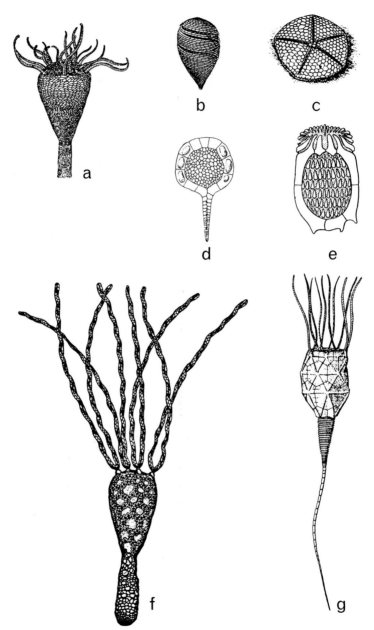

FIG. 3. For legend see opposite.

specialized orifices for feeding and breathing, thus enabling them, so to speak, to live permanently inside the castle. Their respiratory and feeding orifices are analogous to the slit-windows of castles which allow some light in but keep out enemy missiles and it is the same basic requirement of protection which dictated the slit shape in castle windows and, for example, cystoid pectinirhombs.

2. Protection

If we consider protection first, the moist obvious advance in "design" is the development of a skeleton, which appeared first in the lower Cambrian but was by no means ideal as a means of protection. Helicoplacoids (Fig. 3b), camptostromatoids and possibly the earliest edrioasteroids (Fig. 3c) as well, had entirely imbricate skeletons in which adjacent plates could move with respect to each other. Their skeletons were not at all rigid. The fourth known lower Cambrian group, the imbricate eocrinoids (Fig. 3a), show a slight advance in which, although most of the aboral theca is imbricated, there is a rigid oral surface of tesselated plates. Even so, this plated oral surface is weakened by the presence of numerous respiratory epispires, which are holes running through the skeleton along plate sutures.

In the middle Cambrian other advances appeared. First some groups (the Cincta (Fig. 3d), Ctenocystoidea (Fig. 3e), Cyclocystoidea) evolved skeletons with a strong and/or rigid marginal ring of plates between which plated membranes were suspended. These skeletons could resist distortion more than imbricated skeletons could. Secondly, some eocrinoids (*Gogia* (Fig. 3f) now known from the late lower Cambrian, Durham, 1978, and particularly *Lichenoides*) developed skeletons composed entirely of tesselated plates, although still weakened by abundant epispires. One middle Cambrian eocrinoid, *Gogia? radiata* Sprinkle, shows the first development of the "ornament" of ridges or folds running from plate centre to plate centre which forms a rigid triangulated girder structure. This "ornament" is very common among echinoderms, has evolved

---

FIG. 3. Early echinoderms. a, *Kinzercystis*, an imbricate eocrinoid; b, *Helicoplacus*, a helicoplacoid; c, *Stromatocystites*, the earliest edrioasteroid; d, *Trochocystites*, a cinctan; e, *Ctenocystis*, the only known ctenocystoid; f, *Gogia*, a tesselate eocrinoid; g, *Macrocystella*, the earliest rhombiferan cystoid. Note the imbricate plates in the thecae of a and b, the strong marginal rings in d and e, and the triangulated girder system in g. a–c. Lower Cambrian, d–f. Middle Cambrian, g. Upper Cambrian (Tremadoc).

repeatedly in different groups, and is of undoubted protective significance since it makes the theca more rigid. It first appears fully developed in the rhombiferan cystoid *Macrocystella* (Tremadoc) (Fig. 3g). Spines, the most characteristic skeletal protective devices of echinoderms, appear relatively late, in the earliest Ordovician. Nevertheless, virtually all the morphological advances of skeletal protection in echinoderms had appeared by lower Ordovician times. Furthermore the groups which appear to have been less efficient are small and have short stratigraphic ranges. The helicoplacoids and camptostromatoids are confined to the lower Cambrian and have three and one genus, respectively. Of the groups with marginal rings, the Cincta and Ctenocystoidea are confined to the middle Cambrian, the Cyclocystoidea range from the middle Cambrian to the Devonian, but are rare. There are few genera, although undoubtedly more than the single genus, *Cyclocystoides*, recognized in the Treatise (Kesling, 1966). Of these early Cambrian groups only the edrioasteroids (lower Cambrian to Carboniferous) and eocrinoids (lower Cambrian–middle Ordovician) could be considered at all successful and they have about 30 and 10 genera respectively.

3. Respiration

Respiratory gas exchange requires a surface to prevent mixing. The exchange surface may be within or outside the theca, endo- and exo-thecal, respectively. Endothecal exchange surfaces are protected by the theca and tend to be more extensive and delicate than exothecal surfaces. They correspond more closely to the paradigm of an exchange surface but suffer from the difficulty of passing sea water through the respiratory canals without the entrances becoming choked by suspended particles or recycling of the expelled, deoxygenated water. In addition to endothecal and exothecal exchange surfaces many echinoderms exchange gases through their tube-feet which combine both types of surface. Gases are exchanged externally through the tube-foot wall to and from sea water and internally through the ampulla wall to and from coelomic fluids. Tube-feet are of great respiratory significance to all sea urchins (Farmanfarmaian, 1966; Fenner, 1973) and their evolutionary success is probably due to their multiple functions (Nichols, 1972). Tube-feet and exothecal epispires appear first in the fossil record, tube feet in the helicoplacoids and epispires in the eocrinoids. Although epispires are not very efficient since they are blind ending, their construction is very simple and, contrary to my diagram in Hallam's book (Paul, 1977a, fig. 7, p. 144) they survive to the present day in the papulae of asteroids. Epispires showed a slight advance in the middle Cambrian with extensions over the theca that enlarged their surface area, but the real advance of exothecal exchange surfaces came with the appearance of diplopores in the Tremadoc and of rhombic sets of canals (humatirhombs) in the basal Ordovician. Both allow a one way circulation of body fluids so that deoxygenated water entered at one end, became oxygenated outside the theca and returned through the other end. Thus deoxygenated and oxygenated water could be kept separate. From a protective point of view, both diplopores and humatirhombs are an advance over epispires because they penetrate the plates, not the sutures, and therefore do not produce the weakening effect (analogous to the perforations on postage stamps) that epispires had.

The first endothecal pore-structures, pectinirhombs, appear in the Tremadoc, the last, blastoid hydrospires, in the Silurian. In between there was a diverse array of respiratory pore-structures among Ordovician echinoderms. Comparison of the evolution of pectinirhombs in the rhombiferan cystoids and of goniospires in the inadunate crinoid family Porocrinidae, illustrates the protection/respiration compromise particularly well.

The earliest pectinirhombs where not very efficient in comparison with later ones for three reasons. First they had continuous slits which did not separate inhalent and exhalent orifices. Secondly, the ridges of the triangulated girder system run through the axis of the pectinirhomb, dividing it into two demi-rhombs with the loss of some potential respiratory surface in the middle. Finally, the canals were separated from each other by thin strips of plate material which in later forms were replaced by respiratory surface. Undoubtedly the retention of the girder system through the axis of these early pectinirhombs was related to their continuous slits which weakened the theca. The next advance was the development of separate inhalent and exhalent openings in the middle Ordovician, but the axial ridge arrangement with two demi-rhombs was retained initially. Finally the axial ridge was lost, all the canals were connected becoming effectively isoclinal folds in the thecal wall and the rhombs were surrounded by new ridges which strengthened them and probably also helped to prevent recycling of used sea water. Rhombiferans seem to have been under strong selection for efficient respiratory devices which were developed at the expense of the strengthening effects of the rigid girder system.

In contrast, the Porocrinidae seem to have been under strong selection for protection and developed the girder system at the expense of efficient respiratory structures. The earliest porocrinid, *Triboloporus cryptoplicatus*, shows single slits at the plate corners in the three-pronged pattern. Internally, the fold which produces the slit bifurcates. Weak suggestions of the triangulated girder system are present in the beaded plate "ornament". Stratigraphically the next species, *T. xystrotus*, has the pair of folds opening separately in two slits and a more obvious girder system. In *Porocrinus* species the girder system is very strongly developed and the triangular areas in between are progressively filled with a series of folds, the goniospires. In the latest species the girder system carries all the strength of the theca and the goniospire areas are very extensive. Clearly the strength of the theca has been maintained by the girder system, but equally clearly the three-pronged goniospires can never develop an efficient respiratory current system. Furthermore, the growth of goniospires is more complex than that of pectinirhombs. (For a fuller account of these echinoderms see Kesling and Paul, 1968; Paul, 1968, 1975.)

As with protection, all types of respiratory structures in echinoderms had appeared by the middle Silurian at the latest and, in general, those which were least efficient occurred in small groups and/or have short stratigraphic ranges. However, the pattern is not quite as clear and the survival of epispires (papulae) to the present day does not agree with the model. Perhaps it is their sheer simplicity which has caused their repeated independent evolution.

4. Filter-feeding

Finally, let us consider filter-feeding briefly. Modern filter-feeding crinoids capture food particles by trapping them in mucous ejected from special cells in the tube-feet. The arms and pinnules, which bear the tube-feet, are arranged to form a filter in two characteristic patterns which depend on current strength. Crinoids which feed in moderate to strong currents form a filtration fan set perpendicular to the current direction with the food groove downstream. Those feeding in still waters extend their arms with the pinnules held in four radiating series along their length. Either way, the arms, pinnules and tube-feet form a filter, the paradigm for which has a large area and an equidimensional mesh. Many modern crinoids, particularly the rheophilic (current seeking) ones form filtration fans which agree very well with the paradigm (see Magnus, 1963). The efficiency of a filter-feeder may be judged by the area and complexity of its filter. The earliest echinoderms seem to have been filter-feeders. Helicoplacoids and edrioasteroids (lower Cambrian) had just tube-feet extending from epithecal food grooves, while the lower Cambrian eocrinoids were a little more efficient in having their tube-feet born on short brachioles; the blastozoan organization typical of later Cambrian and Ordovician "pelmatozoan" echinoderms. The typical crinoid pattern of branched arms with pinnules that bore tube-feet appeared later in the Ordovician. Thus again we find that the most efficient type of filter appeared fairly early in the fossil record, but that the very earliest forms were less efficient.

The efficiency of a living system ought to be judged in terms of its survival value. After all our lungs are only 20% efficient at extracting oxygen from the air, but they enable us to survive sleeping and sprinting and from depths well below sea level to over 5000 m altitude. In a mature animal which is not growing, food and oxygen consumption rates are directly linked by a simple formula which depends on the type of food metabolized. Although direct measurements of food and oxygen uptake are impossible in fossils, it is possible to make crude estimates of them. Thus the efficiency of filter-feeding and respiratory exchange systems of fossil echinoderms ought to and can, be compared with the food and oxygen *requirements* of the fossils. Very broadly speaking, food and oxygen requirements are proportional to thecal volume, while the food gathering and respiratory capacities are proportional to the areas of filter and respiratory exchange surface respectively. These are, of course, only first order approximations, but they are adequate for the arguments to be presented here. If we re-examine the early echinoderms in terms of their ability to meet their food and oxygen requirements, we find that helicoplacoids have large thecae, but limited food gathering and respiratory capacities, both depending on the tube-feet of the single ambulacrum. Equally almost all blastozoans and the Palaeozoic camerate crinoids have large thecae and hence high food and oxygen requirements. Blastozoans are characterized by specialized respiratory structures, but their food gathering capacity is limited. They have tube-feet on brachioles, but no arms to form extensive fans. Crinoids, on the other hand, typically have no respiratory devices, but extensive food gathering capacity. One may contrast these in terms of modes of life. The blastozoans appear to have been adapted to an environment with abundant food, but a shortage of oxygen (at least temporarily) which necessitated the development of their respiratory structures, while the crinoids are adapted to environments with abundant oxygen, but periodic shortages of food (Paul, 1977b).

Furthermore, the evolutionary success of the inadunate crinoids and particularly their modern descendents, the articulates becomes apparent in these terms. The modern crinoids have very small cups and even lose the stem in the free swimming forms. Thus they are all filter – they have the minimum food and oxygen requirements and the maximum food gathering capacity. No part of the body is more than 1–2 mm from the surface and diffusion alone will meet all their respiratory requirements. The evolutionary success of modern crinoids is due to the reduction of their requirements and concomitant increase in their ability to meet them.

All vital functions are interrelated and the number of relationships to be considered increases very rapidly with additional functions. However, we may finish by considering another aspect of protection in relation to crinoid feeding behaviour. Modern stemmed crinoids live in deep waters which are characterized by stability. Temperature and oxygen concentration do not vary, storms and sudden influxes of sediment are unknown. A fixed crinoid can filter safely and continuously without the risk of burial by sediment, suffocation, overheating, etc. In shallow water the environment is unstable and predators which use vision to locate prey are active in the daytime. A fixed filter-feeder would be subject to all sorts of predators and other physical dangers. Modern shallow water crinoids are free-swimming. They hide in crevices during the day and emerge to feed at night. They are also able to move with food supplies and avoid burial by sediment. Thus this protective behaviour, possible because they have flexible arms to swim with and prehensile cirri to become attached with, enables them to flourish in an environment closed to stalked crinoids. One wonders how the Palaeozoic pelmatozoans survived since most were sessile and lived in shallow waters.

Brief consideration of just these few aspects of the functional morphology of early echinoderms tends to confirm the model. All the short-lived echinoderm classes are confined to the Cambrian and Ordovician and seem to have been relatively inefficient. Clearly more detailed work is required to do a thorough analysis, but the point I wish to make is that the functional significance of morphological changes during evolution can be interpreted and, in turn, can be used to test hypotheses about modes of evolution, particularly if these hypotheses involve ideas about competition and selection.

Finally, and particularly appropriately in a Systematics Association symposium, I wish to consider the taxonomic implications of the colonization–radiation/competition–retrenchment model. Most biological classifications aim to reflect the evolution of the group being classified. These, so-called "natural" classifications in effect summarize knowledge of evolutionary relationships. I see no reason to make echinoderm classification an exception. Having argued that the large number of small short-lived classes demonstrates an early explosive radiation among echinoderms, I shall now come full circle and maintain that the existence of this early radiation justifies the recognition of these small groups as distinct classes. However circular, this is not an unwarranted argument. The hypothesis on which it is based has been tested briefly in the previous section and is open to test in the future. In effect I am arguing that morphological (and hence functional) distinctiveness should be the main criterion on which higher taxa are defined, not taxonomic size or longevity. This idea has been expressed forcibly by at least one other contributor to this symposium, but has met with some resistence among echinoderm workers. The classification of the pelmatozoan, or fixed, echinoderms has been discussed recently by Breimer and Ubaghs (1974) and Sprinkle (1976). The different viewpoints of these authors need consideration since they are relevant to the classification of all echinoderms. Interestingly, however, despite their different emphasis and nomenclature, the classifications of the pelmatozoans which result are remarkably similar. Fossil echinoderms have enjoyed more attention from palaeontologists since 1960 than at any time previously. Not surprisingly new discoveries and new interpretations have had their effects on classification. Breimer and Ubaghs argue against too rapid a change in classification because this may lead to nomenclatural instability. However, these new data must be recognized and, if classification of echinoderms is currently in a state of flux, surely this reflects the healthy state of echinoderm studies at present? No classification can hope to be stable for all time, at least not before we have discovered all the fossils and living organisms that there are to find.

Many of the recently discovered early echinoderms are very distinct from known classes (e.g. helicoplacoids and ctenocystoids) and, more or less inevitably, new classes have been created for them. Equally, new interpretations of known fossils (e.g. *Camptostroma*, originally described as a medusoid) and new formal groupings of known taxa, such as the

Blastozoa, have resulted in other new high level taxa. Breimer and Ubaghs oppose these new taxa for two reasons. First because they feel that the resulting classification is less "natural" and obscures relationships. Secondly because some of the newly proposed taxa ignore earlier names which are available and have historical, if not exactly legal, priority. Although not explicitly stated, it is clearly that Breimer and Ubaghs regard taxonomic size (i.e. number of genera) as an important criterion for the recognition of high level taxa. Hence they tend to reject "enigmatic" groups and leave them unclassified. The Edrioblastoidea, with the single genus *Astrocystites*, suffers this fate at their hands. Surely this practice obscures evolutionary relationships even more than creating new classes for small groups does? All workers not directly familiar with these rare fossils will be unaware of their existence if attention is not drawn to them. Anyone attempting a classification will be doing so on the basis of incomplete data. As Sprinkle (1976, p. 89) put it, these "enigmatic" fossils will not "go away" if they are ignored. I am not advocating the creation of a new class for every puzzling Palaeozoic genus, but at least current practice does advertise the diversity of form among early echinoderms.

As to historical priority, there are no formal rules of nomenclature to cover high level taxa. The proof of the nomenclatural pudding is in the eating and I feel sure that Breimer and Ubaghs resurrection of Burmeister's "Brachiotoidea" for the familiar Crinozoa will stick in most throats. To be sure we should give Burmeister credit for being the first person to emphasize the difference between arm-bearing crinozoans and brachiole-bearing blastozoans, but we do not *have* to accept his names for these groups. Breimer and Ubaghs argument in favour of "Cystoidea" for the Blastozoa is perhaps stronger, but Sprinkle (1973, 1976) has pointed out that the Cystoidea has never really had a fixed composition and had already been rejected as a formal taxon by me (Paul, 1968) when I elevated the Diploporita and Rhombifera to class level. I would personally have preferred Cystozoa for this subphylum, but do not propose to suggest any more names for a group whose constituent parts are very largely agreed upon by everyone.

Turning now to the actual taxa, Breimer and Ubaghs' principal argument is that fixed, or pelmatozoan, echinoderms form a natural unit (fundamental unity, in their terminology), for which the old subphylum Pelmatozoa is retained. By implication, one supposes that eleutherozoan (free-living) echinoderms form an opposed and separate natural group, but Breimer and Ubaghs do not consider the classification of non-pelmatozoan echinoderms. Any character which can only exist in a small number of possible states is a poor criterion on which to base high level taxa. For example, echinoderms may be either fixed or free. What other possibilities exist? It is inherently likely that different members of the same class or subphylum will adopt one or other of these two possibilities, as indeed they have. Holothurians are typically eleutherozoan, but the Psolidae are fixed. Crinoids are typically pelmatozoan, but comatulids are free-living, while the genera *Lichenoides*, *Brockocystis* and *Protocrinites* are examples of free-living eocrinoids, rhombiferans and diploporites, respectively, and all three classes are typically pelmatozoan. Equally, members of the class Edrioasteroidea are typically fixed, but are not classified within the Pelmatozoa by Breimer and Ubaghs, or any other post-war echinoderm worker, although all sessile echinoderms were once included within the Pelmatozoa. Thus it is not fixture itself which characterizes the Pelmatozoa of Breimer and Ubaghs, but the mode of fixture (by a functional stalk) and also the mode of feeding, using exothecal appendages (arms or brachioles). We must therefore examine the stems and subvective systems of pelmatozoans to see if they are truly homologous structures characterizing a natural grouping. Taking first the subvective system, it is already agreed by Breimer and Ubaghs, and most other echinoderm workers, that there is a fundamental difference between arm-bearing crinozoans and brachiole-bearing blastozoans. These two grades of organization are clearly distinct from their first appearance in the Cambrian and I have argued (Paul, 1977b) that they were adapted to different environments. There is hardly a fundamental unity here, especially when one considers that many eleutherozoan ophiuroids capture food particles using tube-feet elevated off the substrate by their arms. The functional stalk is an equally dubious character uniting the pelmatozoan groups together. Sprinkle has argued (1973) that a typical pelmatozoan stem, formed of discrete columnals, evolved independently in the Blastozoa and Crinozoa, but the situation is much more complex than this. The column of glyptocystitid rhombiferans is unique in its construction, has distinct proximal and distal parts of different morphology and may well respresent yet another separate evolution of a stem. The diploporite superfamilies Sphaeronitida and Aristocystitida do not contain a single individual with any vestige of a stem at any known stage in their development. They are typically directly attached by an aboral attachment area. Where the theca is elevated off the substrate, this is achieved by an aboral extension of the theca which is clearly *part* of the theca. It does not even resemble the polyplated holdfasts of eocrinoids like *Gogia*. Equally, the rhombiferan *Echinosphaerites* shows no vestige of a stem, although this might be due to secondary loss as the related genus *Heliocrinites* shows reduction of the relative size of the stem facet during growth. Thus, not only was a true stem evolved independently in different groups, but there are some directly attached groups which apparently never evolved a stem at all. Once again we find that there are only a few ways in which the food gathering surface can be raised off the substrate. The theca may become elongate (as in the sphaeronitid diploporite *Eucystis*, the aristocystitid diploporite *Calix*, or the edrioasteroid *Pyrgocystis*), or a polyplated holdfast may develop (as in the eocrinoid *Gogia*, or the primitive crinoids *Ecmatocrinus* and *Aethocrinus*), or a true stem with discrete, bead-like columnals, may occur (the typical pelmatozoan stem) or finally, the animal may become free-living and climb to an elevated point as comatulid crinoids and filter-feeding ophiuroids often do. Furthermore, it appears that the transition from a polyplated holdfast to a true column may have followed a different course in the Crinozoa and Blastozoa. Ordovician crinoids frequently have columnals composed of five parts, or pentameres, but as far as I am aware, no such intermediate stage has ever been reported for a blastozoan. To summarize, neither the stem nor the subvective system appears to be a good character uniting all pelmatozoan echinoderms into a natural grouping, rather the fundamental difference between the blastozoans and crinozoans is as clear in the stem characters as in the arms and brachioles.

Finally, in considering the pelmatozoan classes, there is the problem of the Paracrinoidea, a group which apparently includes features of both the Blastozoa and Crinozoa. Parsley and Mintz (1976) have created a new subphylum, the Paracrinozoa, for the eight or so genera in this class. The idea of creating a new high level taxon, for a small group which shares characters of two equivalent level taxa, seems to me a somewhat naive approach to taxonomy. The paracrinoids do not represent a group with a fundamentally distinct morphology, merely one which necessitates an awkward decision as to which of two other groups it more closely resembles. As an illustration of this difficulty, I rather arbitrarily attributed them to the Crinozoa last year (Paul, 1977a), but reverse this here and include them in the Blastozoa. Paracrinoids have an enclosed theca, a single internal gonad and often have respiratory pore-structures in the thecal wall; all characters typical of blastozoans. However, they have very distinctive uniserial, branched arms with the branches coming off the left side only of each arm which, if not exactly typical, resemble crinoid arms more than blastozoan structures. The problem of classifying the paracrinoids is one of assessing the importance of their subvective system compared with their other thecal characters. Here I feel that too much emphasis has been placed on uniserial or biserial structures in subvective systems which are in fact highly variable. Subvective appendages may be only uniserial, biserial or polyserial; they may be branched or unbranched. Typically, we are told, blastozoans have biserial unbranched structures, crinozoans branched uniserial structures, but there are numerous exceptions when the whole morphology of the subvective system is considered. The terminal structures of the subvective systems of blastozoans (where known) are biserial brachioles; those of crinozoans, uniserial pinnules. Even this distinction is not universal as some crinoids lack pinnules altogether, while at least three genera of diploporites had uniserial terminal structures. The brachioles of *Lichenoides* (Eocrinoidea) and supposed brachioles (they have never been found preserved) of diploporites such as *Eucystis* and *Glyptosphaerites*, arose from facets developed directly on thecal plates. Epithecal food grooves, excavated in the surface of the thecal plates, carried the food particles to the mouth. In most other blastozoans there are true arm structures, which are recumbent on the thecal surface and the brachioles arose from facets on the arms, each facet being shared by two plates. Such recumbent arms may be branched (*Callocystites*) or unbranched (*Pseudocrinites, Cheirocrinus*, all blastoids), but are pinnate in the sense that the brachioles arose from either side of a central axis. In *Caryocrinites* erect pinnate arms occurred, paralleling those of many crinoids. *Caryocrinites* is not set apart in a new subphylum or class, merely because it has biserial brachioles. Early crinoid arms are not pinnate. They are unbranched in *Ecmatocrinus* (middle Cambrian), isotomously (equally) branched in *Aethocrinus, Compagicrinus, Ramseyocrinus*, etc. (lower Ordovician) and the earliest pinnate (or pinnulate) arms occur in middle Ordovician crinoids such as *Glyptocrinites*. Pinnation was evolved in different ways (using brachioles and pinnules) independently in blastozoans and crinozoans and resulted from the requirements of efficient filtration. Some Ordovician crinoids, like *Porocrinus* and *Palaeocrinus* have unbranched, non-pinnate arms and respiratory pore-structures in the calycal wall. They represent crinoids convergent on blastozoan organization, while *Caryocrinites* represents a blastozoan which mimics crinozoans (Sprinkle, 1975). Viewed against this background, the rather unique arms of paracrinoids are just another pattern of food gathering appendages. Their remarkable feature is not that they are uniserial, but that they are asymmetrical. A similar asymmetry, also with branches predominantly, but not exclusively, to the left, occurs in the two-armed rhombiferan *Schizocystis*. I am, therefore, disinclined to accept the arms of paracrinoids as of fundamental importance.

Of their other morphological features, the presence of respiratory pore-structures in some paracrinoids is not of great significance either. All Ordovician classes of echinoderms have at least one representative with pore-structures, the paracrinoids do not have a single type of pore-structure uniting them as a group, nor do all paracrinoids have pore-structures. The enclosing theca and single internal gonad seem to me more fundamental points, although again there are limited possibilities. Gonads may be either internal or external, but from the point of view of protection, it seems at first sight so illogical for crinoids to develop such vital reproductive structures outside the theca, that this morphology is judged to be very important. Even so, since the gonads are developed in proximal pinnules of living crinoids, one may well ask where they lay in early non-pinnulate fossil crinoids. All blastozoans and paracrinoids have a single internal gonad. Quite literally, they put all their eggs in one basket! The crinoids, on the other hand, dispersed their reproductive tissue which, in the long run, seems to have proved a more successful adaptive strategy. External gonads set crinoids apart from all other echinoderms. Similarly, of all pelmatozoans, only crinoids have separate openings in the cup for food (the mouth) and the external extensions of the internal coelomic systems. The arms of crinoids join the cup well away from the mouth, the food grooves continue to the mouth on what is morphologically the ventral surface of the cup, but the coelomic systems pass into the cup at the point where the arms join it. Such an arrangement is unknown in all blastozoans and the paracrinoids and led Sprinkle (1973) to argue that blastozoans lacked external extensions of the coelomic systems, including the water vascular system, in their arms and brachioles. This idea has been hotly disputed, by Breimer and Ubaghs among others and in this I agree wholeheartedly with the latter two authors.

In summary, the crinoids are set apart from both the blastozoans and paracrinoids by their external gonads and cup morphology. Therefore, I feel the paracrinoids are best viewed as blastozoans with arms somewhat convergent on these of crinozoans, much as *Caryocrinites* is.

Eleutherozoans pose fewer problems, perhaps because I have studied them less and because less has been written on their classification recently. There is reasonable agreement as to the subphyla Echinozoa and Asterozoa, but the Homalozoa and Calcichordata have given rise to some divergent opinions, but there is no need for me to cover these groups here. It therefore remains simply to present the classification which is as follows:

CLASSIFICATION OF THE ECHINODERMS

PHYLUM ECHINODERMATA de Brugière 1789
  Subphylum Homalozoa Whitehouse 1941
    Class Cycloidea Whitehouse 1941
    Class Cyamoidea Whitehouse 1941
    Class Ctenocystoidea Robison and Sprinkle 1969
    Class Soluta Jaekel 1901
    Class Cincta Jaekel 1918

  Subphylum Blastozoa Sprinkle 1973
    Class Eocrinoidea Jaekel 1918
      Order Imbricata Sprinkle 1973
      Two unnamed orders (Sprinkle, 1973)
    Class Diploporita Müller 1854
    Class Rhombifera von Zittel 1879
      Order Dichoporita Jaekel 1899
      Order Fistuliporita Paul 1968
    Class Parablastoidea Hudson 1907
    Class Blastoidea Say 1825
      Order Fissiculata Jaekel 1918
      Order Spiraculata Jaekel 1918
    Class Paracrinoidea Regnéll 1945
    Class Edrioblastoidea Fay 1962

  Subphylum Crinozoa Matsumoto 1929
    Class Crinoidea Miller 1821
      Subclass Camerata Wachsmuth and Springer 1885
      Subclass Inadunata Wachsmuth and Springer 1885
      Subclass Flexibilia von Zittel 1879
      Subclass Articulata Miller 1821

  Subphylum Asterozoa von Zittel 1895
    Class Somasteroidea Spencer 1951
    Class Asteroidea de Blainville 1830
    Class Ophiuroidea Gray 1840

  Subphylum Echinozoa Haeckel in von Zittel 1895
    Class Echinoidea Leske 1778
    Class Edrioasteroidea Billings 1858
    Class Camptostromatoidea Durham 1966
    Class Ophiocystioidea Sollas 1899
    Class Helicoplacoidea Durham and Caster 1963
    Class Cyclocystoidea Miller and Gurley 1895
    Class Holothuroidea de Blainville 1834

REFERENCES

BREIMER, A. and UBAGHS, G. (1974). A critical comment on the classification of the pelmatozoan echinoderms. *Proc. K. ned. Akad. Wet.* **B77**, 398–417.

DURHAM, J. W. (1978). A lower Cambrian eocrinoid. *J. Paleont.* **52**, 195–199.

FARMANFARMAIAN, A. (1966). The respiratory physiology of echinoderms. *In* "Physiology of Echinodermata" (R. A. Boolootian, ed.), pp. 245–265. John Wiley, New York.

FENNER, D. H. (1973). The respiratory adaptations of the podia and ampullae of echinoids (Echinodermata). *Biol. Bull. mar. biol. Lab., Woods Hole*, **145**, 323–339.

KESLING, R. V. (1966). Cyclocystoids. *In* "Treatise on Invertebrate Paleontology" (R. C. Moore, ed.), Part U, pp. U188–U210. Geol. Soc. Am. and Univ. Kansas Press.

KESLING, R. V. and PAUL, C. R. C. (1968). New species of Porocrinidae and brief remarks upon these unusual crinoids. *Contr. Mus. Paleont. Univ. Mich.* **22**, 1–32.

MAGNUS, D. E. B. (1963). Der Federstern *Heterometra savignyi* im Roten Meer. *Natur Mus.* **93**, 355–368.

NICHOLS, D. (1972). The water-vascular system in living and fossil echinoderms. *Palaeontology*, **15**, 519–538.

PARSLEY, R. L. and MINTZ, L. W. (1975). North American Paracrinoidea (Ordovician: Paracrinozoa, new, Echinodermata). *Bull. Am. Paleont.* **68**, 1–115.

PAUL, C. R. C. (1968). The morphology and function of dichoporite pore-structures in cystoids. *Palaeontology*, **11**, 697–730.

PAUL, C. R. C. (1975). A reappraisal of the paradigm method of functional analysis in fossils. *Lethaia*, **8**, 15–21.

PAUL, C. R. C. (1977a). Evolution of primitive echinoderms. *In* "Patterns of evolution as illustrated by The Fossil Record" (A. Hallam, ed.), pp. 123–158. Elsevier Science Publishing Co., Amsterdam.

PAUL, C. R. C. (1977b). Feeding and respiration rates in fossil echinoderms. *J. Paleont.* **51** (Suppl. 2), 20.

SPRINKLE, J. (1973). Morphology and evolution of blastozoan echinoderms. *Spec. Publ. Mus. comp. Zool. Harv.* 284pp.

SPRINKLE, J. (1975). The "arms" of *Caryocrinites*, a rhombiferan cystoid convergent on crinoids. *J. Paleont.* **49**, 1062–1073.

SPRINKLE, J. (1976). Classification and phylogeny of "pelmatozoan" echinoderms. *Syst. Zool.* **25**, 83–91.

16 | Early Evolution of Graptolites and Related Groups

R. B. RICKARDS

Sedgwick Museum, Cambridge, England

Abstract: The stratigraphy of early graptolites and undoubted hemichordates is such that a common ancestor in the Cambrian is reasonable to those who support the idea of close affinity of the two groups. The case for this affinity is briefly outlined before discussion of the nature of early graptolites and a summary of the problems pertaining to them. Further it is suggested that the evolution to graptolites and pterobranchs may have taken place along the following route, if not through each step of it: (i) in the late Precambrian or early Cambrian an infaunal phoronid-like worm existed and gave rise, in the early Cambrian or early middle Cambrian to (ii), a filter-feeding sessile or upright burrow-living form which existed in dense association as an immediate prelude to either (iii), the loose coloniality of an *Eocephalodiscus*-like form, or (iv), a rhabdopleuran-like animal exhibiting true coloniality. A *Chaunograptus*-like form might arise at about the same time as either (iii) or (iv) by bundling the dense associations into upright tubes and connecting them with stolons.

INTRODUCTION

The broad evolution of graptolites is beyond dispute and has recently been outlined by Rickards (1975, 1977) and by Rickards and Palmer (1977). The benthonic dendroid graptolites appeared in the middle Cambrian, became extinct in the upper Carboniferous and gave rise to the planktonic graptolites (graptoloids) in the Tremadoc. The graptoloids diversified strongly but became extinct in the uppermost lower Devonian as far as is known at present. The evolutionary relationships of the other graptolite

Systematics Association Special Volume No. 12, "The Origin of Major Invertebrate Groups", edited by M. R. House, 1979, pp. 435–441. Academic Press, London and New York.

orders, Crustoidea, Camaroidea, Stolonoidea, Tuboidea, Dithecoidea and Archaeodendrida, is far from certain, partly because their stratigraphic record is so sparse. The Dithecoidea and Archaeodendrida, however, appear in the middle Cambrian and must be important in any model which links early graptolites with undoubted hemichordates. Undoubted hemichordates (*Eocephalodiscus*) occur in the Tremadoc, and an evolutionary connection with the graptolites in the middle Cambrian is, therefore, reasonable if the affinities of the two groups are considered very close.

AFFINITY OF GRAPTOLITES AND HEMICHORDATES

It is not intended to discuss the question of graptolite affinities in great detail since this has been dealt with at length in several papers, including review papers (Kozlowski, 1949, 1966; Bohlin, 1950; Bulman, 1955, 1970; Kirk, 1974; Rickards, 1975). Each of these authors considered that, in life, the graptolite skeleton was more or less covered by a secreting layer of soft tissue which was responsible for secreting the cortex or outer peridermal layer. It was thought that the zooid itself occupied the inside of the thecal tube, and various mechanisms were deduced by each author for connecting the zooid with the outer secreting layer (usually termed extrathecal tissue). Bulman (1970), although accepting the concept, failed to represent such extrathecal tissue in his zooidal reconstructions which were constructed by analogy with the rhabdopleuran animals. The presence of this supposed extrathecal layer was considered one of the main differences between graptolites and hemichordates and led Bohlin (1950) and Kirk (1974), in almost identical arguments, to deduce a coelenterate relationship for the graptolites; whereas a majority of researchers opted (and opt) for a hemichordate affinity despite the extrathecal tissue. In some views the presence of extrathecal tissue is taken to indicate that the graptolites should be rather further removed from undoubted hemichordates than the 1970 *Treatise* has them (Urbanek, 1976).

However, recent work by Rickards and Crowther (1979) and Crowther and Rickards (1977) on the ultrastructure of the cortex established that in that layer the whole thickness is composed of narrow bandages which entomb the fusellar layer in a mummy-like wrapping of collagen fibrils. The bandage dimensions and arrangements bear a direct relationship to the nearest thecal tube and it was concluded by these authors that the concept of extrathecal tissue was quite unnecessary to explain the bandage secretion: such a secretion could be achieved by a rhabdopleuran-like zooid leaving the thecal aperture, yet remaining in its proximity, and there depositing with its pre-oral disc the collagen fibrillar bandages. Only the growing tip of the nema (and homologous structures) needs a secreting layer of soft tissue in the fashion outlined by Rickards (1977). It is now known (Dilly, 1976) that *Rhabdopleura* zooids do leave their thecal tubes to lay down a thin external layer of skeletal material over the top of the fusellar layer, whilst in cephalodiscids such activity is crucial to the construction of the coenoecium.

The main differences between rhabdopleurans and graptolites, therefore, can be summarized as follows: (i) rhabdopleurans have a permanent leading bud whereas in graptolites each "stolotheca" in turn becomes the "leading bud"; (ii) there are differences in ultrastructure of the fusellar layers of the two groups. Otherwise the striking feature of the two groups is the remarkable similarity of skeletal construction seen, in fact, in no other groups. As pointed out by Bulman (1970) the budding differences may not be very important since in calyptoblastean hydroids both monopodial and sympodial budding occurs within a single order.

In my opinion, not only are graptolites relatively close to extant hemichordates, but it seems reasonable to include them in the phylum Hemichordata. Having reached this conclusion one of the main questions, when considering the origin and early evolution of the graptolites, becomes one of examining the nature and value of the stratigraphic record in the Cambrian and early Ordovician.

NATURE OF THE EARLY DENDROIDS

Records of dendroids in the middle Cambrian are remarkably rare being restricted to a *Dendrograptus* from the *P. davidis* Zone of Comley, Shropshire, England and a *Dendrograptus* from the *P. davidis* Zone of Norway (see Bulman, 1970). No detailed structure is known in these forms. In the upper Cambrian Ruedemann (1933) described *Dendrograptus*, *Callograptus* and *Dictyonema*; Decker (1945) described a larger fauna from the Wilberns Formation of Texas. All three genera survived into the Carboniferous, joined only by *Desmograptus* and *Ptilograptus* both close relatives of *Dictyonema*. The only other dendroid genera recorded from the upper Cambrian are *Aspidograptus* and possibly *Acanthograptus*. There is a lack of structural information on all these records other than the general similarity of rhabdosome and stipe to later, better known forms, black stolons and the presence in some of autothecae and bithecae. They are undoubted graptolites but some may eventually prove referable to other orders such as the Tuboidea. At present it is assumed that most of these early species have a triad division of a black stolon and the presence of autothecae, bithecae and stolothecae.

NATURE OF EARLY UNDOUBTED HEMICHORDATES

By contrast the early rhabdopleurans and cephalodiscids are quite well known mainly as a result of the work of Kozlowski upon material dissolved from the rock matrix. The earliest form is *Eocephalodiscus polonicus* (Eocephalodiscidae) from the Tremadoc (Kozlowski, 1949) a minute form with a compact, unspined coenoecium. *Pterobranchites* from the lower Ordovician (Kozlowski, 1967) is probably a true member of the Cephalodiscidae and the earliest, therefore, of that extant family. It has a coenoecium of irregularly aggregated tubes and elongate vesicles.

Rhabdopleura-like genera themselves appeared in the low Ordovician with *Rhabdopleurites* and *Rhabdopleuroides* (Kozlowski, 1967; Kozlowski, 1961 respectively). Both are very similar indeed to *Rhabdopleura* but in *Rhabdopleuroides* the coenoecium is attached throughout its length. In these early genera it is possible that the ultrastructure of the fusellar layers is slightly closer to that of the graptolites than to that of the extant hemichordates, although much work is needed on both. The stratigraphic record of the undoubted hemichordates is very sparse and no great weight can be attached to their absence below the Tremadoc. It is possible they originated much earlier, but probably only suitable etching techniques on suitable rocks will detect them. However, they occur sufficiently close and sufficiently early for a reasonable model to be constructed below for their origin and divergence from graptolites.

OTHER MIDDLE CAMBRIAN GRAPTOLITES

A number of little known graptolite genera occur in the middle Cambrian which have been placed under "taxonomic position uncertain" by Bulman (1970) but some of which were placed in the orders Archaeodendrida and Dithecoidea by Obut (1964), my view being in agreement with the latter author; *Protohalecium* Chapman and Thomas (1936); *Sphenecium* Chapman and Thomas (1936); *Archaeolafoea* Chapman (1919); *Archaeocryptolaria* Chapman (1919); *Siberiodendrum* Obut (1964) and *Chaunograptus*

Hall (1883). In addition *Haplograptus* Ruedeman (1933) and *Mastigograptus* Ruedeman (1908) occur in Cambrian strata of uncertain age; *Aellograptus* Obut (1964), *Dithecodendrum* Obut (1964) and *Siberiograptus* Obut (1964) in the upper Cambrian.

Although these are undoubted graptolites with occasional traces of fusellae, traces of stolons, and rare bithecae, the development is only known reasonably well in *Mastiograptus*, a very similar graptolite to *Chaunograptus* recorded by Ruedemann (1947) from the middle Cambrian Burgess Shale. These genera are characterized by a black stem, unquestionably hollow, from which arise at intervals large, more or less transparent autothecae and diminutive bithecae. The autothecae closely resemble the upright coenoecial tubes of *Rhabdopleura* in that the fusellae are ring-like with only occasional zig-zag sutures (Andres, 1961). Obut (personal communication) has noted that the thecorhizal development (holdfast) is similar to that in *Rhabdopleura*: a probable sicula-like theca has spirally coiled adnate stolons budding from it which then grow upwards on the inside of the upright black tube. The upright tube probably houses bundles of stolons and resorption of the containing wall takes place to allow one or more stolons (or ?zooids) through to form a new branch. Each new branch has an enlarged base called a cladorhiza which is not unlike the basal thecorhiza and the branches may occasionally give rise to large conical autothecae. Clearly a great deal more information is needed on these early non-dendroid graptolites but the strong suggestion is that some at least will prove to be intermediate between graptolites and rhabdopleurans. Furthermore, if the development of stolons is an adaptation to coloniality, as seems a reasonable hypothesis, the bundling of several individuals into an upright tube surely suggests an early attempt to raise the aggregated animals above the substrate. *Mastigograptus* continues into the upper Ordovician but most of the early genera are short lived.

SUMMARY

The origin and early diversification of graptolites and other hemichordates may have taken place along the following lines, although not necessarily through each step:

1. an early Cambrian infaunal phoronid-like worm gave rise in the early Cambrian or early middle Cambrian to

2. a filter-feeding sessile or upright burrow-living form which existed in dense association as an immediate prelude to either,

3. the loose coloniality of an *Eocephalodiscus*-like genus or,

4. a rhabdopleuran-like animal exhibiting true coloniality.

5. A *Chaunograptus* or *Mastigograptus*-like form might arise at about the same time by bundling the dense associations into upright tubes and connecting them with stolons; thus achieving in (3) and (4) encrusting colonies and in (5) upright sessile colonies.

6. Further development would have resulted in divergence in that the cephalodiscans and rhabdopleurans had established a successful adaptation to an encrusting niche, whereas the chaunograptids needed to simplify their rather primitive upright colonial structure so that the zooids had a degree of freedom from each other for feeding purposes.

7. At this stage in the evolution of graptolites the rhabdosomes could adopt a high degree of symmetry, to tap more systematically the food supply carried on the currents; and the dendroid habit of *Mastigograptus* and *Dendrograptus* could be replaced by those genera of symmetrical habit notably *Callograptus*, *Dictyonema* and, later, others.

REFERENCES

ANDRES, D. (1961). Die Struktur von Mastigograptus aus einem ordovischen Geschiebe Berlins. *Neues Jb. Geol. Paläont., Mh.* **12**, 636–647.

BOHLIN, B. (1950). The affinities of Graptolites. *Bull. geol. Instn Univ. Uppsala,* **34**, 107–113.

BULMAN, O. M. B. (1955). Graptolithina. *In* "Treatise on Invertebrate Paleontology, Part V" (R. C. Moore, ed.). Geol. Soc. Am. and Univ. Kansas Press.

BULMAN, O. M. B. (1970). Graptolithina with sections on Enteropneusta and Pterobranchia. *In* "Treatise on Invertebrate Paleontology, Part V" (R. . C. Moore, ed.). (2nd edition.) Geol. Soc. Am. and Univ. Kansas Press.

CHAPMAN, F. (1919). On some hydroid remains of lower Palaeozoic age from Monegetta, near Lancefield. *Proc. R. Soc. Vict.,* (2), **31**, 388–393.

CHAPMAN, F. and THOMAS, D. E. (1936). The Cambrian Hydroida of the Heathcote and Monegetta Districts. *Proc. R. Soc. Vict.,* (2), **48**, 193–212.

CROWTHER, P. and RICKARDS, R. B. (1977). Cortical bandages and the graptolite zooid. *Geol. et Palaeont.* **11**, 9–46.

DECKER, C. E. (1945). The Wilbern upper Cambrian graptolites from Mason, Texas. *Univ. Tex. Publs,* 4401, 13–61.

DILLY, P. N. (1976). Some features of the ulstrastructure of the coenoecium of *Rhabdopleura compacta. Cell. Tiss. Res.* **170**, 253–261.

HALL, J. (1883). Descriptions of new species of fossils found in the Niagaran group at Waldron, Indiana. *Geol. Surv. Indiana, Rept.* **11**, 217–346.

KIRK, N. (1974). Some thoughts on convergence and divergence in the Graptolithina. *Univ. Coll. Wales, Aberystwyth, Dept. Geol. Publ.* **5**, 1–29.

KOZLOWSKI, R. (1949). Les graptolithes et quelques nouveauz groupes d'animaux du Tremadoc de la Pologne. *Paleont. Pol.* **3**, 1–235.

KOZLOWSKI, R. (1961). Découverte d'un Rhabdopleuride (Pterobranchia) Ordovicien. *Acta palaeont. pol.* **6**, 3–16.

KOZLOWSKI, R. (1966). On the structure and relationships of graptolites. *J. Paleont.* **40**, 489–501.

KOZLOWSKI, R. (1967). Sur certains fossiles ordoviciens à test organique. *Acta. palaeont. pol.* **12**, 99–132.

OBUT, A. M. (1964). Podtip Stomochordata. Stomokhordovye. *In* "Osnovy paleontogii; Echinodermata, Hemichordata, Pogonophora; Chaetognatha" (Yu. A. Orlov, ed.), pp. 279–337. Nedra Press (Moskva).

RICKARDS, R. B. (1975). Palaeoecology of the Graptolithina, an extinct class of the phylum Hemichordata. *Biol. Rev.* **50**, 397–436.

RICKARDS, R. B. (1977). Patterns of Evolution in the Graptolites. *In* "Patterns of Evolution" (A. Hallam, ed.). Elsevier Sci. Publ. Co., Amsterdam.

RICKARDS, R. B. and CROWTHER, P. (1979). New observations on the mode of life, evolution and ultrastructure of graptolites. *In* "Biology and Systematics of Colonial Organisms" (G. Larwood and B. R. Rosen, eds), The Systematics Association Special volume No. 11, pp. 397–410. Academic Press, London and New York.

RICKARDS, R. B. and PALMER, D. C. (1977). "Graptolites 1." Baldwin Educational Palaeontological Publications.

RUEDEMANN, R. (1904–1908). Graptolites of New York, pts. 1, 11. *N.Y. State Mus. Mem.* **7**, 457–803; *Mem.* **11**, 457–583.

REUDEMANN, R. (1933). The Cambrian of the upper Mississippi Valley, pt. 3, Graptoloidae. *Publ. Mus. Milwaukee, Bull.* **12** (3), 307–348.

RUEDEMANN, R. (1947). Graptolites of North America. *Mem. geol. Soc. Am.* **19**, 1–652.

URBANEK, A. (1976). The problem of graptolite affinities in the light of ultrastructural studies on peridermal derivatives in pterobranchs. *Acta. palaeont. pol.* **21**, 3–36.

17 | The Origin of Chordates—a Methodological Essay

R. P. S. JEFFERIES

Department of Palaeontology, British Museum (Natural History), England

Abstract: The methodology of Hennig is basic to phylogenetic studies. It is discussed especially as it applies to fossils. The term "crown group" (as opposed to "stem group") is proposed for the latest common ancestor of the living members of a monophyletic group, plus all descendants of that ancestor. The criteria of primitiveness are reviewed in the light of Riedl's concept of burden. It is suggested that the stratigraphical criterion of primitiveness only applies when homologies are uncertain.

Within the deuterostomes, the echinoderms plus chordates form a monophyletic group with the rank of sub-superphylum for which the name Dexiothetica is proposed. The synapomorphies of this group, often secondarily lost, are a calcitic skeleton of stereom mesh and signs of descent from a *Cephalodiscus*-like ancestor that lay down on the original right side (*dexios* = right; *thetikos* = suitable for laying down). The hemichordates are probably a paraphyletic group of deuterostomes i.e. it is likely that some of them are more closely related to the dexiothetes than others are.

The term calcichordate is used informally for any chordate which retains a calcitic skeleton. The calcichordates can be divided into a number of groups according to closeness of relationship with living chordates. The traditional group of the cornutes belongs to the stem group of the chordates and can be divided into a number of intermediate categories showing increasing synapomorphy with the crown chordates. The traditional group of the mitrates represents primitive crown chordates. All known mitrates can be seen as stem acraniates, stem tunicates or stem vertebrates. The acraniates, contrary to received opinion, are probably the sister group of vertebrates plus tunicates, on the basis both of fossil and of recent evidence.

Systematics Association Special Volume No. 12, "The Origin of Major Invertebrate Groups", edited by M. R. House, 1979, pp. 443–477. Academic Press, London and New York.

INTRODUCTION

The first part of this paper outlines phylogenetic methodology, with particular reference to the work of Hennig (1966, 1969). In my view this work has not received from invertebrate palaeontologists the attention that it deserves. Systematics will make no real further progress until its methodology becomes explicit and explicitly connected with phylogeny. This justifies a long preliminary discussion.

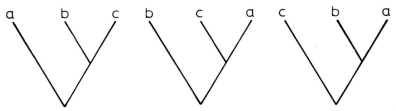

FIG. 1. The three possible relationships between three species *a*, *b* and *c*. These are topological diagrams. Which lines meet which, and the relative heights of the meeting points, shall signify, but nothing else does.

The second part of this paper, which cannot be understood without the first part, will apply Hennig's method to the particular problem of the origin of the chordates. This problem has intrinsic interest, of course, but I hope that it will also be useful as a methodological example.

METHODOLOGY

Hennig maintained that we have not properly understood the phylogeny of a group until we have arranged its members on a dichotomously branching tree of descent.

Hennig defined closeness of relationship exclusively in terms of recency of common ancestry. Any three species, even selected at random, are most unlikely to have arisen by trichotomy of the latest common ancestral species. It can be taken as certain that two of them will be more closely related to each other, in the Hennigian sense of sharing a more recent common ancestor, than they are to the third. If the species are *a*, *b* and *c* then there will be three possibilities (Fig. 1). Either *b* and *c* are more closely related to each other than they are to *a*; or *a* and *c* are more closely related to each other than they are to *b*; or *b* and *a* are more closely related to each other than they are to *c*.

How should we decide which of these possibilities is correct? Hennig suggested that to do this we should look for derived (advanced) character states in common between two of the species but not shown by the third one. Thus Fig. 2 supposes that there is a character I with the character states

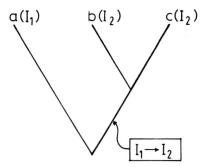

FIG. 2. A shared derived character state I_2 suggests that b and c are more closely related to each other than they are to a.

I_2 and I_2. Furthermore we have deduced, by some form of comparison, that I_1 is more primitive than I_2 and we know that a shows I_1 while b and c show I_2. In this case (Fig. 2) the first and initially most probable assumption is that the change from I_1 to I_2 has only happened once and within the exclusive common ancestry of $b+c$. These two species are therefore more closely related to each other than they are to a, for any other assumption implies either that the change I_1 to I_2 has happened twice in parallel, or else that a reverse change I_2 to I_1 has occurred. Both these complications can happen, but the first, most parsimonious, and initially most probable assumption is that they have not. A homologous shared, derived character state which shows that two taxa are more closely related to each other than they are to a third, because it evolved in the exclusive common ancestry of the two more closely related species, was called by Hennig a synapomorphy of the groups that show it. ("Apomorphous" was Hennig's term of "advanced" or "derived" whereas "plesiomorphous" was his term for primitive.)

Shared primitive characters, on the other hand, tell us nothing about phylogeny—at least they tell us nothing except that they make synapomorphies possible to recognize. The fact that shared primitive characteristics, or symplesiomorphies, are not directly useful in phylogenetic studies is Hennig's central and brilliant insight. For example, if a and b

show I_1, while c shows I_2 then, even without convergence or reversal of evolution, there are three possibilities as shown in Fig. 3. Since none of these possibilities implies more, nor less, convergent evolution than the others, all are *prima facie* equally probable. Because symplesiomorphy is not directly useful it follows that, even if we found the actual ancestor

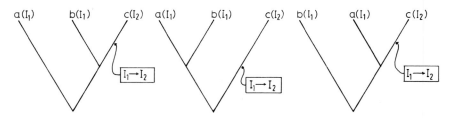

Fig. 3. Symplesiomorphy does not directly help.

of a monophyletic group, we could never, by purely morphological comparison, show that it was such. In my view it may, very rarely, be possible to show that one species is ancestral to another by observing a *gradual* and *continuous* stratigraphical change from ancestor to descendant. Apart from this exception, the question whether one organism is ancestral to another can only be answered when the answer is: "No" (cf. Henning, 1969, p. 39).

Hennig sharpened the concept of monophyly. For him a group was monophyletic if it comprised a latest common ancestral species, *all* descendants of that latest common ancestor and, by implication, nothing else. A group was called "paraphyletic" by Hennig if it comprised a latest common ancestor and some direct descendants of that ancestor but excluded some advanced forms which were descended from members of the group. (A paraphyletic group is necessarily a primitive portion of a larger monophyletic group). Finally a group was polyphyletic for Hennig if the ancestry of some of its members included forms descended from the latest common ancestor which themselves were not included in the group. The commonest type of polyphyletic group excludes the latest common ancestor and the immediate descendants of the latter (Fig. 4c), but this is not necessarily true of all polyphyletic groups (Fig. 4d). Paraphyletic groups would traditionally be accepted as monophyletic. However, paraphyletic groups are like polyphyletic ones in that some members of the group will be more closely related to particular non-members than to some other members. With monophyletic groups, on the other hand, all members are more closely related to all other members than to any non-members.

Monophyletic groups will in practice be recognized as such because all members, or at least all primitive members, share an advanced feature absent from all other members of the monophyletic group of higher rank

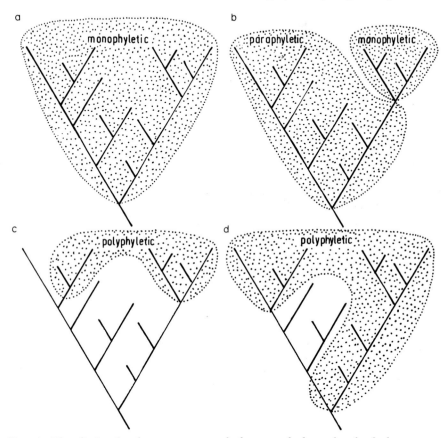

FIG. 4. The distinction between monophyly, paraphyly and polyphyly.

of which it forms part. Thus, within the amniotes, the mammals are characterized by having hair, warm blood and milk glands (for instance), which the primitive amniote would not have had. Paraphyletic groups, on the other hand, in practice are recognized by means of particular combinations of advanced and primitive characteristics. Thus reptiles are characterized by possessing an amnion, which is advanced compared to non-amniotes; but they lack feathers, which is primitive compared to birds; and they lack hair and milk, which is primitive compared to mammals. Finally polyphyletic groups are in practice recognized by means of a defining convergence. Thus the group of "all flying vertebrates" is characterized by wings, but these have certainly evolved independently in flying fish, birds, pterosaurs and bats. Hennig recommends that we should classify living organisms, so far as at all possible, into strictly monophyletic groups. Paraphyletic groupings of extant organisms should be avoided, or at least recognized as such. Paraphyletic groupings of fossils are sometimes unavoidable. Polyphyletic groupings should, of course, always be avoided, as traditional systematics has long recognized. There is a presumption that, within a monophyletic group, any subgroup characterized only by features primitive to that monophyletic group may well prove to be paraphyletic. Thus, within the monophyletic group of the vertebrates, there is a well-founded suspicion that the agnathans are paraphyletic, the lampreys being more closely related to the gnathostomes than the hagfishes are (Løvtrup, 1977).

The sister-group concept was proposed by Hennig as a way of helping to describe a dichotomously branching tree. When a "mother" species splits into two, then its two "daughter" species will be "sisters" of each other. And one "sister" plus all its descendants will together comprise the sister group of the other "sister" plus all *its* descendants. It is important that, when we become convinced that such a dichotomy has occurred, neither of the two sister groups should be given the same name as the mother group—or at least that we should strive to avoid giving them the same name. For, if we do otherwise, we are implying that one of the two daughters is identical to the mother but the other not. This can only lead to confusion if the aim is to arrange all organisms on a dichotomously branching tree of descent. Thus, if echinoderms are the sister group of chordates (Fig. 12), the latest common ancestor of both should not itself be regarded as echinoderm or as chordate but as belonging to the monophyletic group including both. (In the terms explained below, it will be the first crown dexiothete). If we call it an echinoderm, for example, it would be psychologically difficult to avoid ascribing features to it which in fact probably evolved later, within the monophyletic group of the echinoderms. We might be tempted to assume, for example, that it had possessed radial symmetry. Possibly this assumption may prove to be true, but at present there are no grounds for believing it. Obviously the aspiration to name a mother group differently to the two daughter groups may be difficult to achieve, particularly as concerns species within a genus.

Hennig's views on applying his methodology to fossils were expounded in "Die Stammesgeschichte der Insekten" (1969). Suppose we have two monophyletic groups (1 and 2) of extant organisms, each being the sister group of the other (Fig. 5). Then, if we take account of extinct relatives also, there will be two obvious delimitations, wide or narrow, of each of these groups. The narrow delimitation of group 1, for example, would comprise the latest common ancestor of all living members of group 1, *plus* all descendants of this ancestor, whether these are living or extinct. Hennig referred to the subgroup thus narrowly delimited as the ⋆ group, but I propose to call it the crown group. The wide delimitation of group 1 is more difficult to express in words. The latest common ancestor of groups 1 and 2 gave rise to two daughter species, one of them ultimately ancestral to the living members of group 1 and the other ultimately ancestral to the living members of group 2. The wide delimitation of group 1 would include the daughter species ultimately ancestral to the living members of 1, plus all descendants of that daughter, both living and extinct. This wide delimitation can be called the total group. Now, if we subtract the crown group of 1 from the total group 1, we are left with a residue of extinct forms which can be called the stem group of 1. "Stem" and "crown" suggest an obvious comparison with an apple tree, for example—a comparison which holds good for some other languages beside English. Stem groups are extinct by definition and will be paraphyletic in that some members will be more closely related to the crown group than others are. All monophyletic groups containing more than one living species will, in principle, be divisible into a stem group and a crown group.

Within a stem group it is possible to recognize different degrees of relationship to the crown group. In practice this is possible because not all synapomorphies connecting members of a crown group together, and distinguishing it from its sister crown group, will have been acquired at once. Instead they will have been acquired successively within the two stem groups. This allows each stem group to be divided into a series of intermediate categories (Zwischenkategorien) showing increasing relationship to the respective crown group. The minimal limiting case of an intermediate category is the plesion (Patterson and Rosen, 1977). A plesion (Fig. 5) is a group comprising all those members of the stem group which are equally closely related to the crown group: i.e. all members of a plesion

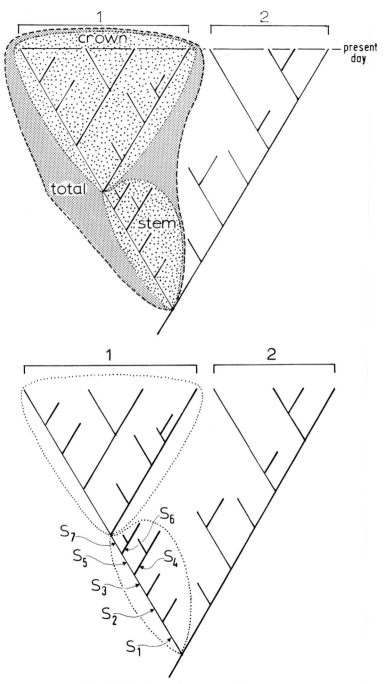

Fig. 5. For legend see opposite

will descend from one and the same member of the lineage leading to the crown group and this member will be the latest common ancestor shared with the crown group. Each plesion will be recognized in practice by possessing a synapomorphy with the crown group which more primitive plesions lack, and by lacking a synapomorphy with the crown group which more advanced plesions possess; or, in the case of the most advanced plesion, by lacking a synapomorphy which the crown group possesses. These synapomorphies *need* be shared only with early, extinct members of the crown group. However, they will be shared with living members of the crown group if the latter have not lost them secondarily. In addition, if the plesion contains more than one known species, it can only be recognized as a plesion if its members are connected by a synapomorphy not shared by plesions more closely related to the crown group—this synapomorphy shows that the members of the plesion are *not* members of the lineage that led direct to the crown group.

I shall finish this summary of Hennigian methodology by discussing how to recognize primitiveness. Here I become somewhat eclectic, drawing on the work of Riedl (1975; 1979) and adding a few thoughts of my own. There are, first, a number of morphological methods whose ultimate logical basis is the recognition of features as being homologous. (I use the word "morphological" here in a wide sense, to cover even histological, cytogenetic or biochemical resemblances). These morphological methods are widely applicable. They can be contrasted with stratigraphical and geographical methods which rest on a different logical basis and, in my view, are much less widely applicable though useful in restricted instances.

The first morphological method can be called in-group comparison. If a homologous feature is universal in a monophyletic group, then it will have been present in the latest common ancestor of the group (e.g. hair in mammals). With infinitesimally less certainty, if a homologous feature is universal in a monophyletic group except where it has been secondarily lost, it will have been present in the latest common ancestor.

FIG. 5. Stem group, crown group, total groups, intermediate categories and plesions. The stem group of 1 is shown as containing four plesions, with increasing relationship to the crown group of 1. Synapomorphous changes S_1, S_2, S_3, S_5, S_7 indicate the evolution of synapomorphies in common with the crown group. S_5 and S_6 are the origins of synapomorphies which show that the two most advanced plesions are both monophyletic. Plesions are minimal intermediate categories. Two or more successive plesions constitute a non-minimal intermediate category.

The second morphological method is out–group comparison. If a feature is present among many relatives of a monophyletic group, or best of all in the sister group, and if its homologue occurs, sometimes or

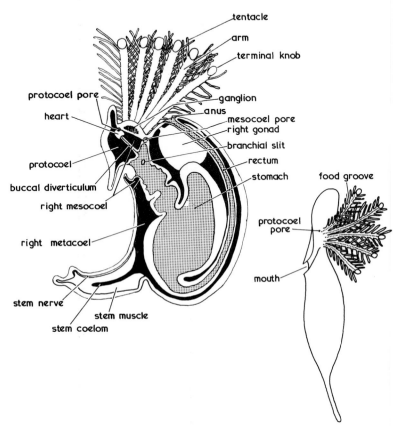

FIG. 6. The pterobranch hemichordate *Cephalodiscus*. From van der Horst, after Schepotieff and Andersson. The animal is shown bisected sagitally in left aspect, and from the left externally. In the bisected animal the projected positions of the protocoel pore and mesocoel pore are shown.

universally, in the members of the monophyletic group, it will have been primitive for the group. An example is the teeth of *Archaeopteryx* which are assumed to be a primitive feature for birds, although not represented in modern birds. This is because teeth are present in crocodiles (the living sister group of birds) and in other amniotes, tetrapods and indeed vertebrates. When in-group and out-group comparisons both point in the same direction we can say with virtual certainty that a feature was primitive for a monophyletic group.

Another morphological method is based on Haeckel's law of recapitulation. New additions to an organ are more easily added to the end of its ontogenetic development than interpolated into that development. For accidental interpolation will usually disrupt the development of the organ and kill the organism. Thus the sequence of embryological inductions in the ontogeny of an organ will usually parallel its phylogenetic development.

All these morphological methods are based on recognizing homologies, and it is necessary to consider how this can be done. Riedl (1975, 1979) has compared the identification of homologues with a recognition that repeated messages are identical. Thus consider a source that transmits the series of numbers: 75783 75783 75783 75783 75783 ..... . Even after the first repetition everyone would suppose that here was orderliness, with the same decisions behind the second group of five numbers as behind the first. After all, the probability of the second occurrence exactly resembling the first by chance alone, assuming that all digits can be chosen with equal probability, is 1:100000. Homologues, in biological practice, are more like not quite identical repetitions of a message, perhaps: 75783 75793 75784 75783. Here again the probability that chance alone is responsible for the repetitiveness is negligible, though exact calculation of that probability will depend on our picture of the decisions behind the choice of numbers. Homologues will normally be obvious to anyone faced with all the relevant data and are a special case of biological identicality. (The human brain mainly functions by recognizing things as identical).

Riedl also points out that all complex homologues are hierarchical in structure. There are minimal homologues at the bottom of the hierarchy and these are grouped into homologues of higher rank and these into homologues of still higher rank. Within a complex homologue, the number of minimal homologues plus the number of homologues of the various ranks higher than minimal (cadre homologues) can be added together to give a measure of the complexity of the complex homologue, in unit homologues. Thus the mammalian vertebral column is a homologue in all mammals (and outside the mammals) with five homologous regions within it; the cervical region is a homologue in all mammals with seven homologous vertebrae within it (with a few exceptions); the second cervical vertebra (axis) is a homologue in all mammals, with a number of homologous features within it, including the odontoid process; the odontoid process is a homologue in all mammals with several smaller homologues within it, including the ventral articular facet; however, the ventral articular facet of the odontoid process is in some ways an end point. If it is subdivided further, we do not meet single homologues, recognizable in related species or individuals. Instead we encounter large numbers of structures such as Haversian columns, no single one of which can be homologized with a particular Haversian column in another individual, or another species. The name "homonom" is current in the German literature for such mass homologues. Riedl estimates the complexity of the mammalian vertebral column at 4381 unit homologues. If each homologue is assumed, conservatively, to depend on a single yes/no decision, then a mammalian vertebral column would require 4381 binary digits for its stipulation. The probability of identical repetition by chance is therefore $(\frac{1}{2})^{4381}$ which is negligible. Riedl points out that minimal homologues will be more subject to change than cadre homologues. This is because, within the system in question, no homologues depend on them. This leads Riedl to propose the concept of "burden". The burden of a unit homologue, in this connection, is the number of other unit homologues included in it. This type of burden I call homological burden. A mammalian vertebral column has a burden of about 4381 unit homologues and the decision, or mutation, not to produce a vertebral column would certainly kill the animal. The ventral articular facet of the odontoid process, on the other hand, has no other homologues included in it. Its homological burden is nothing. The decision not to produce it would probably not be fatal, but merely subvital.

Another sort of hierarchy can be called positional hierarchy. In it the high ranks have a central bodily position, the low ranks a peripheral one. In this connection we can speak of positional burden. Thus the distal phalanges of the fingers have a lower positional burden than the humerus. In fact it is observed that, within mammals for example, the digits show much more taxonomic variety than the humerus. Moreover, in ontogeny there is a hierarchy based on order of origin with early decisions having a higher ontogenetic burden than later ones, and this is the basis for Haeckel's law. Whenever we can recognize a hierarchy, the degree of fixation of a unit homologue will increase with its burden, whether this is homological, positional or ontogenetic.

The certainty with which a homologue can be recognized will depend on its complexity, i.e. on its homological burden. So long as we can recognize homologues, we can deduce that features are primitive for a monophyletic group by using in–group or out–group comparison. When dealing with features of low burden, however, such as distinguishing closely related forms from each other, homologues will be difficult to recognize and convergence will become an increasing nuisance to the phylogeneticist.

The reason why closely related groups are distinguished by homologues of low burden, while high-ranking groups are distinguished by homologues of high burden, is that complex homologues do not originate as such. The highest-ranking constituent homologues of a complex homologue were the first to appear in phylogeny and were minimal homologues when they arose. Their burden subsequently increased by the addition of new homologues upon them, simultaneously raising their rank, their recognizability and their burden.

This brings me to non-morphological methods of recognizing primitiveness. I do not propose to discuss geographical distribution, but the stratigraphical criterion of primitiveness obviously needs attention. Perhaps it is logically the simplest criterion in that known, geologically early members of a group are *prima facie* more likely to be primitive than known, geologically late members. Indeed if the stratigraphical record were complete, there would be no need for any other method and we could follow phylogeny, without benefit of self-conscious methodology, by stepping from one species to the almost indistinguishable next one. Even in this case, phylogeny would still be traceable only by recognizing homologous derived features in common i.e. synapomorphies. With a perfect stratigraphical record, however, the synapomorphies would be common to observed ancestor and descendant, as well between descendants. This naive stratigraphical method is what many palaeontologists have sought to use even when gaps in the record are obvious. Indeed, with real geological records rather than perfect ones, the stratigraphical method has to be used with caution, or rather unceasing distrust, for we can never show it is complete in any relevant aspect. Moreover, in considering the two species b and c which are the sister group of a (Fig. 2), it is evident that the exclusive common ancestry of b and c, during which the synapomorphous change $I_1 \quad I_2$ occurred, may have lasted a very short time indeed. Theoretically it need not have lasted longer than two generations. The prior existence of related forms sharing the primitive condition I_1 may also be very short. On the other hand the subsequent existence of a, b and c may last hundreds of millions of generations. Under these unfavourable circumstances an incomplete stratigraphical distribution will not give the information needed about what is the primitive condition for character *I* and, indeed, will give the wrong answer as often as the right one. Morphological comparisons involving homologies, however, can often allow sure phylogenetic deductions about events that took place hundreds of millions of years ago even in the absence of any fossil record. Thus the presence of notochord, dorsal nerve cord and associated muscles, and also a pharynx with gill slits, endostyle, etc. in tunicates, acraniates and vertebrates showed that the chordates are a monophyletic group long before their fossil record was appreciated.

However, as already mentioned, the certainty with which homologues can be recognized as such depends on their complexity i.e. the number of constituent unit homologues. And complex synapomorphies are characteristic of the differences between high-ranking groups of organisms. Consider for example, the complex synapomorphies which hold living chordates together as a group, compared with the sole synapomorphy of the Giraffidae which is an extra cusp on the lower canine (Hamilton, 1978).

I therefore propose a methodological rule: the stratigraphical criterion of primitiveness is applicable when the relevant homologies (synapomorphies) cannot be certainly recognized without it. Conversely, the stratigraphical criterion is *not* applicable when synapomophies *can* certainly be recognized without it. The stratigraphical criterion will tend to apply when the differences concerned are of low homological, positional or ontogenetic burden i.e. the sorts of differences that separate species and subspecies rather than phyla. Palaeontologists do right, therefore, to emphasize stratigraphy in discussing evolution within a genus of ammonites, for here the differences, in practice, are confined to ribbing, involution and sutures, so that convergence is likely and homologies uncertain. They are also right to ignore stratigraphy in discussing the relationships of high-ranking groups. The adepts of these two different types of study belong to two almost mutually exclusive camps, both somewhat hostile to the other. This hostility exists because|neither camp understands, nor values, what the other is trying to do. The first step in peace-making is to mark a boundary between the two sides. This is what I have tried to do in suggesting that the stratigraphical criterion of primitiveness only applies when homology is otherwise uncertain.

Two thought-experiments will illustrate the varying value of the morphological and stratigraphical criterian. Firstly, it might be supposed that the teeth of *Archaeopteryx* are to be regarded as a primitive feature of birds merely because *Archaeopteryx* is the oldest bird known (upper Jurassic, about 150 million years old). However, this reasoning is false, because, if a toothless bird were discovered tomorrow in the upper Jurassic, nobody would change their opinion about the primitiveness of the teeth of *Archaeopteryx*. In point of fact this primitiveness is deduced by out-group comparison, which is firmly based on the homology of the teeth of *Archaeopteryx* with those of crocodiles and other vertebrates in general. It cannot be shaken by future finds, unless these, inconceivably, made the homology doubtful. Where homology is firmly based, stratigraphical sequence is irrelevant, except to provide a time scale—usually a series of latest possible dates—for the reconstructed dichotomies. However fossils, as opposed to stratigraphical sequence, may still be relevant, to the extent that synapomorphies may be discovered among them that do not still exist in living forms.

On the other hand, consider the echinoid genus *Micraster*, as a converse to the teeth of *Archaeopteryx*. Rowe (1899) demonstrated a famous sequence of phylogenetic changes in this genus in the English Chalk in rocks ranging from Turonian to Santonian in age. The early ones have, for example, no prominent labrum (lower lip) to the mouth and a number of other features are assumed to be primitive purely on stratigraphical grounds. The late ones have a long labrum and a number of other features assumed to be advanced on stratigraphical grounds. But suppose that tomorrow somebody discovered a morphologically "late" *Micraster* contemporary with the early ones; this would throw the gravest doubt on Rowe's phylogeny by destroying the base for supposing that "short labrum", for example, was primitive. All the differences between *Micraster* in the English Chalk are features of low homological and positional burden, separating species within a genus. With them, the stratigraphical criterion of primitiveness is paramount.

To sum up, Hennig's method of reconstructing phylogeny is the search for sister groups by finding synapomophies. This method is more powerful than searching for ancestors because ancestor–descendant relationships will exist within a group of known organisms only by good luck, can exist between contemporaneous organisms only by grace of metaphor, and are logically impossible to prove, by purely morphological methods, between any two known species. Sister-group relationships, on the other hand, must always exist within any group of known organisms and, in favourable cases, can be proven by recognizing synapomorphies. Among taxa of high rank, fossils, but not their stratigraphical sequence, are relevant to the search for sister groups because stem groups may show synapomorphies which living groups have lost all trace of (this is contrary to Løvtrup, 1977, p. 21). After all, the primitive members of a group are in particular danger of becoming extinct by competition with their advanced relatives. Furthermore, fossils, but not their stratigraphical sequence, may allow the order of origin of synapomorphies within a stem group to be established. The stratigraphical sequence of fossils will indicate the time scale of phylogeny. However, when the relevant homologies can be established with certainty on purely morphological grounds, as is often true for the differences between high-ranking groups, stratigraphical sequence will not help in working out phylogeny and indeed will be deceptive. Stratigraphy will be useful in working out phylogeny only when the relevant homologies cannot otherwise be recognized as such, i.e. in dealing with difference of low rank and burden. At that level, stratigraphy is essential—*faute de mieux*. Above that level it is a false guide.

A HENNIGIAN APPROACH TO THE CLASSIFICATION OF THE DEUTEROSTOMES

The deuterostomes contain the conventionally recognized phyla of hemichordates, echinoderms and chordates (Fig. 12). They are characterized as a group by the fact that the mouth appears in ontogeny as a secondary perforation connecting the archenteron with the outside (1 in Fig. 12). The blastopore becomes the anus, or else the anus appears as a secondary perforation in the region of the blastopore. This is contrasted with the protostomatous condition where the blastopore becomes the mouth in ontogeny. The protostomatous condition is likely to be more primitive than the deuterostomatous one, on the basis of Haeckel's comparison of a gastrula with a coelenterate. If this comparison holds, which is inevitably uncertain, the secondary mouth can be regarded as a synapomorphy of the deuterostomes and the latter are a monophyletic group.

The hemichordates are conventionally divided into the tentaculate pterobranchs and the worm-like enteropneusts. However all the features which characterize the hemichordates as a group are primitive deuterostome ones. There is therefore a distinct possibility that the hemichordates are paraphyletic i.e. some of their members are more closely related to the echinoderm–chordate group than others are.

Grobben (1924) compared echinoderms with the pterobranch hemichordate *Cephalodiscus* (Fig. 6). The latter is a very important animal and requires description. It is semi-colonial, living in horny tubes on the sea floor. Each individual zooid is divided into head shield, collar and trunk (protosoma, mesosoma and metasoma). The head shield contains a single unpaired coelom (protocoel). The collar contains a pair of coeloms (mesocels), each of which extends into a group of arms with tentacles. The trunk also contains a pair of coeloms (metacoels). This condition with five coeloms is conventionally, but confusingly, referred to as tricoelomatous. The gut is U-shaped and situated in the trunk. It is suspended in a mesentery formed by the medial walls of the left and right metacoels. A pair of branchial slits penetrates the walls of the pharynx and connects the latter with the outside. Also in the trunk are the paired gonads. The trunk extends into a muscular contractile stem in which the left and right metacoel are not separated from each other. The protocoel opens by a pair of pores and each mesocoel also opens by a pore, but the metacoels are not open to the exterior. Inside the protocoel is a heart, a possibly excretory organ called the glomerulus and an extension from the pharynx called the buccal diverticulum which was once held to be homologous with the notochord. The arms are ciliated and a ciliated groove runs down the ventral side of each one to the mouth. The nervous system of *Cephalodiscus* consists of a meshwork of neurons in the deep parts of the ectoderm, especially thick in the ventral part of the stem and dorsally between the two groups of arms (ganglion). The symmetrical "tricoelomatous" condition seen in *Cephalodiscus* is presumably primitive for deuterostomes, by out-group comparison with the tentaculate proto-stomatous phyla of ectoprocts ("bryozoa") and brachiopods.

A series of homologies can be recognized between *Cephalodiscus* and echinoderms (Table I). As Hyman pointed out (1959), these resemblances suggest a close relationship between *Cephalodiscus* and echinoderms. However, echinoderms are less symmetrical than *Cephalodiscus*, lacking many organs of the right side. Grobben assumed that in this respect *Cephalodiscus* retained the more primitive condition. Out-group comparison with brachiopods and ectoprocts makes it likely that he was right.

The relevant fossil evidence can now be introduced, I have been working since 1964 on a group of strange fossils conventionally regarded as "carpoid" echinoderms but which I see as chordates, broadly in accordance with a suggestion by Gislén (1930). I call this group calci-chordates, but no longer regard it as a subphylum (see Jefferies and Lewis, 1978, p. 209). The calcichordates range geologically from middle Cambrian to middle Devonian. I have reviewed them in some detail in Jefferies (1975), so the present treatment can be shorter and will not consider functional anatomy nor why particular organs are identified as such. The calcichordates are divided into two traditional groups: the cornutes and the mitrates. The cornutes are almost certainly more primitive than the mitrates and, as a group, are ancestral to them.

The cornute *Ceratocystis* is the first calcichordate to be considered (Fig. 9). It comes from the middle Cambrian of Bohemia and is known from

TABLE I. Homologies between *Cephalodiscus* and echinoderms

Cephalodiscus	Echinoderms
Protocoel	axocoel (axial sinus)
Left protocoel pore	hydropore
Right protocoel pore	absent
Glomerulus	axial gland
Pericardium and heart	madreporic vesicle (dorsal sac)
Buccal diverticulum	absent
Left mesocoel	hydrocoel (water vascular system). Note the feeding function which this organ has in common with its homologue.
Right mesocoel	absent in normal adults but present in abnormal larvae and abnormal imagos
Mesocoel pores (left and right)	absent; the left hydrocoel is connected by an internal opening to the axocoel so the water vascular system opens indirectly to the outside by the hydropore.
Left metacoel	left somatocoel
Right metacoel	right somatocoel
Left gonad	gonad
Right gonad	absent
Gill slits (left and right)	absent

one species only (*Ceratocytis perneri* Jaekel, 1900). It has been described in detail by me (Jefferies, 1969) and also, with a different anatomical interpretation, by Ubaghs (1967, 1968). It consists, like all calcichordates, of head and tail. I formerly called these parts body and tail (Jefferies, 1975) or theca and stem (Jefferies, 1967, 1968, 1969, 1973; Jefferies and Prokop, 1972). The terms "head" and "tail", however reflect the best comparison with vertebrates (see Jefferies and Lewis, 1978, p. 211). The terms theca and stem reflected a mistaken homology with crinoids (discussed in

Jefferies and Lewis, 1978, p. 211; see Sprinkle, 1973) and my change of mind on the matter follows criticism by Eaton (1970).

The head of *Ceratocystis* is boot-shaped and its skeleton was made up of large plates of echinoderm type. (There is no direct evidence that it was calcitic in *C. perneri*, but it certainly was so, since the individual plates can be homologized with those of *Nevadaecystis* (Fig. 9) which observably consisted of calcite with each plate a single crystal). The head had a flattish ventral surface that lay on the sea floor and a convex dorsal surface traversed by three ridges which meet at a junction. There are signs that the roof could move slightly up and down relative to the sides of the head, by the action of muscles between the sutures (Jefferies, 1969). This power of movement presumably functioned in coughing, serving to eject unwanted particles from the pharynx. It points towards the condition in *Nevadaecystis* and *Cothurnocystis* where the roof of the head was a flexible integument. The openings of the head consisted of seven gill slits on the left side while on the right side there was a mouth anteriorly, and a hydropore, gonopore and an anus posteriorly (Fig. 9). Inside the head there is evidence for the same five chambers as in *Cothurnocystis* (see below). The tail of *Ceratocystis* was divided into fore, mid and hind parts and ended abruptly. The skeleton of the fore tail had a wide lumen. This was probably filled with muscle and would have contained an anti-compressional notochord along its central axis so that the mid and hind tail could wag from side to side. The tail probably functioned to pull the head backwards across the sea floor. Where the tail joined the head there was a brain with large left and right trigeminal ganglia. The brain had a median eye above it and there was an ear, presumably functioning as lateral line, occupying a groove left of the left trigeminal ganglion (see Jefferies, 1969).

From the classificatory point of view, the most significant feature of *Ceratocystis* is its hydropore (h in Figs 7, 9). This is a symplesiomorphy with echinoderms which all other calcichordates lack. Its loss in all other chordates is a synapomorphy of them, which implies that *Ceratocystis* was the primitive sister group of all other calcichordates plus their descendants. This conclusion, much as with the teeth of *Archaeopteryx*, agrees with the fact that *Ceratocystis* is one of the oldest calcichordates known, but does not depend on it. For a *Cothurnocystis*, in several respects more advanced than *Ceratocystis*, is also known from the middle Cambrian (Sprinkle, 1976).

The extraordinary shape of *Ceratocystis*, with the anus right of the tail

Ceratocystis   Cephalodiscus

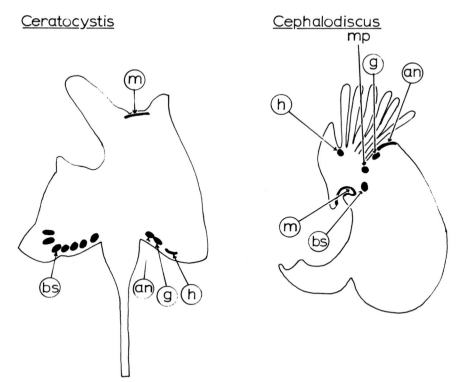

Fig. 7. The figure represents *Ceratocystis*, viewed in dorsal aspect, and *Cephalo-discus*, viewed in left lateral aspect. The sequence of openings encountered in going round the diagrams clockwise is the same in both cases i.e. bs = branchial slit; m = mouth; h = hydropore; g = gonopore; an = anus. The mesocoel pore (mp), is an opening not represented in *Ceratocystis*.

and the boot-shaped head, demands an explanation. It suggests descent from a bilaterally symmetrical ancestor which lay down on one side, flatfish fashion. (This recalls a suggestion by MacBride (1909) who proposed that amphioxus had such a stage in its ancestry.) More particularly it suggests a *Cephalodiscus*-like ancestor that lay down on the right, for the sequence of openings recorded in passing round a diagram of *Cephalodiscus* in left aspect, is like the sequence in passing round *Ceratocystis* in dorsal aspect (Fig. 7).

Such an ancestral phase, lying right-side-downwards on the sea floor, would also explain why echinoderms, as compared with *Cephalodiscus*,

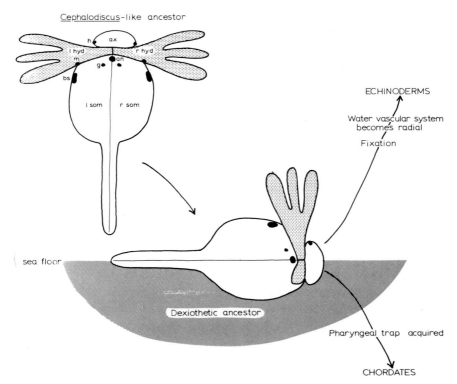

FIG. 8. The dexiothetic condition. The resemblances between *Ceratocystis* and *Cephalodiscus* (cf. Fig. 7) suggest that *Ceratocystis* descended from a *Cephalodiscus*-like ancestor (seen in dorsal aspect) that lay down on its right side. The same origin is likely for echinoderms, suggesting that chordates and echinoderms together form the monophyletic group Dexiothetica. an = anus; ax = axocoel; bs = branchial slit; g = gonopore; l hyd, r hyd = left and right hydrocoel; l som, r som = left and right somatocoel; m = mesocoel pore.

lack the openings and tentacles of the right side (Fig. 8). The body orientation of the phase in question can be called dexiothetic (*dexios* = right, *thetikos* = suitable for laying down) and traces of ancestral dexiothetism can be regarded as a synapomorphy of chordates and echinoderms. I therefore propose the name Dexiothetica, with the rank of sub-superphylum, for the monophyletic group of echinoderms plus chordates. Another synapomorphy that strongly confirms that the dexiothetes are a monophyletic group is the calcite skeleton made of stereom mesh with each plate a single crystal of calcite. Obviously the recognition of Dexiothetica as

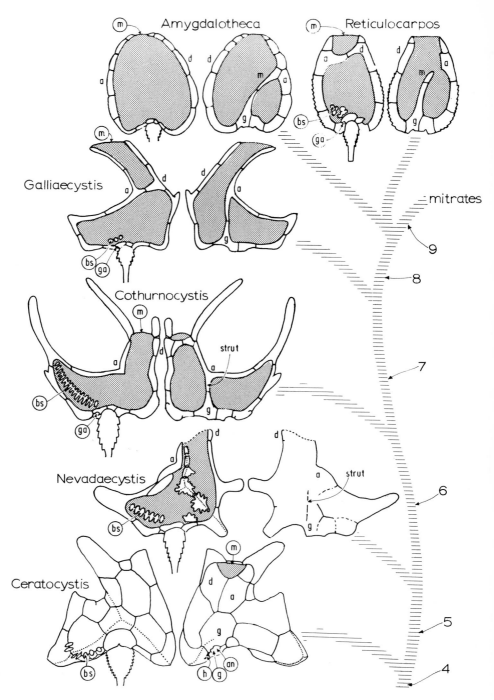

Fig. 9. For legend see opposite.

a monophyletic group depends crucially on fossil evidence. Dexiothetism and the calcite skeleton were acquired at point 2 of Fig. 12.

The next fossil to be described is *Nevadaecystis americana* (Ubaghs) (Fig. 9) known as a single specimen from the upper Cambrian (Tremadoc) of Nevada (Ubaghs, 1963, 1968; Jefferies, 1969). The preservation is tantalisingly incomplete but the head is like *Ceratocystis* in being boot-shaped. The floor of the head must have been rigid for it is made up of a small number of large plates and the same plates are thickened to form a frame round the head. A linear thickening of the floor, passing forward just right of the tail, is homologous with the strut across the ventral integument of *Cothurnocystis*. The most striking difference from *Ceratocystis* is the fact that the roof of the head was obviously flexible. The plates of the roof are mostly star-shaped and often have large gaps between them which presumably lodged muscles. This would allow more vigorous coughing than in *Ceratocystis*. There were seven gill slits in the left dorsal integument and in structure they were like those of *Cothurnocystis*. It seems almost certain that *Nevadaecystis* had no hydropore. The position of the gonopore-anus is uncertain but presumably was in the dorsal integument, intermediate in position between that of *Ceratocystis* (right of the tail), and that of *Cothurnocystis* (left of the tail). The lack of a hydropore is a synapomorphy of *Nevadaecystis* with all other chordates except *Ceratocystis*, as already mentioned. The flexible roof to the head, the fact that the gonopore-anus is not right of the tail and that the gill slits are of *Cothurnocystis* type all distinguish *Nevadaecystis* from *Ceratocystis* and constitute synapomorphies with more advanced cornutes (point 5 in Figs 9, 12). *Nevadaecystis* can be seen as the primitive sister group of all other calcichordates except *Ceratocystis*. It is stratigraphically too late to be ancestral to the forms of which it is the primitive sister group, since *Cothurnocystis* is known already from the middle Cambrian. However, this is a common situation which does not disturb a classification based on Hennigian principles. *Ceratocystis perneri* and *Nevadaecystis americana* constitute separate plesions in the sense of Patterson and Rosen (1977).

The next fossil to be described is *Cothurnocystis elizae*, from the Ashgill

---

FIG. 9. Phylogenetic relations in the cornutes. Dorsal aspect, left; ventral aspect, right. Numbers refer to evolution of synapomorphies (cf. Fig. 12). Integument stippled. Letters without circles: a, d, g, m = homologous plates (for full explanation see Jefferies and Lewis, 1978, p. 223). Letters inside circles, an = anus; bs = branchial slit; g = gonopore; h = hydropore: m = mouth. The trichotomy leading to *Reticulocarpos*, *Amygdalotheca* and the mitrates represents two unresolved dichotomies.

(uppermost Ordovician) of Scotland (Fig. 9). It is the type species of *Cothurnocystis* and, along with *Scotiaecystis curvata*, it is stratigraphically the last cornute known. The head is boot-shaped (*cothurnos* = boot) with a flexible dorsal and ventral integument and a frame of marginal plates. The ventral integument is crossed by a strut homologous with the strut of *Nevadaecystis*.

The openings of the head are: a mouth at the front end, surrounded by a cone of pointed integument plates; a series of about 16 gill slits in the left part of the dorsal integument; and a gonopore-anus situated just left of the tail. The gonopore-anus is connected by a groove inside the skeleton to the approximate position that the gonopore-anus occupies, right of the tail, in *Ceratocystis*. Each gill slit has a frame round it of two U-shaped plates, articulated together by the tips of the U's. Attached to the anterior U was a soft tongue, convex outwards, stiffened by platelets. Such a structure is excellently adapted as an outlet valve (Jefferies, 1968) which confirms the suggestion, made originally by Gislén (1930), that these openings were in fact gill slits.

The embryology of the gill slits of amphioxus is very remarkable, and relevant here. The slits appear in three groups which can be called early, intermediate and late ("primary", "secondary" and "tertiary" of Lankester and Willey, 1890; Willey, 1891). This ontogenetic sequence is totally asymmetrical. The early slits begin to appear when the animal starts to feed and are morphologically left gill slits. They perforate the pharyngeal wall in sequence, beginning at the front end of the series. The intermediate gill slits appear towards the end of larval life and are right gill slits morphologically. They appear more or less simultaneously just right of the early slits. The late slits begin to form in the early imago but continue to form throughout life and are morphologically left and right. They are added posterior to the early and intermediate slits. This history is recapitulatory in Haeckel's sense. The early slits recapitulate the situation seen in *Cothurnocystis* and other cornutes, which all have left gill slits only. The early slits are probably also homologous with the left gill slits of tunicates and vertebrates. The intermediate slits were probably represented in the mitrates and are homologous with the right gill slits of vertebrates and tunicates. And the late gill slits are a peculiarity of amphioxus and other acrianates among living animals. They were present also in the mitrate *Lagynocystis* which, for this reason, can be assigned to the stem group of acraniates as its only known member (Figs 11, 12).

Five chambers probably existed inside the head of *Cothurnocystis elizae* (Fig. 10, c, d). Four of these are based on direct skeletal evidence while the fifth is based purely on comparative evidence. Firstly, there was the buccal cavity, situated just behind the mouth, and occupying the "ankle" part of the boot. Secondly, there was a posterior coelom situated just in front of the tail. Thirdly, there was the pharynx, situated in the foot part of the "boot", with the gill slits emerging through its roof. Fourthly, there was the right anterior coelom, situated also in the "foot" part of the "boot"; it underlay the pharynx and was most capacious in the posterior right portion of the head; it would have contained the non-pharyngeal gut, the heart and pericardium, the gonad and other viscera (see Jefferies and Lewis, 1978). The fifth chamber, the one based on comparative evidence only, is the left anterior coelom. The argument for its existence depends on a comparison with *Cephalodiscus*, imagined as lying on its right side. This comparison suggests that the right metacoel of *Cephalodiscus* would be homologous with the ventrally situated right anterior coelom of *Cothurnocystis*. A homologue of the left metacoel would therefore be expected to exist, overlying the pharynx of *Cothurnocystis*. This hypothetical chamber, which would have had no patent cavity, I call the left anterior coelom. It would be homologous with the left somatocoel of echinoderms (cf. Fig. 8).

Compared with *Nevadaecystis*, *Cothurnocystis elizae* has a number of advanced features in common with mitrates, and the latter, as shown below, represent primitive crown chordates. The most important such features are the flexible floor to the head and the gonopore-anus situated just left of the tail; these would have evolved at point 6 in Figs 9, 12. There are several other species of *Cothurnocystis* and other genera, which share these same features with *Cothurnocystis elizae*, but do not show any additional synapomorphies with the mitrates. All these forms, together with *C. elizae* itself, can be regarded as forming an intermediate category within the stem group of the chordates. This intermediate category would include: *Chauvelicystis*, characterized by a fringe of spines (see Ubaghs, 1970); *Phyllocystis*, with its externally almost symmetrical head (see e.g. Ubaghs, 1970); the Scotiaecystidae, with their chevron-shaped inter-branchial plates and internal gonopore-anus (see Jefferies, 1968, for *Scotiae-cytis* (= "*Cothurnocystis*") *curvata*; see also Ubaghs, 1967, 1970); a number of species grouped under *Cothurnocystis* including *C. elizae*, *C. primaeva* Thoral, *C. fellinensis* Ubaghs, *C. courtessolei* Ubaghs and an unnamed species from the middle Cambrian of Utah (Sprinkle, 1976). It is not yet possible to divide this intermediate category into plesions, nor to state whether it comprises a single plesion.

The next fossil to be described is *Galliaecystis*, known from two specimens from the lower Arenig of the Montagne Noire (*G. lignieresi* Ubaghs, 1970; see also Jefferies and Prokop, 1972). The most notable feature of this fossil is a bar of calcite crossing the dorsal integument just behind the buccal cavity (arising at point 7 in Figs 9, 12). This dorsal bar is almost certainly homologous with the front of the dorsal shield of the primitive mitrate *Peltocystis* for it is made of the same plates (a and d in Figs 9, 11). It can be regarded as a synapomorphy with that genus, which all other known mitrates have lost. There has been a general lightening of the skeleton in *Galliaecystis*, especially as concerns the branchial skeleton. The integument has spread forward in the mouth region, and the ventral mouth bar of *Cothurnocystis* has been lost (Fig. 9). All these changes were probably adaptations to a life on soft mud. *Galliaecystis* constitutes a plesion.

The next fossil to be mentioned is *Reticulocarpos hanusi*, from the Llanvirn of Bohemia (Fig. 9 herein; see Jefferies and Prokop, 1972). It shares the dorsal buccal bar with *Galliaecystis* but has a number of additional synapomorphies in common with the mitrates including: the bilaterally symmetrical shape, the fact that the strut has begun to be reduced, the existence of a peripheral flange round the skeleton, signs that the tail acted by ventral flexion, and small size (in common with primitive mitrates the head is only about 10 mm long). *Reticulocarpos hanusi* seems to be adapted, like *Galliaecystis*, to life on very soft mud, relying only on the strength of the mud to support its weight. Calculations suggest that it would have imposed a load on the sea bottom of only about 10 mg cm^{-2}.

It is not certain that *Reticulocarpos hanusi* constitutes a plesion. For there is another fossil, *Amygdalotheca griffei* Ubaghs (Fig. 9 herein; Ubaghs, 1970), from the L. Arenig of the Montagne Noire, which likewise has a bilaterally symmetrical head and a peripheral flange. Moreover the strut of this form, though complete, shows signs of derivation from the incomplete condition as seen in *Reticulocarpos hanusi* i.e. it contains a separate plate m. *Reticulocarpos hanusi* and *Amygdalotheca griffei* constitute an immediate category between *Galliaecystis* and the mitrates. It may or may not be a plesion.

All the fossils so far described belong to the traditional group of the cornutes, originally set up by Jaekel (1900) as an order. Their most striking characteristic is that they possess left gill slits only, probably homologous with the early slits of amphioxus. This is probably a primitive condition derived from the hypothetical first crown dexiothete which had lost the gill slit (or slits) of the ventral surface. The animals which I shall now discuss belong, by contrast, to the traditional group of the mitrates, set

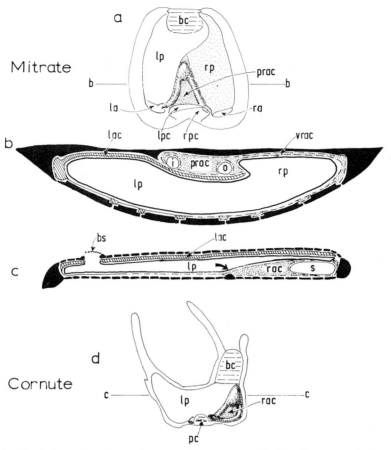

Fig. 10. Head chambers in a mitrate and a cornute. (a) dorsal aspect of mitrate *Mitrocystites*; (b) transverse section through b–b in a; (c) transverse section of cornute *Cothurnocystis* through c–c in d. The arrow in c indicates where the right pharynx pouched out, at the origin of the mitrates. bc = buccal cavity; bs = branchial slit; i = intestine; la = left atrium; lac = left anterior coelom; lp = left pharynx; lpc = left posterior coelom; o = oesophagus; pc = posterior coelom; prac = patent right anterior coelom; ra = right atrium; rac = right anterior coelom; rp = right pharynx; s = stomach; vrac = virtual right anterior coelom (overlying right pharynx).

up by Jaekel (1918) as an order. There is evidence that the mitrates had acquired right gill slits, homologous with the intermediate gill slits of amphioxus, as well as retaining the left ones. These right gill slits are a synapomorphy with all living adult chordates, which cornutes do not share. Moreover, all known mitrates can, by means of synapomorphies, be allocated to one of the three living subphyla of chordates, i.e. they can all be seen as stem acraniates or stem tunicates or stem vertebrates. It follows that, whereas all cornutes are stem chordates, all known mitrates are primitive crown chordates (Figs 11, 12). I shall not describe the mitrates in detail but shall recount their general features and then emphasize the synapomorphies which relate them to the various living chordates.

Mitrates share a number of common features which must be described first. Head and tail are distinct, as in cornutes. The head is always externally almost bilaterally symmetrical, usually dorsally flat and ventrally convex. There was an anterior mouth and right and left posterior gill openings. The presence of right and left gill slits can be inferred in all of them. However, there is direct evidence for gill slits only in *Lagynocystis*, where the gill slits that show best are median slits, not homologous with the right and left gill slits deduced to occur in other mitrates and in *Lagynocystis* itself. The chambers of the head are like those of *Cothurnocystis* but with additions. The most notable addition is the right pharynx which pouched out from the left or primitive pharynx towards the posterior right corner of the head (Fig. 10). In so doing it lifted up the right anterior coelom, squashed it against the ceiling and forced its cavity and contents into a median position. There were left and right atria, and the posterior coelom was probably a double structure, only half of it being equivalent to the posterior coelom of *Cothurnocystis*. The velum, or posterior border of the buccal cavity, was located mainly in the field of the entirely virtual left anterior coelom. This is important because, by way of a comparison with amphioxus, it suggests that the left and right anterior coeloms are homologous with the left and right mandibular somites of vertebrates (Jefferies and Lewis, 1978, p. 267ff.). An organ called the crescentic body was probably homologous with the premandibular somites of vertebrates. The first pair of segments of the tail were homologous with the hyoidean

---

FIG. 11. Phylogenetic relations of the mitrates with extant chordate subphyla, and with cornute *Reticulocarpos* (cf. Figs 9, 12). Dorsal aspect, left; ventral aspect, right. a, d, g, j, n are homologous plates. Letters inside circles are: bs = branchial slits; e = transpharyngeal eye; lbo = left branchial opening; ll = lateral line; m = mouth; mbo = median branchial opening; n_{4+5} = common opening of nerves n_4 and n_5 (dorsal branches of trigeminal) onto dorsal surface of head; rbo = right branchial opening. The numbers indicate the origin of synapomorphies (see Fig. 12). Regular stipple = integuments. Irregular stipple = tail insertion.

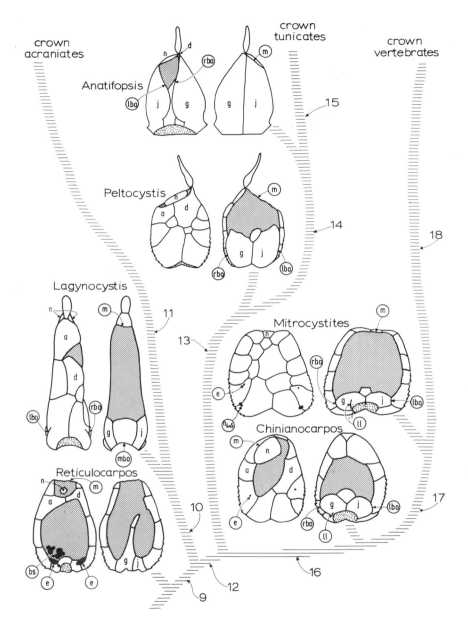

Fig. 11. For legend see opposite.

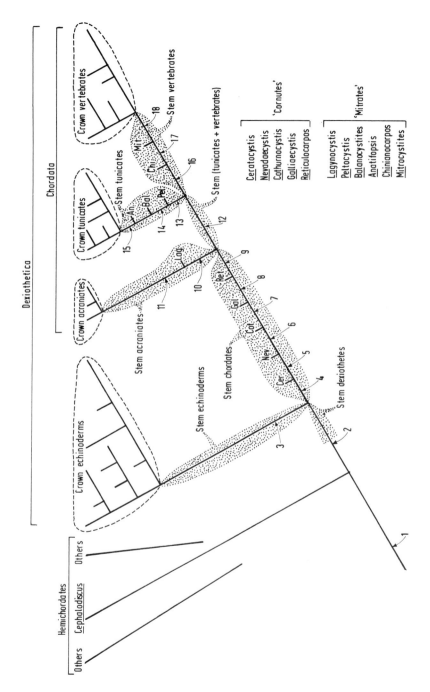

Fig. 12. For legend see opposite.

somites of vertebrates. Many peculiar asymmetries indicate that the pharynx of mitrates functioned very much like the pharynx of tunicates and was homologous with it. There was a singularly complicated cranial nervous system and brain, comparable in detail with those of a fish (Jefferies and Lewis, 1978, p. 271).

The tail of mitrates had fore, mid and hind parts, like the tail of cornutes. However, it is likely that the whole tail of a mitrate was homologous only with the fore tail of a cornute. Autotomy was probably a regular feature of the ontogeny of all calcichordates. In the early history of mitrates it probably broke the tail off more proximally than before, so that the cornute mid and hind tail dropped off. The mitrate fore, mid and hind tail would be regionated from the remaining stump (Jefferies, 1973; Jefferies and Prokop, 1972).

Lagynocystis pyramidalis, from the Llanvirn (lower Ordovician) of Bohemia (Fig. 11), can be regarded as a stem acraniate. It shows good evidence of gill slits homologous with the late gill slits of amphioxus. These constitute a synapomorphy with living acrianates which is not known

---

Fig. 12. Phylogenetic relationships of the deuterostomes. The numbers signify the origin of synapomorphies, as follows: 1 = deuterostomy developed; 2 = dexiothetic condition and calcite stereom-mesh skeleton developed; 3 = fixation and radiality developed; 4 = loss of water vascular system; 5 = loss of hydropore, gonopore-anus shifts from right of tail, roof of head becomes flexible, strut arises as thickening of rigid floor; 6 = flexible floor to head developed (except for strut), gonopore-anus left of tail; 7 = development of dorsal mouth bar formed of plates a and d, loss of ventral mouth bar, integument extends forward in mouth region, lightening of skeleton; 8 = head symmetrical, strut incomplete anteriorly, peripheral flange developed, tail flexes ventrally; 9 = right gill slits and right pharynx developed, right posterior coelom developed, head ventrally convex, old mid and hind tail lost, new mid and hind tail developed on remaining stump, gill slits internal (opening into atria); 10 = late gill slits and median atrium developed; 11 = skeleton lost, notochord extends forward to front end of head, posterior coelom split into nephridia; 12 = dorsal bearing surfaces on tail developed, nerves innervating muscle blocks developed i.e. "muscle tails" lost; 13 = branchial openings elongated; 14 = branchial openings elongated further and moved onto dorsal surface of head; 15 = loss of skeleton, fixation, loss of tail and brain in adult; 16 = lateral line developed; 17 = dorsal branches of trigeminal nerve (n_4, n_5) developed; 18 = loss of skeleton, origin of trunk region, notochordal head and kidneys, elaboration of sense organs and other changes in origin of crown vertebrates.

elsewhere. These gill slits in *Lagynocystis* have the posterior coelom dorsal to them. The situation in other mitrates shows that the posterior coelom is homologous with chambers called epicardia in tunicates. Moreover, the epicardia of tunicates are primitively excretory organs. Consequently, if the gill slits of *Lagynocystis* extended upwards in subsequently phylogeny, they would split the posterior coelom into a number of excretory chambers, each one above a gill bar, like the highly peculiar nephridia of amphioxus. Another peculiarity of *Lagynocystis* is that the buccal cavity was probably mainly innervated from the left, like that of amphioxus. However, this left-sided innervation may have existed in all mitrates and indeed may be the primitive condition for crown chordates. If so it would not be a synapomorphy of *Lagynocystis* with living acraniates though it constitutes a striking and probably homologous resemblance.

Many mitrates can be regarded as stem vertebrates. The most important synapomorphy with living vertebrates appearing at point 16 in Figs 11, 12, is the possession of a lateral line which takes the form of a pit, a straight groove or a cross-shaped groove in a postero-ventral position just right of the tail (ll in Fig. 11).

The stem vertebrate mitrates can be arranged in two intermediate categories of increasing relationship with crown vertebrates. The more primitive category is the plesion of *Chinianocarpos thorali* Ubaghs, from the L. Arenig (lowermost Ordovician) of the Montagne Noire. In it the lateral line is a small circular pit and there seem to be no dorsal branches to the trigeminal nerves. All other known stem vertebrates belong to the more advanced intermediate category of *Mitrocystites* since they show the additional synapomorphy with the crown vertebrates of dorsal branches to the trigeminal (n_{4+5} in Fig. 11). This intermediate category includes *Mitrocystites*, *Mitrocystella*, *Placocystites*, *Enoploura* and others. Some of them, placed in the order Anomalocystitida (e.g. *Placocystites*, *Enoploura*), had lost the lateral line, probably as a result of adopting a half-fossorial mode of life (Jefferies and Lewis, 1978).

Some other mitrates can be regarded as stem tunicates. We are still studying this group in London, but there seem to be three genera involved: *Peltocystis*, *Balanocystites* and *Anatifopsis* (= *Anatiferocystis*) (Figs 11, 12). The last two genera share a remarkable synapomorphy with tunicates in that the gill openings, instead of being posterior and ventro-lateral, have greatly elongated and moved onto the dorsal surface of the head, indicating that the atria extended dorso-lateral to the pharynx. *Peltocystis* is the most primitive member of this group since, though the gill openings have elongated considerably, they still retain their primitive position. It will take some years to understand this group properly.

As concerns the mutual relationships of the living subphyla of chordates, I have argued elsewhere that the acraniates are probably the sister group of tunicates plus vertebrates (Jefferies, 1973). This was based on how the tail muscles are innervated. Some remarkable recent work has shown that amphioxus has no ventral roots i.e. no motor nerves going from the dorsal nerve cord to the muscle blocks. Instead each muscle block has a special strip of muscle extending to the dorsal nerve cord (Flood, 1966). Such "muscle tails" are an echinoderm-like method of innervation, though of course no echinoderm has a dorsal nerve cord. Out-group comparison with echinoderms therefore suggests that "muscle tails" are the primitive method of innervating the muscle blocks. Tunicates and vertebrates, by contrast, innervate the homologous muscles by special nerves with motor end plates at their ends. This is presumably a synapomorphy of tunicates and vertebrates, implying that they together are the sister group of acraniates.

The fossils seem to confirm this conclusion to some extent. The hind tail of *Lagnocystis* has ventral bearing surfaces on the ventral plates. These surfaces were presumably used to push downwards and forwards against the mud, pulling the animal backwards. They are probably homologous with similar surfaces on the fore tail of the cornute *Reticulocarpos* and this out-group comparison suggests that they represent the primitive mitrate condition. A different device is found both in stem tunicates (*Peltocystis*, *Anatifopsis*) and in stem vertebrates (*Chinianocarpos*, *Mitrocystites*, *Mitrocystella*). In these, bearing surfaces are developed on the dorsal ossicles of the hind tail (see Jefferies, 1973). These surfaces should probably be regarded as a synapomorphy of stem-tunicates and stem-vertebrates, supporting the relationships suggested by the living animals.

As concerns dating, I believe that the skeleton of echinoderms is homologous with that of calcichordates. If so, echinoderms and chordates probably separated from each other later than the beginning of the Cambrian, when skeletons began. (This is a rare occasion when it is possible to say: "later than...".). The separation of the plesions of *Ceratocystis* and *Nevadaecystis* from each other, and from the intermediate category of *Cothurnocystis*, must have happened earlier than the middle Cambrian, Spence Shale of Utah, in which the earliest known *Cothurnocystis* occurs. The origin of the plesion of *Galliaecystis*, the intermediate category of *Reticulocarpos hanusi*, the mitrates and therefore the crown chordates, must have happened earlier than basal Ordovician. For at that level the stem tunicate *Peltocystis* and stem vertebrate *Chinianocarpos* are both found. Moreover, the oldest known agnathan fish, *Anatolepis heintzi*, also comes from the lowest Ordovician which helps to show that the crown chordates must already have existed in the Cambrian, although they have not yet been found there.

In summary, the fossils preserve much information about the early history of chordates. Without correct and explicit methodology, however, this information cannot be interpreted. In particular we must be clear about what stratigraphy can, and cannot, do.

REFERENCES

EATON, T. H. (1970). The stem-tail problem and the ancestry of chordates. *J. Paleont.* **44**, 969–979.

FLOOD, P. R. (1966). A peculiar mode of muscular innervation in amphioxus. *J. comp. Neurol.* **126**, 181–218.

GISLÉN, T. (1930). Affinities between the Echinodermata, Enteropneusta and Chordonia. *Zool. Bidr Upps.* **12**, 199–304.

GROBBEN, K. (1924). Theoretische Erörterungen betreffend die phylogenetische Ableitung der Echinodermen. *Sber. Akad. Wiss. Wien,* **132** (1), 263–290.

HAMILTON, W. R. (1978). Fossil giraffes from the Miocene of Africa and a revision of the phylogeny of the Giraffoidea. *Phil. Trans. R. Soc. Lond. B,* **283**, 165–229.

HENNIG, W. (1966). "Phylogenetic systematics." University of Illinois Press, Chicago and London.

HENNIG, W. (1969). "Die Stammesgeschichte der Insekten." Waldemar Kramer, Frankfurt-am-Main.

HYMAN, L. (1959). "The Invertebrates: smaller coelomate groups." Vol. 5. McGraw-Hill, New York and London.

JAEKEL, O. (1900). Über Carpoideen, eine neue Klasse von Pelmatozoen. *Z. dt geol. Ges.* **52**, 661–677.

JAEKEL, O. (1918). Phylogenie und System der Pelmatozoen. *Paläont. Z.* **3**, 1–28.

JEFFERIES, R. P. S. (1967). Some fossil chordates with echinoderm affinities. *Symp. zool. Soc. London,* **20**, 163–208.

JEFFERIES, R. P. S. (1968). The subphylum Calcichordata (Jefferies, 1967)—primitive fossil chordates with echinoderm affinities. *Bull. Br. Mus. Nat. Hist. (Geol.),* **16**, 243–339.

JEFFERIES, R. P. S. (1969). *Ceratocystis perneri* Jaekel—a middle Cambrian chordate with echinoderm affinities. *Palaeontology,* **12**, 494–535.

JEFFERIES, R. P. S. (1973). The Ordovician fossil *Lagynocystis pyramidalis* (Barrande) and the ancestry of amphioxus. *Phil. Trans R. Soc. Lond. B,* **265**, 409–469.

JEFFERIES, R. P. S. (1975). Fossil evidence concerning the origin of the chordates. *Symp. Zool. Soc. Lond.* **36**, 253–318.

JEFFERIES, R. P. S. and LEWIS, D. N. (1978). The English Silurian fossil *Placocystites forbesianus* and the ancestry of the vertebrates. *Phil. Trans. R. Soc. Lond. B,* **282**, 205–323.

JEFFERIES, R. P. S. and PROKOP, R. J. (1972). A new calcichordate from the Ordovician of Bohemia and its anatomy, adaptations and relationships. *Biol. J. Linn. Soc.* **4**, 69–115.

LANKESTER, E. R. and WILLEY, A. (1890). The development of the atrial chamber of amphioxus. *Q. J. microsc. Sci.* **31**, 445–466.

LØVTRUP, S. (1977). "The Phylogeny of Vertebrata." John Wiley. London and New York.

MACBRIDE, E. W. (1909). The formation of the layers in amphioxus, and its bearing on the interpretation of the early ontogenetic processes in other vertebrates. *Q. J. microsc. Sci.* **54**, 279–345.

PATTERSON, C. and ROSEN, D. E. (1977). Review of the ichthyodectiform and other Mesozoic teleost fishes and the theory and practice of classifying fossils. *Bull. Am. Mus. nat. Hist.* **158**, 85–172.

RIEDL, R. (1975). "Die Ordnung des Lebendigen: Systembedingungen der Evolution." Paul Parey, Hamburg.

RIEDL, R. (1979). "Order in living organisms—a systems analysis of evolution." John Wiley, London and New York.

ROWE, A. W. (1899). An analysis of the genus *Micraster*, as determined by rigid zonal collecting from the zone of *Rhynchonella cuvieri* to that of *Micraster cor-anguinum. Q. Jl geol. Soc. Lond.* **55**, 494–544.

SPRINKLE, J. (1973). Morphology and evolution of blastozoan echinoderms. *Spec. Publ. Mus. comp. Zool. Harv.* 1–196.

SPRINKLE, J. (1976). Biostratigraphy and paleoecology of Cambrian echinoderms from the Rocky Mountains. *Brigham Young Univ. geol. Stud.* **23**, 61–73.

UBAGHS, G. (1963). *Cothurnocystis* Bather, *Phyllocystis* Thoral and an undetermined member of the order Soluta (Echinodermata, Carpoidea) in the uppermost Cambrian of Nevada. *J. Paleont.* **37**, 1133–1142.

UBAGHS, G. (1967). Le genre *Ceratocystis* Jaekel (Echinodermata, Stylophora). *Paleont. Contr. Univ. Kans.* Paper 22, 1–16.

UBAGHS, G. (1968). General characters of the Echinodermata. *In* "Treatise on Invertebrate Paleontology" (R. C. Moore, ed.), Part S, Echinodermata, Vol. 1, pp. S3–S60. Geol. Soc. Amer. and Univ. Kansas Press.

UBAGHS, G. (1970). Les echinodermes carpoïdes de l'Ordovicien inférieur de la Montagne Noire (France). *Cahiers de Paléontologie*, 1–112.

WILLEY, A. (1891). The later larval development of amphioxus. *Q. Jl microsc. Sci.* **32**, 183–234.

18 | Discussion on the Origin of Major Invertebrate Groups

M. R. HOUSE

Department of Geology, University of Hull, England

INTRODUCTION

The presence of so broad a spectrum of specialists at the Symposium led to a range of lively and enlightening debate, and it seems of value to record some of this discussion. Whilst the sessions were tape-recorded, and many speakers submitted written copies of their comments, there is an abridgement and selection in the contributions recorded here. Furthermore, in order to give a more logical presentation, the order of a few of the discussions has been rearranged; however the succession of main themes follows that of the symposium as a whole.

ENVIRONMENT AND ORGANISMS OF THE PRECAMBRIAN

Dr K. R. Ashby had asked that if blue-green algae existed some 3200 m.y (million years ago), when did reducing conditions on the earth give place to oxidizing conditions? Mr O. A. Christopherson commented that there has probably been the present amount of oxygen in the atmosphere since 2300 m.y. Roscoe had described sansdstones with detrital uraninite, indicating oxygen poor environments from the lower part of the Huronian whereas the upper part contains haematite pseudomorphs after detrital magnetite indicating that the oxygen content then was relatively high since oxygenation of magnetite, even today, is a slow process in sedimentary environments; he would deduce that the build up of oxygen

---

Systematics Association Special Volume No. 12, "The Origin of Major Invertebrate Groups", edited by M. R. House, 1979, pp. 479–494. Academic Press, London and New York.

was a relatively rapid process although there were factors which would limit the build up. The role of the Varangian glaciation in the late Precambrian, Mr Christopherson thought, might have increased the salinity of the seas, thus inhibiting organic activity in marine waters. Professor Holland remarked that it had been suggested that the Varangian glaciations had limited the possibility of the development of shelled organisms by affecting the relative proportion of magnesian and calcium salts.

The lengthy period preceding the Cambrian radiation was commented on by Dr M. D. Brasier: he thought a series of phases may have been involved and that each may have been controlled by different circumstances, and that whilst Mr Christopherson's suggestion might be applicable to the post Varangian/Vendian radiation of soft-bodied organisms, some 100 m.y. intervened between that and the Cambrian radiation, and there seemed little to suggest a causal relationship. Dr Brasier commented on his recent discovery with Dr A. Perejon of "*Beltanelloides*" from the Pusa Shale of the Toledo Mountains of Spain: both *Beltanelloides* and *Chuaria* are at least an order of magnitude bigger than most extant microplankton. Might they be neither planktonic nor algae, he asked, and what were their chemical affinities? Dr T. D. Ford remarked that he had no information on the two genera mentioned but rather similar forms had a composition similar to sporopollenin.

LATE PRECAMBRIAN AND EARLY CAMBRIAN RADIATION

Dr E. Yochelson said that he preferred Brasier's concept of the Cambrian radiation, but he drew attention to how Sepkowski had taken essentially the same data and suggested that there is a continuous expansion of life forms rather than a dramatic expansion. Mr C. W. Wright, on the other hand, drew attention to the extremely important point from Dr Brasier's paper that the so-called Cambrian Radiation Event in fact covered a period comparable with all subsequent Phanerozoic time: such a period allowed far more protracted a scale of staged evolution than had recently been accepted for the major groups of the Metazoa. Dr Brasier agreed that there had been a long period of time over which invertebrates generally had evolved, nonetheless he regarded the early Cambrian event as an important radiation in which some eight to ten of the major invertebrate phyla appeared. Professor Sleigh commented that it seemed to him likely that this radiation was fuelled by the rise of eukaryote phytoflagellates which may be assumed to have evolved in the late Precambrian: only once these genuine planktonic forms were abundant would sufficient food be present for many of the animal groups to become numerous; such biological features of the environment would contribute as well as physical changes to the explanation of the Cambrian radiation.

Although, Dr A. Bidder said, she was impressed with the idea that the early Metazoa were burrowers, since burrowing involves specialization, these must have had earlier less specialized forms before them. Dr Brasier agreed that there must have been preceding acoelomates, but they would presumably be small and they are not yet known in the early fossil record. Dr Bidder further commented that it was not only a coelom which could provide a hydrostatic skeleton for burrowing, a haemocoel could also provide this.

Dr R. A. Bromley remarked on the supposed appearance of crawling traces (exogenes) in the Vendian, in contrast with the burrows (endogenes) in the Riphaean. He thought it notoriously difficult to distinguish trace fossils produced at interfaces within the sediment from exogene trails, and said that the latter were very rarely preserved and he asked how Dr Brasier made this distinction. In reply Dr Brasier said that his main concern was to emphasize the contrast in the record, he accepted that there were difficulties in interpretation.

On the evidence for the appearance of the Cnidaria in the late Vendian Dr C. Edwards remarked that it was interesting that this was later than the origin of other metazoan forms; he thought this agreed with the fact that they were predatory carnivores and with the probability that they arose with the appearance of zooplankton. He noted Dr Brasier's correlation between the late Vendian/early Cambrian marine transgression and the Cambrian radiation and asked where did the masses of water come from and what geological evidence was there for oceanic water masses (as distinct from shallow seas) before the Vendian? On this no-one was prepared to comment.

Dr K. Vogel agreed that the Cambrian radiation was driven forward by ecological events, but these comprise both external and internal environments: Brasier had only discussed the former, the radiating animal groups. He took the internal environment to be as important; animals could not have made such use of the new environmental possibilities if they had not been preadapted to it; so the external events at the Precambrian/Cambrian boundary may have been preceded by improvements of body construction, biomechanics, physiology and harmonization of all body activities. All this, he said, may have happened without considerable environmental changes, but it would have made the organisms ready to invade the newly appearing habitats.

In reply Dr Brasier said that he had emphasized external environments because these provide great insights into the story. If by "preadapted" Dr Vogel meant something such as the development of a lophophore prior to skeletonization in brachiopods, then this seemed likely, but (he said) there is no evidence to suggest that the transition from the "pre-brachiopod" to the brachiopod took more than a few million years (perhaps because the development stage took place in small biotopes with little potential for preservation); where there appeared to be disagreement was in the relative significance of external and internal environments. Dr Brasier thought it likely that the external changes preceded the internal ones, and acted selectively upon genetic material to encourage the development of new organs and new cavities.

BASIC PHYLOGENETIC RELATIONSHIPS

The several speakers discussing the evolutionary radiation of the lower invertebrates having given a broad conspectus of the facts and hypotheses, the discussion concentrated on certain rather specific inquiries. Dr P. J. Miller observed that if flatworms arose from primitive coelomates, and the Cnidaria from zooflagellates, as Sleigh proposed, this would cut out the Cnidaria as precursors from acoelomate metazoans if, according to Jägersten's views, we wished to continue phyletic association of coelomate and acoelomate metazoa. But Professor Sleigh replied that he thought separate origins seemed most likely.

Dr R. P. S. Jefferies asked Dr Scrutton whether he considered the hydroids or the medusoids the more primitive. Dr Scrutton replied that he did not think the fossil record was good enough to arbitrate between a medusoid or polyploid ancestor for the Cnidaria: of the various theories advanced by zoologists the actinate theory, which takes a hydrozoan medusoid form as the ancestral radicle, he found the most attractive; whilst he thought the fossil record did not conflict with this he could not say that it supported the hypothesis.

An older theory which sought the origin of the coelom from an anthozoan such as *Cerianthus* was described by Dr H. Henson; he thought this provided a theory for the origin of metamerism. The bilaterally produced mesenteries and form of the larva suggested a simple form of annelid structure which might also provide a basis for chordate metamerism; he thought that the trimerous coelomate condition could be derived from the persistent "medusoid" front end of the larva with loss of posterior growth. Professor Clark replied that earlier views of coelenterate origin of coelomates (whether from a medusoid or polyploid) are contained in a modified form in the archicoelomate theories. He did not favour seizing upon similarities of structure or development between existing animals and regarding them as a basis for postulating hypotheses; the enormous number of theories in existence is a consequence of this. A very searching analysis of the implications of a proposal is required before it can be accepted.

Regarding analogies with higher invertebrates Dr P. J. Miller asked whether the evolution of the proboscis coelom (protocoel) in trimerous forms was related to its use in burrowing, as in the enteropneusts, but Professor Clark thought probably not, since the burrowing is incompatible with the mesocoel bearing a tentaculate feeding organ—it is a prehensile organ and used as such in the pterobranchs. *Balanoglossus*, he said, was different and had lost its lophophore and the mesocoel is reduced to remnants inside the collar region suggesting a loss of function. Dr K. R. Ashby asked what was the function of the coelom in *Balanoglossus*, Professor Clark replied that it provided a space in which the gut are free to move; some peristalsis occurs in the tail-end and, in this sense, the metacoel behaves like the individual coelom of a worm.

BRYOZOA

Professor Spjeldnaes said that the model of development and diversification of the Bryozoa presented by Larwood and Taylor was both logical and tempting, but there was one peculiarity: the first occurrence of a diversified bryozoan fauna is the middle Arenig of Esthonia and Øland and in the Oslo area. At one bedding plane, in the *Asaphus expansus* Zone, a rich fauna is met including Trepostomata, Cyclostomata and Cryptostomata: most species are massive and some have specialized colonial shapes. *Croynotrypa*-like bryozoans appear well after this first "burst" of bryozoans. He thought the first fauna had migrated into the area following ecological changes. Dr C. T. Scrutton asked concerning zooecial size in the early bryozoans. Dr Larwood thought there was not much variation and in this respect they were little different from later bryozoans. Dr R. G. Bromley asked why the authors derived the Bryozoa from endofaunal phoronids rather than epifaunal ones. Dr Larwood replied that in this they were following Farmer, and he noted that the epifaunal phoronids were smaller. Dr Larwood thought the radiation was a response to a newly available planktonic food source and that appropriate epifaunal filter-feeding types were best newly developed from infaunal types.

BRACHIOPODA

In view of the evidence for substantial differences between groups of early brachiopods several speakers asked concerning their taxonomy. Dr E. N. K. Clarkson suggested that if it is established that the orders of inarticulate brachiopods arose independently from several simple, infaunal lophophorates, was there not a case for altering their taxonomic rank. Professor Wright was loathe to proliferate a series of classes for the brachiopods. He was satisfied with class rank for the "articulates" which arose from the billingsellacean stock and form a major group, but would be loathe to raise to class rank those "non-articulate" stocks which were either short-lived experiments or at least minor elements in the fauna. Mr C. W. Wright also asked whether, rather than being polyphyletically derived from different pre-brachiopod lophophorates, it was not more probable that all the brachiopods with pedicle attachment were derived monophyletically from a single lophophorate. But Professor Wright re-emphasized the great differences between some of the groups involved, for example, the acrotretids and the lingulids. He thought the ancestral lophophorates were a diverse infaunal group. Notwithstanding this he considered the brachiopods as a whole to form a clear group and a distinct phylum.

Dr R. A. Hewitt commented that Pelemen's work suggested that the oldest known Tommotian brachiopods formed an abundant but low diversity assemblage just above the "Precambrian/Cambrian" boundary of the Siberian section. These forms consist of *Micromitra*-like phosphatic inarticulates; inarticulates such as *Lingulella* are not reported before the late lower Cambrian in that area. Professor Wright commented that Peleman's work being yet unpublished has not been seen by him.

Dr C. Edwards noted that the phylogenetic theories of the evolution of the brachiopods suggested that infaunal forms gave rise to epifaunal ones. He suggested that detritus-feeding became more advantageous with evolution and the development of the zooplankton.

ARTHROPODA

The account of the Burgess Shale Fauna by Professor H. B. Whittington initiated discussion on arthropodan relationships and radiations which was concluded after the address by Dr S. Manton and Professor D. T. Anderson. Several distinct strands ran through the exchanges. Were the arthropods monophyletic or polyphyletic? How should the Burgess Shale fauna be interpreted? What is the relation with the annelids? What is the relative status of groups within the arthropods? How reliable are supposed significant characters? What rigour can, or should, be applied to determining classification?

1. Monophyly versus Polyphyly

The case put by Manton and Anderson for arthropod polyphyly, Dr D. E. G. Briggs saw as reinforced by Whittington who, by implication, would raise the taxon Trilobita to equal rank with the others: he remarked that the Trilobita, Crustacea and Uniramia have representatives in the Cambrian and the chelicerates appear soon after, but there is no fossil evidence for the origin of these four taxa. If these groups are considered phyla (he said) then it implies that they all acquired the "grade of organization" represented by the arthropods independently; on the other hand, Hessler would interpret them as classes, and this implied monophyly. The latter Dr Briggs was still prepared to entertain, at least for the trilobites, Crustacea and chelicerates: he thought that the major arthropod groups were separated by smaller morphological distances than those separating the arthropods as a whole from other phyla. Dr K. Sdzuy was of the opinion that, since he thought most would agree that the many arthropod lines would converge somewhere in the Precambrian, it was a game of words whether the differences are considered as justifying separate phyla or not; he thought it was only a matter of convenience.

2. The Burgess Shale Fauna

Dr R. R. Hessler emphasized that the apparently great diversity of arthropod limb morphologies in the Burgess Shale need be no sure indication of wide taxonomic gaps: if a Recent fauna were similarly preserved, its crustacean components would show an equally broad range of superficial morphologies, yet, from our privileged position of knowing nearly the entire Recent fauna in a condition of perfect preservation, we are able to determine that this broad range of morphologies can be assigned to a single class. Perhaps (he said) the preservation of the Burgess Shale Fauna might not allow a sure determination of affinity which might exist. Dr D. E. G. Briggs interpreted the diversity of the Fauna (Størmer's phylogenetic "grass") as testifying to considerable early radiation which, he felt, could not all be independent, indeed some of the members of the Fauna showed characters of accepted major groups which suggested links between them.

Dr C. J. C. Rees asked whether evidence of trophic analysis and community structure of the Burgess Shale Fauna would help discrimination, but Dr S. Conway Morris pointed out the difficulties of this and said that the Fauna seemed almost entirely adult and wholly benthonic. Professor H. B. Whittington stressed that he could not be really sure that a natural community was represented; the forms may all have been living on a muddy bottom or some may have been washed in from other environments. Whilst numbers are large and there is a good range of size there are never very small forms.

3. Relation with the Annelids

This topic was initiated when Dr K. R. Ashby asked whether, in view of the great range of the Burgess Shale Fauna and the soft-bodied forms it included, there was a case for linking the fauna with the polychaetes as a systematic unit. But Dr S. Conway Morris was convinced that a distinction between the annelids and arthropods was clear, the former possessing setae and the latter not. On the general matter of affinity, Dr S. Manton vigorously affirmed that Anderson's embryological evidence would not allow that Palaeozoic annelids could have given rise to the Crustacea and that this point was developed in their joint paper.

4. Relative Status of Arthropod Groups

The way in which the same data base may be an area of disagreement was illustrated by Dr R. R. Hessler with regard to the constancy of the arthropod head. He agreed with Manton that the crustacean head does have a very constant number of segments—first antenna, second antenna, mandible, maxilulles—but if these crustaceans were fossilized and interpreted without knowing their embryology you would find (he said) the mystacocarids having one fewer segment than the rest; the malacostracans

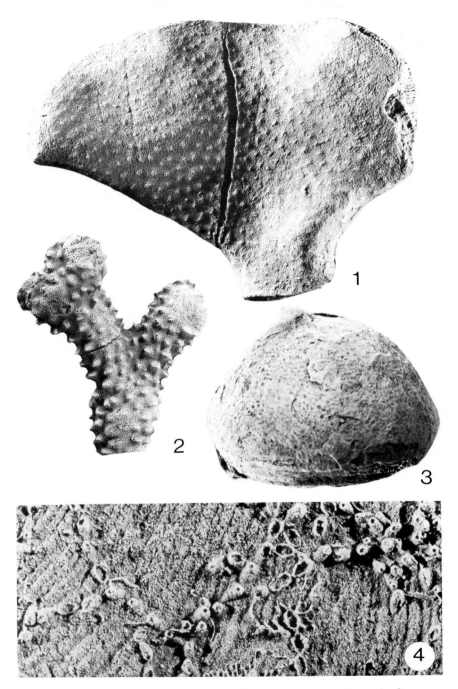

PLATE I. COLONY FORM IN ORDOVICIAN BRYOZOA (see legend overleaf)

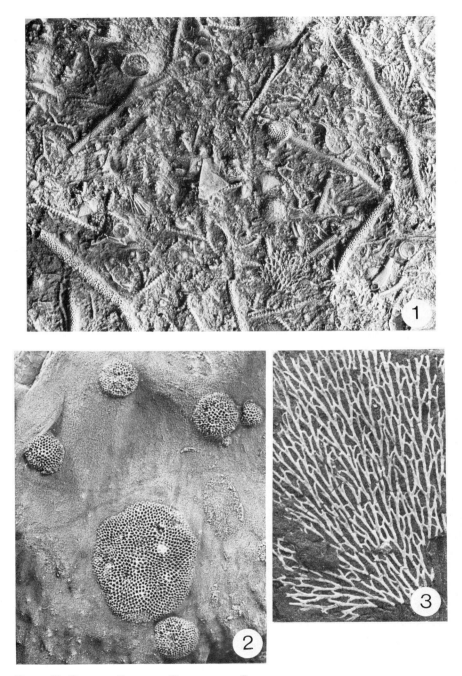

PLATE II. COLONY FORM IN ORDOVICIAN BRYOZOA

Plate I. Colony Form in Ordovician Bryozoa

1. Erect frondose trepostome with monticules (× 1·1). Cincinnati Group, Ohio, USA. BM(NH), D.29759.

2. Erect massive cylindrical branched trepostome with monticules (× 1·8). Cincinnati Group, Ohio, USA. BM(NH), D.1012.

3. Erect domed trepostome (× 2·7) Lower *Chasmops* Shale, Billingstad, Asker, nr. Oslo, Norway. BM(NH), PD4994.

4. Encrusting linear branched cyclostome—*Corynotrypa* sp. (× 16·5). Cincinnati Group, Ohio, USA. BM(NH), D.1253.

Plate II. Colony Form in Ordovician Bryozoa

1. Many fragments of narrow cylindrical branched cyclostomes and cryptostomes (× 2·1). Trenton Limestone, Trenton Falls, New York, USA. BM(NH), D.809.

2. Encrusting sheet-like trepostomes (× 4·6). Cincinnati Group, Ohio, USA. BM(NH), PD.5404–5.

3. Erect fenestrate cryptostome (× 2·3). Middle Ordovician, C2 horizon, Koht-lavärvi, Estonia. BM(NH), PD.4192.

Plate III. Articulate Inarticulate brachiopods

1–5. *Spondylotreta parva* Wright 1963, Portrane Limestone (Ashgill), Portrane, Co. Dublin. × 20. **1**, interior of brachial valve (BB29873) showing hinge teeth. **2, 3**, lateral and ventral views of a brachial valve interior (BB29872). **4, 5**, posterior and lateral views of a marginally damaged pedicle valve (BB29874).

6–9. *Dictyonella capewelli* (Davidson, 1848). × 5. **6, 9**, dorsal and postero-ventral views of conjoined valves (BB93771). J. F. Walker Colln., 1903, Wenlock Shales, Buildwas. **7, 8**, exterior and interior of broken brachial valve (BB93772), showing the long, postero-laterally sited articulation ridge and groove. Disarticulated valves are uncommon, a pointer to the efficient articulation mechanism. John Gray Colln., 1869, Wenlock Shales, Walsall.

10, 11. *Eodinobolus* sp. nov. Lower Drummuck (Ashgill), Glenmard Quarry, Girvan; coll. D. A. T. Harper, 1977. × 5. **10, 11**, internal mould and latex cast of a pedicle valve (H.M. L11958).

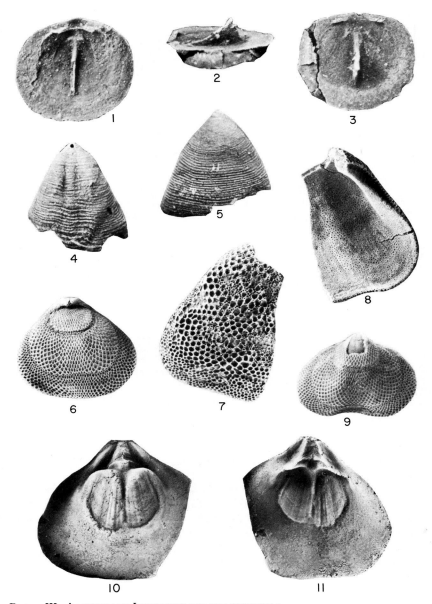

PLATE III. ARTICULATE INARTICULATE BRACHIOPODS

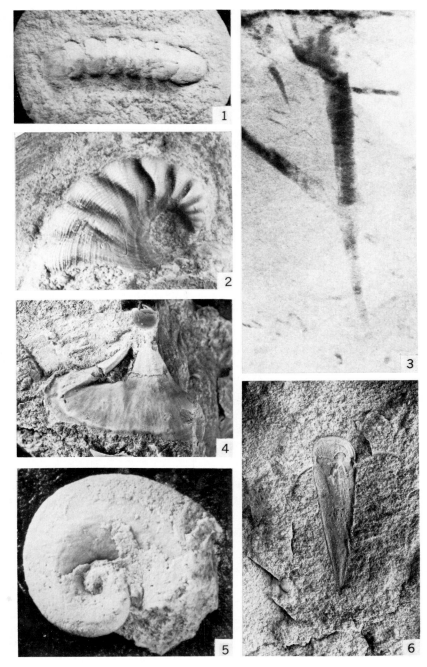

Plate IV. (see legend overleaf)

PLATE IV. MOLLUSCA AND MOLLUSC-LIKE GROUPS

1, *Pterochiton concinnus* Richardson (× 2). Francis Creek Shale of Savage (1927) (upper Carboniferous), Mazon Creek, Illinois. This specimen has all eight valves held in place by the girdle; the head valve is to the right. Field Museum of Natural History PE31950.

2, *Helcionella buttsi* Resser (× 5). "Austinville Formation" (middle Cambrian), Austinville, Virginia. (U.S. Geological Survey locality 8089–CO). The aperture is not complete, but the specimen is well preserved; in the writer's interpretation the head would be to the left. U.S. National Museum 257274.

3, *Viriatellina fuchsi* Kutscher (× 5). Hunsrück Shale (lower Devonian) Gemiinden, Germany. This X-ray shows soft parts extended; adjacent to this specimen are others less well preserved. Private collection of G. Brassel.

4, *Matthevia variabilis* Walcott (× 2). Hoyt Limestone (upper Cambrian), Saratoga, New York. (This paralectotype is the original of plate 33, figure 1a of Walcott, 1886.) In my interpretation, this steinkern would be at the anterior or posterior of the animal, with powerful muscles in the two cavities and presumably with the left side of the photo as ventral; the change in texture of the steinkern near the margin may be comparable to a pallial line. U.S. National Museum 257275.

5, *Pelagiella primaevum* Billings (× 5). (Lower Cambrian), near Greenwich, New York (U.S.N.M. locality 35 green). This top view conveys the size but not the profile; an apparent muscle track occurs near the middle of this surface, but is poorly shown. U.S. National Museum 8257276.

6, *Hyolithes carinatus* Matthew (× 2). Steven Formation (middle Cambrian) near Field, British Columbia. This is a view of the presumed ventral surface with "whiskers" (helens of Runnegar and Pojeta) to the right and left of the conch in approximately life position; the anterior of the conch is eroded and the interior of the operculum is seen in this area. U.S. National Museum 257277.

PLATE V. MOLLUSCA AND MOLLUSC-LIKE GROUPS

1, 2, Two specimens of unnamed "worm" tubes, × 55. Valhallfona Formation (lower Ordovician) north NY Firesland, Spitsbergen. The first resembles a small snail whereas the second does not; shapes intergrading from bilaterally symmetry to straight occur in the same collection. Paleontological Museum of Oslo NF 3283, NF 3284.

3, *Pterotheca undulata* Hall (× 1). Black River Group (middle Ordovician), Poland, New York. This is a representative species, though others are a bit less oval; in my interpretation the head is to the top and the main body mass mostly below the obvious "wedge", with a broad foot below the expanded shell. U.S. National Museum 79288.

4, *Dentalium martini* Whitfield (× 1½). Columbus Limestone (middle Devonian), Dublin, Ohio. This might be an ancient scaphopod, but it also might be a worm tube; the shell is about half the thickness of Permian scaphopods of the same size. Orton Museum, Ohio State University 11488.

5, *Proplina cornutaforme* Walcott (× 1½). Cotter Dolomite (lower Ordovician), Christian County, Missouri (U.S. Geological Survey locality 101o). This steinkern is not especially well preserved but it is one of the oldest showing the large discrete paired muscle scars laterally; in my opinion, the head was to the right. U.S. National Museum 252278.

6, *Scenella reticulata* Billings ('2). Lower Cambrian, Conception Bay, Newfoundland. This oblique view of a steinkern shows the rugosities well; I have no idea of which direction is anterior. U.S. National Museum 257279.

7, *Bellerophon (Bellerophon) scissile* conrad (× 1). Ste. Genevieve Limestone (lower Carboniferous), near Ste. Genevieve, Missouri. The slight irregularity in profile during the lower 2 cm to the left is the site of the short slit; in my interpretation the head is to the left. U.S. National Museum 99034.

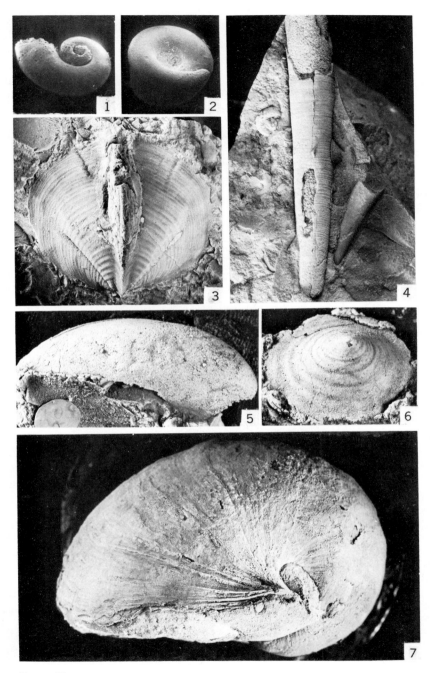

PLATE V.

PLATE VI.

1. *Ribeiria pholadiformis* Sharpe. Lectotype, natural internal cast, a, dorsal view, b, right lateral view showing muscle attachment scars. GSM7798 and BM(NH)4177, L. Ordovician (?Llandeillo), nr. Coimbra, Portugal (× 2). **2.** *Ribeiria apusoides* Shubert and Waagen, anterior view of rubber mould with growth increments crossing the dorsal margin, BM(NH)66362 (× 2). **3.** *Conocardium longipennis* Brown, right lateral view of cast of the interior of the central part of the shell showing muscle attachment scars, PPR, posterior pedal retractor, B, other body attachment scars, G, possible position of the gill, PL, line of mantle attachment (see also Fig. 6), BM(NH)L45406 (× 5·2), Viséan, Northumberland, England. **4.** *Conocardium aliforme* (J. de C. Sowerby), 4a, anterior view showing large gape and internal shell plates, BM(NH)L19055 (× 3), Tournasian, Tournai, Belgium, 4b, view of natural internal cast, B, body attachment muscle scars, PPR, posterior pedal retractor scar (× 2·3), BM(NH)L45403, Viséan, Northumberland, England. **5.** *Helcionella rugosa* (Hall), apex with initial shell attached, a, anterior view, b, view of right side, BM(NH)G17337, SEM (× 21), lower Cambrian, Comely, England. **6.** ?*Euchasma* sp. presumed larval shell, a, anterior, b, view of right side, BM(NH)LL31372, SEM (× 133), L. Ordovician (L. Llanvirn), Spitzbergen. **7.** *Antalis* cf. *panormum* (Chenu), young shell with larval shell attached, SEM (× 98), Pantelleria Island, Italy, 720 m, Porcupine Expedition, M, Martin of larval shell, F, fold in larval shell formed by subsequent growth, O, early shell ornament. **8.** *Pseudoconocardium lanterna* (Branson), cellulose acetate peel of transverse section of dorsum showing structure of the "ligament", BM(NH)LL31373 (× 16), U. Pennsylvanian, Fort Jackson, Texas, L, organic rich ligament-like material, EL, early deposited layer bent during growth, N, nacreous layer.

and decapods, having all of the thoracic segments fused to the head, might be interpreted as having thirteen rather than five segments, and even if it was decided that the posterior thoracic segments were not part of the head, almost certainly it would be concluded that the most anterior three thoracic segments were part of the head because they were firmly cephalized. Thus, Dr Hessler argued, as seen in the fossil record, the "crustacean" head would not appear constant.

In the light of Dr Hessler's comments Dr D. E. G. Briggs remarked that in the fossil record it is striking that the few trilobites of which the appendages are known appear to show a constant arrangement; *Olenoides*, *Naraoia*, *Triarthrus* and *Phacops* all appear to have had a pair of antennae followed by three pairs of biramous appendages in the cephalon; an apparently similar arrangement in the cephalon of the non-trilobite *Burgessia* and possibly *Aglaspis*, however, shows (he said) that cephalic segmentation in Palaeozoic arthropods must be used with caution: the evidence of *Olenoides* and the crustacean *Canadaspis* suggests that stabilization of cephalic segmentation in the major groups might have taken place at an early stage although they coexisted with a variety of "experimental" forms exemplified by some of the Burgess Shale arthropods.

Dr H. Henson observed that there were curious similarities which are difficult to ascribe to adaptation in the three main living groups which are under discussion: for instance, in the amphipods, Insecta and Arachnida there are outgrowths of the midgut, the Malpighian tubules, which are found in these three groups and in no other; they are strictly homologous and are endodermal gut outgrowths. There is another peculiarity; in the amphipods and Insecta the first abdominal segment is atrophied. Does one accept (he asked) this evidence of affinity or that of limb structure? Another peculiarity of insects, occurring in pairs of genera in almost any family or subfamily, is the presence of species without a fourth anal vein; these forms cannot be derived from a common ancestor and the character may reflect some tendency of the inherited genetic composition. Dr J. Sudd confirmed that, in the last case, such rather closely coordinated changes might be due to their possessing special attributes at genetic loci.

5. Reliability of Characters

Developing a similar theme, Dr A. Bidder made the general point that an earlier generation might have taken too readily the conjunction of primitive characters as indicating a common ancestor: three characters were regarded as fundamental; the cuticular exoskeleton, metameric segmentation, and the coelom. What reason, she asked, was there for thinking these only happened once? Is it not time we were less fundamentalist and more agnostic about common ancestors? At this Dr R. R. Hessler argued that if one no longer gave reliance on similarity as an indication of affinity then taxonomy should be given up: similarity as a criterion should only be abandoned after proof. In reply, Dr A. Bidder did not deny this, but still favoured more agnosticism.

6. Rigour in Arthropod Classification

The question of Hennigian analysis was raised by Dr R. P. S. Jefferies in relation to the four main groups of arthropods. If there are four monophyletic groups involved (he said) then there ought to be 15 ways of combining them onto a dichotomously branching tree, but this can only be done using synapomorphies; he thought it was terrible that Whittington and his group had presented an interpretation requiring either agnosticism or Special Creation. Mr C. W. Wright, on the other hand, did not regard this as terrible at all, he commented that "Palaeontology is a vast area of ignorance modified by small areas of illumination: it is unwise to jump before you can crawl."

MOLLUSCA

With four speakers covering aspects of the radiation of the Molluscan groups a reasonable consensus on the factual background and the general interpretation emerged. The discussion was not very much concerned, therefore, with problems of interpretation, but with relevant aspects which had not been covered in detail and from this several points of general interest arose.

1. Early Radiation

Dr K. Vogel commented that a comparison of the Ediacara fauna with Cambrian faunas from a functional morphology point of view shows that the Ediacara animals had large exterior surfaces which they used for respiration and food acquisition: the majority of new Metazoa in the Cambrian, on the other hand, possessed special cavities for food and oxygen acquisition, and this seemed a clear adaptive improvement giving more independence from the environment and the sudden appearance of exoskeletons would appear to be related to the evolution of these cavities.

2. Gastropoda

The geological record, said Dr E. Yochelson, totally supports Graham's notion of the feeding habits and the evolution of the radula. He added that there is a very important point that the stromatolites are essentially thought to have been driven almost to extinction by grazing organisms: most stromatolites disappear, or become greatly reduced, at the close of the early Ordovician, and it is in the late Cambrian and early Ordovician that we see the real rise of the gastropods if we ignore the small shells in the lower Cambrian.

3. Aplacophora

Dr R. R. Hessler asked Dr E. Yochelson to comment on the Aplacophora. He was answered that Dr Yochelson had no experience with Aplacophora, only prejudices; he thought the Aplacophora may be "derived" rather than "primitive" since they were grazing carnivores, were somewhat specialized in diet, and lived in moderately deep water: because they have a radula and a small mantle cavity he thought they were molluscs but he thought that palaeontologists have no basis for making comments on groups which do not have a fossil record.

4. Cephalopoda

Dr A. Bidder commented on the diagrams of early Mollusca showing pedal retractors going down to the foot: generally (she said) the mantle of the Nautiloidea (s.l.) is not concerned with locomotion; in the coleoids there are pedal and funnel retractors, but in *Nautilus*, since the funnel is within the mantle cavity, there is nothing homologous with the funnel retractors; however she emphasized how tricky it is to extrapolate from *Nautilus*.

On the statement of Holland that the cephalopods were not herbivorous, Professor M. R. House remarked that he was not wholly convinced of this for some ammonoids: the records of radulae in a few Palaeozoic and Mesozoic ammonoids suggested otherwise, and the existence of shell apertures, in several groups, which very closely abut preceding whorls could be interpreted as suggesting a suspension-feeding habit.

5. Bivalvia

Dr M. D. Brasier remarked on the reference by Morris to *Fordilla* as the only known Cambrian bivalve apart from recently discovered middle Cambrian forms: he asked concerning the phosphatic internal cast reported by Poulsen in 1967 from the Atdabanian of Bornholm which appeared to be taxodont. Dr N. Morris replied that he had not studied that material but that the possibility of it representing other groups could not be discounted.

Dr K. Vogel said that he did not believe in Pojeta's derivation of bivalves from the Rostroconchia, a group which had neither functional hinge nor adductor muscles; without the latter a hinge would have had no use and he did not see how the gap could be closed by step-by-step functional improvement. He suggested a model which he thought had much greater probability: a univalved mollusc invading fine soft sediment "required" new methods for keeping the mantle cavity free from pollution; this might be acquired by drawing down the marginal part of its yet unmineralized "shell"; this might be achieved by a combined contraction of transverse and rorso-ventral muscles of its foot. Later calcification might start from at least two lateral points, leaving a median strip of organic shell to become the hinge; a dorsal shift of the transverse pedal muscles, and their insertion onto the mineralized shell, would produce the bivalve adductor muscles.

6. Shell Composition

In reply to questions by Dr R. P. S. Jefferies on the nature of molluscan shell composition and whether he thought calcification had only happened once, Dr E. Yochelson replied that exhaustive studies of shell structure and ultrastructure in the last few years had shown that the molluscs can do a number of things, but he was not convinced that there is necessarily any phylogenetic significance in shell structure: on the other hand, there was a pervasiveness about the mollusc shell; some groups have calcitic shells, but on the whole, the impression one gets from breaking molluscs out of rock is that one is dealing primarily with an aragonitic shell. He thought that molluscs had developed an organic integument and it was the common type of calcification of this· which gave them similarity. He thought that it was the matrix upon which calcification took place that was the important thing, and that mollusca and hyolithids, shared a similar pattern. Dr R. A. Hewitt commented on a record of apatite in the siphuncle of *Spirula* and asked if phosphatic molluscs were known. Dr E. Yochelson replied if a shell was primarily rather than secondarily phosphatic then it could not be a mollusc; the record of apatite was no reason to infer otherwise any more than the record of magnetite in certain radulae could lead to the conclusion that iron secreting bacteria belong to the molluscs.

ECHINODERMATA

In the general discussion following Dr Paul's contribution Dr E. N. K. Clarkson commented that one of the striking factors of the lower Palaeozoic echinoderms is the common tendency for structures or characters to turn up in groups that have already become distinct; thus there are folded membrane structures in cystoids, blastoids and porocrinoids, and paired pores in diploporite cystoids and echinoids, edrioblastoids that look like blastoids with lancet-shaped ambulacra but have the characters of edrioasteroids, and *Aulechinus* in some ways resembles *Edrioaster*, but has a mouth and lantern facing downwards; furthermore pentamery may have been independently acquired in cystoids. Whilst it is reasonable to assume that there are only a very few ways of fulfilling particular functions, hence the same kind of structure is going to arise more than once, is there, he asked, a deeper evolutionary significance? Was there a limitation in the genetic potential of echinoderms so that only certain kinds of structure could be produced?

On the particular example of folded membrances, Dr Paul replied that the blastoid hydrospire, cystoid pectinirhombs and the parablastoid and porocrinoid structures are due to functional constraints even if the porocrinoid structure does not seem to be particularly efficient. He considered that there are only two or three ways in which some of these functions could be achieved, but that this applied to other groups as well. He agreed there must be some limit in the genetic potential and that the recurrence of five-fold structures was particularly diagnostic. However, he thought nothing operates under the requirements of *one* particular function. When Rudwick erected the paradigm he discussed the possibility of conflicting requirements producing a compromise structure. Dr Paul did not think this was a possibility alone, but that it happened all the time: all the real structures were compromises between the ideal situation for several factors.

Mr C. W. Wright confirmed, from an asterozoan case, what had been said about morphology constraints: *Podosphaerites* was a perfectly spherical asteroid; in the process of evolving to greater globularity, skeletal restraints led to a close resemblance to many globular Mesozoic echinoids. He thought such constraints were more important than "genetic tendencies". Dr K. R. Ashby asked what was the significance of the skeleton in Cambrian echinoderms and were there predators present which could have attacked animals of their size. Dr Paul thought that the early echinoderms may have required protection from predators but that the skeleton also provided protection against the physical energy of the environment, especially if they were living in shallow marine environments. He thought that the appearance of phosphatic, calcareous and agglutinating skeletons as well as dwelling burrows around the Precambrian/Cambrian boundary, all of which are protective structures to some extent, may reflect the appearance of predation or some other increased need for protection such as colonization of higher energy environments.

HEMICHORDATA

On the question of the earliest graptoloids, Dr Sdzuy remarked that middle Cambrian graptolites have been described from southern Australia and from Spain and that the latter might be the oldest found so far and they show large free thecae and no evidence of thecal dimorphism. Dr R. B. Rickards commented that a few earlier than middle Cambrian may yet be found.

Dr K. M. Towe asked whether the cortex secreted by the living *Rhabdopleura* had any ultrastructural comparison with the graptolite cortex. Dr Rickards replied that details were awaited. On the nature of the bandaging he said that this might be the thickness of one collagen fibre, for example on the inside of a skeletal tube, but that on the outside it went up to 18 layers of collagen fibres in some dendroids.

On the matter of the lophophore Dr P. J. Miller asked whether the "lophophore" of hemichordates could have been evolved separately from that of the lophophorate phyla, Dr Rickards agreed that this might be so. He also agreed that the enteropneusts might represent a more primitive type of hemichordate than the pterobranchs. Dr R. P. S. Jefferies remarked that the earliest enteropneusta might have retained the lophophore.

THE ORIGIN OF THE CHORDATES

The contribution of Dr R. P. S. Jefferies aroused considerable discussion relating to the methodology of elucidating phylogeny, and he develops this at length in his paper: few doubted the logic and rigour of the Hennigian method if all the facts were available, but substantial doubts arose in those many cases in palaeontology where characters cannot be unambiguously assessed as "primitive" or "advanced" and where iterative evolution may have applied. Dr E. N. K. Clarkson thought the basic Hennigian bifurcation itself was unusual and Dr Jefferies agreed that the system worked better at high taxon levels than at the species level. Mr C. W. Wright commented that the Hennigian methodology has a number of difficulties, but thought it was still the best logical tool available and is the discipline which makes one think the hardest.

Commenting on the asymmetric pattern of a number of the forms described by Jefferies, Dr P. J. Miller asked whether similar asymmetrical features had been discovered in vertebrate development. Dr Jefferies replied that they had not yet been, but that there were many examples of such development known from the tunicates. Dr K. Vogel asked why forms such as *Cephalodiscus* should have laid down on one side, but there seemed no clearer reason for this than for the similar habit of scallops and flatfish.

Professor N. Spjeldnaes remarked that, as he understood Jefferies' scheme, the calcitic skeleton was lost three times independently—in amphioxus, in tunicates and in the vertebrates. He thought it should be possible to see whether some or all of these groups have vestiges of the appropriate skeletal physiological and molecular mechanisms. He added that this had been done in the primitive vertebrates, where the petromyzontids and myxinids have entirely different physiology: the petromyzontids have a phosphatic skeleton which has been phylogenetically reduced whereas the myxinids had never had a skeleton in their history. Dr Jefferies was uncertain who could attempt such a comparison; he added that he thought it possible that the myxinids were the primitive sister group of all other vertebrates and agreed that they had not had a skeleton. He thought there were signs of resorption in all of the stem groups of mitrates—the stem tunicates, stem vertebrates and stem achraniates.

Dr P. J. Miller asked whether the calcified otolith of the fish inner ear might be a vestige of the calcichordate exoskeleton in the vertebrates. Dr Jefferies agreed that this might be so but that in cyclostomes it was phosphatic. In reply to a further question on the nephridia of amphioxus, Dr Jefferies said that the nephridia may well be derived from segmentation of the posterior coelom of *Lagynocystis* above the gill slits.

EPILOGUE

In a concluding vote of thanks, Mr C. W. Wright reminded members that this was the Jubilee Year of the University of Hull and he said that he had known many of the founders at the time when his father was the first Honorary Treasurer of the University College. It was a privilege, he said, on behalf of participants to thank their host and through him the University for the admirable facilities provided. He ventured to suggest also that the founders of Hull University would have regarded the proceedings during the two days as being in the tradition of distinction which they had inaugurated.

Index

The numbers in *italics* refer to pages where illustrations occur, roman numerals in **bold** refer to plate illustrations.

A

495

The Systematics Association Publications

LONDON. Published by the Association

Systematics Association Special Volumes

★Published by Academic Press for the Systematics Association
†Published by the Palaeontological Association in conjunction with the
Systematics Association

Systematics Association Special Volumes

★Published by Academic Press for the Systematics Association